Recent Titles in This Series

169 **William Abikoff, Joan S. Birman, and Kathryn Kuiken, Editors,** The mathematical legacy of Wilhelm Magnus, 1994

168 **Gary L. Mullen and Peter Jau-Shyong Shiue, Editors,** Finite fields: Theory, applications, and algorithms, 1994

167 **Robert S. Doran, Editor,** C^*-algebras: 1943–1993, 1994

166 **George E. Andrews, David M. Bressoud, and L. Alayne Parson, Editors,** The Rademacher legacy to mathematics, 1994

165 **Barry Mazur and Glenn Stevens, Editors,** p-adic monodromy and the Birch and Swinnerton-Dyer conjecture, 1994

164 **Cameron Gordon, Yoav Moriah, and Bronislaw Wajnryb, Editors,** Geometric topology, 1994

163 **Zhong-Ci Shi and Chung-Chun Yang, Editors,** Computational mathematics in China, 1994

162 **Ciro Ciliberto, E. Laura Livorni, and Andrew J. Sommese, Editors,** Classification of algebraic varieties, 1994

161 **Paul A. Schweitzer, S. J., Steven Hurder, Nathan Moreira dos Santos, and José Luis Arraut, Editors,** Differential topology, foliations, and group actions, 1994

160 **Niky Kamran and Peter J. Olver, Editors,** Lie algebras, cohomology, and new applications to quantum mechanics, 1994

159 **William J. Heinzer, Craig L. Huneke, and Judith D. Sally, Editors,** Commutative algebra: Syzygies, multiplicities, and birational algebra, 1994

158 **Eric M. Friedlander and Mark E. Mahowald, Editors,** Topology and representation theory, 1994

157 **Alfio Quarteroni, Jacques Periaux, Yuri A. Kuznetsov, and Olof B. Widlund, Editors,** Domain decomposition methods in science and engineering, 1994

156 **Steven R. Givant,** The structure of relation algebras generated by relativizations, 1994

155 **William B. Jacob, Tsit-Yuen Lam, and Robert O. Robson, Editors,** Recent advances in real algebraic geometry and quadratic forms, 1994

154 **Michael Eastwood, Joseph Wolf, and Roger Zierau, Editors,** The Penrose transform and analytic cohomology in representation theory, 1993

153 **Richard S. Elman, Murray M. Schacher, and V. S. Varadarajan, Editors,** Linear algebraic groups and their representations, 1993

152 **Christopher K. McCord, Editor,** Nielsen theory and dynamical systems, 1993

151 **Matatyahu Rubin,** The reconstruction of trees from their automorphism groups, 1993

150 **Carl-Friedrich Bödigheimer and Richard M. Hain, Editors,** Mapping class groups and moduli spaces of Riemann surfaces, 1993

149 **Harry Cohn, Editor,** Doeblin and modern probability, 1993

148 **Jeffrey Fox and Peter Haskell, Editors,** Index theory and operator algebras, 1993

147 **Neil Robertson and Paul Seymour, Editors,** Graph structure theory, 1993

146 **Martin C. Tangora, Editor,** Algebraic topology, 1993

145 **Jeffrey Adams, Rebecca Herb, Stephen Kudla, Jian-Shu Li, Ron Lipsman, and Jonathan Rosenberg, Editors,** Representation theory of groups and algebras, 1993

144 **Bor-Luh Lin and William B. Johnson, Editors,** Banach spaces, 1993

143 **Marvin Knopp and Mark Sheingorn, Editors,** A tribute to Emil Grosswald: Number theory and related analysis, 1993

142 **Chung-Chun Yang and Sheng Gong, Editors,** Several complex variables in China, 1993

141 **A. Y. Cheer and C. P. van Dam, Editors,** Fluid dynamics in biology, 1993

140 **Eric L. Grinberg, Editor,** Geometric analysis, 1992

(Continued in the back of this publication)

The Mathematical Legacy
of Wilhelm Magnus

Groups, Geometry
and Special Functions

CONTEMPORARY MATHEMATICS

169

The Mathematical Legacy
of Wilhelm Magnus

Groups, Geometry
and Special Functions

Conference on the
Legacy of Wilhelm Magnus
May 1–3, 1992
Polytechnic University
Brooklyn, New York

William Abikoff
Joan S. Birman
Kathryn Kuiken
Editors

American Mathematical Society
Providence, Rhode Island

The Conference on the Legacy of Wilhelm Magnus was held at Polytechnic University, Brooklyn, New York, May 1–3, 1992. The conference was sponsored by the Courant Institute of Mathematical Sciences of New York University, Merrill Lynch and Co., Metropolitan Life Insurance Company, Polytechnic University, and Teleport Communications Group, with grant support from the American Mathematical Society.

1991 *Mathematics Subject Classification.* Primary 20Exx, 20Fxx, 30–XX, 33–XX.

Library of Congress Cataloging-in-Publication Data

Conference on the Legacy of Wilhelm Magnus (1992: Brooklyn, New York, N.Y.)
  The mathematical legacy of Wilhelm Magnus: groups, geometry and special functions: proceedings of the Conference on the Legacy of Wilhelm Magnus, May 1–3, 1992, Polytechnic University, Brooklyn, New York/William Abikoff, Joan S. Birman, Kathryn Kuiken, editors.
    p. cm. — (Contemporary mathematics; v. 169)
  Includes bibliographical references.
  ISBN 0-8218-5156-X (acid-free)
  1. Group theory—Congresses. 2. Functions of complex variables—Congresses. 3. Functions, Special—Congresses. I. Abikoff, William. II. Birman, Joan S., date. III. Kuiken, Kathryn. IV. Title. V. Title: Groups, geometry and special functions. VI. Series: Contemporary mathematics (American Mathematical Society); v. 169.
QA174.C66   1992
512′.2—dc20
                                                                                            94-11625
                                                                                            CIP

Contents

Preface   ix

Inequalities for plane quasiconformal mappings
G. D. ANDERSON, M. K. VAMANAMURTHY, AND M. VUORINEN   1

A look at the Bateman project
RICHARD ASKEY   29

Linear-central filtrations on groups
HYMAN BASS AND ALEXANDER LUBOTZKY   45

Musings on Magnus
GILBERT BAUMSLAG   99

The monodromy group of a transcendental function
KEVIN BERRY AND MARVIN TRETKOFF   107

Finite-dimensional representations of Artin's braid group
J. S. BIRMAN, D. D. LONG, AND J. A. MOODY   123

Squaring rectangles: The finite Riemann mapping theorem
J. W. CANNON, W. J. FLOYD, AND W. R. PARRY   133

The Freiheitssatz and its extensions
BENJAMIN FINE AND GERHARD ROSENBERGER   213

A Rodrigues-type formula for the q-Racah polynomials and some related results
ISMOR FISCHER   253

Air on the Dirac strings
GEORGE K. FRANCIS AND LOUIS H. KAUFFMAN   261

Does Lyndon's length function imply the universal theory of free groups?
ANTHONY M. GAGLIONE AND DENNIS SPELLMAN   277

Schottky groups and the boundary of Teichmüller space: Genus 2
DANIEL M. GALLO   283

Lacunary series as quadratic differentials in conformal dynamics
FREDERICK P. GARDINER AND DENNIS SULLIVAN   307

The geometry of cycles in the Cayley diagram of a group
ROBERT H. GILMAN 331

Braids, Riemann surfaces and moduli
W. J. HARVEY 341

Some remarks on J replacement in direct products
R. HIRSHON 353

Wilhelm Magnus, applied mathematician
HARRY HOCHSTADT 365

On the combinatorial curvature of groups of F-type and other one-relator free products
A. JUHÁSZ AND G. ROSENBERGER 373

Branched dihedral structures on Riemann surfaces
KATHRYN KUIKEN AND JOHN T. MASTERSON 385

Groups and Lie algebras: The Magnus theory
JOHN P. LABUTE 397

Semiregular continued fractions whose partial denominators are 1 or 2
JOSEPH LEHNER 407

Testing for the center of a one-relator group
FRANK LEVIN 411

Generalizing the Baer-Stallings pregroup
SEYMOUR LIPSCHUTZ 415

On binary σ-invariant words in a group
O. MACEDONSKA AND DONALD M. SOLITAR 431

Explicit matrices for Fuchsian groups
BERNARD MASKIT 451

Levi-properties generated by varieties
ROBERT FITZGERALD MORSE 467

Chains of primitive ideals
D. S. PASSMAN 475

Families of closed geodesics on hyperbolic surfaces with common self-intersections
THEA PIGNATARO AND HANNA SANDLER 481

On the isometry groups of hyperbolic manifolds
JOHN G. RATCLIFFE 491

A generalization of Lazard's theorem on modular dimension subgroups
VLADIMIR TASIĆ 497

Preface

This volume constitutes the Proceedings of the conference held on May 1-3, 1992 at Polytechnic University to celebrate the mathematical legacy of Wilhelm Magnus and to honor his memory. Many of Wilhelm's former students and colleagues were present. Also in attendance were members of his New York Group Theory Seminar and mathematicians who did not know him personally but had been inspired by his work. The Magnus family was also present.

The conference was sponsored by the American Mathematical Society, the Courant Institute of Mathematical Sciences of New York University, Merrill Lynch and Co., Metropolitan Life Insurance Company, Polytechnic University, and Teleport Communications Group.

A brief glance at the diversity of the papers in this volume gives an indication of Magnus' invaluable contributions to mathematics as well as their continuing relevance and inspiration. In the course of editing this volume, we were led to a new appreciation of the many interconnected themes in Magnus' work. We were reminded that his papers contain many difficult, profound and original ideas and that, even at this writing, there are many unexplored themes in them.

Magnus' Ph.D. dissertation was published in 1930. In it, he proved a theorem which is commonly known as the Freiheitssatz or the Freedom Theorem. It gives a remarkably simple criterion which assures that certain subgroups of a 1-relator group are free. It was discovered at a time when combinatorial group theory was in its infancy, and the simple yet delicate arguments which he used had a profound effect on the development of the area. Indeed, there were many attempts to simplify his argument or to find a new proof cast in a geometric framework, but Magnus' approach has survived to this day. In another aspect of his work on the Freiheitssatz, Magnus showed that it is possible to learn about the structure of a group from its presentations. This idea of Magnus was new in 1930 and is an idea whose depth we have come to appreciate many times in the intervening years. In this volume, the reader will find a review article by Fine and Rosenberger as well as new and original research, all on the work which has grown out of Magnus' Freiheitssatz.

A second and far-reaching discovery of Magnus was that there is an embedding e of a free group F into the group of units of an appropriate ring R. Elements of F map to elements of the form $1 + \phi_x \in R$, where ϕ_x belongs to a particular ideal $I \subset R$. Magnus proved that the kernel of e is trivial, that the image of F has a natural structure permitting it to be viewed as a metric space, and that certain subideals $I^n \subset I$ are precisely the images under e of the groups F^n of the lower central series, where $F^0 = F$ and $F^n = [F, F^{n-1}]$. This volume includes two expository papers, one by Labute, the other by Baumslag, which explore the *Magnus embedding* from different viewpoints. Many other authors use the Magnus embedding in interesting ways.

The spiritual roots of Magnus' work on discrete groups may be found in earlier studies by Euler, Gauss and Riemann on the hypergeometric equation, as is pointed out in the articles by Harvey and by Berry and Tretkoff. These beginnings were extended by the contributions of Beltrami, Klein, and Poincaré. Collectively, these giants merged important aspects of the theories of special functions, uniformization and conformal mapping, hyperbolic geometry, and complex algebraic curves into a magnificent edifice. Harvey's paper explains how Magnus was led from these ideas to the studies of group actions on the hyperbolic plane, monodromy groups of branched coverings of surfaces, and braids. The interested reader will find many papers in this volume which relate to the ideas discussed by Harvey. In an expanding set of developments which is loosely tied to Magnus' work, some of these papers connect directly to Magnus' work while others range further afield, passing from Riemann surface theory to Teichmüller theory and thence to dynamical systems.

We note that in initially developing the relationship between hyperbolic geometry and complex algebraic curves, Klein and his student Dyck established a connecting link between these subjects and combinatorial group theory. Magnus' book *Non-euclidean Tessellations and Their Groups* presents this connection to the modern reader. It contains reproductions of many nineteenth century drawings associated with groups of motions. Referring to the classic work on automorphic functions first published in 1897, Wilhelm often remarked that *Pictures from Fricke-Klein* would have been a more appropriate title for his book. Several papers in this volume are directly related to the ideas in Magnus' book.

Wilhelm was particularly fascinated by the special functions of mathematical physics. His first work in this area was the famous handbook, written jointly with F. Oberhettinger, entitled *Formulas and Theorems for the Special Functions of Mathematical Physics*. Somewhat later, there followed the three volume set with Erdélyi, et al., *Higher Transcendental Functions*, commonly known as the *Bateman Manuscript Project*. For precise references, see the review articles by Askey and by Hochstadt. Magnus loved examples and had a special delight in facts; these books are crammed to the brim with examples and facts. In the essay *Mathematical Recollections*, Magnus' introduction to his *Collected Papers*, one finds Wilhelm's description of how he came to be interested in special functions. After a brief discussion of his work during World War II, he goes on to describe how he met the physicist Sommerfeld, whose work on diffraction problems fascinated him for its ingenuity and mathematical sophistication. Sommerfeld introduced him to physical applications of special functions, and as Magnus continues "I started collecting them. I am not sure how far these functions may have appealed to my collector's instinct, an instinct which manifests itself in many people with application to diverse objects, regardless of any consideration of usefulness." The flavor of this remark evokes for the editors sharp memories of Wilhelm's wit, love of conversation, and sparkling intellect.

The mathematical legacy of Magnus was growing even as it was being celebrated at the conference which this volume records. T. Stanford, a young graduate student who was working on his Ph.D. thesis at the time, was in the audience. He learned about the Magnus embedding of a free group for the first time at this conference; by the day's end, he had used it in yet another way to solve a problem in geometric topology.

On a more personal note, Wilhelm was truly a special man – warm, gentle and self-effacing. His creative intellect was evident to those whose lives he touched. It was always shared in a spirit of love, caring and enthusiasm, thus greatly enriching the lives, both personal and professional, of those fortunate enough to know him.

William Abikoff, University of Connecticut at Storrs
Joan S. Birman, Columbia University
Kathryn Kuiken, Polytechnic University

Contemporary Mathematics
Volume 169, 1994

Inequalities for
Plane Quasiconformal Mappings

G. D. ANDERSON, M. K. VAMANAMURTHY, AND M. VUORINEN

ABSTRACT. Let f be a K-quasiconformal mapping of the unit disk B into itself fixing the origin. A generalization, due to Hersch and Pfluger, of the classical Schwarz Lemma, gives a sharp upper bound $\varphi_K(r)$ for the distance of the image $f(z)$ from the origin in terms of $r = |z|$. In this paper the function $\varphi_K(r)$ is estimated, and applications are found to the Hölder continuity of quasiconformal mappings of the unit disk. This function also occurs in number theory in the study of modular equations. The function $\varphi_K(r)$ has an expression in terms of the complete elliptic integral $\mathcal{K}(r)$. The proofs make use of the properties of $\mathcal{K}(r)$ and some related special functions.

1. Introduction

The classical Schwarz Lemma for analytic functions was generalized in 1952 by J. Hersch and A. Pfluger [14] to the class of quasiconformal mappings of the unit disk. They showed that there is a strictly increasing distortion function $\varphi_K : (0,1) \to (0,1)$ such that $|f(z)| \leq \varphi_K(|z|)$ for each K-quasiconformal mapping of the unit disk B into itself with $f(0) = 0$. This distortion function is defined by

$$(1.1) \qquad \varphi_K(r) = \mu^{-1}\left(\frac{1}{K}\mu(r)\right),$$

where $\mu(r)$ denotes the modulus of the plane Grötzsch ring $B \setminus [0, r]$, $0 < r < 1$, and has the explicit expression

1991 Mathematics Subject Classification. Primary 30C62. Secondary 33E05, 26D07.

This research was completed during the second author's visit to Michigan State University and the third author's visit to the University of Michigan. The research of the third author was supported in part by the Academy of Finland and by U. S. National Science Foundation Grant NSF-DMS-9003438 (Prof. F. W. Gehring).

This paper is in final form and no version of it will be submitted for publication elsewhere.

(1.2)
$$\mu(r) = \frac{\pi \mathcal{K}'(r)}{2\mathcal{K}(r)}.$$

Here $\mathcal{K}$, $\mathcal{K}'$ are the complete elliptic integrals [17,pp.358-359]

(1.3)
$$\mathcal{K}(r) = \int_0^{\pi/2} \frac{dt}{\sqrt{1 - r^2 \sin^2 t}}, \quad \mathcal{K}'(r) = \mathcal{K}(r'), \quad r' = \sqrt{1 - r^2}.$$

We extend the definition of φ_K to $[0,1]$ by assigning it the values $0, 1$ at $0, 1$, respectively. We shall also later need the complementary elliptic integrals

(1.4)
$$\mathcal{E}(r) = \int_0^{\pi/2} \sqrt{1 - r^2 \sin^2 t}\, dt, \quad \mathcal{E}'(r) = \mathcal{E}(r'), \quad r' = \sqrt{1 - r^2}.$$

The equation

(1.5)
$$\frac{\mathcal{K}'(s)}{\mathcal{K}(s)} = p \frac{\mathcal{K}'(r)}{\mathcal{K}(r)}$$

is called the modular equation of degree p [10, Ch. IV]. The unique solution of (1.5) is given by $s = \varphi_{1/p}(r)$. For positive integral values of p the modular equation (1.5) has been studied extensively in number theory [10], [8], [9].

Our work is motivated mainly by applications to quasiconformal theory, and it is a continuation of earlier work in [5] and [23].

Section 2 gives various inequalities for $\varphi_K(r)$, $\mu(r)$, and $\mathcal{K}(r)$. For instance, we prove the sharp inequality

$$\mathcal{K}(r) \le \mathcal{K}(\varphi_K(r)) \le K\mathcal{K}(r)$$

for $K \ge 1$ and $r \in (0,1)$. Section 3 reduces the problem of finding bounds for the rather complicated function $\varphi_K(r)$ to that of studying monotone properties of $\mu(r)$. Additional estimates for these functions are obtained in Section 5. Our main results appear in Section 4, and they provide bounds for the change of distance under quasiconformal automorphisms of the unit disk.

Throughout this paper, for $t \in [0,1]$, t' will denote $\sqrt{1 - t^2}$, as in (1.3) and (1.4). When the argument of the function is clear, we sometimes write $\mathcal{K}, \mathcal{K}'$ and $\mathcal{E}, \mathcal{E}'$ instead of $\mathcal{K}(r), \mathcal{K}'(r)$ and $\mathcal{E}(r), \mathcal{E}'(r)$. The hyperbolic sine, cosine, tangent functions and their inverses are denoted by sh, ch, th, arsh, arch, arth, respectively. If E, F, G are subsets of $\bar{\mathbb{R}}^2$, we let $\Delta(E, F; G)$ denote the family of curves joining E and F in G. If $G = \mathbb{R}^2$ or $\bar{\mathbb{R}}^2$, we write simply $\Delta(E, F)$. The conformal capacity and modulus of a plane ring R will be denoted by cap R and mod R, respectively. The capacities of the Grötzsch and Teichmüller extremal rings are denoted by $\gamma(s)$, $s > 1$, and $\tau(t)$, $t > 0$, respectively; these quantities are connected by the relations

(1.6)
$$\gamma(s) = 2\tau(s^2 - 1) = 2\pi/\mu(1/s)$$

[16, Theorem 1.1, p.55]. The open unit disk in the plane is denoted by B. If a, b, c, d are distinct points in $\bar{\mathbb{R}}^2$, the symbol $|a, b, c, d|$ denotes the *absolute ratio*, which is the absolute value of the cross ratio (a, b, c, d). For each $K \ge 1$, we sometimes denote by $QC_K(B)$ the class of all K-quasiconformal mappings of B into itself. In general, we follow the relatively standard notation of [16].

1.7. ACKNOWLEDGEMENT. We are grateful to S.-L. Qiu for reading the manuscript and offering helpful suggestions.

2. Properties of the distortion functions

It is easy to see from the definitions that $\varphi_K(r)$ is strictly increasing and concave [3, Theorem 3.27] on $[0,1]$ if $K > 1$. We here obtain similar results for functions related to φ_K.

2.1. LEMMA. *If f is positive and concave on an interval $I \subset \mathbb{R}$, then f is log-concave on I.*

PROOF. For $x, y \in I$ and $a \in (0,1)$, we have

$$\log f(ax + (1-a)y) \geq \log(af(x) + (1-a)f(y)) \geq a \log f(x) + (1-a) \log f(y),$$

since $\log$ is concave and increasing on $(0, \infty)$. $\qquad\square$

2.2. COROLLARY. *Each of the following functions is log-concave:*

(1) $K \mapsto \varphi_K(r)$, for fixed $r \in (0,1)$, on $[1, \infty)$,
(2) $r \mapsto \varphi_K(r)$, for fixed $K \geq 1$, on $(0,1)$,
(3) $r \mapsto \mu(1/r)$, $r > 1$.

Each of the following is log-convex:

(4) *the Grötzsch capacity $\gamma(s)$, $s > 1$,*
(5) *the Teichmüller capacity $\tau(t)$, $t > 0$.*

PROOF. For (1), with $s = \varphi_K(r)$, $f(K) = \log \varphi_K(r)$, we have

$$f'(K) = \frac{1}{s}\frac{ds}{dK} = \frac{1}{s}\frac{4}{\pi^2}\frac{ss'^2 K^2(s)\mu(r)}{K^2} = \frac{4\mu(r)}{\pi^2}\frac{s'^2 K^2(s)}{K^2},$$

which is positive and decreasing by [4, Theorem 2.2]. Parts (2), (3) follow from Lemma 2.1 and [3, Theorem 3.27], [4, Theorem 4.5], respectively. Part (4) follows from Lemma 2.1 and [4, Corollary 4.7]. Part (5) follows from (1.6), part (4), and the relation

$$\frac{d}{dt}\log\tau(t) = \frac{\frac{d}{ds}\log\gamma(s)}{2\sqrt{t+1}}, \quad s = \sqrt{t+1}. \qquad\square$$

2.3. COROLLARY. *For $a, b, r \in (0,1)$, $a+b = 1$, $K_1, K_2 \in (1,\infty)$,*

$$\varphi_{aK_1 + bK_2}(r) \geq \varphi_{K_1}^a(r)\varphi_{K_2}^b(r).$$

2.4. COROLLARY. *For $r, s, a, b \in (0,1)$, $a+b = 1$, $K \in [1,\infty)$,*

$$\varphi_K(ar + bs) \geq \varphi_K^a(r)\varphi_K^b(s).$$

For $K = 1$, Corollary 2.4 reduces to the well-known Young's inequality [7, p. 15].

The inequality in the next theorem appears in [15].

2.5. THEOREM. *For $K \in (1, \infty)$, $r \in (0, 1)$, let*

$$f(K, r) = r^{-1/K} \varphi_K(r) \exp[(\frac{1}{K} - 1)(m(r) + \log r)],$$

where $m(r) = (2/\pi) r'^2 \mathcal{K}(r) \mathcal{K}'(r)$. Then, for fixed r, f is a decreasing function of K from $(1, \infty)$ onto $(e^{-m(r)}/r, 1)$. In particular,

$$\varphi_K(r) \le r \exp((1 - (1/K)) m(r)),$$

with equality if and only if $K = 1$ or $r = 0$ or $r = 1$.

PROOF. With $s = \varphi_K(r)$, define

$$g(K) = \log f(K, r) = \log s - \log r - (1 - \frac{1}{K}) m(r).$$

Then

$$
\begin{aligned}
g'(K) &= \frac{1}{s} \frac{ds}{dK} - \frac{1}{K^2} m(r) = \frac{4}{\pi^2 s} \frac{s s'^2 \mathcal{K}^2(s) \mu(r)}{K^2} - \frac{2 r'^2 \mathcal{K} \mathcal{K}'}{\pi K^2} \\
&= \frac{2\mathcal{K}'(r)}{\pi K^2 \mathcal{K}(r)} [s'^2 \mathcal{K}^2(s) - r'^2 \mathcal{K}^2(r)],
\end{aligned}
$$

which is negative by [4, Theorem 2.2(3)]. The limiting values are clear. □

2.6. THEOREM. *For each $r \in (0, 1)$ the function*

$$f(K) \equiv \log \varphi_K(r) - \log r - (1 - \frac{1}{K}) r' \mu(r)$$

is decreasing from $[1, \infty)$ onto $(-r' \mu(r) - \log r, 0]$. In particular,

$$\varphi_K(r) \le r \exp((1 - (1/K)) r' \mu(r)) \le (2(1 + r'))^{r'(1-(1/K))} r^{1/K} \le 4^{r'(1-(1/K))} r^{1/K},$$

with equality if and only if $K = 1$ or $r = 0$ or $r = 1$.

PROOF. Let $s = \varphi_K(r)$. Then $\mu(s) = \mu(r)/K$, so that

$$\frac{ds}{dK} = \frac{2\mathcal{K}'(r) s s'^2 \mathcal{K}^2(s)}{\pi K^2 \mathcal{K}(r)}$$

(cf. [4, Lemma 2.1(2)]). Hence

$$f'(K) = \frac{2\mathcal{K}'(r)}{\pi K^2 \mathcal{K}(r)} \left[s'^2 \mathcal{K}^2(s) - \frac{\pi^2}{4} r' \right] < \frac{2\mathcal{K}'(r)}{\pi K^2 \mathcal{K}(r)} r' \left[s' \mathcal{K}^2(s) - \frac{\pi^2}{4} \right] < 0$$

by [4, Theorem 2.2(3)], and the monotoneity follows. Hence, in particular, for $K > 1$ and $0 < r < 1$, $f(K) < f(1)$ yields by [16, p.62]

$$
\begin{aligned}
\varphi_K(r) \;&<\; r\exp((1 - \tfrac{1}{K})r'\mu(r)) < r\exp\left((1 - \tfrac{1}{K})r'\log\frac{2(1+r')}{r}\right) \\
&=\; r\left(\frac{2(1+r')}{r}\right)^{(1-(1/K))r'} < (2(1+r'))^{(1-(1/K))r'}r^{1/K} \\
&<\; 4^{(1-(1/K))r'}r^{1/K}.
\end{aligned}
$$

$\square$

2.7. REMARK. Theorem 2.6 is best possible in the sense that the inequality

$$
\varphi_K(r) < r\exp((1 - \tfrac{1}{K})r'^2\mu(r))
$$

fails for large K. For suppose

$$
\log\varphi_K(r) - \log r - (1 - \tfrac{1}{K})r'^2\mu(r) \le 0
$$

for $K \ge 1$. Then, since $\varphi_K(r) \to 1$ as $K \to \infty$, in the limit we get $\mu(r) \ge r'^{-2}\log(1/r)$ for $r \in (0,1)$. Letting $r \to 1$ and using l'Hôpital's rule we obtain the contradiction $0 = \lim_{r\to 1}\mu(r) \ge 1/2$.

2.8. COROLLARY. *For $1 < K < \infty$, $r' = \sqrt{1 - r^2}$, $0 < r < 1$,*

(1) $\varphi_K(r)\varphi_K(r') < 4^{\sqrt{2}(1-(1/K))}(rr')^{1/K} < 8^{1-(1/K)}(rr')^{1/K}.$

In particular,

(2) $\varphi_K(1/\sqrt{2}) < 4^{(3/4)-(1/K)}.$

PROOF. By Theorem 2.6,

$$
\varphi_K(r)\varphi_K(r') < 4^{(1-(1/K))(r+r')}(rr')^{1/K}.
$$

Now by elementary calculus, $1 \le r + r' \le \sqrt{2}$, while $4^{\sqrt{2}} < 8$, and (1) follows. For (2) substitute $r = 1/\sqrt{2}$ in (1). $\square$

2.9. THEOREM. *For $1 \le K < \infty$, $r \in [0,1]$,*

(1) $\mu(r^K) \le K\mu(r),$

(2) $\varphi_K(r) \le \varphi_{K^2}(r^K),$

(3) $\varphi_K(a)\varphi_K(b) \le \varphi_{K^2}^2((ab)^{K/2}).$

PROOF. Applying μ^{-1} on both sides of (1), we see that it is equivalent to $\varphi_{1/K}(r) \le r^K$, which is true (cf. [3, Theorem 3.18]). Part (2) is true iff

$\mu(\varphi_K(r)) \geq \mu(\varphi_{K^2}(r^K))$ iff $\mu(r)/K \geq \mu(r^K)/K^2$, which is (1). Finally, (3) follows from [23, Theorem 1.7(1)] and part (2). □

2.10. THEOREM. For $1 < K < \infty$, $0 \leq r < 1$, let $f(r) = \mathcal{K}(s)/\mathcal{K}(r)$, $s = \varphi_K(r)$. Then f is increasing from $[0,1)$ onto $[1, K)$. In particular,

$$1 \leq \frac{\mathcal{K}(s)}{\mathcal{K}(r)} \leq K.$$

PROOF. By logarithmic differentiation we obtain

$$rr'^2 \mathcal{K}(r)\mathcal{K}'(r)\frac{f'(r)}{f(r)} = g(s) - g(r),$$

where

$$g(x) = x\mathcal{K}'(x) \cdot \frac{\mathcal{E}(x) - x'^2\mathcal{K}(x)}{x}$$

is a product of two positive increasing functions [4, Theorem 2.2(3),(7)]. Hence $g(s) > g(r)$, so that $f'(r) > 0$. Finally, $\lim_{r\to 1} f(r) = \lim_{r\to 1} K\mathcal{K}'(s)/\mathcal{K}'(r) = K$ by (1.2) and (1.1). □

We recall the following result from [5, Theorem 1.11(1)].

2.11. THEOREM. For $K \geq 1$ let $L = 2^{1-(1/K)}$, $\ell = 2^{1-K}$, $\ell_1 = 2\ell/(1 + \ell^2)$, $A(r) = r/(1 + r')$. Then $\varphi_K(r) \leq \text{th}(2 \text{ arth}(LA^{1/K}(r)))$ for $r \in (0, \ell_1)$.

3. A class of inequalities for $\varphi_K(r)$

Assume that g and h are continuous injective functions from $(0,1)$ into $(0,\infty)$ such that

(3.1) $g(r) < \mu(r) < h(r)$.

We shall now study the conditions under which (3.1) implies the estimates

(3.2) $g(s) < \frac{1}{K}g(r) < \frac{1}{K}h(r) < h(s)$,

where $s = \varphi_K(r)$, $K > 1$, $0 < r < 1$.

3.3. THEOREM. Let f, g, and h be strictly decreasing continuous functions from $(0,1)$ into $(0,\infty)$ such that $f(1^-) = 0$ and $g(r) < f(r) < h(r)$. Then

$$g(s) < \frac{1}{C}g(r) < \frac{1}{C}h(r) < h(s)$$

for all $C > 1, 0 < r < 1$, where $f(s) = (1/C)f(r)$, if and only if $g(r)/f(r)$ and $f(r)/h(r)$ are strictly decreasing on $(0,1)$.

PROOF. First, suppose the monotoneity properties hold. Then $r < s$, so that

$$\frac{f(s)}{g(s)} > \frac{f(r)}{g(r)} \quad \text{and} \quad \frac{f(s)}{h(s)} < \frac{f(r)}{h(r)}.$$

These give $g(r) > Cg(s)$ and $h(r) < Ch(s)$, so the desired inequalities follow.

For the converse, let $0 < r < s < 1$, and define $C = f(r)/f(s)$. Then $C > 1$. Hence

$$g(s) < \frac{1}{C}g(r) < \frac{1}{C}h(r) < h(s).$$

These give $f(r)/g(r) < f(s)/g(s)$ and $f(r)/h(r) > f(s)/h(s)$. $\qquad\square$

3.4. EXAMPLE. We let $f(r) = \mu(r)$ in Theorem 3.3. If $g(r) = \text{arch}(1/r)$ and $h(r) = \log(4/r)$, then $g(r) < \mu(r) < h(r)$ by [16, (2.10), p.61] and [5, Theorem 1.9]. Since $h(r)$ is obviously decreasing and $\mu(r)/h(r)$ is strictly decreasing by [4, Theorem 4.3(6)], it follows from Theorem 3.3 that, with $s = \varphi_K(r)$,

$$\frac{1}{K}\log\frac{4}{r} < \log\frac{4}{s},$$

so that $s < 4^{1-(1/K)}r^{1/K}$. This upper bound is well known [16, (3.6), p. 65]. In [3, Theorem 1.3] it was shown that

$$\varphi_K(r) > g^{-1}\left(\frac{1}{K}g(r)\right) = \frac{2r^{1/K}}{(1+r')^{1/K} + (1-r')^{1/K}}.$$

Since g is strictly decreasing, it follows from Theorem 3.3 that $\mu(r)/g(r)$ is strictly increasing.

3.5. THEOREM. For $1 < K < \infty$, $0 < r < 1$, let $s = \varphi_K(r)$, and let $f(r) = s(1+r')^{1/K}/((1+s')r^{1/K})$. Then f is decreasing from $(0,1)$ onto $(1, 2^{1-(1/K)})$. In particular,

$$\left(\frac{r}{1+r'}\right)^{1/K} < \frac{s}{1+s'} < 2^{1-(1/K)}\left(\frac{r}{1+r'}\right)^{1/K},$$

and hence

$$\varphi_K(r) < (2(1+r'))^{1-(1/K)}r^{1/K} < 4^{1-(1/K)}r^{1/K}.$$

PROOF. By logarithmic differentiation,

$$\begin{aligned}
\frac{f'(r)}{f(r)} &= \left(\frac{1}{s} + \frac{s}{s'(1+s')}\right)\frac{ds}{dr} - \frac{1}{K}\left(\frac{1}{r} + \frac{r}{r'(1+r')}\right) \\
&= \frac{1}{Krr'}\left(\frac{s'}{r'}\left(\frac{\mathcal{K}(s)}{\mathcal{K}(r)}\right)^2 - 1\right) < 0,
\end{aligned}$$

since $s > r$ and the function $g(x) \equiv x'\mathcal{K}^2(x)$ is decreasing on $(0,1)$ by [4, Theorem 2.2(3)]. $\qquad\square$

4. Improvements in Mori's theorem

If f is a K-quasiconformal automorphism of the unit disk B with $f(0) = 0$ then, by a theorem of Mori [18], [16, p.66], f is Hölder continuous with

$$(4.1) \qquad |f(x) - f(y)| \leq 16|x - y|^{1/K}$$

for all $x, y \in B$, and the number 16 cannot be replaced by a smaller one independent of K. It was shown by Lehto and Virtanen [16, p.68] that if the above conditions hold and if x, y lie on ∂B then

$$(4.2) \qquad |f(x) - f(y)| \leq 16^{1-(1/K)}|x - y|^{1/K},$$

where the constant $16^{1-(1/K)}$ is the smallest possible. Thus, the best possible constant C for (4.1) must satisfy $C \in [16^{1-(1/K)}, 16]$, and it is natural to expect (4.2) to hold for all $x, y \in B$.

Since 1985 several refinements of Mori's theorem have appeared; see [21], [24, p.150], where further references are given. For related results see [22]. The present authors [6, Theorem 5.8] have recently improved the constant to $64^{1-(1/K)}$, and Qiu Songliang has informed us in private correspondence that he has done the same [21]. In this section we obtain a further improvement of the constant 16 that is sharp as K tends to 1. First, we recall a theorem of Agard and Gehring [2, Theorem 2] and obtain some of its consequences.

4.3. THEOREM [2]. *Suppose f is a K-quasiconformal mapping of the extended plane with $f(\infty) = \infty$. Then for each triple of distinct finite points x, y, z,*

$$\varphi_{1/K}\left(\sin \frac{\alpha}{2}\right) \leq \sin \frac{\beta}{2} \leq \varphi_K\left(\sin \frac{\alpha}{2}\right),$$

where

$$\alpha = \arcsin\left(\frac{|x - y|}{|x - z| + |y - z|}\right), \quad \beta = \arcsin\left(\frac{|f(x) - f(y)|}{|f(x) - f(z)| + |f(y) - f(z)|}\right).$$

The next result is due to Qiu [21].

4.4. COROLLARY. *Under the hypotheses of Theorem 4.3,*

$$(4\delta)^{1-K}(\sin \alpha)^K \leq \sin \beta \leq (4\delta)^{1-(1/K)}(\sin \alpha)^{1/K},$$

where $\delta = \cos(\alpha/2) + \cos^2(\alpha/2)$.

PROOF. We prove only the second inequality, since the other proof is similar. If $\beta \in (0, \alpha]$ there is nothing to prove. Hence we may assume $\beta > \alpha$. Now by Theorem 3.5,

$$
\begin{aligned}
\sin \beta &= 2\sin(\beta/2)\cos(\beta/2) \leq 2\varphi_K(\sin(\alpha/2))\cos(\alpha/2) \\
&\leq 2 \cdot 2^{1-(1/K)}(1 + \cos(\alpha/2))^{1-(1/K)}(\sin(\alpha/2))^{1/K}\cos(\alpha/2) \\
&= (4\delta)^{1-(1/K)}(\sin \alpha)^{1/K},
\end{aligned}
$$

which proves the second inequality. $\qquad \square$

4.5. COROLLARY. *Let f be a K-quasiconformal automorphism of B with $f(0) = 0$. Then for $x, y \in B$*

$$|f(x) - f(y)| \le (8\delta)^{1-(1/K)} r |x - y|^{1/K} \le 64^{1-(1/K)} |x - y|^{1/K}$$

and

$$|f(x) - f(y)| \ge (8\delta)^{1-K} s |x - y|^K \ge 64^{1-K} |x - y|^K,$$

where

$$\delta = \cos(\alpha/2) + \cos^2(\alpha/2), \quad \sin \alpha = |x - y|/(|x| + |y|),$$

$$r = \frac{|x|^{1/K} + |y|^{1/K}}{(|x| + |y|)^{1/K}} (\max\{1 + |x|', 1 + |y|'\})^{1-(1/K)},$$

$$s = \frac{|x|^K + |y|^K}{(|x| + |y|)^K} (\max\{1 + |x|', 1 + |y|'\})^{1-K}.$$

PROOF. By reflection [16, p.47] f can be extended to a K-quasiconformal self-mapping of $\bar{\mathbb{R}}^2$ with $f(\infty) = \infty$. Hence Corollary 4.4 and Theorem 3.5 together with [7,Ch.I, Section 16] yield

$$
\begin{aligned}
|f(x) - f(y)| &\le (|f(x)| + |f(y)|)(4\delta)^{1-(1/K)} (\sin \alpha)^{1/K} \\
&\le (8\delta)^{1-(1/K)} (\sin \alpha)^{1/K} [(1 + |x|')^{1-(1/K)} |x|^{1/K} + \\
&\qquad\qquad (1 + |y|')^{1-(1/K)} |y|^{1/K}] \\
&\le (8\delta)^{1-(1/K)} r |x - y|^{1/K} \le 64^{1-(1/K)} |x - y|^{1/K},
\end{aligned}
$$

which proves the first inequality. The proof of the second inequality is similar. $\square$

We now prove the following result related to Theorem 4.3.

4.6. LEMMA. *Let f be a K-quasiconformal mapping of the extended plane with $f(\infty) = \infty$. Then*

$$\varphi_{1/K}\left(\frac{\sin \alpha}{2}\right) \left(\sec \frac{\alpha}{2}\right)^{K-1} \le \frac{\sin \beta}{2} \le \varphi_K\left(\frac{\sin \alpha}{2}\right) \left(\cos \frac{\alpha}{2}\right)^{1-(1/K)},$$

where α, β are as in Theorem 4.3.

PROOF. First note that $\alpha, \beta \in [0, \pi/2]$. The proof is divided into two parts.
Case 1. If $\alpha \le \beta$, then by Theorem 4.3,

$$\frac{\sin \beta}{2} = \sin \frac{\beta}{2} \cos \frac{\beta}{2} \le \varphi_K\left(\sin \frac{\alpha}{2}\right) \cos \frac{\beta}{2} \le \varphi_K\left(\sin \frac{\alpha}{2}\right) \cos \frac{\alpha}{2}.$$

By [13, Lemma 1], $r^{-1/K} \varphi_K(r) \le (rs)^{-1/K} \varphi_K(rs)$ for $r, s \in (0, 1)$, $K \ge 1$. Hence with $r = \sin(\alpha/2)$, $s = \cos(\alpha/2)$, we get

$$\varphi_K\left(\sin \frac{\alpha}{2}\right) \left(\cos \frac{\alpha}{2}\right)^{1/K} \le \varphi_K\left(\sin \frac{\alpha}{2} \cos \frac{\alpha}{2}\right) = \varphi_K\left(\frac{\sin \alpha}{2}\right),$$

and the second inequality follows.

Next, since $\varphi_{1/K}(r) < r^K$ for $K > 1$ (cf. [3, Theorem 3.18]),

$$\varphi_{1/K}\left(\frac{\sin\alpha}{2}\right)\left(\sec\frac{\alpha}{2}\right)^{K-1} \leq \left(\frac{\sin\alpha}{2}\right)^K \left(\sec\frac{\alpha}{2}\right)^{K-1}$$

$$= \left(\sin\frac{\alpha}{2}\right)^K \cos\frac{\alpha}{2}$$

$$\leq \sin\frac{\alpha}{2}\cos\frac{\alpha}{2} = \frac{\sin\alpha}{2} \leq \frac{\sin\beta}{2},$$

and the first inequality follows.

Case 2. If $\beta < \alpha$, then

$$\frac{\sin\beta}{2}\left(\sec\frac{\alpha}{2}\right)^{1-(1/K)} \leq \frac{\sin\alpha}{2}\left(\sec\frac{\alpha}{2}\right)^{1-(1/K)} = \sin\frac{\alpha}{2}\left(\cos\frac{\alpha}{2}\right)^{1/K}$$

$$\leq \left(\sin\frac{\alpha}{2}\cos\frac{\alpha}{2}\right)^{1/K} = \left(\frac{\sin\alpha}{2}\right)^{1/K} \leq \varphi_K\left(\frac{\sin\alpha}{2}\right),$$

proving the second inequality.

Finally, by [13, Lemma 1],

$$(rs)^{-K}\varphi_{1/K}(rs) \leq r^{-K}\varphi_{1/K}(s)$$

for $r, s \in (0, 1)$, hence with

$$r = \sin\frac{\alpha}{2}, \quad s = \cos\frac{\alpha}{2}$$

we get

$$\varphi_{1/K}\left(\frac{\sin\alpha}{2}\right)\left(\sec\frac{\alpha}{2}\right)^K \leq \varphi_{1/K}\left(\sin\frac{\alpha}{2}\right) \leq \sin\frac{\beta}{2}.$$

Therefore

$$\varphi_{1/K}\left(\frac{\sin\alpha}{2}\right)\left(\sec\frac{\alpha}{2}\right)^{K-1} \leq \sin\frac{\beta}{2}\cos\frac{\alpha}{2} \leq \sin\frac{\beta}{2}\cos\frac{\beta}{2} = \frac{\sin\beta}{2}. \qquad \square$$

We now derive an invariant form of a consequence of Lemma 4.6.

4.7. COROLLARY. *For a quadruple a, b, c, d of distinct points in $\mathbb{R}^2$, let $\sigma(a, b, c, d) = |a, b, c, d| + |a, b, d, c|$. If $f : \bar{\mathbb{R}}^2 \to \bar{\mathbb{R}}^2$ is a K-quasiconformal map, then*

$$\varphi_{1/K}\left(\frac{1}{2\sigma(a, b, c, d)}\right) \leq \frac{1}{2\sigma(f(a), f(b), f(c), f(d))} \leq \varphi_K\left(\frac{1}{2\sigma(a, b, c, d)}\right).$$

PROOF. By Möbius invariance we may assume that $b = f(b) = \infty$. Then the inequality is equivalent to

$$\varphi_{1/K}\left(\frac{|c-d|}{2(|a-c|+|a-d|)}\right) \leq \frac{|f(c)-f(d)|}{2(|f(a)-f(c)|+|f(a)-f(d)|)}$$

$$\leq \varphi_K\left(\frac{|c-d|}{2(|a-c|+|a-d|)}\right),$$

which is equivalent to

$$\varphi_{1/K}(\frac{\sin\alpha}{2}) \leq \frac{\sin\beta}{2} \leq \varphi_K(\frac{\sin\alpha}{2}),$$

where

$$\alpha = \arcsin\left(\frac{|c-d|}{|a-c|+|a-d|}\right), \ \beta = \arcsin\left(\frac{|f(c)-f(d)|}{|f(a)-f(c)|+|f(a)-f(d)|}\right). \ \square$$

4.8. COROLLARY. With notation as in Lemma 4.6, if $f(0) = 0$ and $|x-y|/(|x|+|y|) \geq \sqrt{3}/2$ (i.e., $\alpha \geq \pi/3$) then

$$|f(x) - f(y)| \leq (32\sqrt{3})^{1-1/K}|x-y|^{1/K}.$$

PROOF. By Lemma 4.6,

$$\frac{\sin\beta}{2} \leq \left(\cos\frac{\alpha}{2}\right)^{1-(1/K)} \varphi_K\left(\frac{\sin\alpha}{2}\right) \leq \left(\frac{\sqrt{3}}{2}\right)^{1-(1/K)} \varphi_K\left(\frac{\sin\alpha}{2}\right).$$

Hence by Theorem 4.3 and [16, (3.6), p.65]

$$|f(x) - f(y)| \leq 2(|f(x)|+|f(y)|)\left(\frac{\sqrt{3}}{2}\right)^{1-(1/K)} \varphi_K\left(\frac{|x-y|}{2(|x|+|y|)}\right)$$

$$\leq 2 \cdot 8^{1-(1/K)}\left(\frac{\sqrt{3}}{2}\right)^{1-(1/K)}\left(\frac{|x-y|}{2}\right)^{1/K} 4^{1-(1/K)}$$

$$= (32\sqrt{3})^{1-(1/K)}|x-y|^{1/K}. \hspace{2cm} \square$$

4.9. THEOREM. Let $f : \bar{\mathbb{R}}^2 \to \bar{\mathbb{R}}^2$ be K-quasiconformal, with $f(0) = 0$, $f(\infty)=\infty$, and $f(B) \subset B$. Then

(1) $$|f(x) - f(y)| \leq C(K)|x-y|^{1/K}$$

for all $x, y \in B$, where $C(K) = \min\{16^{1-(1/4K)}, 64^{1-(1/K)}\}$,

(2) $$|f(x) - f(y)| \leq 16^{1-(1/K)}\left[1 + \sqrt{1 - \frac{|x-y|}{4}}\right]^{2-(2/K)} |x-y|^{1/K}$$

for all $x, y \in B$.

PROOF. The second bound in (1) was obtained in [6, Theorem 5.8], where its proof was essentially the same as that of Corollary 4.5 above. For the first bound we consider two cases.

Case 1. If $|x - y| \geq 1/4$ then

$$
\begin{aligned}
|f(x) - f(y)| &\leq 2 \leq 2(4|x - y|)^{1/K} \\
&= 2^{1+(2/K)}|x - y|^{1/K} \leq 2^{4-(1/K)}|x - y|^{1/K}
\end{aligned}
$$

Case 2. $|x - y| < 1/4$.

a) If $|x + y| < 1$ then by [16, (3.9), (3.6), p.65]

$$
\begin{aligned}
|f(x) - f(y)| &\leq 2\varphi_K\left(\frac{|x - y|}{|1 - \bar{x}y|}\right) \leq 2\varphi_K\left(\frac{64}{47}|x - y|\right) \\
&\leq 2 \cdot 4^{1-(1/K)}\left(\frac{64}{47}\right)^{1/K}|x - y|^{1/K} \leq 16^{1-(1/4K)}|x - y|^{1/K}.
\end{aligned}
$$

b) If $|x + y| \geq 1$ we may assume that $0 < |x| \leq |y|$ and $x \neq y$. Let $R = R(C_0, C_1)$ be the plane ring with $C_0 = [x, y]$ and $C_1 = \{-t(x + y) : t \in [0, \infty)\}$, and let $R' = f(R)$. Then

$$
\text{cap } R \leq \tau\left(\frac{1 - a}{2a}\right),
$$

where $a = |x - y|/|x + y|$, so that

$$
\text{mod } R \geq \mu\left(\frac{a}{1 + a'}\right)
$$

by [24, Lemma 5.27] and Landen's transformation [11]. Next, by [16, Theorem 1.2, p.57],

$$
\begin{aligned}
\text{cap } R' &\geq 2\tau(|-\sqrt{f(x)}, -\sqrt{f(y)}, \sqrt{f(y)}, \sqrt{f(x)}|) \\
&= 2\tau\left(\frac{|\sqrt{f(x)} + \sqrt{f(y)}|^2}{|\sqrt{f(x)} - \sqrt{f(y)}|^2}\right) \\
&= \gamma\left(\frac{\sqrt{2(|f(x)| + |f(y)|)} \, |\sqrt{f(x)} + \sqrt{f(y)}|}{|f(x) - f(y)|}\right) \geq \gamma\left(\frac{4}{|f(x) - f(y)|}\right).
\end{aligned}
$$

Thus

$$
\text{mod } R' \leq \mu\left(\frac{|f(x) - f(y)|}{4}\right).
$$

Hence by quasi-invariance for the modulus [16, p. 38] we get

$$
\mu\left(\frac{a}{1 + a'}\right) \leq K\mu\left(\frac{|f(x) - f(y)|}{4}\right).
$$

Finally,

$$|f(x) - f(y)| \leq 4\varphi_K\left(\frac{a}{1+a'}\right)$$

$$\leq 4^{2-(1/K)}\left(\frac{a}{1+a'}\right)^{1/K} \leq 16(4+\sqrt{15})^{-1/K}|x-y|^{1/K}.$$

Next, for (2), let $r = (1/2)\sqrt{|x-y|}$. Then, by Lemma 4.6, [3, Theorem 3.27], [23, Theorem 1.7(1)], and [5, Theorem 3.1(3)] with $n = 2$, we get

$$|f(x) - f(y)| \leq 2(|f(x)| + |f(y)|)\varphi_K\left(\frac{|x-y|}{2(|x|+|y|)}\right)$$

$$\leq 2(\varphi_K(|x|) + \varphi_K(|y|))\varphi_K\left(\frac{|x-y|}{2(|x|+|y|)}\right)$$

$$\leq 4\varphi_K\left(\frac{|x|+|y|}{2}\right)\varphi_K\left(\frac{|x-y|}{2(|x|+|y|)}\right)$$

$$\leq 4\varphi_K^2(r) \leq 4(2(1+r'))^{2(1-\frac{1}{K})}r^{\frac{2}{K}}$$

$$= 16^{1-\frac{1}{K}}(1+r')^{2(1-\frac{1}{K})}|x-y|^{\frac{1}{K}}. \qquad \square$$

For points on the boundary of the unit disk we have the following improvements in Mori's theorem.

4.10. THEOREM. Let $f : \bar{\mathbb{R}}^2 \to \bar{\mathbb{R}}^2$ be K-quasiconformal with $f(\bar{B}) \subset \bar{B}$, $f(0) = 0$, $f(\infty) = \infty$. Suppose $x = e^{i\alpha}$, $y = e^{-i\alpha}$, $0 < \alpha \leq \pi/2$, and $|f(x)| = |f(y)| = 1$. Then

$$|f(x) - f(y)| \leq \left[8\left(1+\sqrt{1-\frac{\sin^2\alpha}{4}}\right)\cos\frac{\alpha}{2}\right]^{1-(1/K)}|x-y|^{1/K}$$

$$\leq 16^{1-(1/K)}|x-y|^{1/K}.$$

PROOF. We may assume that $f(x) = e^{i\beta}$, $0 < \beta < \pi/2$, $f(y) = e^{-i\beta}$.

Case 1. $\beta \leq \alpha$.

Then $\sin\beta \leq \sin\alpha$, and

$$|f(x) - f(y)| = 2\sin\beta \leq 2\sin\alpha = |x-y| = |x-y|^{1-(1/K)}|x-y|^{1/K}$$

$$\leq 2^{1-(1/K)}|x-y|^{1/K}.$$

Case 2. $\beta > \alpha$.

Then by Theorems 3.5, 4.3 and [13, Lemma 1],

$$
\begin{aligned}
|f(x) - f(y)| &= 2\sin\beta = 4\sin\frac{\beta}{2}\cos\frac{\beta}{2} \\
&\le 4\varphi_K\left(\sin\frac{\alpha}{2}\right)\cos\frac{\alpha}{2} \\
&= 4\varphi_K\left(\sin\frac{\alpha}{2}\right)\left(\cos\frac{\alpha}{2}\right)^{1/K}\left(\cos\frac{\alpha}{2}\right)^{1-(1/K)} \\
&\le 4\varphi_K\left(\sin\frac{\alpha}{2}\cos\frac{\alpha}{2}\right)\left(\cos\frac{\alpha}{2}\right)^{1-(1/K)} \\
&\le 4\varphi_K\left(\frac{\sin\alpha}{2}\right)\left(\cos\frac{\alpha}{2}\right)^{1-(1/K)} \\
&\le \left[8\left(1+\sqrt{1-\frac{\sin^2\alpha}{4}}\right)\cos\frac{\alpha}{2}\right]^{1-(1/K)}|x-y|^{1/K}. \qquad \square
\end{aligned}
$$

The next result is an extension of Mori's theorem to a more general class of domains.

4.11. THEOREM. (1) Let $f : \bar{\mathbb{R}}^2 \to \bar{\mathbb{R}}^2$ be K-quasiconformal, with $f(0) = 0$, $f(\infty) = \infty$. Then

$$
\frac{|f(x) - f(y)|}{|x-y|^{1/K}} \le 8^{1-(1/K)}\frac{|f(x)| + |f(y)|}{(|x| + |y|)^{1/K}}
$$

for all distinct $x, y \in \mathbb{R}^2$.

(2) Further, suppose $|f(x)| \le C^{1-(1/K)}|x|^{1/K}$ for all $x \in D$, where $0 \in D \subset \mathbb{R}^2$. Then

$$
|f(x) - f(y)| \le (16C)^{1-(1/K)}|x-y|^{1/K}
$$

for all $x, y \in D$.

PROOF. By Lemma 4.6,

$$
\frac{1}{2}\sin\beta \le \varphi_K(\frac{1}{2}\sin\alpha),
$$

where

$$
\alpha = \arcsin\left(\frac{|x-y|}{|x|+|y|}\right), \quad \beta = \arcsin\left(\frac{|f(x)-f(y)|}{|f(x)|+|f(y)|}\right).
$$

Thus

$$
\frac{|f(x)-f(y)|}{2(|f(x)|+|f(y)|)} \le \varphi_K\left(\frac{|x-y|}{2(|x|+|y|)}\right),
$$

so that

$$
|f(x)-f(y)| \le 2(|f(x)|+|f(y)|)\varphi_K\left(\frac{|x-y|}{2(|x|+|y|)}\right)
$$

and

$$|f(x) - f(y)| \ \leq \ 2(|f(x)| + |f(y)|)4^{1-(1/K)}\left(\frac{|x-y|}{2(|x|+|y|)}\right)^{1/K}$$

$$= \ 8^{1-(1/K)}\frac{|f(x)| + |f(y)|}{(|x|+|y|)^{1/K}}|x - y|^{1/K},$$

proving the first part.

For the second part, for $x \neq y$ we have

$$\frac{|f(x) - f(y)|}{|x-y|^{1/K}} \ \leq \ 8^{1-(1/K)}C^{1-(1/K)}\left[\left(\frac{|x|}{|x|+|y|}\right)^{1/K} + \left(\frac{|y|}{|x|+|y|}\right)^{1/K}\right]$$

$$\leq \ 8^{1-(1/K)}C^{1-(1/K)}2^{1-(1/K)} = (16C)^{1-(1/K)}. \qquad \square$$

4.12. COROLLARY. *If $f(B) \subset B$ in Theorem 4.11, then*

$$|f(x) - f(y)| \leq 64^{1-(1/K)}|x - y|^{1/K}.$$

PROOF. By the quasiconformal Schwarz Lemma [16, pp.63-65] we have $C \leq 4$. $\qquad \square$

4.13. THEOREM. *Let f be a K-quasiconformal automorphism of $\bar{\mathbb{R}}^2$ with $f(0) = 0$, $f(\infty) = \infty$, and $f(B) \subset B$. Then*

$$|f(x) - f(y)| \leq 16^{1-(1/K^2)}|x - y|^{1/K^2} \ for \ all \ x, y \in B.$$

PROOF. By [2, Theorem 2] and [23, Theorem 1.7]

$$|f(x) - f(y)| \ \leq \ 2(|f(x)| + |f(y)|)\varphi_K\left(\frac{|x-y|}{2(|x|+|y|)}\right)$$

$$\leq \ 2(\varphi_K(|x|) + \varphi_K(|y|))\varphi_K\left(\frac{|x-y|}{2(|x|+|y|)}\right)$$

$$\leq \ 4\varphi_K\left(\frac{|x|+|y|}{2}\right)\varphi_K\left(\frac{|x-y|}{2(|x|+|y|)}\right)$$

$$\leq \ 4\varphi_{K^2}\left(\frac{|x-y|}{4}\right)$$

$$\leq \ 4^{2-(1/K^2)}\left(\frac{|x-y|}{4}\right)^{1/K^2} = 16^{1-(1/K^2)}|x - y|^{1/K^2}. \qquad \square$$

4.14. THEOREM. *Let f be a K-quasiconformal automorphism of B with $f(0) = 0$. Then*

$$(4.15) \qquad |(|f(x)| - |f(y)|)| \le 32^{1-(1/K)} |(|x| - |y|)|^{1/K}$$

for all $x, y \in B$ if and only if f maps the concentric circles $|x| = r \in (0,1)$ onto concentric circles $|f(x)| = s \in (0,1)$.

PROOF. For the necessity, suppose there is $r \in (0,1)$ such that the image of $|x| = r$ is not a circle with center 0. Then there exist x, y with $|x| = |y| = r$ but $|f(x)| \ne |f(y)|$, in contradiction to (4.15).

Next, for the sufficiency, if $|x| = |y|$, then $|f(x)| = |f(y)|$, and we are done. Let $|x| = a$, $|y| = b$, $0 < a < b < 1$. Then $|f(x)| = c$, $|f(y)| = d$, $0 < c < d < 1$. Hence, by [23, Theorem 1.4],

$$\frac{|f(y)|}{d} - \frac{|f(x)|}{d} \le 8^{1-(1/K)} \left(\frac{|y|}{b} - \frac{|x|}{b} \right)^{1/K},$$

and

$$0 < |f(y)| - |f(x)| \le 8^{1-(1/K)} \frac{d}{b^{1/K}} (|y| - |x|)^{1/K}$$

$$\le 32^{1-(1/K)} (|y| - |x|)^{1/K}. \qquad \square$$

We next obtain results of Mori type in the hyperbolic metric [19]. We begin by recalling a result from [24] in a slightly modified form.

4.16. THEOREM. *For $K \ge 1$, let $f : B \to B$ be K-quasiconformal, $x, y \in B$, and let $\rho = \rho(x,y)$, $\rho' = \rho(f(x), f(y))$. Then*

$$(1) \qquad \operatorname{sh}^2 \frac{\rho'}{2} \le \tau^{-1} \left(\frac{1}{K} \tau \left(\operatorname{sh}^2 \frac{\rho}{2} \right) \right)$$

and

$$(2) \qquad \operatorname{th} \frac{\rho'}{4} \le \sqrt{\frac{1-a}{1+a}}, \quad a = \varphi_{1/K} \left(\frac{1}{\operatorname{ch}(\rho/2)} \right).$$

For $K = 1$ and $f(B) = B$ both (1) and (2) reduce to equality.

PROOF. A proof of inequality (1) is given in [24, Theorem 11.19]. For (2) we observe that

$$\tau^{-1} \left(\frac{1}{K} \tau \left(\operatorname{sh}^2 \frac{\rho}{2} \right) \right) = \frac{1-a^2}{a^2}$$

by virtue of [24,7.53, 5.61(2)] and (1.1). This identity together with (1) and

$$\operatorname{th} \frac{\rho'}{4} = \frac{b}{1 + \sqrt{1+b^2}}, \quad b = \operatorname{sh} \frac{\rho'}{2},$$

yield (2). Finally, the equality statement follows if we apply the results to f and f^{-1}. $\qquad \square$

4.17. COROLLARY. *Let the notation be as in Theorem 4.16. Then for* $x, y \in B$,

$$|f(x) - f(y)| \le 2\sqrt{\varphi_K(\text{th}^2(\rho/4))} \le 2\varphi_K(\text{th}(\tfrac{\rho}{4})).$$

PROOF. By [5, (3.10)], Theorem 4.13, and [3, (3.5), (3.13)], we have

$$|f(x) - f(y)| \le 2\,\text{th}\left(\frac{\rho'}{4}\right) \le 2\sqrt{\frac{1-a}{1+a}} = 2\sqrt{\varphi_K\left(\text{th}^2\left(\frac{\rho}{4}\right)\right)} \le 2\varphi_K\left(\text{th}\left(\frac{\rho}{4}\right)\right).$$
$\square$

4.18. COROLLARY. *Let notation be as in Theorem 4.16. Then for* $x, y \in B$,

$$|f(x) - f(y)| \le 2\sqrt{\varphi_{K/2}(\text{th}(\rho/2))}.$$

PROOF. By [3, (3.8)] we have

$$\varphi_K(\text{th}^2(\rho/4)) = \varphi_{K/2}(\varphi_2(\text{th}^2(\rho/4))) = \varphi_{K/2}\left(\frac{2\,\text{th}(\rho/4)}{1+\text{th}^2(\rho/4)}\right) = \varphi_{K/2}(\text{th}(\rho/2)).$$

The result now follows from Corollary 4.17. $\square$

4.19. THEOREM. *For* $K \ge 1$ *let* $f \in QC_K(B)$, $x, y \in B$, $\rho = \rho(x, y)$, $\rho' = \rho(f(x), f(y))$, $\ell = 2^{1-K}$, $L = 2^{1-(1/K)}$. *Then, for* $\rho < 4\,\text{arth}\,\ell$,

$$\rho' \le 4\,\text{arth}\left(\varphi_K\left(\text{th}\frac{\rho}{4}\right)\right) \le 8\,\text{arth}(Lt), \quad t \equiv \left(A\left(\text{th}(\tfrac{\rho}{4})\right)\right)^{1/K},$$

where $A(r) = r/(1+r'), r \in [0, 1]$.

PROOF. The first inequality follows from Theorem 4.16 (2) and [3, (3.5)], while the second one follows from [5, Theorem 1.11(1)]. $\square$

4.20. THEOREM. *For* $r \in (0, 1)$, $K \ge 1$,

$$\varphi_K(r) \le \left(\frac{1 - \ell a^K}{1 + \ell a^K}\right)^2,$$

where $\ell = 2^{1-K}$ *and* $a = (1 - \sqrt{r})/(1 + \sqrt{r})$.

PROOF. This follows from [5, Theorem 1.11(2)] and the identity

$$\varphi_K(r) = \frac{1 - \varphi_{1/K}(s)}{1 + \varphi_{1/K}(s)}, \quad s = \frac{1 - r}{1 + r}.$$
$\square$

4.21. REMARK. For $K > 0, 0 < r < 1$, let $T_K(r) = \text{th}(2\,\text{arth}(LA(r)^{1/K}))$, where $A(r) = r/(1 + r')$ and $L = 2^{1-(1/K)}$. Then it is easy to show that $T_K(T_N(r)) = T_{KN}(r)$ when r is so small that both sides are defined. If $t_K(r) = \text{th}(2\,\text{arth}(\ell A(r)^K))$, where $\ell = 2^{1-K}$, then $t_K(t_N(r)) = t_{KN}(r)$ and $T_K^{-1} = t_K$.

A drawback of Theorem 2.11 is the fact that the majorant is valid only for $r \in (0, \ell_1)$. We shall now use an iteration to show how one can apply this result for all $r \in (0, 1)$.

4.22. THEOREM. *For $K > 1$ let $\ell = 2^{1-K}$ and $L = 2^{1-(1/K)}$. For each integer j let $B_j(r) = \text{th}(2^j \text{ arth } r)$, $0 \leq r < 1$. Denote $B_j(\ell)$ by ℓ_j. If p is a positive integer and $r \in (0, \ell_p)$ then*

$$\varphi_K(r) \leq B_p(LB_{-p}(r)^{1/K}).$$

PROOF. For the induction proof we note first that by [5, Theorem 1.11(1)] the inequality holds for $p = 1$. Next we observe that by [16, p.64]

$$\varphi_2(r) = \frac{2\sqrt{r}}{1+r} = \text{th}(2 \text{ arth } \sqrt{r}),$$

$$\varphi_{1/2}(r) = \frac{1-r'}{1+r'} = \varphi_2^{-1}(r) = (\text{th}(\frac{1}{2} \text{arth } r))^2$$

and that $\varphi_K(r) = \varphi_2(\varphi_K(\varphi_{1/2}(r)))$. Suppose the inequality holds for some fixed $p \geq 1$. Then for $r \in (0, \ell_{p+1})$ we have

$$\varphi_K(r) = \text{th}(2 \text{ arth}(\sqrt{\varphi_K(\varphi_{1/2}(r))})) \leq \text{th}(2 \text{ arth}(\varphi_K(r/(1+r'))))$$

since $\varphi_K(s^2) \leq \varphi_K^2(s)$ [3, 3.13]. Since $r \in (0, \ell_{p+1})$ we have $r/(1+r') \in (0, \ell_p)$ and thus the induction hypothesis gives

$$\varphi_K(r) \leq \text{th}(2 \text{ arth}(B_p(L(B_{-p}(r/(1+r')))^{1/K}))) = B_{p+1}(L(B_{-p-1}(r))^{1/K}),$$

as desired.   □

For a given $r \in (0, 1)$ we can apply Theorem 4.22 with p as the smallest integer such that $r \in (0, \ell_p)$, that is with arth $r \leq 2^p$ arth ℓ. Note also that $B_{-p} = B_p^{-1}$.

4.23. REMARK. In a simplified form the method of proof of Theorem 4.22 yields the following technique of modification of the bounds for $\varphi_K(r)$. Since

$$\varphi_K(r) = \varphi_2(\varphi_K(\varphi_{1/2}(r))) = \varphi_{1/2}(\varphi_K(\varphi_2(r)))$$

and

$$\varphi_2(r) = \frac{2\sqrt{r}}{1+r}, \ \varphi_{1/2}(r) = \frac{1-r'}{1+r'},$$

the inequality

(a) $$f_1(r) \leq \varphi_K(r) \leq f_2(r) < 1$$

yields

(b) $$\frac{2\sqrt{f_1(u)}}{1+f_1(u)} \leq \varphi_K(r) \leq \frac{2\sqrt{f_2(u)}}{1+f_2(u)}; \ u = \varphi_{1/2}(r),$$

(c) $$g(f_1(v)) \leq \varphi_K(r) \leq g(f_2(v)),$$

where $g(t) = \varphi_{1/2}(t)$ and $v = \varphi_2(r)$.

It is possible that (a) – (c) together give better bounds for $\varphi_K(r)$ than (a) alone. We have received a preprint from D. Partyka [20], in which he elaborates on bounds similar to (a) – (c), obtaining interesting convergence properties. He shows, for example, that

$$\varphi_K(r) = \lim_{t \to \infty} \varphi_t(\varphi_{1/t}^{1/K}(r)).$$

5. Properties of μ

In this section we study some properties of $\mu(r)$, the basic function in (1.2) on which the distortion function $\varphi_K(r)$ is built. In the proof of our first theorem we shall need the following technical lemma.

5.1. LEMMA. *Let* $f : (a, b) \to (1, \infty)$ *be a log-convex monotone function. Then the function*

$$g(x) \equiv \frac{1 + f(x)}{\sqrt{f(x)}}$$

is also log-convex on (a, b).

PROOF. The derivative of $\log g(x)$ can be written as

(5.2) $$\frac{d}{dx} \log g(x) = \frac{1}{2} \frac{f'(x)}{f(x)} \frac{f(x) - 1}{f(x) + 1}.$$

If f is increasing, this is the product of two positive increasing functions. If f is decreasing, then

(5.3) $$-\frac{d}{dx} \log g(x) = -\frac{1}{2} \frac{f'(x)}{f(x)} \frac{f(x) - 1}{f(x) + 1}$$

is the product of two positive decreasing functions. $\square$

5.4. THEOREM. *The function* $f(x) \equiv \mu(x) + \mu(x')$ *is convex from* $(0, 1)$ *onto* $[\pi, \infty)$. *It is decreasing on* $(0, 1/\sqrt{2}]$ *and increasing on* $[1/\sqrt{2}, 1)$.

PROOF. By Jacobi's series [10, p.52],

(5.5) $$2f(x) = 8 \log 2 - \log(xx') + 3 \sum_{n=1}^{\infty} 2^{-n} \log \frac{a_n c_n}{b_n d_n},$$

where $a_0 = 1 = c_0$, $b_0 = x'$, $d_0 = x$,

$$a_{n+1} = \frac{a_n + b_n}{2}, \quad b_{n+1} = \sqrt{a_n b_n}, \quad c_{n+1} = \frac{c_n + d_n}{2}, \quad d_{n+1} = \sqrt{c_n d_n}.$$

It will be sufficient to prove that each term of (5.5) is convex. The proof is by induction. First,

$$\frac{d}{dx}\log\frac{1}{x} = -\frac{1}{x} \quad \text{and} \quad \frac{d}{dx}\log\frac{1}{x'} = \frac{x}{x'^2}$$

are increasing on $(0, 1)$. That is, a_0/b_0 and c_0/d_0 are log-convex. Hence, by Lemma 5.1,

$$\frac{a_1}{b_1} = \frac{1+x'}{\sqrt{x'}} \quad \text{and} \quad \frac{c_1}{d_1} = \frac{1+x}{\sqrt{x}}$$

are log-convex. If we assume that a_n/b_n and c_n/d_n are log-convex, then, by the same lemma,

$$\frac{a_{n+1}}{b_{n+1}} = \frac{1+a_n/b_n}{\sqrt{a_n/b_n}}, \quad \frac{c_{n+1}}{d_{n+1}} = \frac{1+c_n/d_n}{\sqrt{c_n/d_n}}$$

are log-convex. Hence each term of (5.5) is convex.

Finally,

$$f'(x) = \frac{\pi^2}{4xx'^2}\left[\frac{1}{\mathcal{K}^2(x')} - \frac{1}{\mathcal{K}^2(x)}\right],$$

which is negative on $(0, 1/\sqrt{2})$ and positive on $(1/\sqrt{2}, 1)$.   □

In our next theorem we shall need the following technical lemma.

5.6. LEMMA. *The function $g(r) \equiv \sqrt{1+r'^2}\mathcal{K}(r)$ is strictly increasing from $(0, 1)$ onto $(\pi/\sqrt{2}, \infty)$.*

PROOF. By Landen's transformation [11,164.02] we may write

$$g(r) = \frac{2\sqrt{1+r'^2}}{1+r'}\mathcal{K}\left(\frac{1-r'}{1+r'}\right),$$

and the result follows.   □

5.7. THEOREM. *The function $f(x) \equiv \mu(x) + \log x - \frac{1}{2}\log x'$ is strictly increasing and convex from $(0, 1)$ onto $(\log 4, \infty)$.*

PROOF. By [4, Lemma 2.1],

$$f'(x) = \frac{1}{2xx'^2\mathcal{K}^2(r)}\left[(1+x'^2)\mathcal{K}^2(x) - \frac{\pi^2}{2}\right],$$

which is positive by Lemma 5.6.

Next, by Jacobi's series [10, p.52]

$$f(x) = \log 4 + \frac{3}{2}\sum_{n=1}^{\infty}2^{-n}\log\frac{a_n}{b_n},$$

where $a_0 = 1$, $b_0 = x'$, $a_{n+1} = (a_n + b_n)/2$, $b_{n+1} = \sqrt{a_n b_n}$. By the proof of Theorem 5.4 each term of this series is convex.   □

5.8. COROLLARY. *For $K \geq 1$, $f(r) \equiv r^{-1/K} \varphi_K(r)/\sqrt{\varphi_{1/K}(r')}$ is increasing from $(0,1)$ onto $(4^{1-(1/K)}, \infty)$.*

PROOF. With $s = \varphi_K(r)$,

$$
\begin{aligned}
\log f(r) &= \log s - \frac{1}{K}\log r - \frac{1}{2}\log s' \\
&= (\mu(s) + \log s - \frac{1}{2}\log s') - \frac{1}{K}(\mu(r) + \log r)
\end{aligned}
$$

is increasing by Theorem 5.7 and [12, Lemma 6, p.514].   □

In [4, Theorem 4.3(2)] it was shown that the function $f : r \mapsto \mu(r) + \log r$ is concave on $(0,1)$. We shall now prove a stronger result, namely that $r \mapsto [r/(1+r')]\exp(\mu(r))$ is concave. For this we need a lemma.

5.9. LEMMA. *Suppose f and g are positive functions.*
  (1) *If f increasing and concave and g decreasing and concave, then $f \circ g$ is decreasing and concave.*
  (2) *If f is decreasing and concave and g increasing and convex, then $f \circ g$ is decreasing and concave.*

PROOF. By the Chain Rule,

$$
-(f \circ g)'(x) = f'(g(x))(-g'(x)) = -f'(g(x))g'(x).
$$

In both cases this is the product of two positive increasing functions.   □

5.10. LEMMA. *The function $f(r) \equiv r\exp(\mu(r))$ is decreasing and concave from $(0,1)$ onto $(1,4)$.*

PROOF. The monotoneity and range follow from [4, Theorem 4.3(2)]. Next, by [4, Lemma 2.1], we may write

$$
-f'(r) = g(r)h(r),
$$

where

$$
g(r) \equiv \frac{r}{r'}\exp(\mu(r)), \quad h(r) = \frac{1}{r}\left[\frac{\pi^2}{4r'\mathcal{K}^2(r)} - r'\right].
$$

Now g is increasing by [4, Theorem 4.3(3)]. To show that h is increasing it will be sufficient by [5, Lemma 2.2] to show that

$$
F(r) \equiv \frac{\pi^2}{4r'\mathcal{K}^2(r)} - r'
$$

has an increasing derivative. Now

$$
F'(r) = \frac{\pi^2}{4r'^3\mathcal{K}^3}\frac{(\mathcal{K} - \mathcal{E}) - (\mathcal{E} - r'^2\mathcal{K})}{r} + \frac{r}{r'}.
$$

Clearly r/r' is increasing, while $r'\mathcal{K}$ is decreasing by [4, Theorem 2.2(3)]. Next

$$\frac{d}{dr}[(\mathcal{K} - \mathcal{E}) - (\mathcal{E} - r'^2\mathcal{K})] = \frac{r}{r'^2}(\mathcal{E} - r'^2\mathcal{K})$$

is increasing by [4, Theorem 2.2(7)]. Hence F' is increasing by [5, Lemma 2.2], $-f'$ is increasing, and finally f' is decreasing. □

5.11. THEOREM. *The function*

$$g(r) \equiv \frac{r}{1+r'}\exp(\mu(r))$$

is decreasing and concave from $(0,1)$ *onto* $(1,2)$.

PROOF. Let $t = (1 - r')/(1 + r')$. Then

$$g^2(r) = f\left(\frac{1-r'}{1+r'}\right),$$

where f is the function in Lemma 5.10. The result then follows from Lemma 5.9. □

5.12. THEOREM. *The function* $f(r) \equiv \exp(-c\mu(r))$, $0 < r < 1$, *is convex if and only if* $c \geq 1$. *If* $c < 1$ *it is neither convex nor concave.*

PROOF. If $c = 1$,

$$f'(r) = e^{-\mu(r)}\frac{\pi^2}{4rr'^2\mathcal{K}^2(r)} = \frac{\pi^2}{4}\frac{1}{r'^2\mathcal{K}^2(r)}\exp(-(\mu(r) + \log r)).$$

By [4, Theorem 2.2(2),(3)] this is the product of two positive increasing functions, hence increasing. If $c > 1$ then $f(r) = (e^{-\mu(r)})^c$ is the composition of two convex functions, hence convex. If $c < 1$, then

$$\frac{4}{\pi^2}f'(r) = \frac{ce^{-c(\mu(r)+\log r)}}{r'^2\mathcal{K}^2(r)} \cdot \frac{1}{r^{1-c}}.$$

Hence $f'(1^-) = f'(0^+) = \infty$, so that f' cannot be monotone. □

We conclude this section by proving some additional inequalities for μ.

5.13. LEMMA. *For* $0 < r < 1$, $r' = \sqrt{1 - r^2}$, $c = \exp(\pi/2)$, *we have*

$$\log\frac{4}{r'} < \mathcal{K}(r) < \log\frac{c}{r'}.$$

PROOF. The lower bound is well known (see, e.g., [4, 2.2(5)]). The upper bound was proved in [4, 2.2(2)]. □

5.14. THEOREM. *For* $0 < r < 1$, $r' = \sqrt{1-r^2}$, $c = \exp(\pi/2)$,

(1)
$$(\log 4)\frac{\log(4/r)}{\log(4/r')} < \frac{\pi}{2}\frac{\log(4/r)}{\log(c/r')} < \mu(r),$$

(2)
$$\mu(r) < \frac{\pi}{2}\frac{\log(c/r)}{\log(4/r')} < \frac{\pi^2}{4\log 4}\frac{\log(4/r)}{\log(4/r')}.$$

PROOF. In (1) the first inequality is elementary and the second one follows from (1.2) and Lemma 5.13. Inequality (2) follows from (1) and the functional identity $\mu(r)\mu(r') = \pi^2/4$ for μ [16, (2.7), p.61]. □

The next result improves [4, Lemma 4.2 (5)].

5.15. THEOREM. *Let* $m(r) = (2/\pi)r'^2\mathcal{K}(r)\mathcal{K}'(r)$, $0 < r < 1$. *Then the functions* $r'\mu(r)/m(r)$, $m(r)/\log(1/r)$, *and* $r'\mu(r)/\log(1/r)$ *are strictly increasing from* $(0,1)$ *onto* $(1,\infty)$. *In particular,*

$$\mu(r) > (2/\pi)r'\mathcal{K}(r)\mathcal{K}'(r) > (1/r')\log(1/r).$$

PROOF. The monotoneity of $r'\mu(r)/m(r) = \pi^2/(4r'\mathcal{K}^2(r))$ follows from the fact that $r'\mathcal{K}^2(r)$ is strictly decreasing from $(0,1)$ onto $(0,\pi^2/4)$ [4, Theorem 2.2 (3)], and the monotoneity of $m(r)/\log(1/r)$ follows from [5, Lemma 2.2] and the Legendre identity [11, 110.10]. Then the monotoneity of $r'\mu(r)/\log(1/r)$ follows immediately. □

The next result gives another pair of lower bounds for $\mu(r)$.

5.16. THEOREM. *For* $0 < r < 1$, $r' = (1-r^2)^{1/2}$,

$$\mu(r) > r'\log 4 + \log(1/r) > (1+r)^2\log(1/r).$$

PROOF. From [16, (2.10), p.61] and the fact that $(1/x)\log(1+x)$ is decreasing on $(0,1)$ we have

$$\mu(r) > 2\log(1+r') + \log(1/r) > 2(\log 2)r' + \log(1/r) = r'\log 4 + \log(1/r).$$

For the second inequality we need to prove that $r'\log 4 > r(r+2)\log(1/r)$. We let

$$f(r) = \frac{r'\log 4}{r^2 + 2r} - \log\frac{1}{r}.$$

Then $f(1) = 0$, $f(0+) = \infty$. Thus it is enough to prove that $f'(r) < 0$ or, equivalently, $rr'(r+2)^2 < (r+2+rr'^2)\log 4$. For this let $g(r) = rr'(r+2)$. If $0 \le r \le 1/\sqrt{2}$ then $g(r) \le g(1/\sqrt{2}) < 1.36 < \log 4$. If $1/\sqrt{2} < r < 1$, then $g'(r) = 0$ iff $h(r) \equiv 3r^3 + 4r^2 - 2r - 2 = 0$. But $h(r) = 0$ has exactly one root, say r_0, in $(1/\sqrt{2}, 1)$ and $g(r) \le g(r_0)$ there. Clearly $1/\sqrt{2} < r_0 < 0.75$. Since rr' is decreasing and $r+2$ is increasing on $(1/\sqrt{2}, 1)$, we get $r_0r_0' < 1/2$ and

$r_0 + 2 < 2.75$. Hence $g(r_0) < (1/2)(2.75) = 1.375 < \log 4$, so that $1/\sqrt{2} < r < 1 \Rightarrow rr'(r+2) \le g(r_0) < \log 4$. Thus

$$rr'(r+2)^2 \le (r+2)\log 4 < (r+2+rr'^2)\log 4,$$

and we have $f'(r) < 0$.   □

5.17. REMARK. It is easy to show that Theorem 5.15 cannot be improved to $\mu(r) > (1/r')^2 \log(1/r)$ and Theorem 5.16 cannot be improved to $\mu(r) > (1+r)^3 \log(1/r)$.

5.18. THEOREM. For $0 < r < 1$,

$$\frac{\pi}{2\mathcal{K}(r)} < \mathrm{th}\,\mu(r) < \frac{2\pi}{2\mathcal{K}(r) + \pi}.$$

There is equality throughout when r tends to 0 or 1.

PROOF. Since $\mathcal{K}(0) = \pi/2$, $\mu(1-) = 0$, and $\mathcal{K}(1-) = \mu(0+) = \infty$, equality is clear when r tends to 0 or 1. Next, let $q = \exp(-2\mu(r))$ for $0 < r < 1$. Then $0 < q < 1$ and

$$\mathrm{th}(\mu(r)) = \frac{1 - e^{-2\mu(r)}}{1 + e^{-2\mu(r)}} = \frac{1-q}{1+q},$$

so that

$$\frac{1}{\mathrm{th}(\mu(r))} = \frac{1+q}{1-q} = \frac{2}{1-q} - 1 = 1 + 2\sum_{n=1}^{\infty} q^n,$$

and by [11, 900.04]

$$\mathcal{K}(r) = \frac{\pi}{2}\left[1 + 4\sum_{n=1}^{\infty}\frac{q^n}{1+q^{2n}}\right] > \frac{\pi}{2}\left[1 + 2\sum_{n=1}^{\infty}q^n\right] = \frac{\pi}{2}\frac{1}{\mathrm{th}\,\mu(r)}.$$

For the upper bound we write

$$\mathcal{K}(r) = \frac{\pi}{2}\left[1 + 4\sum_{n=1}^{\infty}\frac{q^n}{1+q^{2n}}\right] < \frac{\pi}{2}\left[1 + 4\sum_{n=1}^{\infty}q^n\right]$$

$$= \frac{\pi}{2}\left[1 + 2\left(\frac{1}{\mathrm{th}(\mu(r))} - 1\right)\right] = -\frac{\pi}{2} + \frac{\pi}{\mathrm{th}(\mu(r))}.$$ □

In conclusion we obtain the following improvement of He's theorem [13].

5.19. THEOREM. *For each $K > 1$ the function*

$$f(r) \equiv \varphi_K(r)\mathrm{ch}((1/K)\mathrm{arch}(1/r))$$

is decreasing from $(0,1)$ onto $(1, 2^{1-(1/K)})$. In particular,

$$\frac{2r^{1/K}}{(1+r')^{1/K} + (1-r')^{1/K}} < \varphi_K(r) < \frac{2^{2-(1/K)}r^{1/K}}{(1+r')^{1/K} + (1-r')^{1/K}}$$

for $K > 1$ and $0 < r < 1$.

PROOF. Let $s = \varphi_K(r)$ and $t = 1/\mathrm{ch}((1/K)\mathrm{arch}((1/r)))$. Then $s' = \varphi_{1/K}(r')$ and $t' = \mathrm{th}((1/K)\mathrm{arth}(r'))$, so that $s' < t'$ by [3,Lemma 3.15]. Next, by [4, Lemma 2.1(2)],

$$t^2 f'(r) = t\frac{ds}{dr} - s\frac{dt}{dr} = \frac{t}{K}\frac{ss'^2\mathcal{K}^2(s)}{rr'^2\mathcal{K}^2(r)} - \frac{s}{K}\frac{tt'}{rr'} = \frac{stt'}{Krr'}\left[\frac{s'}{t'}\frac{s'\mathcal{K}^2(s)}{r'\mathcal{K}^2(r)} - 1\right],$$

which is negative by [4,Theorem 2.2(3)]. The inequalities follow by the explicit expression

$$\mathrm{ch}((1/K)\mathrm{arch}(1/r)) = \frac{(1+r')^{1/K} + (1-r')^{1/K}}{2r^{1/K}}$$

and the fact that $\lim_{r\to 0} r^{-1/K}\varphi_K(r) = 4^{1-(1/K)}$ [16, p.65]. □

6. Conjectures

(1) $\mathrm{th}\,\varphi_K(r) \le \varphi_K(\mathrm{th}\,r)$ for $K \in [1,\infty)$, $r \in [0,1]$.

(2) For fixed $K > 1$ define

$$f_K(r) = r^{-1/K}\varphi_K(r)\frac{\mathcal{K}(\varphi_K(r))}{\mathcal{K}(r)}.$$

For $1 < K < 2$, $f_K(r)$ is strictly decreasing, $f_2(r)$ is the constant function 2, and for $K > 2$, $f_K(r)$ is strictly increasing.

(3) For $K > 2$, $g_K(r) \equiv \varphi_{1/K}^{1/K}(r)\mathcal{K}(\varphi_{1/K}(r))/(r\mathcal{K}(r))$ is strictly decreasing from $(0,1)$ onto $(1/K, 4^{(1/K)-1})$.

(4) For $K \ge 1$, $r \in (0,1)$,

$$\frac{\varphi_K^2(\sqrt{r})}{\varphi_K(r)} \le (1+r')^{2(1-1/K)}.$$

(5) $f(K,r) \equiv \log\varphi_K(r)$ is jointly concave as a function of two variables (K,r) on $(1,\infty) \times (0,1)$.

6.1. REMARK. The authors have verified conjectures 2 and 4 for the special values of $K = 2^n$, $n \in \mathbb{N}$. S.-L. Qiu informs us that he has proved (1)–(4) of the above conjectures.

6.2. OPEN PROBLEM. Find counterparts of Theorem 2.1 and Corollary 2.4 for $K \in (0, 1)$.

REFERENCES

[1] S. Agard, *Distortion theorems for quasiconformal mappings*, Ann. Acad. Sci. Fenn. Ser. AI **413** (1968), 1-12.

[2] S. Agard and F. W. Gehring, *Angles and quasiconformal mappings*, Proc. London Math. Soc., **14** (1965), 1-21.

[3] G. D. Anderson, M. K. Vamanamurthy, and M. Vuorinen, *Distortion functions for plane quasiconformal mappings*, Israel J. Math., **62** (1988), 1-16.

[4] G. D. Anderson, M. K. Vamanamurthy, and M. Vuorinen, *Functional inequalities for complete elliptic integrals and their ratios*, SIAM J. Math. Anal. **21** (1990), 536-549.

[5] G. D. Anderson, M. K. Vamanamurthy, and M. Vuorinen, *Inequalities for quasiconformal mappings in space*, Pacific J. Math. **160** (1993), 1-18.

[6] G. D. Anderson, M. K. Vamanamurthy, and M. Vuorinen, *Conformal invariants, quasiconformal maps, and special functions*, in Quasiconformal Space Mappings: A Collection of Surveys, Lecture Notes in Math., Vol. **1508**, Springer-Verlag, 1992, 1-19.

[7] E. F. Beckenbach and R. Bellman, *Inequalities*, Ergebnisse der Mathematik und ihrer Grenzgebiete, Vol. **30**, Springer-Verlag, Berlin – Heidelberg – New York, 1961.

[8] B. C. Berndt, *Ramanujan's Notebooks*, Vol. I, Springer-Verlag, Berlin – Heidelberg – New York, 1985.

[9] B. C. Berndt, *Ramanujan's Notebooks*, Vol. III, Springer-Verlag, Berlin – Heidelberg – New York, 1991.

[10] J. M. Borwein and P. B. Borwein, *Pi and the AGM*, John Wiley & Sons, New York, 1987.

[11] P. F. Byrd and M. D. Friedman, *Handbook of Elliptic Integrals for Engineers and Physicists*, Die Grundlehren der math. Wissenschaften Vol. **57**, Springer-Verlag, Berlin – Göttingen – Heidelberg, 1954.

[12] F. W. Gehring, *Symmetrization of rings in space*, Trans. Amer. Math. Soc. **101** (1961), 499-519.

[13] C.-Q. He, *Distortion estimates of quasiconformal mappings*, Sci. Sinica Ser. A **27** (1984), 225-232.

[14] J. Hersch and A. Pfluger, *Généralisation du lemme de Schwarz et du principe de la mesure harmonique pour les fonctions pseudo-analytiques*, C. R. Acad. Sci. Paris **234** (1952), 43-45.

[15] O. Hübner, *Remarks on a paper by Lawrynowicz on quasiconformal mappings*, Bull. de l'Acad. Polon. des Sci. **18** (1970), 183-186.

[16] O. Lehto and K. I. Virtanen, *Quasiconformal Mappings in the Plane*, 2nd ed., Die Grundlehren der math. Wissenschaften, Band **126**, Springer-Verlag, New York – Heidelberg – Berlin, 1973.

[17] W. Magnus, F. Oberhettinger, and R. P. Soni, *Formulas and Theorems for the Special Functions of Mathematical Physics*, Die Grundlehren der math. Wissenschaften, Band **52**, Springer-Verlag, New York, 1966.

[18] A. Mori, *On quasi-conformality and pseudo-analyticity*, Trans. Amer. Math. Soc. **84** (1957), 56-77.

[19] R. Nevanlinna, *Analytic Functions*, Die Grundlehren der math. Wissenschaften, Band **162**, Springer-Verlag, New York – Heidelberg – Berlin, 1970.

[20] D. Partyka, *Approximation of the Hersch-Pfluger distortion function. Applications,* Ann. Univ. Mariae Curie – Sklodowska Sect. A. XLV, **12** (1992), 99-111.

[21] S.-L. Qiu, *Distortion properties of $K-qc$ maps and better estimate of Mori's constant,* Acta Math. Sinica **35** (1992), 492-504 (Chinese).

[22] V. I. Semenov, *Certain applications of the quasiconformal and quasiisometric deformations,* Rev. Roumaine Math. Pures Appl. **36** (1991), 503-511.

[23] M. K. Vamanamurthy and M. Vuorinen, *Functional inequalities, Jacobi products, and quasiconformal maps,* Illinois J. Math. (to appear).

[24] M. Vuorinen, *Conformal Geometry and Quasiregular Mappings,* Lecture Notes in Math., Vol. **1319**, Springer-Verlag, 1988.

G. D. Anderson
Michigan State University
East Lansing MI 48824 USA

M. K. Vamanamurthy
University of Auckland
Auckland, New Zealand

M. Vuorinen
University of Helsinki
Helsinki, Finland

Contemporary Mathematics
Volume **169**, 1994

A LOOK AT THE BATEMAN PROJECT

RICHARD ASKEY

ABSTRACT. General comments on the three volumes of Higher Transcendental Functions, which Magnus coauthored, are given. This is followed by a look at quadratic and cubic transformations for hypergeometric functions.

1. GENERAL COMMENTS

First, I want to thank the committee for the invitation to speak at this meeting. Also, I want to thank Joan Birman for the very interesting comments on Magnus as a teacher and friend which she asked for from some of you and from others. These were published in the AWM Newsletter. I did not know Magnus, but I remember Creighton Buck telling me that Magnus had three new Ph.D. students each year. After looking at the list in his "Collected Works", I see this estimate was essentially correct. I still do not understand how he found the time or the ideas for this, but the comments of some of his students help a little.

My connection with Magnus is through his work on handbooks of special functions, see [19, 20, 21, 22,23]. This was a small part of his work, but an important part. While most mathematicians take these books for granted, they are both important and hard to write. The authors need to know not only what to include, but also what to leave out. A broad knowledge of applications is essential, but that is not enough. It is also very useful to be able to predict what will happen in the future, since these books are not written very often and what they contain can influence what will be discovered. I will look at some of the chapters in the three volumes of "Higher Transcendental Functions"; (or HTF), pointing out some of the important results which have been found in the last forty years, and also pointing out some results which were known then but were either not fully appreciated or were known in the sense that they had been published but were not really known. About fifteen years ago I wrote a short appreciation [4] of the volumes which are sometimes referred to as the "Bateman Project", but are better referred to as "Higher Transcendental Functions", and sent a copy of it to Magnus to see if I had made any obvious misstatements. This paper was written for an issue of a journal dedicated to Arthur Erdélyi's memory. His immediate response was generous in his praise of Erdélyi, which I now see was characteristic of him. Some of this response was printed in [5, pp. 381–382]. The most forward-looking chapter in these volumes is the one on spherical harmonics, which Magnus wrote using unpublished notes of Herglotz as a basis.

1991 *Mathematics Subject Classification.* Primary 33–00, 33C20.

Supported by NSF Grant DMS–8922990.

The first chapter deals with gamma and beta functions, or as I would now say, with the gamma function and beta integrals. One nice feature of the chapter is the inclusion of Mellin-Barnes type integrals, including what I call Barnes's beta integral, which is 1.19(8).

(1.1)
$$\int_{-i\infty}^{i\infty} \Gamma(a+x)\Gamma(b+x)\Gamma(c-x)\Gamma(d-x)dx$$
$$= \frac{2\pi i \Gamma(a+c)\Gamma(a+d)\Gamma(b+c)\Gamma(b+d)}{\Gamma(a+b+c+d)}.$$

Here the contour is the imaginary axis if the parameters have positive real parts and is indented appropriately in other allowable cases. It is not clear from reading this section if the authors thought this integral belonged in this chapter because it contained gamma functions in the integrand and in the value, or because it is an extension of Euler's beta integral. I suspect they thought the first was so, although if it were pointed out to them that the second was also true, they would have thought a few minutes and then said "of course". That was my reaction when I saw the paper [10] which introduced a new orthogonality relation for a known set of polynomials and realized that these polynomials extended Jacobi polynomials, which are the polynomials orthogonal with respect to Euler's beta integral. Since this was true, the weight functions had to be related in the same way. This description is not quite correct since Jim Wilson and I had earlier [7] found the symmetric case of these polynomials; yet I still remember my annoyance at not finding the general case. I would have if I had really understood (1.1).

There are important types of beta integrals which are omitted. The one which was well-known then is Dirichlet's multidimensional beta integral. This is important, especially in multivariate statistics, but it is really an iterated one-dimensional integral. In the early 1940's, A. Selberg [45, 46] found a very important multidimensional beta integral which is not just an iteration of one-dimensional integrals.

(1.2)
$$\int_0^1 \cdots \int_0^1 \prod_{1\leq i<j\leq n} |t_i - t_j|^{2c} \prod_{j=1}^{n} t_j^{a-1}(1-t_j)^{b-1}dt_j$$
$$= \prod_{j=1}^{n} \frac{\Gamma(a+(n-j)c)\Gamma(b+(n-j)c)\Gamma(jc+1)}{\Gamma(a+b+(2n-j-1)c)\Gamma(c+1)}.$$

The paper [46], which contained a derivation of this result, was reviewed very nicely in Mathematical Reviews and then almost completely forgotten for forty years. Recently, an extension was found by Aomoto [2].

(1.3)
$$\int_0^1 \cdots \int_0^1 \prod_{j=1}^{k} t_j \prod_{1\leq i<j\leq n} |t_i - t_j|^{2c} \prod_{j=1}^{n} t_j^{a-1}(1-t_j)^{b-1}dt_j$$
$$= \prod_{j=1}^{k} \frac{(a+(n-j)c)}{(a+b+(2n-j-1)c)} \prod_{j=1}^{n} \frac{\Gamma(a+(n-j)c)\Gamma(b+(n-j)c)\Gamma(jc+1)}{\Gamma(a+b+(2n-j-1)c)\Gamma(c+1)}.$$

Not only is Aomoto's extension useful in some settings where Selberg's integral is not quite general enough, but also his derivation of (1.3) is simple enough so it can be taught to students in advanced calculus. Selberg's integral has a number of important applications. One came from work on large nuclei by M. L. Mehta and Freeman Dyson [36]. They needed to evaluate the derivative of

$$(1.4) \qquad M_n(z) = \frac{1}{(2\pi)^{n/2}} \int_{-\infty}^{\infty} \cdots \int_{-\infty}^{\infty} \prod_{1 \le i < j \le n} |t_i - t_j|^{2z} \prod_{j=1}^{n} e^{-t_j^2/2} dt_j$$

when $z = 1/2$. They were able to add the case $z = 2$ to the cases $z = 1$ and $z = 1/2$ where the value of the integral in (1.4) was known. They could also evaluate (1.4) when $n = 2$, and later someone else did the case $n = 3$. There was a natural conjectured value for (1.4), which they stated, but the problem then remained open for many years before it was realized that their conjecture followed from Selberg's integral. A natural setting for these integrals, and a related one of Dyson [18], has been found in the last ten years. This setting is root systems, i.e., sets of vectors with a large group of symmetries which arise in many parts of mathematics. For example, they can be used to generate semi-simple Lie algebras. There are five infinite families of root systems and five exceptional ones. To each of these is associated a beta integral with free parameters which distinguish roots of different lengths. Selberg's integral is attached to $BC(n)$ and by inclusion also to $B(n)$, $C(n)$ and $D(n)$. $A(n)$ is a different object, and it has two beta type integrals attached to it. One is that of Dyson which was mentioned above, and the other is a second beta integral of Selberg which corresponds to his integral (1.2) in the same way that Cauchy's beta integral corresponds to Euler's integral. For $A(n)$, all the roots have the same lengths, but it is possible to add extra parameters. Thus root lengths do not tell the whole story.

For Selberg's case, there are two parameters in one dimension and three for all higher dimensions. Dyson's integral has two parameters in one dimension, and a new parameter is added for each new dimension.

To give you a taste of this, the case of $BC(n)$ is equivalent to the following.

$$(1.5) \quad \text{C.T.} \prod_{j=1}^{n} h(x_j, a) h(x_j^2, b) \prod_{1 \le i < j \le n} h(x_i/x_j, c) h(x_i x_j, c)$$

$$= \prod_{j=0}^{n-1} \frac{(2a + 2b + 2cj)!(2b + 2cj)!(c(j+1))!}{(a + 2b + (n + j - 1)c)!(a + b + cj)!(b + cj)!c!} .$$

where

$$(1.6) \qquad h(x, a) = (1 - x)^a (1 - x^{-1})^a .$$

Here C.T. means to take the constant term of the Laurent polynomial, and the x_j come from the roots r_j by taking a formal exponential,

$$x_j = \exp(r_j) .$$

The realization that Selberg's integral can be given as this constant term identity is due to Macdonald [34] as is the connection with root systems. Dyson [18] was the first to state one of these identities as a constant term identity.

We now know that it would have been useful to include a q version of the gamma function in this chapter and some of the many q versions of beta integrals. I will say more about this later but just remark that the q gamma function is the natural extension of the q factorial, which is defined by

$$(1.7) \quad n!_q = 1(1+q)(1+q+q^2)\cdots(1+q+\cdots+q^{n-1}) = \prod_{j=1}^{n}(1-q^j)/(1-q)^n.$$

It would also have been useful to include a section on Gauss and Jacobi sums. These are finite field analogues of the gamma function and a beta integral and have applications in number theory which are old but constantly being added onto, and applications in some part of combinatorics. There are now p-adic versions of the gamma function and of beta integrals. This should be included in future handbooks. Most mathematicians would be surprised to know that some physicists are studying p-adic strings and also q-strings.

Chapter 2, which was primarily written by Magnus, is the most important chapter. There were books on hypergeometric functions and books like Whittaker and Watson [53] which included a chapter on hypergeometric functions, but there was nothing as useful as this chapter. Bailey's book [13] primarily deals with higher hypergeometric functions, a subject treated in Chapter 4 of HTF. Since hypergeometric functions are not familiar to many mathematicians, here are a few definitions and comments.

A generalized hypergeometric series is a series Σa_n with the term ratio a_{n+1}/a_n equal to a rational function of n. When the ratio a_{n+1}/a_n is

$$\frac{(n+a)(n+b)x}{(n+c)(n+1)}$$

the series is called a hypergeometric series. This series is usually written as

$$(1.8) \qquad {}_2F_1\left(\begin{matrix} a, b; x \\ c \end{matrix}\right) = \sum_{n=0}^{\infty} \frac{(a)_n(b)_n}{(c)_n n!} x^n$$

where the shifted factorial $(a)_n$ is defined by

$$(1.9) \qquad (a)_n = \Gamma(n+a)/\Gamma(a) = a(a+1)\cdots(a+n-1).$$

One reason these hypergeometric functions are important is that they satisfy second order linear differential equations with at most three regular singular points. Riemann [41] showed that this property is very strong, and the differential equations are determined up to a few constants by the location of the singular points. These constants determine the singularities of the solutions. However, there are also other reasons these functions are important. They satisfy important difference equations in the parameters. Gauss made a systematic study of these in connection with continued fractions, but they are even more important in connection with orthogonal polynomials. The singular points of the differential equation can be mapped to each other, and this induces transformations in the functions. There is a special case of these functions where a quadratic transformation exits. These functions are Legendre functions, the subject of Chapter 3 of HTF. Hobson [30] had written a long book on these functions and remarked that the methods of Barnes using integrals of gamma functions "leads to considerable economy of labour in obtaining various transformations requisite for the investigation of various forms in which the functions can be represented", but he never mentioned another key observation of Barnes that Legendre functions are just hypergeometric functions which have a quadratic transformation, and so the transformations found by Kummer are the new transformations which need to be added to the linear fractional transformations to give all the necessary transformations. That is said explicitly in [19, page 121].

Chapter 4 deals with generalized hypergeometric series in one variable. Again this is a chapter Magnus wrote, or at least he wrote most of it. Here the connection with differential equations is not as important, so it was not clear forty years ago which results would be very important and which more incidental. Most of the material included is important, but some very important material was only briefly mentioned. The best illustration of this is basic hypergeometric series, a topic covered in three pages, which we now see should have had a full chapter. Basic hypergeometric series are series of the form

$$(1.10) \qquad \Sigma a_n \text{ with } a_{n+1}/a_n = \text{a rational function of } q^n.$$

These have been studied since 1750 or so when Euler discovered some important identities, but they were usually thought to be a minor branch of number theory with some connections to elliptic functions. With the recent development of quantum groups, it is now recognized that they play a role in many other areas where combinatorial problems arise. In the last century, Heine [29] and Rogers [42, 43, 44] found very important results. Heine found extensions of say half of the material in Chapter 2 of HTF, and Rogers built on this to discover some very important results, some of which had not been found for hypergeometric functions yet. Two of Rogers's identities, those known as the Rogers-Ramanujan identities, are listed. That is the only reference to Rogers I have found in these books and is one of the few references to Ramanujan. One type of hypergeometric result which should have been mentioned is a set of infinite series found by Ramanujan [39, 40] giving a representation of π as a rapidly convergent series. Jonathan and Peter Borwein and David and Gregory Chudnovsky finally gave complete proofs of these identities of Ramanujan [14, 15, 16], and found some he had missed; the Chudnovskys have a very important paper explaining the new ideas Ramanujan had which led to these series. In addition, the Chudnovskys found p-adic versions of these results and used them to compute π to a very large number of places. An interesting general account of these calculations appears in the March 2nd, 1992 issue of "The New Yorker" [38]. When reading this, don't forget that these calculations are only a very small part of the work the Chudnovskys are doing. It is just the part of their work which can be most easily written about for a general audience.

The few references to Ramanujan's work are one of the shortcomings in HTF, but this can be understood since we now have a much better idea where some of his works fits with the rest of mathematics than anyone had forty years ago.

The next chapter is one of three which deals with problems in several variables. Forty years ago it was only barely possible to write more on multiple hypergeometric functions than was contained in the classic book of Appell and Kampé de Fériet [3]. Now it is possible to write much more, but it is still not clear what the final form of multiple hypergeometric functions will be. The one thing we know is that they will be very important.

The next two chapters deal with confluent hypergeometric functions. The first of these deals with them in general and the second deals with Bessel functions. Bessel functions are to general confluent hypergeometric functions what Legendre functions are to the general $_2F_1$ in the sense that there is a quadratic transformation which connects the first of these to a special case of the second when one parameter of the second is appropriately specialized. These are classical subjects, and Bessel functions were treated extensively in Watson's book [51]. More is known now, but the basic framework is the same. This is also true for the next two chapters but changes in the chapter on orthogonal polynomials.

Most of the chapter on orthogonal polynomials deals with the classical polynomials of Jacobi, Laguerre and Hermite. These are orthogonal with respect to the beta, gamma and normal distributions respectively. A look toward the future was given near the end of this chapter when other sets of polynomials were introduced. There is a short section on polynomials of a discrete variable which extend the classical polynomials. The general classes are named after Charlier, Meixner, Krawtchouk and Hahn, with a special case of the last being named after Tchebichef. Actually, Tchebichef had the most general of these polynomials [48]. He used them for data fitting. Later, a special case was rediscovered by Fisher [24], the statistician, and also used for data fitting. Fisher's use of them was practical, so he looked for and found a different representation for the first few polynomials, and later F. Allan worked out this representation for the polynomials of arbitrary degree [1]. These polynomials arise in angular momentum theory, where they are not given as polynomials, and are known as 3-j symbols, Wigner coefficients or Clebsch-Gordan coefficients. For mathematicians, they are the coefficients which arise when the tensor product of two representations of SU(2) is decomposed as the sum of representations. Special cases also arise in coding theory. Krawtchouk polynomials also arise in coding theory. Allan's representation eventually led to a more general set of orthogonal polynomials which are balanced $_4F_3$'s. The discrete case of these orthogonal polynomials is equivalent to the orthogonality of the Racah coefficients from quantum angular momentum theory. This work of Racah dates to the early 1940's, but the connection with orthogonal polynomials was missed for 35 years since a series transformation needs to be done to Racah's representation before the polynomial aspect becomes clear.

There is one paragraph which had great impact. This is the last paragraph on page 165. To describe it, start with the classical polynomials of Jacobi, Laguerre and Hermite. Each of these classes of polynomials has the property that the derivative of each set consists of the same set of polynomials, with the parameters shifted by one for Jacobi and Laguerre polynomials. Sonine and, much later, Hahn proved that these are the only sets of orthogonal polynomials whose derivatives are also orthogonal. Hahn [27] asked and answered the question of finding all sets of orthogonal polynomials for which the operator

$$\text{(1.11)} \qquad \qquad \text{Lf}(x) = \frac{f(qx + w) - f(x)}{(q - 1)x + w}$$

maps orthogonal polynomials to orthogonal polynomials. When $q = 1$, these are the discrete polynomials mentioned earlier. When $w = 0$, these are basic hypergeometric extensions of both the classical continuous and the classical discrete polynomials. This theorem of Hahn turns out not to be the final story since there is a divided difference operator which can also be used. For hypergeometric functions the most general of these orthogonal polynomials are the Racah and Wilson polynomials mentioned above. There are three term recurrence relations which every set of orthogonal polynomials must satisfy. The recurrence relation found by Apéry (see [49]) in his proof of the irrationality of $\zeta(3)$ is an easy consequence of the three term recurrence relation for these most general polynomials (see [8]). Many of the basic hypergeometric orthogonal polynomials arise in the representation theory of $\text{SU}_q(2)$. Surprisingly, some of these polynomials can be used to study braids, a subject Magnus worked on. See [31]. I think he would be surprised by this since he seems to have primarily thought of special functions as arising from partial differential equations, at least most of the important ones. It is clear he would have been pleased by this unexpected development.

Magnus's chapter on spherical harmonics has been very influential. One of the early Springer Lecture Notes in Mathematics [37] is essentially a redoing of this chapter with many more details. Here, very interesting new developments have arisen. What makes it possible to do spherical harmonics is the existence of a space with a group acting on it in a transitive way, a distance function is defined on the space and is preserved under the group action, and the space is two-point homogeneous in the sense that any point can be mapped to any other, and if two pairs of points have the same distance between them, they can be mapped to each other by the group. One example is the unit sphere in R^n. The spherical functions here are symmetric Jacobi polynomials. This case is covered in Magnus's chapter. A second example is the unit cube in R^N, with the Hamming distance. The spherical functions here are symmetric Krawtchouk polynomials. Other Jacobi polynomials are spherical functions on real, complex and quaternionic projective spaces. For complex projective spaces this was first observed by E. Cartan in the late 1920's. The case of the unit cube was first considered by S. Lloyd in the setting of coding theory, and for many years the people doing this work were unaware of Krawtchouk's earlier work, so the polynomials were called Lloyd polynomials. It was only after this work became known to people who recognized the connection with the classical work on spherical harmonics that anyone thought to find an addition formula for Krawtchouk polynomials. This was discovered by Dunkl [17]. Some of the Hahn polynomials and some of the q versions of Krawtchouk and Hahn polynomials also arise in similar settings (see Stanton [47]).

There is a very surprising connection between some of the q orthogonal polynomials and the Rogers-Ramanujan identities which were stated in Chapter 4. Rogers derived these identities by means of a set of polynomials which are now seen to be q analogues of Hermite polynomials. There are at least five different q versions of Hermite polynomials, so one can see that the world of q is surprisingly rich. Each of these leads to a q analogue of the harmonic oscillator, and physicists are now looking at these [11].

The next chapter deals with orthogonal polynomials in several variables. Here it was too early to write this chapter and include very useful sets of polynomials. Much more can be written now, but again the full theory is far from clear. It seems likely there are a number of different extensions of the classical polynomials to several variables, but some seemingly different functions are connected, so it is far too early to predict what will eventually develop.

The last chapter in the second volume deals with elliptic functions and integrals. It is a standard treatment, not given in complete detail since many books give extensive developments.

The third volume was less successful than the first two. Partly this was because some of the functions were not well understood then and still are not. Others, like the last chapter on generating functions, do not work very well because the subject does not really fit into this type of format. A book like the one of Hansen [28] listing many series might be useful, but there is now a better way to treat generating functions and some of the material listed earlier. There are methods which use computers to do massive calculations, and in many cases new theories which tell one what computations to do. Thus it is now possible to consider putting some of this material in a computer algebra system and have it do more than just let you look at a massive array of formulas. See Wilf and Zeilberger [54] for some new developments.

Here are two examples where handbooks have been used to teach someone mathematics. The first deals with a different handbook: "Tables of Integrals, Series, and Products" by Gradshteyn and Ryzhik. Rodney Baxter has solved a number of important two-dimensional models in statistical mechanics. Often, the solution involves elliptic functions. Baxter told me he learned elliptic functions from Gradshteyn and Ryzhik's book. This seems improbable after looking at it, but one should realize that Baxter rediscovered the Rogers-Ramanujan identites while solving the hard hexagon model and was able to prove them, unlike what happened in 1914 when Ramanujan told them to Hardy, and none of the people Hardy wrote to were able to prove them. The proof then came when Ramanujan discovered Rogers's forgotten paper of 1894. The second case deals with me. When I started to do research, these volumes were recently published or about to be published. For a number of years they contained all I needed to know about special functions. Then, some new problems arose where the formulas I needed were not there, so I had to learn something serious about these functions. It took a while, and HTF was an invaluable aid. I now know some of the drawbacks of these volumes, but they were an excellent summary of what was known about hypergeometric functions in the middle of the 20th century.

When a new version is done, there will have to be much more algebra in it. While Lie groups are not the final answer to special functions as some think, they are an important setting and tool to discover very useful formulas. However, we must remember the old Indian legend of the six blind men and an elephant. Any rich subject should be looked at from many points of view. Special functions have been studied extensively by Euler, Gauss, Jacobi, Riemann, and Ramanujan and they all knew what they were doing. I have regularly been surprised by their usefulness in many areas of mathematics and expect to be surprised many more times.

2. Specific comments

The section on quadratic and cubic transformations of $_2F_1$'s, Section 2.1.5, is a bit misleading, and the conditions for quadratic and cubic transformations are not given correctly. Here are some comments which the reader might find useful.

For quadratic transformations of

$$(2.1) \qquad {}_2F_1\left(\begin{matrix} a, b; z \\ c \end{matrix}\right)$$

it is claimed that they exist if and only if the numbers

$$(2.2) \qquad \pm(1-c) \qquad \pm(a-b) \qquad \pm(a+b-c)$$

have the property that one of them equals $\frac{1}{2}$ or that two of them are equal. Then four examples are given:

$$(2.3) \qquad {}_2F_1\left(\begin{matrix} a, b \\ 2b \end{matrix}; \frac{4z}{(1+z)^2}\right) = (1+z)^{2a}\,{}_2F_1\left(\begin{matrix} a, a+\frac{1}{2}-b \\ b+\frac{1}{2} \end{matrix}; z^2\right)$$

$$(2.4) \qquad {}_2F_1\left(\begin{matrix} a, b \\ 1+a-b \end{matrix}; z\right) = (1-z)^a\,{}_2F_1\left(\begin{matrix} \frac{a}{2}, \frac{a}{2}+\frac{1}{2}-b \\ 1+a-b \end{matrix}; \frac{-4z}{(1-z)^2}\right).$$

(The denominator parameter was listed incorrectly on the right as $1-a+b$)

$$(2.5) \qquad {}_2F_1\left(\begin{matrix} a, a+\frac{1}{2} \\ b \end{matrix}; z\right) = 2^{2a}\left[1+(1-z)^{\frac{1}{2}}\right]^{-2a}$$
$$\cdot\,{}_2F_1\left(\begin{matrix} 2a, 2a-b+1 \\ b \end{matrix}; \frac{1-(1-z)^{\frac{1}{2}}}{1+(1-z)^{\frac{1}{2}}}\right)$$

$$(2.6) \qquad {}_2F_1\left(\begin{matrix} a, b \\ a+b+\frac{1}{2} \end{matrix}; 4z(1-z)\right) = {}_2F_1\left(\begin{matrix} 2a, 2b \\ a+b+\frac{1}{2} \end{matrix}; z\right).$$

These are introduced by the sentence "The fundamental formulas are those of Gauss and Kummer."

The first thing wrong with this is the way the conditions for a quadratic transformation to exist are stated. The condition involving $\frac{1}{2}$ is almost correct, but the other condition that two of the numbers being equal misses the restriction that this does not involve one of them being equal to its negative. I would state the conditions in the following way.

A quadratic transformation of a $_2F_1$ is a transformation between two $_2F_1$'s where the variable in one of the $_2F_1$'s is a quadratic function of the other, possibly combined with a linear fractional transformation. With this definition (2.4), (2.5) and (2.6) are quadratic transformations but (2.3) is not.

On the linear side the conditions come in two groups. These are

$$a + b = 1$$

(2.7)
$$c = \frac{a + b + 1}{2}$$

$$c = a + 1 - b$$

$$c = b + 1 - a$$

and

(2.8)
$$c = 2a$$

$$c = 2b$$

These come from the equality of two of the parameters in (2.2) when they are taken from two different pairs of $\pm$ (parameter).

On the quadratic side, setting one of these numbers to $\frac{1}{2}$ gives two groups

$$a = b + \frac{1}{2}$$

$$b = a + \frac{1}{2}$$

(2.9)
$$c = a + b + \frac{1}{2}$$

$$c = a + b - \frac{1}{2}$$

and

(2.10)
$$c = \frac{1}{2}$$

$$c = \frac{3}{2}.$$

If quadratic transformations are restricted as I restricted them above and the series do not terminate, then there are two fundamental quadratic transformations. One is (2.6), or equivalently (2.4) or (2.5), since these are all equivalent after using Pfaff's linear (fractional) transformation

(2.11)
$$\,_2F_1\left(\begin{matrix} a, b \\ c \end{matrix}; x\right) = (1 - x)^{-a} \,_2F_1\left(\begin{matrix} a, c - b \\ c \end{matrix}; \frac{x}{x - 1}\right).$$

Any of these three transformations connects a $\,_2F_1$ with a restriction in (2.7) with another restriction in (2.9). Notice the denominator parameters in (2.4), (2.5) and (2.6) are the same on both sides.

Formula (2.3) connects a series with parameters which satisfy (2.7) with another whose parameters satisfy (2.8). This is not one of the fundamental transformations but an iterate of one of the other three with a second fundamental quadratic transformation which connects (2.9) and (2.8). One of these is

(2.12)
$$\,_2F_1\left(\begin{matrix} a, b \\ 2b \end{matrix}; z\right) = \left(1 - \frac{z}{2}\right)^{-a} \,_2F_1\left(\begin{matrix} \frac{a}{2}, \frac{a+1}{2} \\ b + \frac{1}{2} \end{matrix}; \left(\frac{z}{2 - z}\right)^2\right).$$

Notice that the denominator parameters here differ. That is how I remember the difference between these two cases. I remember the conditions (2.7) and (2.8) from Euler's integral

$$(2.13) \qquad {}_2F_1\left(\begin{matrix} a, b \\ c \end{matrix}; z\right) = \frac{\Gamma(c)}{\Gamma(b)\Gamma(c-b)} \int_0^1 (1-zt)^{-a} t^{b-1} (1-t)^{c-b-1} dt.$$

The conditions (2.7) and (2.8) come from equating two of the exponents in the integrand along with the symmetry in a and b and Euler's transformation

$$(2.14) \qquad {}_2F_1\left(\begin{matrix} a, b \\ c \end{matrix}; z\right) = (1-z)^{c-a-b} {}_2F_1\left(\begin{matrix} c-a, c-b \\ c \end{matrix}; z\right).$$

The other conditions come from the parameters a and b differing by $\frac{1}{2}$, using

$$(2.15) \qquad (a)_{2n} = \left(\frac{a}{2}\right)_n \left(\frac{a+1}{2}\right)_n 2^{2n}$$

or c and 1 differing by $\frac{1}{2}$, and then Pfaff's transformation (2.11) combined with the numerator parameters differing by $\frac{1}{2}$. The conditions $c = \frac{1}{2}$ and $c = \frac{3}{2}$ play a different role than the other conditions. When one of the series terminates, it is possible to read the series backwards. When the linear side terminates, (2.12) is equivalent to any of (2.4), (2.5) or (2.6). When the quadratic side terminates and $b = a + \frac{1}{2}$, reading the series backwards leads to $c = \frac{1}{2}$ or $c = \frac{3}{2}$. In the nonterminating case, to get a quadratic transformation using $c = \frac{1}{2}$ or $c = \frac{3}{2}$, three series are needed. Transformations connecting (2.7) and (2.10) are given in HTF, Section 2.11.

These transformations with $c = \frac{1}{2}$ and $c = \frac{3}{2}$ are important, so it is clear that the definition of a quadratic transformation I gave above, connecting just two series, is too special. However, once you allow three series, there is no reason to restrict these transformations to three series, and then the conditions (2.7)–(2.10) can be changed. Mourad Ismail and I worked out other examples about 15 years ago and will eventually publish what we did.

I feel I understand quadratic transformations, but do not feel I understand cubic transformations. The fundamental one for ${}_2F_1$'s was found by Riemann [41] as a general cubic transformation between two hypergeometric differential equations. Goursat [26] worked out many explicit examples, and Watson [50] and Fowler [25] worked out others. One interesting special case of Watson's formula is the evaluation of

$$ {}_2F_1\left(\begin{matrix} a, 3a-1 \\ 2a \end{matrix}, e^{\pi i/3}\right), $$

as he observed.

Watson's cubic transformation and some of Goursat's are given in [19, Section 2.11]. When they are examined, it is seen that the conditions for a cubic transformation as given in Section 2.15 do not tell the full story. These conditions are given as either

$$(2.16) \qquad 1 - c = \pm(a-b) = \pm(c-a-b)$$

or two of the numbers

(2.17) $\pm(1-c)$ $\pm(a-b)$ $\pm(c-a-b)$

are equal to one-third.
    Here is one of Goursat's cubic transformations:

(2.18) $$_2F_1\left(\begin{matrix} 3a, a + \frac{1}{6} \\ 4a + \frac{2}{3} \end{matrix}; z\right) = \left(1 - \frac{z}{4}\right)^{-3a}$$

$$_2F_1\left(\begin{matrix} a, a + \frac{1}{3} \\ 2a + \frac{5}{6} \end{matrix}; \frac{-27z^2}{(z-4)^3}\right).$$

On the right hand side, $b - a = \frac{1}{3}$ is satisfied, but $c - a - b = \frac{1}{2}$ rather than $\frac{1}{3}$. Thus the conditions for a quadratic transformation play a role. On the left, $c = a + b + \frac{1}{2}$, which is a condition for a quadratic transformation, is one of the restrictions. The other restriction does not seem to be among those for either a quadratic or a cubic transformation. So where does (2.18) come from? Bailey discovered this in [12]. Both quadratic transformations and cubic transformations exist at the $_3F_2$ level. There is one quadratic transformation due to Whipple [52]:

(2.19) $$_3F_2\left(\begin{matrix} a, b, c \\ a + 1 - b, a + 1 - c \end{matrix}; z\right) = (1 - z)^{-a}$$

$$_3F_2\left(\begin{matrix} a + 1 - b - c, \frac{a}{2}, (a+1)/2 \\ a + 1 - b, a + 1 - c \end{matrix}; \frac{-4a}{(1-z)^2}\right).$$

For cubic transformations, there are two which were found by Bailey [12]:

(2.20) $$_3F_2\left(\begin{matrix} a, 2b - a - 1, a + 2 - 2b \\ b, a - b + \frac{3}{2} \end{matrix}; \frac{z}{4}\right) = (1 - z)^{-a}$$

$$_3F_2\left(\begin{matrix} \frac{a}{3}, (a+1)/3, (a+2)/3 \\ b, a - b + \frac{3}{2} \end{matrix}; \frac{-27z}{4(1-z)^3}\right)$$

and

(2.21) $$_3F_2\left(\begin{matrix} a, b - \frac{1}{2}, a + 1 - b \\ 2b - 1, 2a + 2 - 2b \end{matrix}; z\right) = \left(1 - \frac{z}{4}\right)^{-a}$$

$$_3F_2\left(\begin{matrix} \frac{a}{3}, \frac{a+1}{3}, \frac{a+2}{3} \\ b, a + \frac{3}{2} - b \end{matrix}; \frac{27z^2}{(4-z)^3}\right).$$

    Formulas (2.19) and (2.20) are included in HTF as (4.5.1) and (4.5.2). When $b = (a + 2)/3$ in (2.20) and a is then replaced by $3a$, formula (2.21) reduces to (2.18).
    There is a way of changing the left-hand side of (2.18) to another series so that two of the conditions above are satisfied. Use the quadratic transformation

(2.22) $$_2F_1\left(\begin{matrix} a, b \\ a + b + \frac{1}{2} \end{matrix}; z\right) = \left[\frac{1 + (1 - z)^{1/2}}{2}\right]^{-2a}$$

$$\cdot\, _2F_1\left(\begin{matrix} 2a, a - b + \frac{1}{2} \\ a + b + \frac{1}{2} \end{matrix}; \frac{(1-z)^{1/2} - 1}{(1-z)^{1/2} + 1}\right)$$

to get a series whose parameters are

$$
{}_2F_1\left(\begin{array}{c} 6a, 2a + \frac{1}{3} \\ 4a + \frac{2}{3} \end{array}; -\right).
$$

Then $c = 2b$ and $c = a - b + 1$, so two conditions for a quadratic transformation hold. The second condition on the parameters on the left-hand side of (2.18) can be stated as $c = 4b$. Clearly cubic transformations are a bit more complicated than the conditions (2.16) and (2.17).

While Bailey found both (2.20) and (2.21), he did not explain why there are two cubic transformations at the ${}_3F_2$ level but only one quadratic transformation. I can explain why there is a second cubic transformation which can be obtained from the first and why the same method does not lead to a second quadratic transformation. Take $a = -n$, a negative integer, in (2.20) and read the series backwards. The result is equivalent to (2.21) after some algebra is done. When the same is tried on the left-hand side of (2.19), the same series arises. This is one of the characteristic properties of well-poised series. The only change is that z is replaced by z^{-1}, which is another characteristic property of well-poised series. Observe that $-4z/(1 - z)^2$ remains invariant when $z \to z^{-1}$ while, when z is replaced by $4z$ in (2.20) and then $z \to z^{-1}$, the resulting power series variable on the right is $27z^2/(4 - z)^3$ as it should be to match the variable in (2.21). This is an argument which shows it is likely there is a second cubic transformation once there is one, but does not explain why quadratic and cubic transformations exist at the ${}_3F_2$ level. I think we have not gotten to the heart of cubic transformations and that it would be worthwhile trying to really figure out what is happening.

There are three conditions on both sides of (2.20) and (2.21). Two are conditions from the ${}_2F_1$ level, while the third condition is the sum of the numerator parameters plus $\frac{1}{2}$ is the sum of the denominator parameters.

In his thesis, Levelt developed a theory for higher hypergeometric differential equations and functions. In the fourth part of the published version [33, IV.3] [IV.3], he showed how to obtain Whipple's quadratic transformation between two ${}_3F_2$'s, [52]. Cubic transformations of ${}_3F_2$'s seem to be more numerous since there are two from reading one series backwards and the other side can be read backwards as well. It is worthwhile seeing what Levelt's work implies about other cubic transformations. My guess is that the root system G_2 underlies all of this since it has both quadratic and cubic symmetry. However, I have no idea how to use these symmetries to get the above results and possible double series extensions. That is likely where cubic transformations can really be explained.

ACKNOWLEDGEMENTS

I hope the comments in Section 2 give some idea how hard it is to write a set of books like this. While these three volumes are not perfect, they are much better than anything else we have. I want to thank Magnus, Erdélyi, Oberhettinger and Tricomi for the work they did. I have mentioned on other occasions that I have had to get Volumes 1 and 2 rebound. Now I can add that the rebound Volume 2 is showing wear and will need a new cover in a few years. That is an indication of how useful I have found them. Because of a kidney stone, I was unable to be present at this meeting. Doron Zeilberger has my thanks for presenting much of what is now Section 1, which is all that was written at the time of the meeting.

REFERENCES

[1] F. E. Allan, *The general form of the orthogonal polynomials for simple series, with proofs of their simple properties*, Proc. Royal Soc. Edinburgh **50** (1930), 310–320.

[2] K. Aomoto, *Jacobi polynomials associated with Selberg's integral*, SIAM J. Math. Anal. **18** (1987), 545–549.

[3] P. Appell and J. Kampé de Fériet, *Fonctions hypergéométrique et hypersphériques*, Gauthier Villars, Paris, 1926.

[4] R. Askey, *Retrospective on the Bateman project*, Applicable Analysis **8** (1978), 5–10.

[5] R. Askey, *Handbooks of special functions*, A Century of Mathematics in America, Part III (P. Duren et al, eds.), Amer. Math. Soc., Providence, 1989, pp. 369–391.

[6] R. Askey and J. Wilson, *A set of orthogonal polynomials that generalize the Racah coefficients or 6-j symbols*, SIAM J. Math. Anal. **10** (1979), 1008–1016.

[7] R. Askey and J. Wilson, *A set of hypergeometric orthogonal polynomials*, SIAM J. Math. Anal. **13** (1982), 651–655.

[8] R. Askey and J. Wilson, *A recurrence relation generalizing those of Apéry*, J. Austral. Math. Soc. (Series A) **36** (1984), 267–278.

[9] R. Askey and J. Wilson, *Some basic hypergeometric orthogonal polynomials that generalize Jacobi polynomials*, Memoirs of Amer. Math. Soc. **319** (1985), 55.

[10] N. M. Atakishiyev and S. K. Suslov, *The Hahn and Meixner polynomials of an imaginary argument and some of their applications*, J. Phys. A. Math. Gen. **18** (1985), 1583–1596.

[11] N. M. Atakishiyev and S. K. Suslov, *A realization of the q-harmonic oscillator*, Theoretical and Mathematical Physics **87 (1)** (1991), 442–442.

[12] W. N. Bailey, *Transformations of generalized hypergeometric series*, Proc. London Math. Soc. **(2) 29** (1929), 495–502.

[13] W. N. Bailey, *Generalized Hypergeometric Series*, Reprinted Hafner, New York, (1972), Camb. U. Press, Cambridge, 1935.

[14] J. Borwein and P. Borwein, P_i *and the AGM; A Study in Analytic Number Theory and Computational Complexity*, Wiley, 1987.

[15] J. M. Borwein and P. B. Borwein, *More Ramanujan-type series for* $1/\pi$, Ramanujan Revisited (G. Andrews et al, eds.), Academic Press, San Diego, 1988, pp. 357–374.

[16] D. V. Chudnovsky and G. V. Chudnovsky, *Approximations and complex multiplication according to Ramanujan*, Ramanujan Revisited (G. Andrews et al, eds.), Academic Press, San Diego, 1988, pp. 375–472.

[17] C. Dunkl, *A Krawtchouk polynomial addition theorem and wreath products of symmetric groups*, Indiana Univ. Math. J. **25** (1976), 335–358.

[18] F. J. Dyson, *Statistical theory of the energy level of complex systems I*, J. Math. Phys. **3** (1962), 140–156.

[19] A. Erdélyi et al, *Higher Transcendental Functions*, volume I, McGraw Hill, New York, 1953.

[20] A. Erdélyi et al, *Higher Transcendental Functions*, volume II, McGraw Hill, New York, 1953.

[21] A. Erdélyi et al, *Higher Transcendental Functions*, volume III, McGraw Hill, New York, 1955.

[22] A. Erdélyi et al, *Tables of Integral Transforms*, volume I, McGraw Hill, New York, 1954.

[23] A. Erdélyi et al, *Tables of Integral Transforms*, volume II, McGraw Hill, New York, 1954.

[24] R. A. Fisher, *Studies in crop variation, I. An examination of the yield of dressed grain from Broadbalk*, J. Agri. Sci. **11** (1920), 107–135; Reprinted in R. A. Fisher, Contributions to Mathematical Statistics, Wiley, New York, 1950, pp. 3.106a–3.135.

[25] R. H. Fowler, *The cubic transformations of Riemann's P-functions*, Quarterly J. Pure and Applied Math. **44** (1913), 205–218.

[26] E. Goursat, *Sur l'equation différentielle qui admet pour intégrale la série hypergéométrique*, Annales de l'École Normale **(2) 10** (1881); 3–142, Supplement.

[27] W. Hahn, *Über Orthogonalpolynome, die q-Differenzengleichungen Genügen*, Math. Nachr. **2** (1949), 4–34.

[28] Eldon Hansen, *A Table of Series and Products*, Prentice-Hall, Englewood Cliffs, NJ, 1975.

[29] E. Heine, *Theorie der Kugelfunctionen*, Erster Band, Reimer, Berlin, 1878.

[30] E. W. Hobson, *The Theory of Spherical and Ellipsoidal Harmonics*, Reprinted Chelsea, New York, 1955, Cambridge Univ. Press, 1931.

[31] A. N. Kirillov and N. Yu. Reshetikhin, *Representations of the algebra $U_q(sl(2))$ q-orthogonal polynomials and invariants of links*, Infinite-Dimensional Lie Algebras and Groups, World Sci. Pub., Teaneck, NJ, 1989, pp. 285–339.

[32] T. H. Koornwinder, *Orthogonal polynomials in connection with quantum groups*, Orthogonal Polynomials: Theory and Practice (P. Nevai, ed.), Kluwer, Dordrecht, 1990, pp. 257–292.

[33] A. H. M. Levelt, *Hypergeometric functions, I, II, III, IV*, Indag. Math. **23** (1961), 361–372, 373–385, 386–396, 397–403.

[34] I. G. Macdonald, *Some conjectures for root systems*, SIAM J. Math. Anal. **13** (1982), 988–1007.

[35] M. L. Mehta, *Random Matrices*, second edition, Academic Press, San Diego, 1991.

[36] M. L. Mehta and F. J. Dyson, *Statistical theory of the energy levels of complex systems*, J. Math. Phys. **4** (1963), 713–719.

[37] C. Mueller, *Spherical Harmonics*, Lecture Notes in Math. **17**, Springer, Berlin, 1966.

[38] R. Preston, *The mountains of π*, The New Yorker, March 2 (1992), 36–67.

[39] S. Ramanujan, *Modular equations and approximations to π*, Quarterly Jour. Math. **45** (1914), 350–372.

[40] S. Ramanujan, *Collected Papers*, edited by G. H. Hardy, P. V. Seshu Aiyer and B. M. Wilson, Cambridge U. Press, 1927, Reprinted Chelsea, New York, 1962.

[41] B. Riemann, *Beiträge zur Theorie der durch die Gauss'sche Reine $F(\alpha, \beta, \gamma, x)$ darstellbaren Functionen*, Abhand. Koniglichen Gesell. der Wiss. zu Göttingen **17** (1857); Reprinted in Gesammelte Math. Werke, New edition edited by R. Narasimhan, Springer, Berlin, 1990, pp. 99–115.

[42] L. J. Rogers, *On the expansion of certain infinite products*, Proc. London Math. Soc. **24** (1893), 337–352.

[43] L. J. Rogers, *Second memoir on the expansion of certain infinite products*, Proc. London Math. Soc. **25** (1894), 318–343.

[44] L. J. Rogers, *Third memoir on the expansion of certain infinite products*, Proc. London Math. Soc. **26** (1895), 15–32.

[45] A. Selberg, *Über einen Satz von A. Gelfond*, Arch. Math. Naturvid. **44** (1941), 159–170; Collected Papers, vol. 1, Springer, Berlin, 1989, pp. 62–73.

[46] A. Selberg, *Bemerkninger om et Multipelt Integral*, Norsk. Mat. Tidsskr. **26** (1944), 71–78; Collected Works, Springer, Berlin, 1989, pp. 204–211.

[47] D. Stanton, *Orthogonal polynomials and Chevalley groups*, Special Functions: Group Theoretic Aspects and Applications (R. Askey, T. H. Koornwinder and W. Schempp, eds.), Reidel, New York, 1984, pp. 87–128.

[48] P. L. Tchebycheff, *Sur l'interpolation des valeurs equidistantes*, Zapiski Imperatorskoi Akad. Nauk (Russia) **25** (1875); suppl. 5; French trans. in Oeuvres, tome 2, Chelsea, New York, 1961, pp. 219–242.

[49] A. van der Poorten, *A proof that Euler missed — Apéry's proof of the irrationality of $\zeta(3)$*, Math. Intelligencer **1** (1979), 195–203.

[50] G. N. Watson, *The cubic transformation of the hypergeometric function*, Quart. J. Pure and Applied Math. **40** (1909), 70–79.

[51] G. N. Watson, *A Treatise on the Theory of Bessel Functions*, 2nd edition, Cambridge Univ. Press, Cambridge, 1944.

[52] F. J. W. Whipple, *Some transformations of generalized hypergeometric series*, Proc. London Math. Soc. **(2) 26** (1927), 257–272.

[53] E. T. Whittaker and G. N. Watson, *A Course in Modern Analysis*, 2nd edition (1915), third edition (1920), fourth edition (1927), Cambridge Univ. Press, Cambridge.

[54] H. S. Wilf and D. Zeilberger, *Rational functions certify combinatorial identities*, J. Amer. Math. Soc. **3** (1990), 147–158.

DEPARTMENT OF MATHEMATICS, UNIV. WISCONSIN, MADISON, WI 53706
E-mail address: askey@math.wisc.edu

Contemporary Mathematics
Volume 169, 1994

Linear - Central Filtrations on Groups

Hyman Bass and Alexander Lubotzky

ABSTRACT. We introduce and develop the notion of a linear-central filtration on a group. Among the applications is a proof that every solvable group of outer automorphism of a free group is polycyclic, thus answering a question of M. Bestvina.

Dedicated to the memory of
Wilhelm Magnus, who pioneered
the methods used here.

0. Introduction

Let F be a group, $G = Aut(F)$, $H = Out(F)$, and

$$1 \to Z(F) \to F \xrightarrow{ad} G \xrightarrow{\pi} H \to 1$$

the canonical exact sequence.

Suppose first that $F = \pi_1(S_g)$, the fundamental group of a closed orientable surface of genus $g \geq 2$. The mapping class group of S_g is an index 2 subgroup of H. The following properties have been established by geometric methods.

I. Every abelian subgroup A of H is finitely generated. (Moreover $rank(A) \leq 3g - 3$). ([BLM], Theorem A).

II. Every (virtually) solvable subgroup B of H is virtually abelian. (In fact B has an abelian subgroup of index bounded in terms only of g; [BLM], Theorem B.)

III. H satisfies the "Tits alternative", i.e. a subgroup K of H is either virtually solvable, or else K contains a non abelian free group. ([Mc]).

1991 Mathematics Subject Classification. Primary 20F14, 20F28, 20F40. Secondary 20E05, 20E26.

Partially supported by US-Israel Binational Science Foundation grant 89-0423

Consider next the case when F is a free group on $r \geq 2$ generators. A great deal of progress has been made in developing geometric methods for studying G and H in this case, notably in the work of Culler-Vogtmann [CV] and Bestvina-Handel [BH]. At the MSRI Workshop on Arboreal Group Theory (September, 1988) M. Bestvina, posed the problem of deciding, for free F as above, whether $H = Out(F)$ satisfies the analogues of I, II, and III above.

This paper grew from the observation that when F is free, property I for G and H can be deduced from some more or less standard "nilpotent technology", using the classical work of Magnus [Mag 1,2]. Since the method offers a potential approach to properties II and III as well, it seemed worthwhile to give a more systematic exposition than is usefully available in the literature. That is what we offer here. Our methods also give a new proof of I for surface groups.

The central concept that we introduce and use is that of a k-linear-central filtration on a group G. Here k is a commutative ring, and the filtration is a chain of subgroups $G = G_0 \geq G_1 \geq G_2 \geq \ldots$ such that: (a) $(G_d, G_e) \leq G_{d+e} \forall\, d, e \geq 0$, so that we can form the groups $L_d(G) = G_d/G_{d+1}$, with $L_+(G) = \oplus_{d \geq 1} L_d(G)$ a graded Lie $\mathbf{Z}$-algebra; (b) $L_0(G)$ is embeddable in some $GL_{N_0}(k)$; and (c) For each $d \geq 1, L_d(G)$ is embeddable in some k^{N_d}. We call the filtration separating if $\cap G_d = \{1\}$, central if $G = G_1$, and we call G $(k-)$linear if $G_1 = \{1\}$.

Linear groups have many useful properties – good control on solvable subgroups, residual properties, cohomological dimension, Tits alternative, ... cf [W], [AS]. However even if F is linear, $G = Aut(F)$ and $H = Out(F)$ need not be (see, for example, [FP] for the case of free groups). Nonetheless G and H often carry separating $\mathbf{Z}$-linear-central filtrations, for example when F is free (Theorem (10.4) below), or when $F = \pi_1(S_g)$ with $g \geq 3$ (Theorem 11.5), and this suffices to provide many of the properties otherwise supplied by $\mathbf{Z}$-linearity. (See Theorem (6.1).) For example solvable subgroups K of G or H of finite Hirsch rank are polycyclic. If K has finite virtual cohomological dimension, vcd, (as in the case when F is free, thanks to [CV]) then K automatically has finite Hirsch rank, and so we obtain property I above for G and H.

Following is a summary of some of the main results. Much of this, though not all, is assembly and organization of folklore. Let F be a finitely generated group, $G = Aut(F), H = Out(F)$, and

$$1 \to Z(F) \to F \xrightarrow{ad} G \xrightarrow{\pi} H \to 1$$

as above.

(A) F admits a G-invariant $\mathbf{Z}$-central filtration $(F_n)_{n \geq 1}$, say with Lie algebra $L(F) = \oplus_{n \geq 1} L_n(F)$. It is separating iff F is residually-torsion-free-nilpotent. (See (7.2).)

(B) G admits a $\mathbf{Z}$-linear-central filtration $(G_d)_{d \geq 0}$ with $L_0(G) \leq Aut_{\mathbf{Z}}(L_1(F))$, and $L_+(G)$ embedded in the graded Lie algebra of Lie derivations of $L(F)$. If $(F_n)_{n \geq 1}$ is separating then so also is $(G_d)_{d \geq 0}$. (See (8.2)).

(C) Filter H by $H_d = \pi(G_d)$. Then $(H_d, H_e) \leq H_{d+e}$ $\forall d, e \geq 0$, and $L_0(H) = L_0(G)$. If $Z(L(F)) = \{0\}$ then we have an exact sequence of graded Lie algebras

$$0 \to L(F) \xrightarrow{L(ad)} L_+(G) \xrightarrow{L_+(\pi)} L_+(H) \to 0,$$

with $L_+(H)$ embedded in the graded Lie algebra of outer derivations of $L(F)$. If $Z(\mathbf{F}_p \otimes L(F)) = \{0\}$ for all primes p then $(H_d)_{d \geq 0}$ is a $\mathbf{Z}$-linear-central filtration. The filtration $(H_d)_{d \geq 0}$ is separating iff F satisfies the condition,

$$IN(F) : \begin{cases} \text{If } g \in G \text{ induces an inner automorphism} \\ \text{on } F/F_n \text{ for all } n \geq 1, \text{ then } g \text{ is inner} \end{cases}$$

(See (9.8).)

(D) If F is a non abelian free group then we know from the work of Magnus ([Mag 1,2]) that F is residually-torsion-free-nilpotent, $L(F)$ is a free Lie algebra, with $Z(k \otimes L(F)) = \{0\}$ for all commutative rings k. Moreover results of Stebe [St] and of Edna Grossman [EKG] imply condition $IN(F)$. Finally results of Culler Vogtmann [CV] and Gersten show that G and H have finite vcd. Thus we obtain all of the conclusions of (A), (B), (C) above, and solvable subgroups of G or H are polycyclic.

(E) Let $F = \pi_1(S_g), g \geq 2$, as above. Then results of Labute (cf. [Lab 1]) furnish an explicit description of $L(F)$. We thus derive $\mathbf{Z}$-linear-central filtrations $(G_d)_{d \geq 0}$ on G and $(H_d)_{d \geq 0}$ on H. Baumslag (cf. [GB2]) has shown that F is residually-free, hence residually-torsion-free-nilpotent, and so $(G_d)_{d \geq 0}$ is separating. Using actions on character varieties ((14.5)) and results of Macbeath and Singerman [MS], we are able to establish $IN(F)$ when $g \geq 3$, whence $(H_d)_{d \geq 0}$ is also separating in this case.

In part III we consider an associative algebra A over a commutative ring k, with a filtration

$$A = A_0 \geq A_1 \geq A_2 \geq \cdots : \quad A_n \cdot A_m \leq A_{n+m} \ \forall \, d, e \geq 0.$$

Let $\bar{A} = \oplus_{n \geq 0} \bar{A}_{(n)}$, $\bar{A}_{(n)} = A_n/A_{n+1}$, be the associated graded algebra.

(F) The group $G = A^\times$ of units of A is filtered by $G_d = Ker(A^\times \to (A/A_d)^\times), (G_d, G_e) \leq G_{d+e} \ \forall d, e \geq 0, L_0(G) \leq \bar{A}^\times_{(0)}$, and $L_+(G)$ embeds in $\bar{A}$, as a graded sub Lie $\mathbf{Z}$-algebra. If $\cap_n A_n = \{0\}$ then $\cap_d G_d = \{1\}$. This result can be found, for example, in Bourbaki [NB2], Ch II, §4.5. (See (13.4) below.) Under suitable finiteness assumptions on $\bar{A}$ as an algebra over the image $\bar{k}_{(0)}$ of k in $\bar{A}_{(0)}$, we conclude (cf. (13.5)) that $(G_d)_{d \geq 0}$ is a $\bar{k}_{(0)}$-linear-central filtration.

(G) As an application, suppose that k is a noetherian integral domain, P is a prime ideal, $P^{(d)} = k \cap P^d k_P$ ("symbolic powers"), and we filter $GL_N(k)$ by $GL_N(P^{(d)}) = Ker(GL_N(k) \to GL_N(k/P^{(d)}))$. Then $(GL_N(P^{(d)}))_{d \geq 0}$ is a separating (k/P)-linear-central filtration on $GL_N(k)$. (See (13.7).)

(H) The latter result has a nice application to representation varieties. Suppose that, in the representation variety $Hom(F, GL_N(\mathbf{C}))$, (F a finitely generated group), there is an irreducible subvariety V containing a faithful representation and the trivial representation. Then F is residually-torsion-free-nilpotent. (See (14.4).)

(I) Consider the group $G = Aut^{filt}_{k-alg}(A)$ of filtration preserving k-algebra automorphisms of A. Then one has an induced filtration $(G_d)_{d\geq 0}$ with $(G_d, G_e) \leq$ G_{d+e} $\forall d, c \geq 0$, $L_0(G) \leq Aut_{gr\ k-alg}(\bar{A})$, $L_+(G) \leq gr Der_{k-alg}(\bar{A})$, the graded Lie algebra of derivations of $\bar{A}$, and $\cap G_d = \{1\}$ if $\cap A_n = \{0\}$. (See (15.2).) Under rather strong finiteness conditions on $\bar{A}$ as a $\bar{k}_{(0)}$-algebra we conclude that $(G_d)_{d\geq 0}$ is a $\bar{k}_{(0)}$-linear-central filtration.

(J) For a finitely generated group F, the latter result can be applied to the actions of $G = Aut(F)$ on the affine algebra of the representation variety $Hom(F, GL_N(\mathbf{C}))$, and the action of $H = Out(F)$ on the affine algebra of the character variety $Hom(F, GL_N(\mathbf{C}))\ modGL_N(\mathbf{C})$. These algebras are filtered by symbolic powers of the prime ideals corresponding to the trivial representation (resp., character). (See section 16.) The idea of exploiting this structure goes back to Magnus [Mag 3].

Our perspective here emphasizes the role of the graded Lie $\mathbf{Z}$-algebra $L(F)$ of a finitely generated group F, as in (A) above. The structure of $L(F)$ and of its Lie algebra of derivations could be a useful source of group theoretic information. For example, when F is free, the question of the Tits alternative for $G = Aut(F)$ naturally raises the question of a Lie-algebra Tits alternative for $L_+(G) \leq gr Der(L(F))$, which seems to be an interesting question in its own right. (Cf. (10.4) and (10.5)).

The most far reaching results on the Lie algebras $L(F)$ are due to John Labute (cf [Lab 1,2]). There seems to have been little study of the Lie algebras $gr Der(L(F))$.

Acknowledgement. We thank G. Bergman for some examples of derivations of free algebras (used in connection with (10.7)), John Labute for much valuable information on the Lie algebras $L(F)$, especially supplying (11.1)(4), Nikolas Vonessen for helpful discussions about representation varieties, and Karen Vogtmann for references and background on $Out(F)$ when F is free.

I. Background

1. Polycyclic groups, Hirsch rank, and vcd.

(1.1) DEFINITION. Let P be a group theoretic property. A group G is called *poly-P* if there is a finite filtration $G = G_0 \geq G_1 \geq G_2 \geq \ldots \geq G_n = \{1\}$ such that $G_i \triangleleft G_{i-1}$ and G_{i-1}/G_i is a P-group, $i = 1, \ldots, n$.

Thus, for example, poly-abelian = solvable, and poly-finitely-generated-abelian = poly-cyclic. We quote the following result.

(1.2) THEOREM. *The following conditions on a group G are equivalent.*

(a) G is polycyclic

(b) G is solvable and all subgroups are finitely generated.

(c) G is solvable and all abelian subgroups are finitely generated.

(d) G is isomorphic to a solvable subgroup of $GL_N(\mathbf{Z})$ for some $N \geq 0$.

The implications (a) $\Leftrightarrow$ (b) $\Rightarrow$ (c) are easy. The implications (c) $\Rightarrow$ (a) $\Leftarrow$ (d) are due to Malcev (cf. [Mal], or [Seg], Ch. 2, B Theorems 1 and 2). The implication (a) $\Rightarrow$ (d) is due L. Auslander (cf. [Aus], or [Seg], Ch 5, C, Theorem 5).

(1.3) The *Hirsch rank* is the unique cardinal valued function defined on virtually solvable groups satisfying:

(0) Hirsch rank $(G) = 0$ iff G is torsion;

(1) Hirsch rank $G = dim_{\mathbf{Q}}(\mathbf{Q} \otimes G)$ if G is abelian;

and

(2) For any exact sequence $1 \rightarrow G' \rightarrow G \rightarrow G'' \rightarrow 1$,

$$\text{Hirsch rank } (G) = \text{Hirsch rank } (G') + \text{Hirsch rank } (G).$$

(1.4) The *cohomological dimension* of a group G, denoted $cd(G)$, is the supremum of all n such that $H^n(G, M) \neq 0$ for some G-module M. For the following basic facts we refer to [Br] or [Gru]. If $cd(G) < \infty$ then G is torsion free. If $H \leq G$ then $cd(H) \leq cd(G)$. If $[G : H] < \infty$ and G is torsion free then $cd(H) = cd(G)$. If follows that, in general, all torsion free finite index subgroups $H \leq G$ have the same $cd(H)$, denoted $vcd(G)$, and called the *virtual cohomological dimension* of G; it is defined when G is virtually torison free. If G is virtually solvable then

$$Hirsch \ rank(G) \leq vcd(G).$$

If, further, $vcd(G) < \infty$, then $vcd(G) \leq Hirsch \ rank(G) + 1$ (cf. [Gru] §8.8).

2. Linear groups

(2.1) **DEFINITION.** We call a group G *linear* if there is an injective homomorphism $\rho : G \to GL_N(k)$, for some commutative integral domain k, and some $N \geq 0$.

If G has a finite set $S = \{s_1, \ldots, s_d\}$ of generators then $\rho G \leq GL_N(k')$, where k' is the finitely generated subring of k generated by the entries of $\rho(s_i), \rho(s_i)^{-1}$ ($i = 1, \ldots, d$).

Linear groups have a number of useful properties, some of which we list below. A general reference for this is Wehrfitz [W]. Assume, for the rest of this section, that

$$G \leq GL_N(k)$$

where k is a commutative integral domain with field of fractions K, and algebraic closure $\bar{K}$.

(2.2) **Solvable subgroups.** Let $H \leq G$ be a solvable subgroup. Then there are normal subgroups $H_u \leq H^0 \leq H$ such that H/H^0 is finite, H^0/H_u is abelian, and H_u is nilpotent (of class $\leq N - 1$).

This follows from the Lie-Kolchin Theorem, applied to the Zariski closure $\bar{H}$ of H in $GL_N(\bar{K})$, which asserts that the connected component $\bar{H}^0$ can be conjugated into triangular form. Then the unipotent radical $\bar{H}_u$ is the kernel of the projection of $\bar{H}^0$ into the (abelian) diagonal group. We take $H^0 = H \cap \bar{H}^0$ and $H_u = H \cap \bar{H}_u$.

In case $k = \mathbf{Z}$ then H is polycyclic ((1.2)).

In case k is finitely generated and algebraic over $\mathbf{Z}$ (so that $[K : \mathbf{Q}] < \infty$) then the Hirsch rank of H is bounded in terms of k and N alone.

(2.3) Residual properties. (Cf [W], Ch. 4). Suppose that k is a finitely generated ring. (As observed above, we can reduce to this case when G is finitely generated.) Write $char(k) = p \geq 0$.

If $p > 0$ then G is virtually residually-finite-p.

If $p = 0$ then for almost all primes q, G is virtually residually-finite-q. Moreover G is virtually torsion free.

If $k = \mathbf{Z}$ (or is integral over $\mathbf{Z}$) then, for all primes q, G is virtually residually-finite-q.

These properties result from the following observations. For J an ideal of k put $GL_N(J) = Ker(GL_N(k) \to GL_N(k/J))$. Let M be a maximal ideal of k with residue field $F = k/M$ of characteristic q. Then F is finite, each $GL_N(k)/GL_N(M^r)$ is finite, and $GL_N(M)/GL_N(M^r)$ is a q-group. Moreover $\cap_r GL_N(M^r) = \{I\}$ since $\cap_r M^r = \{0\}$ (k is a noetherian integral domain). It follows that all torsion in $GL_N(M)$ is q-torsion. Let M' be another maximal ideal and $char(k/M') = q'$. Then $GL_N(M) \cap GL_N(M')$ must be torsion free (of finite index) if $q \neq q'$. If $p = 0$ then all but finitely many primes q occur as above.

(2.4) Tits alternative. (cf [T]) Suppose that either G is finitely generated or that $char(k) = 0$. Let $H \leq G$. Then either H is virtually solvable or H contains a non abelian free group.

(2.5) vcd (cf. [AS]). Assume that k is finitely generated and $char(k) = 0$. Then $vcd(G) < \infty$ iff the Hirsch ranks of the nilpotent (even unipotent) subgroups of G are bounded above.

3. Commutators and central series

(3.1) Commutators. Let G be a group. For $x, y \in G$ we write $^{x}y = xyx^{-1} = y^{x^{-1}}$, and $(x, y) = xyx^{-1}y^{-1} = (y, x)^{-1}$. We have the following identities (cf [NB1], Ch 1, §6, no.2).

(1) $$(x, yz) = (x, y) \cdot{}^{y}(x, z)$$

(2) $$(xy, z) = {}^{x}(y, z) \cdot (x, z)$$

(3) $$((x, y), {}^{y}z)((y, z), {}^{z}x)((z, x), {}^{x}y) = 1.$$

For subgroups $A, B \leq G$,

$$(A, B) = (B, A) = \text{the subgroup generated by all}$$
$$(a, b) \quad (a \in A, b \in B).$$

It is normalized by A and B (loc. cit.). If (A, B) centralizes A (resp., B) then the commutator map $(\ , \ ) : A \times B \to (A, B)$ is linear in the $A-$ (resp., $B-$) variable. If (A, B) normalizes $C \leq G$ then $((A, B), C)$ is generated by all $((a, b), c)$ $(a \in A, b \in B, c \in C)$ (loc. cit.)

For the following discussion, cf. [Seg], Ch 1, B.

(3.2) Descending central series: $C^{n}(G) \leq G$ is defined inductively by $C^{1}(G) = G$ and $C^{n+1}(G) = (G, C^{n}(G))$. Thus $C^{n}(G) \triangleleft G$ and $C^{n}(G)/C^{n+1}(G) \leq Z(G/C^{n+1}(G))$. If $G = G_1 \geq G_2 \geq G_3 \geq \cdots$ and $(G, G_n) \leq G_{n+1}$ for all n then it follows inductively that $G_n \triangleleft G$ and $C^{n}(G) \leq G_n$ for all n.

(3.3) Ascending central series: $Z^{n}(G) \triangleleft G$ is defined inductively by $Z^{0}(G) = \{1\}$ and $Z^{n+1}(G)/Z^{n}(G) = Z(G/Z^{n}(G))$. If $\{1\} = H_0 \leq H_1 \leq H_2 \leq \cdots$ and

$(G, H_n) \leq H_{n-1}$ for all n then it follows inductively that $H_n \triangleleft G$ and $H_n \leq Z^n(G)$ for all n.

The commutator map $G \times Z^{n+1}(G) \to Z^n(G)$ descends to a bilinear map

$$G^{ab} \times (Z^{n+1}(G)/Z^n(G)) \to Z^n(G)/Z^{n-1}(G),$$

whence a homomorphism

(1) $\delta : Z^{n+1}(G)/Z^n(G) \to Hom(G^{ab}, Z^n(G)/Z^{n-1}(G)),$

which is easily seen to be injective.

(3.4) Nilpotent groups. The following conditions are equivalent, in which case we say that G is nilpotent of class $\leq n$.

(a) $C^n(G) = \{1\}$

(b) $Z^{n-1}(G) = G$

(c) There is a chain $G = G_1 \geq G_2 \geq \ldots \geq G_n = \{1\}$ with $(G, G_i) \leq G_{i+1}$ $(1 \leq i < n)$.

(3.5) Finite generation. If G^{ab} is finitely generated then so also is each $C^n(G)/C^{n+1}(G)$. (cf. (4.6) below). If G^{ab} and $Z(G)$ are finitely generated then so also is each $Z^n(G)/Z^{n-1}(G)$ (cf. (3.3)(1)).

(3.6) Torsion. If $x^m = 1$ for all $x \epsilon G^{ab}$ (resp., for all $x \epsilon Z(G)$) then the same holds for each $C^n(G)/C^{n+1}(G)$ (resp., each $Z^n(G)/Z^{n-1}(G)$), (Cf. (4.6), resp., (3.3)(1)). If G is nilpotent then the set $Tors(G)$ of elements of finite order is a subgroup of G, and it is the weak direct product of p-groups, for primes p.

(3.7) Torsion freeness. If $Z(G)$ is torsion free then so also is each $Z^n(G)/Z^{n-1}(G)$. (Cf. (3.3(1)). If G is nilpotent then G is torsion free iff $Z(G)$ is torsion free. If G is finitely generated torsion free nilpotent then G is residually-finite-p for all primes p (cf. [Seg], Ch 1, C, Theorem 4; this is a result of Gruenberg.).

4. Central filtrations and Lie algebras

(4.1) The set-up. Throughout this section we fix the following data and notation.

- F is a group, with a family $(F_n)_{n \in \mathbf{Z}}$ of normal subgroups $F_n \triangleleft F$, with $F_{n+1} \leq F_n \ \forall n$.

- G is a group acting on F as group automorphisms, and leaving each F_n invariant.

- For $d \geq 0$ we put

$$G_d = \cap_n Ker(\rho_{n,d} : G \to Aut(F_n/F_{n+d})),$$

where $\rho_{n,d}$ gives the induced G-action on F_n/F_{n+d}.

When convenient we embed F and G in the semi-direct product $H = F \rtimes G$, defined using the action of G on F, which then becomes conjugation in H. Then in H we see that

$$G_d = \{x \epsilon G | (x, F_n) \subset F_{n+d} \ \forall n\},$$

and so G_d is the largest subgroup of G such that, for all n,

$$(G_d, F_n) \leq F_{n+d}.$$

(4.2) PROPOSITION. (P. Hall [PH], §3, Theorem 2.3). *For all $d, e \geq 0$, $(G_d, G_e) \leq G_{d+e}$. Hence $C^d(G_1) \leq G_d$ for all $d \geq 1$.*

(4.3) COROLLARY. *If*

(1) $$(F, F_n) \leq F_{n+1} \quad \forall n \epsilon \mathbf{Z}$$

then

(2) $$(C^d(F), F_n) \leq F_{n+d} \quad \forall n \epsilon \mathbf{Z}, \ d \geq 1.$$

Proof. Apply (4.2) to $G = F$, acting by conjugation on itself. Then (1) implies that $G = G_1$, and so $C^d(G) \leq G_d$, whence $(C^d(G), F_n) \leq (G_d, F_n) \leq F_{n+d}$, by (4.2).

(4.4) COROLLARY. *For* $n \leq 0$ *define* $C^n(F) = F$ *and* $Z^n(F) = \{1\}$. *Then we have* $(F, C^n(F)) \leq C^{n+1}(F)$ *and* $(F, Z^n(F)) \leq Z^{n-1}(F)$ $\forall n \epsilon \mathbf{Z}$, *and so, by (4.3),*

$$(C^d(F), C^n(F)) \leq C^{n+d}(F)$$

and

$$(C^d(F), Z^n(F)) \leq Z^{n-d}(F)$$

$\forall n \epsilon \mathbf{Z}, d \geq 1$.

(4.5) DEFINITION. We call the sequence

$$\Phi = (F_1 \geq F_2 \geq F_3 \geq \ldots)$$

a *central filtration of* F if $F_1 = F$ and

$$(F_n, F_m) \leq F_{n+m} \quad \forall n, m \geq 1.$$

Then we put

$$L_n(F, \Phi) = F_n/F_{n+1}, \quad \text{written additively,}$$

and

$$L(F, \Phi) = \bigoplus_{n \geq 1} L_n(F, \Phi).$$

The commutator $(,): F_n \times F_m \to F_{n+m}$ descends to a bilinear map (cf. (3.1))

$$[,]: L_n(F, \Phi) \times L_m(F, \Phi) \to L_{n+m}(F, \Phi)$$

making $L(F, \Phi)$ a graded Lie $\mathbf{Z}$-algebra. (The Jacobi identity follows from (3.1)(3).) (Cf. [NB2], Ch II, §4.4)

(4.6) EXAMPLE. It follows from (4.4) that

$$\Phi^C = (C^n(F))_{n \geq 1}$$

is a central filtration on F. In this case we write simply $L^C(F) = \bigoplus_{n \geq 1} L_n^C(F)$
for $L(F, \Phi^C)$. From the formula $C^{n+1}(F) = (F, C^n(F))$ if follows that $[\ ,\ ]$:
$L_1^C(F) \otimes L_n^C(F) \to L_{n+1}^C(F)$ is surjective. It follows by induction on n that the Lie
algebra $L^C(F)$ is generated by $L_1^C(F) = F^{ab}$. Therefore if F^{ab} is finitely generated
(or torsion, or of exponent dividing m, or finite, or ...) then each $L_n^C(F)$ has the
same property. Consequently, if F is nilpotent and F^{ab} is finitely generated then F
is polycyclic.

(4.7) COROLLARY. *The condition*

(1) $$(G_d, G_e) \leq G_{d+e} \quad \forall d, e \geq 0$$

of (4.2) implies that $\Phi = (G_d)_{d \geq 1}$ is a central filtration, so that

$$L_+(G) := L(G_1, \Phi) = \bigoplus_{d \geq 1} L_d(G), \quad L_d(G) = G_d/G_{d+1},$$

is a graded Lie $\mathbf{Z}$-algebra. The conjugation action of G induces an action of $L_0(G) = G_0/G_1$ on $L_+(G)$ as graded Lie algebra automorphisms.

We denote the action of $g \epsilon L_0(G)$ on a $a \epsilon L_+(G)$ by $^g a$.

(4.8) DERIVATIONS. Let k be a commutative ring and A a k-algebra (not
necessarily associative), i.e. A is a k-module with a k-bilinear product $a \cdot b$. A k-
derivation of A is a k-linear map $D : A \to A$ such that $D(a \cdot b) = Da \cdot b + a \cdot Db$. The
k-derivations of A form a Lie k-algebra $Der_k(A)$, with $[D, D'] = D \circ D' - D' \circ D$.

Suppose that $A = \bigoplus_{n \epsilon \mathbf{Z}} A_n$ is a graded k-algebra: $A_n \cdot A_n \subset A_{n+m}$. For $d \epsilon \mathbf{Z}$
let
$$Der_k(A)_d = \{D \epsilon Der_k(A) | DA_n \leq A_{n+d} \ \forall n\}.$$

Then
$$gr Der_k(A) := \bigoplus_{d \epsilon \mathbf{Z}} Der_k(A)_d$$

is a graded Lie subalgebra of $Der_k(A)$.

(4.9) PROPOSITION. *In the setting of (4.1), assume further that*

(1) $$(F, F_n) \leq F_{n+1} \quad n \epsilon \mathbf{Z}$$

and put $M(F) = \bigoplus_{n \epsilon \mathbf{Z}} M_n(F), \quad M_n(F) = F_n / F_{n+1}.$

(a) *For $d \geq 1$ the commutator maps $(\ ,\ ) : G_d \times F_n \to F_{n+d}$ descend to bilinear maps $L_d(G) \times M_n(F) \to M_{n+d}(F)$ making $M(F)$ a faithful graded module over the graded Lie $\mathbf{Z}$-algebra $L_+(G)$. Write $a \cdot x$ for the action of $a \epsilon L(G)$ on $x \epsilon M(F)$.*

(b) *The conjugation action of G on F defines a faithful action, denoted $g : x \mapsto {}^g x$, of $L_0(G)$ on $M(F)$, and we have ${}^g(a \cdot x) = {}^g a \cdot {}^g x$ for $g \in L_0(G)$, $a \in L_+(G), x \in M(F)$.*

(c) *Suppose further that $F = F_1$ and that $(F_n)_{n \geq 1}$ is a central filtration, so that $M(F) = \bigoplus_{n \geq 1} M_n(F)$ is a graded Lie $\mathbf{Z}$-algebra. Then the above actions define an injective group homomorphism*

$$L_0(G) \to Aut_{gr.Lie\ alg}(M(F))$$

and an injective graded Lie $\mathbf{Z}$-algebra homomorphism

$$L_+(G) \to gr Der_{\mathbf{Z}}(M(F)).$$

Proof. For $g \epsilon G = G_0, a \epsilon G_d(d > 0)$ and $x \epsilon F_n(n > 0)$, let $g_0 \epsilon L_0(G)$, $a_d \epsilon L_d(G)$, and $x_n \epsilon M_n(F)$ denote their respective classes. The actions of $L_0(G)$ and $L_d(G)$, calculated in $H = F \rtimes G$, are defined by

$$^{g_0} x_n = (gxg^{-1})_n$$

$$a_d \cdot x_n = (a, x)_{n+d} .$$

Clearly $^{g_0}(a_d \cdot x_n) = (^{g_0}a) \cdot (^{g_0}x_n)$. Thus $^{g_0}x_n = x_n$ iff $(g, x) \epsilon F_{n+1}$. It follows that g_0 acts trivially on $M(F)$ iff $g \epsilon G_1$, i.e. iff $g_0 = 1$; thus $L_0(G)$ acts faithfully on $M(F)$. Similarly a_d acts trivially on $M(F)$ iff $(a, F_n) \leq F_{n+d+1}$ for all n, i.e. iff $a \epsilon G_{d+1}$, i.e. iff $a_d = 0$; thus $L_+(G)$ acts faithfully on $M(F)$.

If $y \in F_m (m > 0)$ then $(a, xy) = (a, x) \cdot^x (a, y)$. Since $(F, F_r) \le F_{r+1}$ for all r it follows that, if $n = m, a_d \cdot (x_n + y_n) = a_d \cdot x_n + a_d \cdot y_n$. If $b \in G_e$ then $(ab, x) = {}^a(b, x) \cdot (a, x)$. Since $(G_d, F_r) \le F_{r+1}$ (as $d > 0$) it follows, when $e = d$, that

$$(a_d + b_d) \cdot x_n = a_d \cdot x_n + b_d \cdot x_n.$$

Next put $x' = b^{-1} x b \in F_n$, and note that $x'_n = x_n$ since $(G_e, F_n) \le F_{n+1} (e > 0)$. From (3.1)(3) we have

(1)
$$\begin{aligned}((a, b), x) &= ((a, b), {}^b x') \\ &= [((b, x'), {}^{x'} a)((x', a), {}^a b)]^{-1}\end{aligned}$$

Now

$$\begin{aligned}((b, x'), {}^{x'} a)_{n+d+e} &= [{}^{x'}({}^{x'^{-1}}(b, x'), a)]_{n+d+e} \\ &= -a_d \cdot^{x'^{-1}} (b, x')_{n+e} \\ &= -a_d \cdot (b_e \cdot x'_n) = -a_d \cdot (b_e \cdot x_n).\end{aligned}$$

Further

$$\begin{aligned}((x', a), {}^a b)_{n+d+e} &= -({}^a b)_e \cdot (-a_d \cdot x'_n) \\ &= b_e \cdot (a_d \cdot x_n).\end{aligned}$$

Thus it follows from (1) that

$$[a_d, b_e] \cdot x_n = (a_d \circ b_e - b_e \circ a_d) \cdot x_n \ .$$

This shows that $M(F)$ is a graded module over the graded Lie algebra $L_+(G)$.

Suppose finally that $F = F_1$ and $(F_n, F_m) \le F_{n+m}$, so that $M(F) = \bigoplus_{n \ge 1} M_n(F)$ is itself a graded Lie $\mathbf{Z}$-algebra. Since ${}^g(x, y) = {}^g(x, {}^g y)$ we see that $L_0(G)$ acts on $M(F)$ as graded Lie algebra automorphisms. It remains to see that the linear map $a_d \cdot$, of degree d, is a Lie algebra derivation. For $x \in F_n, y \in F_m$, $a_d \cdot [x_n, y_m] = (a, (x, y))_{n+m+d}$. Put $a' = y^{-1} ay$. Then, using (3.1)(3),

(2)
$$\begin{aligned}(a, (x, y)) &= ((x, y), a)^{-1} = ((x, y), {}^y a')^{-1} \\ &= ((y, a'), {}^{a'} x)((a', x), {}^x y).\end{aligned}$$

Since both (F, F_r) and (G_d, F_r) lie in F_{r+1} for all r, we have $(^{a'}x)_n = x_n$, and $(y, a')_{m+d} = {}^{y^{-1}}(y, a)_{m+d} = -a_d \cdot y_m$.

$$((y, a'), {}^{a'}x)_{d+n+m} = -[a_d \cdot y_m, x_n]$$
$$= [x_n, a_d \cdot y_m].$$

Similarly

$$((a', x), {}^{x}y)_{d+n+m} = [a_d \cdot x_n, y_m].$$

Now it follows from (2) that

$$a_d \cdot [x_n, y_m] = [a_d \cdot x_n, y_m] + [x_n, a_d \cdot y_m],$$

thus showing that a_d is a derivation, as claimed.

II k-linear central filtrations

In what follows, k denotes a commutative ring.

5. Definition

(5.1) DEFINITIONS. By a *filtration* on a group G we shall mean a chain

$$\Phi = (G = G_0 \geq G_1 \geq G_2 \geq \ldots)$$

of subgroups of G. We call it *separating* if $\cap_d G_d = \{1\}$. We call it a *k-linear-central filtration* if the following conditions are satisfied.

(a) $(G_d, G_e) \leq G_{d+e}$ for all $d, e \geq 0$.

We then put $L_d(G)$ (or $L_d(G, \Phi)$) $= G_d/G_{d+1}$, and $L_+(G)$ (or $L_+(G, \Phi)) = \bigoplus_{d \geq 1} L_d(G)$, a graded Lie **Z**-algebra.

(b) For some N_0 there is an injective group homomorphism $L_0(G) \to GL_{N_0}(k)$.

(c) For $d \geq 1$, there is an $N_d \geq 0$ and an injective group homomorphism $L_d(G) \to k^{N_d}$.

If further $G = G_1$ we call Φ a *k-central filtration*.

6. Consequences of a separating Z-linear-central filtration

(6.1) THEOREM *Let G be a group with a separating $\mathbf{Z}$-linear-central filtra-tion,*

$$G = G_0 \geq G_1 \geq G_2 \geq \cdots$$

and quotients $L_d(G) = G_d/G_{d+1}$.

(a) G_1 is residually-finitely-generated-torsion-free-nilpotent, and hence ((3.7)) residually-finite-p for all primes p.

(b) If G_1^{ab} is finitely generated and $Z_G(G_1) = \{1\}$ then, for all primes p, G is virtually residually-finite-p.

(c) G is virtually torsion free.

(d) Let $H \leq G$ be a solvable subgroup. If Hirsch rank $(H) < \infty$ then H is polycyclic. If $vcd(G) = r < \infty$ then Hirch rank $(H) \leq r$, and so H is polycyclic.

Proof of (a). Since $G_1 \geq G_2 \geq G_3 \geq \ldots$ is a **Z**-central filtration of G_1, each G_1/G_d is, by (3.5) or (4.6), finitely generated torsion free nilpotent, and hence also, by (3.7), residually-finite-p for all primes p.

Proof of (b). The hypothesis $Z_G(G_1) = \{1\}$ implies that G embeds in $Aut(G_1)$. It is shown in (8.2) below that if a group H satisfies H^{ab} is finitely generated and H is residually-torsion-free-nilpotent then, for all primes p, $Aut(H)$ is virtually residually-finite-p.

Proof of (c). Since G_1 is torsion free and G/G_1 embeds in some $GL_{N_0}(\mathbf{Z})$, which is virtually torsion free (cf. (2.3)), it follows that G is virtually torsion free.

Proof of (d). Let Hirsch rank $(H) = h$. If $vcd(G) = r < \infty$ then $h \leq r$, by (1.4). The image of H in G/G_1 is isomorphic to a solvable subgroup of $GL_{N_0}(\mathbf{Z})$, and hence is polycyclic, by (1.2). Putting $H_d = H \cap G_d$ we see then that H/H_1

is polycyclic. For $d \geq 1$, $H_d/H_{d+1} \leq L_d(G) \leq \mathbf{Z}^{N_d}$ (cf. (5.1)), so $H_d/H_{d+1} \cong \mathbf{Z}^{h_d}$ for some $h_d \leq N_d$. Clearly $h_1 + h_2 + \ldots + h_d \leq h = Hirsch\ rank(H)$. It follows that $h_d = 0$ for all $d \geq some\ d_0$. This means that $H_d = H_{d_0}$ for all $d \geq d_0$. But $\cap_{d \geq d_0} H_d = H \cap (\cap_{d \geq d_0} G_d) = \{1\}$ since the filtration of G is separating. Thus $H_{d_0} = \{1\}$. If follows that $H_1 = H_1/H_{d_0}$ embeds in G_1/G_{d_0}, a finitely generated nilpotent, hence, polycyclic group, and so H_1 is polycyclic.

(6.2) QUESTION. Must a group G with a separating $\mathbf{Z}$-linear-central filtration satisfy the Tits alternative (cf. (2.4))? We suspect that the answer is negative. (Cf. Example (15.4) below.)

7. Canonical Z-central filtrations

Let F be a group.

(7.1) DEFINITION. The subgroups

$$(1) \qquad\qquad F = C_{\mathbf{Z}}^1(F) \geq C_{\mathbf{Z}}^2(F) \geq C_{\mathbf{Z}}^3(F) \geq \ldots$$

are defined inductively by the conditions

$$(2) \qquad\qquad (F, C_{\mathbf{Z}}^n(F)) \leq C_{\mathbf{Z}}^{n+1}(F)$$

(and so $C_{\mathbf{Z}}^n(F) \triangleleft F$ for all n), and

$$(3) \qquad \begin{aligned} &C_{\mathbf{Z}}^{n+1}(F)/(F, C_{\mathbf{Z}}^n(F)) = \\ &\text{the torsion subgroup of } C_{\mathbf{Z}}^n(F)/(F, C_{\mathbf{Z}}^n(F)). \end{aligned}$$

Equivalently,

$$(3') \qquad \begin{aligned} &C_{\mathbf{Z}}^n(F)/C_{\mathbf{Z}}^{n+1}(F) = \\ &\text{the torsion free quotient of } C_{\mathbf{Z}}^n(F)/(F, C_{\mathbf{Z}}^n(F)). \end{aligned}$$

It follows inductively that if we have a filtration $F = F_1 \geq F_2 \geq F_3 \geq \ldots$ with $(F, F_n) \leq F_{n+1}$ and F_n/F_{n+1} torsion free for all n, then $C_{\mathbf{Z}}^n(F) \leq F_n$ for all n.

(7.2) PROPOSITION. *(a) For all $n, m \geq 1$,*

$$(C_{\mathbf{Z}}^n(F), C_{\mathbf{Z}}^m(F)) \leq C_{\mathbf{Z}}^{n+m}(F).$$

Thus $(C_{\mathbf{Z}}^n(F))_{n \geq 1}$ is a central filtration, with associated Lie algebra $L^{\mathbf{Z}}(F) = \oplus_{n \geq 1} L_n^{\mathbf{Z}}(F)$, where $L_n^{\mathbf{Z}}(F) = C_{\mathbf{Z}}^n(F)/C_{\mathbf{Z}}^{n+1}(F)$.

(b) For all $n \geq 1$, $C^n(F) \leq C_{\mathbf{Z}}^n(F)$, and

$$C_{\mathbf{Z}}^n(F)/C^n(F) = Tors(F/C^n(F)),$$

the torsion subgroup of the nilpotent group $F/C^n(F)$. Thus the induced homomorphism of graded Lie algebra

$$L(F) \to L^{\mathbf{Z}}(F)$$

has torsion kernel and cokernel.

(c) The graded Lie $\mathbf{Q}$-algebra $L^{\mathbf{Z}}(F)_{\mathbf{Q}} = \mathbf{Q} \otimes_{\mathbf{Z}} L^{\mathbf{Z}}(F)$ is generated by $L_1^{\mathbf{Z}}(F)_{\mathbf{Q}}$.

(d) If F^{ab} is finitely generated then $(C_{\mathbf{Z}}^n(F))_{n \geq 1}$ is a $\mathbf{Z}$-central filtration.

(e) $\bigcap_{n \geq 1} C_{\mathbf{Z}}^n(F) = \{1\}$ iff F is residually torsion-free-nilpotent.

For the proof we shall abbreviate $C^n = C^n(F)$ and $D^n = C_{\mathbf{Z}}^n(F)$.

Proof of (a). By induction on $d \geq 1$ we show that

(1)$_d$ $\qquad\qquad\qquad (D^d, D^n) \leq D^{d+n}$ for all $n \geq 1$.

The case $d = 1$ follows from Definition (7.1). Assuming (1)$_d$, we now show (1)$_{d+1}$. Since $D^{d+n}/D^{d+n+1} \leq Z(F/D^{d+n+1})$ the commutator pairing

(2) $\qquad\qquad\qquad (\cdot, \cdot) : D^d \times D^n \to D^{d+n} \to D^{d+n}/D^{d+n+1}$

is bilinear. For $x \epsilon F, y \epsilon D^d, z \epsilon D^n$, writing $z' = y^{-1} z y \in D^n$ we have, using (3.1)(3)

$$((x, y), z) = ((x, y), {}^y z') = [((y, z'), {}^{z'} x)((z', x), {}^x y)]^{-1}.$$

Now, using $(1)_d$,

$$((y, z'), {}^{z'}x) \in (F, (D^d, D^n)) \leq (F, D^{d+n}) \leq D^{d+n+1},$$

and

$$((z', x), {}^x y) \in (D^d, (F, D^n)) \leq (D^d, D^{n+1}) \leq D^{d+n+1}.$$

This, and the analogous calculation for $(y, (x, z))$, shows that the pairing (2) factors through a bilinear pairing

(3) $$\frac{D^d}{(F, D^d)} \times \frac{D^n}{(F, D^n)} \rightarrow \frac{D^{d+n}}{D^{d+n+1}}.$$

Since D^{n+d}/D^{n+d+1} is, by construction, torsion free, the pairing (3) kills torsion in both variables, and so factors through $(D^d/D^{d+1}) \times (D^n/D^{n+1}) \rightarrow D^{d+n}/D^{d+n+1}$, whence $(1)_{d+1}$.

Proof of (b). Since $(F, D^n) \leq D^{n+1}$ we have $C^n \leq D^n$, for all n. Since F/D^n is torsion free nilpotent, we have

(4) $$Tors(F/C^n) \leq D^n/C^n.$$

Writing $Tors(F/C^n) = T/C^n, C^n \leq T \leq D^n$, we see that $H = F/T$ is torsion free and nilpotent of class $\leq n$. Define

$$H = H_1 \geq H_2 \geq \ldots \geq H_n = \{1\}$$

by $H_r = Z^{n-r}(H)$ $(r = 1, \ldots, n)$, and put $H_r = \{1\}$ for all $r > n$. Then, by (3.7) $(F, H_r) \leq H_{r+1}$, and H_r/H_{r-1} is torsion free for all $r \geq 1$. Let F_r denote the inverse image of H_r in $F : H_r = F_r/T$. Then $(F, F_r) \leq F_{r+1}$, F_r/F_{r+1} is torsion free, and $F_r = T$ for $r \geq n$. It follows from definition (7.1) that $D^r \leq F_r$ for all r. In particular $D^n \leq F_n = T$, and so, in view of (4), $D^n = T$, as claimed.

Now with $L_n(F) = C^n/C^{n+1}$ and $L_n^{\mathbf{Z}}(F) = D^n/D^{n+1}$, we have an exact sequence

$$0 \rightarrow \frac{C^n \cap D^{n+1}}{C^{n+1}} \rightarrow L_n(F) \rightarrow L_n^{\mathbf{Z}}(F) \rightarrow \frac{D^n}{C^n D^{n+1}} \rightarrow 0.$$

Since D^n/C^n and D^{n+1}/C^{n+1} are torsion, so also are the cokernel and kernel above.

Proof of (c). This follows from the isomorphism

$$L(F)_{\mathbf{Q}} \to L^{\mathbf{Z}}(F)_{\mathbf{Q}},$$

given by (b), plus the fact that $L_1(F)$ generates the Lie $\mathbf{Z}$-algebra $L(F)$ (cf. (4.6)).

Proof of (d). If F^{ab} is finitely generated, then the quotient F/D^{n+1} of F/C^{n+1} is polycyclic, by (3.5), and so D^n/D^{n+1} is finitely generated torsion free abelian, hence isomorphic to $\mathbf{Z}^{N_d}$ for some $N_d \geq 0$. Thus, in view of (a), $(D^d)_{d \geq 1}$ is a $\mathbf{Z}$-central filtration.

Proof of (e). One implication follows since each F/D^n is torsion-free-nil-potent. Suppose, conversely, that F is residually-torsion-free-nilpotent. We must show that $\cap_{d \geq 1} D^d = \{1\}$. Let $x \neq 1$. Choose $N \triangleleft F, x \notin N$, with $H = F/N$ torsion-free nilpotent. As in the proof of (b) above, the reverse indexed ascending central series of H produces a filtration $H = H_1 \geq H_2 \geq H_3 \geq \ldots$ such that $(H, H_n) \leq H_{n+1}$ and H_n/H_{n+1} is torsion free, and $H_n = \{1\}$ for all $n \geq$ some n_0. Writing $H_n = F_n/N$ we obtain $F = F_1 \geq F_2 \geq F_3 \geq \ldots$ with $(F, F_n) \geq F_{n+1}$ and F_n/F_{n+1} torsion free for all $n \geq 1$, and $F_n = N$ for $n \geq n_0$. It follows that $D^n \leq F_n$ for all n, and so $D^n \leq N$ for $n \geq n_0$, in which case $x \notin D^n$.

8. Filtered automorphism groups

(8.1) The set-up. Consider a group F with its central filtration (cf. (7.1))

(1) $$F_n = C_{\mathbf{Z}}^n(F) \qquad (n \geq 1)$$

and graded Lie $\mathbf{Z}$-algebra

(2) $$L^{\mathbf{Z}}(F) = \bigoplus_{n \geq 1} L_n^{\mathbf{Z}}(F), \quad L_n^{\mathbf{Z}}(F) = F_n/F_{n+1}.$$

Let

$$G = Aut(F),$$

which acts on F, leaving each F_n invariant. As in (4.1) we put

$$(3) \qquad G_d = \bigcap_{n \geq 1} Ker(\rho_{n,d} : G \to Aut(F_n/F_{n+d})).$$

Thus

$$(4) \qquad G = G_0 \geq G_1 \geq G_2 \geq \ldots$$

and, according to (4.2),

$$(5) \qquad (G_d, G_e) \leq G_{d+e} \quad \forall\, d, e \geq 0.$$

We put $L_d(G) = G_d/G_{d+1}$, so that

$$(6) \qquad L_+(G) = \bigoplus_{d \geq 1} L_d(G)$$

is a graded Lie $\mathbf{Z}$-algebra. According to (4.9) we have an injective group homomorphism

$$(7) \qquad L_0(G) \xrightarrow{\delta_0} Aut_{gr\ Lie\ alg.}(L^{\mathbf{Z}}(F))$$

and an injective homomorphism of graded Lie $\mathbf{Z}$-algebras

$$(8) \qquad L_+(G) \xrightarrow{\delta_+} gr\,Der_{\mathbf{Z}}(L^{\mathbf{Z}}(F)).$$

Since $L^{\mathbf{Z}}(F)$ is torsion free, and $L^{\mathbf{Z}}(F)_{\mathbf{Q}}$ is generated by $L_1^{\mathbf{Z}}(F)_{\mathbf{Q}}$, it follows that a Lie algebra automorphism or derivation of $L^{\mathbf{Z}}(F)$ is determined by its restriction to $L_1^{\mathbf{Z}}(F) = F^{ab}/(torsion)$. Consequently we have

$$(9) \qquad Aut_{gr\ Lie\ alg.}(L^{\mathbf{Z}}(F)) \hookrightarrow Aut_{\mathbf{Z}}(L_1^{\mathbf{Z}}(F))$$

and

$$(10) \qquad Der_{\mathbf{Z}}(L^{\mathbf{Z}}(F))_d \hookrightarrow Hom_{\mathbf{Z}}(L_1^{\mathbf{Z}}(F), L_{d+1}^{\mathbf{Z}}(F)).$$

Combining (7) with (9) and (8) with (10) we obtain embeddings

$$(11) \qquad L_0(G) \hookrightarrow Aut_{\mathbf{Z}}(L_1^{\mathbf{Z}}(F))$$

and, for $d \geq 1$,

(12) $L_d(G) \hookrightarrow Hom_{\mathbf{Z}}(L_1^{\mathbf{Z}}(F), L_{d+1}^{\mathbf{Z}}(F))$, hence $L_d(G)$ is torsion free.

Finally we remark that

(13) $$\bigcap_{n \geq 1} F_n = \{1\} \ \Rightarrow \ \bigcap_{d \geq 0} G_d = \{1\}.$$

For suppose that $g \in G$ and $g \neq 1$, so that $g(x) \neq x$ for some $x \in F$. By assumption, $g(x)x^{-1} \notin F_{d+1}$ for some $d \geq 0$. Hence, by definition, $g \notin G_d$.

We now draw some conclusions from this discussion.

(8.2) **THEOREM.** Let $F_n = C_{\mathbf{Z}}^n(F)$ $(n \geq 1)$, and $(G_d)_{d \geq 0}$ be as in (8.1). Assume that F^{ab} is finitely generated.

(a) $(F_n)_{n \geq 1}$ is a $\mathbf{Z}$-central filtration; say $L_n^{\mathbf{Z}}(F)(= F_n/F_{n+1}) \cong \mathbf{Z}^{N_n}$ $(n \geq 1)$.

(b) $(G_d)_{d \geq 0}$ is a $\mathbf{Z}$-linear-central filtration on $G = Aut(F)$. We have embeddings

$$L_0(G) \hookrightarrow Aut_{\mathbf{Z}}(L_1^{\mathbf{Z}}(F)) \cong GL_{N_1}(\mathbf{Z})$$

and, for $d \geq 1$,

$$L_d(G) \hookrightarrow Hom_{\mathbf{Z}}(L_1^{\mathbf{Z}}(F), L_{d+1}^{\mathbf{Z}}(F)) \cong \mathbf{Z}^{N_1 \cdot N_{d+1}}.$$

(c) If F is residually-torsion-free-nilpotent then $(G_d)_{d \geq 0}$ is separating, i.e. $\bigcap_d G_d = \{1\}$.

Part (a) follows from (7.2)(d). Then (b) follows from (8.1) (5), (11), and (12). Finally (c) follows from (7.2)(e) plus (8.1)(13).

(8.3) **COROLLARY.** If F is finitely generated and residually-torsion-free-nilpotent, then $G = Aut(F)$ admits a separating $\mathbf{Z}$-linear-central filtration, hence satifies the conditions of Theorem (6.1).

For later use we record the following observation.

(8.4) LEMMA. *For all $d \geq 1$,*

$$G_d = Ker(\rho_{1,d} : G \to Aut(F/F_{1+d}))$$

Proof. By definition,

$$G_d = \bigcap_{n \geq 1} Ker(\rho_{n,d}) \leq Ker(\rho_{1,d}).$$

Let $g \in Ker(\rho_{1,d})$. Operating in $F \rtimes G$, we must show that

$(1)_n$ $\qquad\qquad\qquad\qquad (g, F_n) \subset F_{n+d}$

for all $n \geq 1$, this being our assumption for $n = 1$. Assuming $(1)_n$, we show $(1)_{n+1}$. Let $x \in F, y \in F_n$, and put $g' = y^{-1}gy = g^y$. Then, by (3.1)(3),

$$((x, y), g) = ((x, y), {}^y g')$$
$$= [((y, g'), {}^{g'} x)((g', x), {}^x y)]^{-1}.$$

Now $(y, g') = (y, g)^y \in (g, F_n)^y \subset F_{n+d}$, by $(1)_n$, so $((y, g'), {}^{g'} x) \in (F, F_{n+d}) \leq F_{n+1+d}$. Similarly $(g', x) = (g, {}^y x)^y \in (g, F)^y \subset F_{1+d}$, by $(1)_1$, so $((g', x), {}^x y) \in (F_{1+d}, F_n) \leq F_{n+1+d}$. This shows that

(2) $\qquad\qquad\qquad\qquad (g, (F, F_n)) \subset F_{n+1+d}$

Thus $(g, \cdot)$ induces a homomorphism

$$F_n/(F, F_n) \to F_{n+d}/F_{n+1+d} = L_{n+d}^{\mathbf{Z}}(F).$$

since $L_{n+d}^{\mathbf{Z}}(F)$ is torsion free, this factors through $L_n^{\mathbf{Z}}(F) = F_n/F_{n+1}$, thus showing, as required, that $(g, F_{n+1}) \subset F_{n+1+d}$.

9. Filtered outer automorphism groups

(9.1) The set-up. We keep the setting and notation of (8.1), and put

$$H = Out(F),$$

the group of outer automorphisms of F. Thus we have a canonical exact sequence

(1) $$1 \to Z(F) \to F \xrightarrow{ad} G \xrightarrow{\pi} H \to 1.$$

We put

(2) $$H_d = \pi(G_d), \quad \text{and} \quad F'_d = ad^{-1}(G_d).$$

It follows from (8.1) (4) and (5) that

(3) $$H = H_0 \geq H_1 \geq H_2 \geq \dots$$

and

(4) $$(H_d, H_e) \leq H_{d+e} \ \forall \ d, e \geq 0.$$

Put $L_d(H) = H_d/H_{d+1}$ and $L_+(H) = \oplus_{d \geq 1} L_d(H)$. Since $(F, F_n) \leq F_{n+1} \forall n$ we have

(5) $$ad(F) \leq G_1, \quad \text{hence} \quad L_0(G) \xrightarrow{\cong} L_0(H).$$

Further (5) and (8.1)(5) imply that,

(6) $$(F'_n)_{n \geq 1} \text{ is a central filtration of } F.$$

We put

(7) $$L'(F) = \bigoplus_{n \geq 1} L'_n(F), \qquad L'_n(F) = F'_n/F'_{n+1}.$$

It follows from the definition of F'_n that the inclusion induced homomorphism $L'_n(F) \xrightarrow{\varepsilon} L_n(G)$ is injective for $n \geq 1$, and so, by (8.1)(12), $L'_n(F)$ is torsion free. Consequently, by (7.1),

(8) $$F_n \leq F'_n \ \forall n \geq 1$$

(9.2) LEMMA. *The sequence of graded Lie algebra homomorphisms*

$$0 \to L'(F) \xrightarrow{\varepsilon} L_+(G) \xrightarrow{L_+(\pi)} L_+(H) \to 0$$

is exact.

Proof. In degree $d \geq 1$ we have the exact sequence

$$
\begin{array}{ccccc}
 & L_d(G) & & L_d(H) & \\
 & \| & & \| & \\
1 \rightarrow \dfrac{(G_{d+1} \cdot ad(F)) \cap G_d}{G_{d+1}} & \rightarrow & \dfrac{G_d}{G_{d+1}} \xrightarrow{L_d(\pi)} \dfrac{G_d \cdot ad(F)}{G_{d+1} \cdot ad(F)} & \rightarrow 1.
\end{array}
$$

Since $(G_{d+1} \cdot ad(F)) \cap G_d = G_{d+1} \cdot (ad(F) \cap G_d) = G_{d+1} \cdot ad(F'_d)$ we have

$$
\frac{(G_{d+1} \cdot ad(F)) \cap G_d}{G_{d+1}} = \frac{ad(F'_d)}{G_{d+1} \cap ad(F'_d)} = \frac{ad(F'_d)}{ad(F'_{d+1})} \cong \frac{F'_d}{F'_{d+1}} = L'_d(F).
$$

(9.3) LEMMA. *The following diagram of graded Lie algebra homomorphisms is commutative.*

$$
\begin{array}{ccc}
L^{\mathbf{Z}}(F) & \xrightarrow{L(ad)} & L_+(G) \\
\\
Ad \searrow & & \downarrow \delta_+ \\
\\
& gr\, Der_{\mathbf{Z}}(L^{\mathbf{Z}}(F)) &
\end{array}
$$

(Here δ_+ is the homomorphism given by (8.1)(8), or (4.9)(c).)

Proof. Let $a \in F_d, g = ad(a) \in G_d$, and $x \in F_n$. Denote their classes in $L_d^{\mathbf{Z}}(F), L_d(G)$, and $L_n^{\mathbf{Z}}(F)$ by $\bar{a}, \bar{g}$, and $\bar{x}$, respectively. Then

$$
Ad(\bar{a})(\bar{x}) = [\bar{a}, \bar{x}] = \overline{(a, x)} \in L_{n+d}^{\mathbf{Z}}(F).
$$

On the other hand $L(ad)(\bar{a}) = \bar{g}$, and, calculating in $F \rtimes G$,

$$
\delta_d(\bar{g})(\bar{x}) = \overline{(g, x)} = \overline{g(x)x^{-1}}
$$
$$
= \overline{axa^{-1}x^{-1}} = \overline{(a, x)},
$$

whence the assertion of the Lemma.

(9.4) It follows that we have the following chain of graded Lie algebras

$$gr\,Out\,Der(L^{\mathbf{Z}}(F))\begin{cases} gr\,Der(L^{\mathbf{Z}}(F)) \\[4pt] \quad\mid \\[4pt] \delta_+L_+(G)\cong L_+(G) \\[4pt] \quad\mid \\[4pt] \delta_+\varepsilon L'(F)\cong L'(F) \\[4pt] \quad\mid \\[4pt] Ad(L^{\mathbf{Z}}(F))\cong L^{\mathbf{Z}}(F)/Z(L^{\mathbf{Z}}(F)) \\[4pt] \quad\mid \\[4pt] \quad 0 \end{cases} \qquad L_+(H)$$

(9.5) PROPOSITION: *Assume that,*

(1) $$Z(L^{\mathbf{Z}}(F)) = \{0\}.$$

Then $F'_d = F_d$ for all $d \geq 1$, and we have an exact sequence of graded Lie algebras,

(2) $$0 \to L^{\mathbf{Z}}(F) \xrightarrow{L(ad)} L_+(G) \xrightarrow{L(\pi)} L_+(H) \to 0$$

If F is residually-torsion-free-nilpotent then condition (1) implies that

(1') $$Z(F) = \{1\}.$$

Proof. Say $a \in F'_d$. If $a \notin F_d$ then let $e(1 \leq e < d)$ be the largest index such that $a \in F_e$, and let $\bar{a} \neq 0$ denote the class of a in $L^{\mathbf{Z}}_e(F)$. For $x \in F_n$, with class $\bar{x} \in L^{\mathbf{Z}}_n(F)$, we have $[\bar{a}, \bar{x}] = \overline{(a,x)} \in L^{\mathbf{Z}}_{n+e}(F)$. But $(a,x) \in F_{n+d} \leq F_{n+e+1}$ since $e < d$, so $[\bar{a}, \bar{x}] = 0$. Thus $\bar{a} \in Z(L^{\mathbf{Z}}(F)) = \{0\}$; contradiction. Thus $F'_d = F_d$ for all $d \geq 1$, and so the exact sequence (2) follows from (9.2) and (9.3).

If $a \in Z(F) \cap F_d$ then the class $\bar{a}$ of a in $L^{\mathbf{Z}}_d(F)$ belongs to $Z(L^{\mathbf{Z}}(F))$. Thus (1) implies that $\bar{a} = 0$, i.e. that $a \in F_{d+1}$. By induction, (1) implies that $Z(F) \leq \cap_d F_d$. If F is residually-torsion-free-nilpotent then $\cap_d F_d = \{1\}$, by (7.2)(e), and so (1) implies that $Z(F) = \{1\}$.

(9.6) Torsion freeness. Assuming that

(1) $$Z(L^{\mathbf{Z}}(F)) = \{0\}$$

we have a commutative diagram of graded Lie algebras, with exact rows

$$0 \to L^{\mathbf{Z}}(F) \xrightarrow{L(ad)} L_+(G) \qquad\qquad \to L_+(H) \qquad\qquad \to 0$$

$$(2) \qquad\qquad \| \qquad\qquad \downarrow \delta_+ \qquad\qquad \downarrow$$

$$0 \to L^{\mathbf{Z}}(F) \xrightarrow[Ad]{} grDer(L^{\mathbf{Z}}(F)) \to grOutDer(L^{\mathbf{Z}}(F)) \to 0$$

We now investigate conditions for $L_+(H)$ to be torsion free.

Let p be a prime. For any abelian group homomorphism $f : A \to B$, we shall provisionally write $f_p : A_p \to B_p$ for the induced homomorphism $(f \bmod p)$: $A/pA \to B/pB$. Then the diagram (2) $mod \ p$ furnishes part of the following commutative diagram with exact rows.

$$L^{\mathbf{Z}}(F)_p \xrightarrow{L(ad)_p} L_+(G)_p \qquad \to \qquad L_+(H)_p \qquad \to \ 0$$

$$\| \qquad\qquad \downarrow (\delta_+)_p \qquad\qquad \downarrow$$

$$L^{\mathbf{Z}}(F)_p \xrightarrow[Ad_p]{} grDer(L^{\mathbf{Z}}(F))_p \ \to \ grOutDer(L^{\mathbf{Z}}(F))_p \ \to \ 0$$

$$\| \qquad\qquad \downarrow \alpha \qquad\qquad \downarrow \alpha'$$

$$L^{\mathbf{Z}}(F)_p \xrightarrow[Ad]{} grDer(L^{\mathbf{Z}}(F)_p) \ \to \ grOutDer(L^{\mathbf{Z}}(F)_p) \ \to \ 0$$

Here the canonical maps α and α' are easily seen to be injective, since, $L^{\mathbf{Z}}(F)$ being torsion free, a derivation $D : L^{\mathbf{Z}}(F) \to L^{\mathbf{Z}}(F)$ with $DL^{\mathbf{Z}}(F) \le pL^{\mathbf{Z}}(F)$ is of the form $D = pD'$ for a unique derivation $D' : L^{\mathbf{Z}}(F) \to L^{\mathbf{Z}}(F)$. Thus we have the equivalence of the conditions

$$(1)_p \qquad \begin{cases} (a) & Ad_p : L^{\mathbf{Z}}(F)_p \to grDer(L^{\mathbf{Z}}(F))_p \text{ is injective.} \\ (b) & Ad : L^{\mathbf{Z}}(F)_p \to grDer(L^{\mathbf{Z}}(F)_p) \text{ is injective.} \\ (c) & Z(L^{\mathbf{Z}}(F)_p) = \{0\}. \end{cases}$$

Further these conditions imply that

$$(1')_p \qquad\qquad L(ad)_p : L^{\mathbf{Z}}(F)_p \to L^{\mathbf{Z}}_+(G)_p \ \text{ is injective.}$$

Since $L_+(G)$ is torsion free it follows from the exact sequence (2) that

$$(3) \qquad \begin{cases} L_+(H) \text{ is torsion free iff, } \forall \text{ primes } p, \\ (1')_p : L(ad)_p : L^{\mathbf{Z}}(F)_p \to L_+(G)_p \text{ is injective.} \end{cases}$$

From $(1)_p \Rightarrow (1')_p$ above we thus conclude that:

(4) $\qquad \left\{ \begin{array}{l} \text{If } Z(L^{\mathbf{Z}}(F)_p) = \{0\} \text{ for all primes } p \quad \text{then} \\ L_+(H) \text{ is torsion free.} \end{array} \right.$

Note that if $L^{\mathbf{Z}}(F)$ is $\mathbf{Z}$-free, for example if F^{ab} is finitely generated (cf. (7.2)(d)), then $Z(L^{\mathbf{Z}}(F)) = \{0\}$ as soon as $Z(L^{\mathbf{Z}}(F)_p) = \{0\}$ for even a single prime p.

(9.7) Separability. We have $H = G/ad(F)$ and

$$H_d = \frac{G_d \cdot ad(F)}{ad(F)},$$

hence

(1) $\qquad \bigcap_d H_d = (\bigcap_d G_d \cdot ad(F))/ad(F).$

If follows that

The filtration $(H_d)_{d \geq 0}$ is separating,

(2) $\qquad$ i.e. $\bigcap_d H_d = \{1\}$, iff $\bigcap_d G_d \cdot ad(F) = ad(F).$

Recall from (8.4) that

(3) $\qquad G_d = Ker(\rho_{1,d} : G \to Aut(F/F_{1+d})).$

Thus we see that, for $g \in G$,

$$g \in \bigcap_d G_d \cdot ad(F)$$

$$\Leftrightarrow \left\{ \begin{array}{l} \forall d \geq 1 \ \exists \ x_d \in F \text{ such that} \\ g = ad(x_d) \text{ on } F/F_d \end{array} \right.$$

$$\Leftrightarrow \left\{ \begin{array}{l} \forall d \geq 1, g \text{ induces an inner} \\ \text{automorphism on } F/F_d \end{array} \right. .$$

Thus $\bigcap_d H_d = \{1\}$ iff F satisfies:

$$IN(F) : \left\{ \begin{array}{l} \text{If } g \in G = Aut(F) \text{ induces an inner automorphism} \\ \text{of } F/F_d \ \forall d \geq 1, \text{ then } g \text{ is inner.} \end{array} \right.$$

We now summarize our conlusions.

(9.8) THEOREM. *As in (9.1), let F be a group, $G = Aut(F)$, and $H = Out(F)$, with exact sequence*

$$1 \to Z(F) \to F \xrightarrow{ad} G \xrightarrow{\pi} H \to 1.$$

Let $F_n = C_{\mathbf{Z}}^n(F)$, as in (7.1) and, for $d \geq 0$, put

$$G_d = Ker(G \to Aut(F/F_{d+1})) \quad (cf.(8.4))$$

and

$$H_d = \pi(G_d).$$

(a) $(H_d, H_e) \leq H_{d+e}$ $\forall d, e \geq 0$. *Putting* $L_d(H) = H_d/H_{d+1}$ *we have* $L_0(G) \xrightarrow{\cong} L_0(H)$.

(b) *If* $Z(L(F)) = 0$ *then the sequence of graded Lie algebras*

$$0 \to L^{\mathbf{Z}}(F) \xrightarrow{L(ad)} L_+(G) \xrightarrow{L_+(\pi)} L_+(H) \to 0$$

is exact. If $Z(\mathbf{F}_p \otimes L^{\mathbf{Z}}(F)) = 0$ *for all primes p then $L_+(H)$ is torsion free.*

(c) $\bigcap_d H_d = \{1\}$ *iff F satisfies, $IN(F)$: If $g \in G = Aut(F)$ and g on F/F_d is inner for all $d \geq 1$, then g is inner on F.*

(9.9) COROLLARY. *If F^{ab} is finitely generated and $Z(\mathbf{F}_p \otimes L^{\mathbf{Z}}(F)) = \{0\}$ for all primes p then $(H_d)_{d \geq 0}$ is a $\mathbf{Z}$-linear-central filtration on $H = Out(F)$. It is separating iff F satisfies $IN(F)$.*

This follows from (9.8) (a), (b) and (c) together with Theorem (8.2).

10. Free groups

(10.1) NOTATION. Let F be the free group on $r \geq 2$ generators $x_1, \ldots, x_r$, with classes $\bar{x}_1, \ldots, \bar{x}_r$ in $F^{ab} = L_1(F)$. Put

(1)
$$\begin{cases} F_n = C^n(F), \ L_n(F) = F_n/F_{n+1}, \ \text{and} \\ L(F) = \bigoplus_{n \geq 1} L_n(F). \end{cases}$$

It follows from the work of Magnus ([Mag 1,2], cf. also [MKS], Ch 5, or [NB2], Ch II) that

(2) $L(F)$ is the free Lie $\mathbf{Z}$-algebra on $\bar{x}_1, \ldots, \bar{x}_r$,

and, for $d \geq 1$, we have Witt's formula,

$$L_d(F) \cong \mathbf{Z}^{N_d}, \quad \text{where}$$
(3)
$$N_d = \frac{1}{d} \sum_{e \mid d} \mu(e) r^{d/e};$$

here μ denotes the Möbius function. In particular it follows that

(4) $F_n = C_{\mathbf{Z}}^n(F)$, and $L(F) = L^{\mathbf{Z}}(F)$.

As a result of (2) we see that

(5) $Z(\mathbf{F}_p \otimes L(F)) = \{0\}$ for all primes p.

Further

(6) $$\bigcap_{n \geq 1} F_n = \{1\}.$$

We next quote some results of Baumslag-Taylor and of Edna Grossman.

(10.2) PROPOSITION. *(a) If $x, y \in F$ become conjugate in F/F_n $\forall n \geq 1$ then they are conjugate in F.*

(b) If $g \in G = \mathrm{Aut}(F)$ and $g(x)$ is conjugate to x for all elements x of length ≤ 2 relative to $x_1, \ldots, x_r$ then g is inner.

(c) If $g \in \mathrm{Aut}(F)$ is inner on F/F_n for all $n \geq 1$ then g is inner.

Assertion (a) is Proposition 4.9 in [LS]. Assertion (b) is Lemma 1 in [EKG]. Assertion (c) follows immediately from (a) and (b).

(10.3)   PROPOSITION   (Culler-Vogtmann   [CV]   and   Gersten)
$vcd(G) = 2r - 2$ and $vcd(H) = 2r - 3$.

Combining the above observations with (8.3), (9.9) and (6.1)(d) we obtain:

(10.4) THEOREM. *Let F be the free group on $r \geq 2$ generators $x_1, \ldots, x_r$,*
$G = Aut(F), H = Out(F)$, *and*

$$1 \to F \xrightarrow{ad} G \xrightarrow{\pi} H \to 1.$$

Put $F_n = C^n(F), G_d = Ker(G \to Aut(F/F_{d+1}))$ and $H_d = \pi(G_d)$. Then $(G_d)_{d \geq 0}$ and $(H_d)_{d \geq 0}$ are separating $\mathbf{Z}$-linear-central filtrations. We have $L_0(G) \cong L_0(H) \cong GL_r(\mathbf{Z})$, and graded Lie $\mathbf{Z}$-algebra embeddings

$$
\begin{array}{ccc}
L_+(G) & \longrightarrow & gr\,Der(L(F)) \\
L_+(\pi) \downarrow & & \downarrow \\
L_+(H) & \longrightarrow & gr\,OutDer(L(F)),
\end{array}
$$

with $L(F)$ the free Lie $\mathbf{Z}$-algebra on $\bar{x}_1, \ldots, \bar{x}_r \in L_1(F) = F^{ab}$. Further $vcd(G) = 2r - 2$ and $vcd(H) = 2r - 3$, so each solvable subgroup of G or of H is polycyclic.

(10.4) QUESTION. Does $G = Aut(F)$ satisfy the Tits alternative? I.e. if $K \leq G$ is not virtually solvable, must K contain a non abelian free group? Filter K by $K_d = K \cap G_d$. Then $L_0(K) \leq L_0(G) = GL_r(\mathbf{Z})$. If $L_0(K)$ contains a non abelian free group then so also does K. Otherwise $L_0(K) = K/K_1$ is virtually solvable, by Tits' Theorem (2.4). Then K_1 is not virtually solvable, so it follows that $L_+(K) \leq L_+(G) \leq gr\,Der(L(F))$ is a graded subalgebra of infinite rank. If $L_+(K)$ contains a free non abelian Lie subalgebra of $L_+(G)$ then K contains a non abelian free group. This follows from the next Lemma, which thus gives a Lie algebra approach to the Tits alternative for G.

(10.5) LEMMA. *Let $x \in G_d, y \in G_e(d, e \geq 1)$ have classes $\bar{x} \in L_d(G)$ and $\bar{y} \in L_e(G)$. If the Lie algebra generated by $\bar{x}$ and $\bar{y}$ is free on $\bar{x}, \bar{y}$, then the group $K =< x, y >$ is free on x, y.*

Proof. Let $\pi : T \to K$, where T is a free group on X, Y, $\pi(X) = x$ and

$\pi(Y) = y$. Define $\tau(X) = d$ and $\tau(Y) = e$ and give T the $(\{X,Y\}, \tau)$ filtration $(T_n)_{n \geq 1}$, as in Labute [Lab], §2. Then (loc.cit.) $\Phi = (T_n)$ is a central filtration and $L(T) = L(T, \Phi)$ is a free graded Lie algebra based on $\bar{X}$ of degree d and $\bar{Y}$ of degree e. By hypothesis, $L(T) \to L(G)$ is injective, hence so also is $L(\pi) : L(T) \to L(K)$, where we filter K by $K_n = K \cap G_n$. It follows inductively that $T_n = \pi^{-1}(K_n)$. Hence

$$Ker(\pi) \leq \bigcap_n \pi^{-1}(K_n) = \bigcap_n T_n = \{1\},$$

so $\pi : T \to K$ is injective, as claimed.

(10.6) QUESTIONS. Let $L = \oplus_{d \geq 1} L_d$ be a graded Lie subalgebra of $grDer(L(F))$. If L is not nilpotent (hence not solvable) then must L contain a free non abelian graded Lie algebra? If the answer is affirmative then we obtain the Tits alternative for $G = Aut(F)$, by the discussion above.

A similar question for $grOutDer(L(F))$ relates to the Tits alternative for $H = Out(F)$.

We can also try to use the Lie algebras $L_+(G)$ and $L_+(H)$ to study the structure of virtually solvable subgroups of G and H. The following example, based on a related example of George Bergman, shows that this does not furnish a direct approach to testing the question of whether solvable subgroups of G or H are virtually abelian.

(10.7) EXAMPLE. Let $F = F(x, y, z, w)$ be free with basis $\{x, y, z, w\}$. Define $\alpha, \beta \in G = Aut(F)$ by

$$
\begin{array}{llcllc}
\alpha(x) & = & x(y, z) & \qquad \beta(x) & = & x \\
\alpha(y) & = & y & \qquad \beta(y) & = & y(z, w) \\
\alpha(z) & = & z & \qquad \beta(z) & = & z \\
\alpha(w) & = & w & \qquad \beta(w) & = & w.
\end{array}
$$

Put $\gamma = (\alpha, \beta)$; then direct calculation shows that,

$$\gamma(x) = x(y, z)(z, y(z, w)) = x\,^y(z, (z, w))$$

$$\gamma(y) = y$$

$$\gamma(z) = z$$

$$\gamma(w) = w.$$

We have, evidently, $\alpha, \beta \in G_1$ and $\gamma \in G_2$. Let $\bar{\alpha}, \bar{\beta} \in L_1(G)$ and $\bar{\gamma} = [\bar{\alpha}, \bar{\beta}] \in L_2(G)$ denote their classes. On $L(F) =$ the free Lie $\mathbf{Z}$-algebra on $\bar{x}, \bar{y}, \bar{z}, \bar{w}$, the derivations $\bar{\alpha}, \bar{\beta}, \bar{\gamma}$ are given by

$$
\begin{array}{llllll}
\bar{\alpha}(\bar{x}) & = & [\bar{y}, \bar{z}] & \bar{\beta}(\bar{x}) & = & 0 \\
\bar{\alpha}(\bar{y}) & = & 0 & \bar{\beta}(\bar{y}) & = & [\bar{z}, \bar{w}] \\
\bar{\alpha}(\bar{z}) & = & 0 & \bar{\beta}(\bar{z}) & = & 0 \\
\bar{\alpha}(\bar{w}) & = & 0 & \bar{\beta}(\bar{w}) & = & 0 \\
\end{array}
\qquad
\begin{array}{lll}
\bar{\gamma}(\bar{x}) & = & [\bar{z}, [\bar{z}, \bar{w}]] \\
\bar{\gamma}(\bar{y}) & = & 0 \\
\bar{\gamma}(\bar{z}) & = & 0 \\
\bar{\gamma}(\bar{w}) & = & 0 \\
\end{array}
$$

It is easily checked that $[\bar{\alpha}, \bar{\gamma}] = 0 = [\bar{\beta}, \bar{\gamma}]$. Thus $\mathbf{Z}\bar{\alpha} + \mathbf{Z}\bar{\beta} + \mathbf{Z}\bar{\gamma}$ is a nilpotent, but not abelian, sub Lie algebra of $L_+(G) \leq gr Der(L(F))$, and it projects injectively into $L_+(H) \leq gr OutDer(L(F))$.

Consider, on the other hand, the group $K = < \alpha, \beta > \leq G$. Let $F' = < y, z, w >$ and let G' denote the subgroup of G fixing y, z, and w. There is an isomorphism $\sigma : F' \times F' \to G'$ given by $\sigma(u, v)(x) = uxv^{-1}$. Evidently we have $\alpha, \gamma \in \sigma(\{1\} \times F') \cong F'$; in fact

$$\alpha = \sigma(1, (y, z)^{-1}) \text{ and } \gamma = \sigma(1, {}^y(z, (z, w))^{-1})$$

It follows that $< \alpha, \gamma >$ is free on α, γ.

The idea for this example came from some derivations on the free associative algebra similar to $\bar{\alpha}$ and $\bar{\beta}$ above, which were furnished to us by George Bergman.

11. Surface groups

(11.1) **NOTATION.** let S_r be a closed orientable surface of genus $r \geq 2$, with fundamental group

(1)
$$F = \pi_1(S_r)$$

$$= < x_1, y_1, \ldots, x_r, y_r | (x_1, y_1) \ldots (x_r, y_r) = 1 >$$

Put $F_n = C^n(F), L_n(F) = F_n/F_{n+1}$. Then (cf. [Lab1])

(2)
$$
\begin{cases}
L(F) = \oplus_{n\geq1} L_n(F) \text{ is the graded Lie } \mathbf{Z}\text{-algebra} \\
\text{presented by generators } \bar{x}_1, \bar{y}_1, \ldots, \bar{x}_r, \bar{y}_r \in L_1(F) = F^{ab}, \\
\text{subject to the single relation} \\
[\bar{x}_1, \bar{y}_1] + \ldots + [\bar{x}_r, \bar{y}_r] = 0.
\end{cases}
$$

Moreover (loc.cit.)

(3)
$$
\begin{cases}
L_n(F) \cong \mathbf{Z}^{N_n}, \text{ where} \\
N_n = \frac{1}{n}\sum_{d/n} \mu(n/d)[\sum_{0\leq i\leq[d/2]}(-1)^i \frac{d}{d-i}\binom{d-i}{i}(2r)^{d-2i}]
\end{cases}
$$

It follows that $C^n(F) = C_{\mathbf{Z}}^n(F)$, so $L(F) = L^{\mathbf{Z}}(F)$. Further, Labute has shown (private correspondence) that:

(4)
$$
\begin{cases}
\text{For all commutative rings k,} \\
Z(k \otimes_{\mathbf{Z}} L(F)) = \{0\}.
\end{cases}
$$

As pointed out by the referee, this result is also quoted in [As1]; in the preprint [As2] a complete proof is given, as well as of the existence of $\mathbf{Z}$-central filtrations as above associated with outer automorphism groups of free groups and surface groups.

(11.2) Let $G = Aut(F)$ and $H = Out(F)$, with the exact sequence

(1)
$$
1 \to F \xrightarrow{ad} G \xrightarrow{\pi} H \to 1.
$$

We use $(F_n)_{n\geq1}$ to filter G and H as in (8.1) and (9.1) (cf. (8.4) also). In view of (11.1)(3),(4), (8.2), (9.5), and (9.6) (4), we have a commutative diagram of graded Lie $\mathbf{Z}$-algebras

(2)
$$
\begin{array}{ccccc}
0 \to L(F) & \xrightarrow{L(ad)} & L_+(G) & \xrightarrow{L_+(\pi)} & L_+(H) & \to 0 \\
& \| & & \downarrow & & \downarrow \\
0 \to L(F) & \xrightarrow[Ad]{} & gr\,Der(L(F)) \to & grOutDer(L(F)) \to 0
\end{array}
$$

in which the rows are exact, the vertical arrows are injective, and each $L_n(G)$ and $L_n(H)$ is a finitely generated free $\mathbf{Z}$-module.

(11.3) PROPOSITION. (a) (G. Baumslag [GB2]) *F is residually free, hence residually-torsion-free-nilpotent.*

(b) (Edna Grossman [EKG]) *If $g \in G$ and $g(x)$ is conjugate to x for all $x \in F$ then g is inner.*

(c) (P. Stebe [St]). *Let $x, y \in F$. If x and y become conjugate in all finite quotients of F then x and y are conjugate in F.*

(11.4) REMARKS. In view of $(8.1)(13)$, $(11.3)(a)$ implies that

(1) $$\bigcap_d G_d = \{1\}.$$

To obtain the condition, $\bigcap_d H_d = \{1\}$, we need condition $IN(F)$ of $(9.8)(c)$:

$$IN(F) : \begin{cases} \text{If } g \in F \text{ induces an inner automorphism} \\ \text{of } F/F_n \text{ for all } n, \text{ then } g \text{ is inner on } F. \end{cases}$$

This would follow from $(11.3)(b)$ if we know that:

(2) $$\begin{cases} \text{If } x, y \in F \text{ become conjugate in } F/F_n \\ \text{for all } n \text{ then } x \text{ and } y \text{ are conjugate in } F. \end{cases}$$

A result of Blackburn [Bl] (cf. also [GB1], no. 4) implies that (2) is equivalent to the analogue of Stebe's result $(11.3)(c)$, but with finite groups of prime power order in place of all finite groups. However we are nonetheless able to obtain $IN(F)$ by other methods, at least for $g \geq 3$. This follows from Example $(14.6)2$ below:

(3) $$IN(F) \text{ holds for } g \geq 3.$$

We have not yet been able to determine whether $IN(F)$ holds when $g = 2$.

(11.5) THEOREM. *Let $F = \pi_1(S_g), g \geq 2$, and let $(F_n)_{n \geq 1}$ be the descending central series of F, with Lie algebra $L(F)$. Filter $G = Aut(F)$ and $H = Out(F)$ by $G_d = Ker(G \to Aut(F/F_{d+1}))$, and $H_d = \pi(G_d)$, as in (11.2). Then $(F_n)_{n \geq 1}$ is a separating $\mathbf{Z}$-central filtration, $(G_d)_{d \geq o}$ and $(H_d)_{d \geq 0}$ are $\mathbf{Z}$-linear-central filtrations, we have a group embedding $L_0(G) = L_0(H) \leq Aut_{\mathbf{Z}}(L_1(F))$, and graded Lie algebra* *embeddings* $L_+(G) \leq gr\,Der(L(F))$ *and* $L_+(H) \leq grOutDer(L(F))$. *The filtration* $(G_d)_{d \geq 0}$ *is separating, and* $(H_d)_{d \geq 0}$ *is separating if* $g \geq 3$.

12. *p*-filtrations and Lizard algebras

This section is not used elsewhere. We simply indicate how to formulate our methods with $\mathbf{F}_p$ in place of $\mathbf{Z}$, where p is a prime.

(12.1) A *p-central filtration* on a group F is a central filtration

$$(1) \qquad \Phi = (F = F_1 \geq F_2 \geq F_3 \geq \dots)$$

such that $F_n^p \leq F_{n+1} \forall n$, where F_n^p denotes the subgroup generated by all $x^p, x \in F_n$.

Putting

$$(2) \qquad L(F) = L(F, \Phi) = \oplus L_n(F)$$

we then have well defined maps

$$(3) \qquad P : L_n(F) \to L_{n+1}(F), \quad P(x_n) = (x^p)_{n+1}$$

for $x \in F_n$ with class $x_n \in L_n(F)$. According to Lazard [Laz] Ch.II, (1.2), we have the following properties.

$$(4) \qquad \begin{aligned} &\text{If } x, y \in L_n(F) \text{ then } P(x+y) = Px + Py, \\ &\text{unless } p = 2 \text{ and } n = 1, \text{ when } P(x+y) = Px + Py + [x,y] \ . \end{aligned}$$

$$(5) \qquad \begin{aligned} &\text{If } x \in L_n(F) \text{ and } y \in L_m(F) \text{ then } P[x,y] = [Px, y], \\ &\text{unless } p = 2 \text{ and } n = 1, \text{ when } P(x,y) = [Px, Py] + [x, [x,y]] \ . \end{aligned}$$

(12.2) Lizard algebras. A graded Lie $\mathbf{F}_p$-algebra $L = \oplus_{n \geq 1} L_n$, with maps $P : L_n \to L_{n+1}$ satisfying (12.1)(4) and (5) will be called a *Lizard algebra*. These are called "mixed Lie algebras" in [Laz]. More generally Lazard considers algebras graded by positive real numbers in place of positive integers, and appropriately generalized versions of the above axioms. There are natural notions of homomorphism, derivations, etc. for Lizard algebras.

(12.3) A group F has a canonical *p*-central filtration

$(C^n_{\mathbf{F}_p}(F))_{n\geq 1}$, defined inductively by

$$C^{n+1}_{\mathbf{F}_p}(F) = C^n_{\mathbf{F}_p}(F)^p \cdot (F, C^n_{\mathbf{F}_p}(F)).$$

The associated graded Lizard algebra, denoted

$$L^{\mathbf{F}_p}(F) = \bigoplus_{n\geq 1} L^{\mathbf{F}_p}_n(F)$$

is generated by $L^{\mathbf{F}_p}_1(F) = \mathbf{F}_p \otimes F^{ab}$ as a Lizard algebra.

We have

$$\bigcap_n C^n_{\mathbf{F}_p}(F) = \{1\} \qquad \text{iff } F \text{ is residually-finite-}p.$$

(12.4) p-filtered automorphism groups. Let $(F_n)_{n\geq 1}$ be a p-central filtration, and let G be a group acting on F and preserving the filtration. As in (4.1) we filter G by

$$G_d = \bigcap_n Ker(G \to Aut(F_n/F_{n+d}))$$

Then, in (4.9), $(G_d)_{d\geq 1}$ is a p-central filtration of G_1, $L_0(G)$ acts as Lizard algebra automorphisms of $L(F)$, and we have an injective Lizard algebra homomorphism from $L_+(G)$ to the Lizard algebra of graded Lizard derivations of $L(F)$.

III Filtered algebras

Here we indicate how a filtration on an algebra A gives associated filtrations on its group $A^\times$ of units, and on its group of filtration preserving algebra automorphisms. The idea of exploiting these structures can be found, for example, in [Mag 3], [NB2], [BL], [McCo 1,2],....

13. Units.

(13.1) NOTATION. Let k denote a commutative ring, and A a k-algebra (associative with unit), with unit group $A^\times$. For any (2-sided) ideal J we put

$$(1+J)^\times = Ker(A^\times \to (A/J)^\times)$$

(13.2) LEMMA. (a) *If J and K are ideals in A then*

$$((1+J)^\times, (1+K)^\times) \le (1 + (JK + KJ))^\times.$$

(b) *If $J^2 = 0$ then $x \mapsto 1 + x$ is a group isomorphism $J \to 1 + J = (1+J)^\times$.*

Proof. For (a) it suffices to observe that, when $JK = 0 = KJ$, $1 + J$ and $1 + K$ commute, which is obvious. Likewise (b) is immediate.

(13.3) FILTERED ALGEBRAS. Consider a filtration on A,

$$A = A_0 \ge A_1 \ge A_2 \ge \ldots$$

$$A_d \cdot A_e \le A_{d+e} \qquad \forall\, d, e \ge 0.$$

Then each A_d is a 2-sided ideal, and we have the associated graded algebra

$$\bar{A} = \bigoplus_{d \ge 0} \bar{A}_{(d)}, \qquad \bar{A}_{(d)} = A_d/A_{d+1}.$$

For $a \in A_d$ we denote by $a_{(d)}$ its class in $\bar{A}_{(d)}$. $\bar{A}$ is an algebra over

$$\bar{k}_{(0)} = \text{the image of } k \text{ in } \bar{A}_{(0)}.$$

The following result can be found, for example, in [NB2], Ch.II, §4.5.

(13.4) THEOREM. *Let $G = A^\times$, filtered by the subgroups $G_d = (1 + A_d)^\times, d \ge 0$.*

(a) $(G_d, G_e) \le G_{d+e} \;\forall d, e \ge 0$.

Put $L_d(G) = G_d/G_{d+1}$ and $L_+(G) = \oplus_{d \ge 1} L_d(G)$.

(b) *There is an injective group homomorphism $L_0(G) \to \bar{A}_{(0)}^\times$, induced by $g \mapsto g_{(0)}$.*

(c) *There is an injective homomorphism of graded Lie $\mathbf{Z}$-algebras $L_+(G) \to \bar{A}$, induced by $g \to (g-1)_{(d)}$ for $g \in G_d$.*

(d) *If $\bigcap_d A_d = \{0\}$ then $\bigcap_d G_d = \{1\}$.*

Note that (a) results from (13.2)(a).

In case $\bar{k}_{(0)}$ has prime characteristic p then $(G_d)_{d \geq 1}$ is a p-central filtration, in the sense of (12.1).

(13.5) COROLLARY. *Assume that:*

(i) $\bar{A}_{(0)}$ is a finitely generated projective $\bar{k}_{(0)}$-module; and

(ii) For $d \geq 1, \bar{A}_{(d)}$ is embeddable in a finitely generated free $\bar{k}_{(0)}$-module.

Then $(G_d)_{d \geq 0}$ is a $\bar{k}_{(0)}$-linear-central filtration on $G = A^\times$.

Proof. For $d \geq 1$ we have inclusions

$$L_d(G) \to \bar{A}_{(d)} \to \bar{k}_{(0)}^{N_d}$$

for some N_d, by (13.4)(c) and hypothesis (ii). We further have $L_0(G)$ embedded in $\bar{A}_{(0)}^\times$, which embeds in $Aut_{\bar{k}}(\bar{A}_{(0)})$. By hypothesis (i) we can write $\bar{A}_{(0)} \oplus Q \cong \bar{k}_{(0)}^{N_0}$ for some $\bar{k}$-module Q and $N_0 \geq 0$. Then $g \mapsto g \oplus Id_Q$ embeds $Aut_{\bar{k}_{(0)}}(\bar{A}_{(0)})$ into $Aut_{\bar{k}_{(0)}}(\bar{k}^{N_0}) = GL_{N_0}(\bar{k}_{(0)})$.

(13.6) EXAMPLE. Let P be a prime ideal of k, with quotient $\bar{k}_{(0)} = k/P$, and local ring k_P. Define the "symbolic powers" of P by

$$P^{(d)} = \text{``}k \cap P^d k_P\text{''}$$

$$= Ker(k \to k_P/P^d k_P).$$

Then we have a natural inclusion

$$\bar{k}_{(d)} = P^{(d)}/P^{(d+1)} \hookrightarrow P^d k_P/P^{d+1} k_P,$$

which shows that $\bar{k} = \oplus_{d \geq 0} \bar{k}_{(d)}$ is a torsion free $\bar{k}_{(0)}$-module. If, further, k is noetherian, then each $\bar{k}_{(d)}$ is a finitely generated $\bar{k}_{(0)}$-module.

Now let $A = M_N(k)$, the k-algebra of $N \times N$-matrices, filtered by

$$A_d = M_N(P^{(d)}) = P^{(d)} \cdot A.$$

Then we have

$$\bar{A} = \bigoplus_{d \geq 0} \bar{A}_{(d)} = M_N(\bar{k})$$

with $\bar{A}_{(d)} = M_N(\bar{k}_{(d)}) \cong \bar{k}_{(d)}^{N^2}$ as $\bar{k}_{(0)}$-module.

With $G = A^\times = GL_N(k)$, we have

$$G_d = (I + A_d)^\times = GL_N(P^{(d)}) = Ker(GL_N(k) \to GL_N(k/P^{(d)})),$$

the congruence group of level $P^{(d)}$. Now we can apply Theorem (13.4) and Corollary (13.5) to conclude:

(13.7) COROLLARY. *Let P be a prime ideal of k, and filter $G = GL_N(k)$ by $G_d = GL_N(P^{(d)})$, as above. If k is noetherian then $(G_d)_{d \geq 0}$ is a (k/P)- linear-central filtration on G. If further k is an integral domain then $\bigcap G_d = \{1\}$.*

Only the last assertion remains to be justified. It suffices to show that $\bigcap_d P^{(d)} = \{0\}$. Since k is an integral domain, it embeds into k_P, so it suffices to observe that $\bigcap_d P^d k_P = \{0\}$, this being Krull's Intersection Theorem (cf. [NB3], III, §3.2).

14. Representation Varieties.

(14.1) Presentations and representations. Let Γ be a finitely generated group with a presentation

(1) $$\Gamma = < s_1, \ldots, s_m | W >$$

Here W is a subset of the free group $F = F(s_1, \ldots, s_m)$ based on $s_1, \ldots, s_m$, and Γ is defined by by relations $w(s_1, \ldots, s_m) = 1 \; \forall w \in W$. If G is a group and $\rho \in Hom(\Gamma, G)$ then ρ is determined by $(\rho(s_1), \ldots, \rho(s_m)) \in G^m$. In this way we can identify $Hom(\Gamma, G)$ with

(2) $$\{x = (x_1, \ldots, x_m) \in G^m | w(x_1, \ldots, x_m) = 1 \; \forall w \in W\}.$$

When A is a commutative ring and $N \geq 0$ this gives a description of the N-dimensional representations of Γ over A,

(3) $$R_N(\Gamma, A) = Hom(\Gamma, GL_N(A)).$$

In particular it shows how $R_N(\Gamma, \mathbf{C})$ is an affine variety in $GL_N(\mathbf{C})^m$, called the *N-dimensional representation variety of* Γ.

The algebraic quotient

$$X_N(\Gamma, \mathbf{C}) = GL_N(\mathbf{C}) \backslash\backslash R_N(\Gamma, \mathbf{C})$$

by the conjugation action of $GL_N(\mathbf{C})$ is called the *N-dimensional character variety* (cf. §16 below). If $[\rho]$ denotes the class in $X_N(\Gamma, \mathbf{C})$ of $\rho \in R_N(\Gamma, \mathbf{C})$ then $[\rho] = [\rho']$ iff $\chi_\rho = \chi_{\rho'}$, where $\chi_\rho = Trace \circ \rho$ denotes the character of ρ.

The automorphism group $Aut(\Gamma)$ acts on $R_N(\Gamma, \mathbf{C})$, commuting with the action of $GL_N(\mathbf{C})$. The induced action on $X_N(\Gamma, \mathbf{C})$ factors through an action of $Out(\Gamma)$ on $X_N(\Gamma, \mathbf{C})$.

(14.2) The universal N-dimensional representation. Let $x_{\nu ij}(1 \leq \nu \leq m; 1 \leq i, j \leq N)$ be mN^2 indeterminates, and put $A' = \mathbf{Z}[x_{\nu ij}(1 \leq \nu \leq m; 1 \leq i, j \leq N)]$, a polynomial $\mathbf{Z}$-algebra. Then we have the matrices

$$X_\nu = [x_{\nu ij}]_{1 \leq i,j \leq N} \in M_N(A') \quad (1 \leq \nu \leq m).$$

Put $D = \Pi_\nu det(X_\nu)$, and

$$A_N = A'[D^{-1}] \quad (\leq \mathbf{Q}(x_{\nu ij})).$$

Then we have $X_1, \ldots, X_m \in GL_N(A_N)$. Define

$$J_N(\Gamma) = \text{the ideal in } A_N \text{ generated by all entries of}$$

$$w(X_1, \ldots, X_m) - I_N, \quad \forall w \in W.$$

and put

(1) $$\mathbf{Z}_N(\Gamma) = A_N / J_N(\Gamma).$$

The projection $\pi : A_N \to \mathbf{Z}_N(\Gamma)$ also defines $\pi : M_N(A_N) \to M_N(\mathbf{Z}_N(\Gamma))$. Putting $y_{\nu ij} = \pi x_{\nu ij}$, we have

$$\pi X_\nu = Y_\nu = [y_{\nu ij}]_{1 \leq i,j \leq N} \in GL_N(\mathbf{Z}_N(\Gamma))$$

By construction, we have a representation

$$(2) \qquad \rho_{N,\Gamma} : \Gamma \to GL_N(\mathbf{Z}_N(\Gamma)).$$

Evidently it has the following universal property: Given any commutative ring B and any representation $\rho : \Gamma \to GL_N(B)$, there in a unique ring homomorphism $\varphi : \mathbf{Z}_N(\Gamma) \to B$ such that the following diagram commutes:

$$(3) \qquad \begin{array}{ccc} \Gamma & \xrightarrow[\rho_{N,\Gamma}]{} & GL_N(\mathbf{Z}_N(\Gamma)) \\ & & \downarrow \varphi \ (= GL_N(\varphi)) \\ & \rho \searrow & \\ & & GL_N(B) \end{array}$$

In fact $\varphi(y_{\nu ij}) = z_{\nu ij}$, where $\rho(s_\nu) = [z_{\nu ij}]_{1 \le i,j \le N}$ $(1 \le \nu \le m)$. We can reformulate this property as an identification

$$(4) \qquad \begin{aligned} R_N(\Gamma, B) &= Hom_{\mathbf{Z}-alg}(\mathbf{Z}_N(\Gamma), B) \\ &= Hom_{B-alg}(B_N(\Gamma), B), \end{aligned}$$

where $B_N(\Gamma) = B \otimes_{\mathbf{Z}} \mathbf{Z}_N(\Gamma)$.

It follows that $\mathbf{Z}_N(\Gamma)$ depends only on N and Γ, and not on the presentation used for Γ. Further $\mathbf{C}_N(\Gamma)$ is the affine algebra of the affine variety $R_N(\Gamma, \mathbf{C})$, or rather the affine scheme, since $\mathbf{C}_N(\Gamma)$ could have nilpotent elements.

(14.3) The prime ideal of the trivial representation. The trivial representation $\tau_N : \Gamma \to GL_N(\mathbf{Z})$, $\tau_N(\Gamma) = \{I_N\}$, corresponds to a homomorphism $\mathbf{Z}_N(\Gamma) \to \mathbf{Z}$. Let P_0 denote its kernel. Then P_0 is a prime ideal, $\mathbf{Z}_N(\Gamma)/P_0 = \mathbf{Z}$, and we have

$$\rho_{N,\Gamma}(\Gamma) \le GL_N(P_0) = Ker(GL_N(\mathbf{Z}_N(\Gamma)) \to GL_N(\mathbf{Z}_N(\Gamma)/P_0)).$$

(14.4) PROPOSITION. *Suppose that there is an irreducible subvariety $V \subset R_N(\Gamma, \mathbf{C})$ that contains the trivial representation and a faithful representation.*

(a) Γ is residually-torsion-free-nilpotent, i.e. $\bigcap_d C_{\mathbf{Z}}^d(\Gamma) = \{1\}$.

(b) Let $\alpha \in Aut(\Gamma)$ be such that, for all $x \in \Gamma$, $\alpha(x)$ and x become conjugate in $\Gamma/C_{\mathbf{Z}}^d(\Gamma)$ for all d. Then $\chi_\rho = \chi_\rho \circ \alpha$ $(= \chi_{\rho\circ\alpha})$ for all $\rho \in V$.

Proof of (a). The affine algebra A_V of V is an integral domain which is a quotient of $\mathbf{C}_N(\Gamma)$, and the kernel of the canonical homomorphism $\varphi : \mathbf{Z}_N(\Gamma) \to A_V$ is contained in the ideal P_0 of the trivial representation. Put $A = \varphi(\mathbf{Z}_N(\Gamma))$ and $P = \varphi(P_0)$. Let

$$\rho = \varphi \circ \rho_{N,\Gamma} : \Gamma \to GL_N(A).$$

Then $\rho\Gamma \leq GL_N(P)$, and $A/P = \mathbf{Z}_N(\Gamma)/P_0 = \mathbf{Z}$. Since V contains a faithful representation it follows that ρ is faithful.

Now the filtration of $GL_N(A)$ by $GL_N(P^{(d)})$ $(d \geq 0)$ is a separating $\mathbf{Z}$-linear-central filtration, by (13.7). Since $\Gamma \cong \rho\Gamma \leq GL_N(P^{(1)})$ we conclude that Γ is residually-torsion-free-nilpotent, as claimed.

Proof of (b). Putting $\Gamma_d = \rho^{-1}(GL_N(P^{(d)}))$, we just saw that $(\Gamma_d)_{d\geq 1}$ is a separating $\mathbf{Z}$-central filtration of Γ, and so $C_{\mathbf{Z}}^d(\Gamma) \leq \Gamma_d$ $\forall d$. Let $\alpha \in Aut(\Gamma)$ be as in (b), and $x \in \Gamma$. Since, by assumption, $\alpha(x)$ and x are conjugate in $\Gamma/C_{\mathbf{Z}}^d(\Gamma)$, they are also conjugate in Γ/Γ_d, so $\rho\alpha(x)$ and $\rho(x)$ are conjugate mod $P^{(d)}$, whence $\chi_\rho(\alpha(x)) - \chi_\rho(x) \in P^{(d)}$. This holds for all d, and $\bigcap_d P^{(d)} = 0$, whence $\chi_\rho(\alpha(x)) = \chi_\rho(x)$. Now ρ is the generic representation, $\rho : \Gamma \to GL_N(A_V)$, and so $\chi_\sigma \circ \alpha = \chi_\sigma$ $\forall \sigma \in V$.

(14.5) COROLLARY. *Assume that:*

(i) $R_N(\Gamma, \mathbf{C})$ is an irreducible variety and contains a faithful representation; and

(ii) $Out(\Gamma)$ acts faithfully on $X_N(\Gamma, \mathbf{C})$.

Then:

(a) Γ is residually-torsion-free-nilpotent; and

(b) If $\alpha \in Aut(\Gamma)$ and, for all $x \in \Gamma$, $\alpha(x)$ is conjugate to x in $\Gamma/C_{\mathbf{Z}}^d(\Gamma)$ for all d, then α is inner. In particular condition $IN(\Gamma)$ of (9.8)(c) holds.

This follows from (14.4) with $V = R_N(\Gamma, \mathbf{C})$. For assertion (b), it follows from (14.4)(b) that α acts trivially on $X_N(\Gamma, \mathbf{C})$, and so, by hypothesis (ii), α is inner.

(14.6) **EXAMPLES.** 1.   If F is free on $s_1, \ldots, s_m$ then $R_N(F, \mathbf{C}) = GL_N(\mathbf{C})^m$ is irreducible, and F has faithful representations for all $N \geq 2$. Moreover, $Out(F)$ acts faithfully on $X_N(F, \mathbf{C})$, as follows from (10.2).

2. Let $F = < x_1, y_1, \ldots x_g, y_g | [x_1, y_1] \ldots [x_g, y_g] = 1 >$, the fundamental group of a closed surface of genus $g \geq 2$. Then F has faithful representations for all $N \geq 2$. Moreover Rapinchuk and Benyash-Krivetz [RB-K] have recently shown that $R_N(F, \mathbf{C})$ is irreducible for $N \leq 4$. (They conjecture that this is true for all N.) In view of (14.4)(a), this implies that F is residually-torsion-free-nilpotent. Results of Macbeath and Singerman [MS] (cf. also [BL], (4.2)) imply that $Out(F)$ acts faithfully on $X_2(\Gamma, \mathbf{C})$ if $g \geq 3$, and with kernel of order 2 for $g = 2$. Hence (14.4)(b) gives $IN(F)$ for $g \geq 3$.

(14.7) **DEFINITION.** Call a representation $\rho : \Gamma \to GL_N(\mathbf{C})$ *super-faithful* if it is faithful and irreducible and if further the normalizer of $\rho\Gamma$ in $GL_N(\mathbf{C})$ is $(\rho\Gamma) \cdot (\mathbf{C}^\times I)$.

(14.8) **PROPOSITION.** Let $\rho : \Gamma \to GL_N(\mathbf{C})$ be a super-faithful represen-tation, with character $\chi_\rho = Trace \circ \rho$.

(a) If $\alpha \in Aut(\Gamma)$ and $\chi_\rho \circ \alpha = \chi_\rho$ then α is inner. In particular, α is inner if α preserves conjugacy classes.

(b) Suppose that ρ belongs to an irreducible subvariety

$V \subset R_N(\Gamma, \mathbf{C})$ which contains also the trivial representation. Then the canonical $\mathbf{Z}$-central filtration $C_{\mathbf{Z}}^n(\Gamma)$ $(n \geq 1)$ is separating, and if $\alpha \in Aut(\Gamma)$ is inner on $\Gamma/C_{\mathbf{Z}}^n(\Gamma)$ $\forall n$, then α is inner.

Proof of (a). If $\chi_\rho = \chi_\rho \circ \alpha \ (= \chi_{\rho \circ \alpha})$ then, since ρ is irreducible, $\rho \circ \alpha \cong \rho$, i.e. for some $g \in GL_N(\mathbf{C})$ we have $\rho(\alpha(x)) = g\rho(x)g^{-1}$ for all $x \in \Gamma$. Since $g\rho(\Gamma)g^{-1} = \rho(\alpha\Gamma) = \rho(\Gamma)$ we see that g normalizes $\rho(\Gamma)$. By definition of superfaithful it follows that $g = \rho(y) \cdot c$ for some $y \in \Gamma$ and $c \in \mathbf{C}^\times$. Hence we have $\rho(\alpha(x)) = \rho(y)\rho(x)\rho(y)^{-1} = \rho(yxy^{-1})$, i.e., $\rho \circ \alpha = \rho \circ ad(y)$. Since ρ is faithful we have $\alpha = ad(y)$.

If α preserves conjugacy classes, i.e. if $\alpha(x)$ is conjugate to x for all $x \in \Gamma$, then $\chi_\rho \circ \alpha = \chi_\rho$, so α is inner.

Assertion (b) follows from (a) plus (14.5).

15. Automorphisms

(15.1) NOTATION. As in (13.3) let

$$A = A_0 \geq A_1 \geq A_2 \geq \ldots$$

be a filtered k - algebra. We now consider the group

$$G = Aut_{k-alg}^{filt}(A) = \{g \in Aut_{k-alg}(A) \mid gA_d = A_d \ \forall d\}$$

of filtration preserving k-algebra automorphisms of A. We have induced actions

$$\rho_{n,d} : G \to Aut_{k-mod}(A_n/A_{n+d}),$$

and we filter G by $G_d = \bigcap_n Ker(\rho_{n,d})$.

(15.2) THEOREM. (a) $(G_d, G_e) \leq G_{d+e} \ \forall \ d, c \geq 0$. Put $L_d(G) = G_d/G_{d+1}$ and $L_+(G) = \oplus_{d \geq 1} L_d(G)$.

(b) There is an injective group homomorphism $L_0(G) \to Aut_{gr.alg}(\bar{A})$.

(c) There is an injective homomorphism of graded Lie $\mathbf{Z}$-algebras $L_+(G) \to gr Der_{k-alg.}(\bar{A})$

(d) If $\bigcap_n A_n = \{0\}$ then $\bigcap_d G_d = \{1\}$.

Proof. Consider the k-algebra

$$E = End^{filt}_{k-mod}(A) = \{u \in End_{k-mod}(A) | \ uA_d \leq A_d \ \forall d\}.$$

We have induced k-algebra homomorphisms $\rho_{n,d} : E \to End_{k-mod}(A_n/A_{n+d})$, and we filter E by $E_d = \bigcap_n Ker(\rho_{n,d})$; evidently $E_d \cdot E_e \leq E_{d+e}$. Note further that

(1)   If $\bigcap_n A_n = \{0\}$ then $\bigcap_d E_d = \{0\}$.

We have the associated graded algebras

$$\begin{array}{rclcrcl}
\bar{A} & = & \bigoplus_{n \geq 0} \bar{A}_{(n)} & , & \bar{A}_{(n)} & = & A_n/A_{n+1} \\
\bar{E} & = & \bigoplus_{d \geq 0} \bar{E}_{(d)} & , & \bar{E}_{(d)} & = & E_d/E_{d+1}.
\end{array}$$

For $a \in A_n$ (resp., $u \in E_d$) denote by $a_{(n)}$ (resp., $u_{(d)}$) its class in $\bar{A}_{(n)}$ (resp., $\bar{E}_{(d)}$). The k-linear map $u : A_n \to A_{n+d}$ descends to a map $\bar{A}_{(n)} \to \bar{A}_{(n+d)}$, that depends only on $u_{(d)}$. In this way $u_{(d)}$ defines a k-linear map $\bar{A} \to \bar{A}$ that is homomorphism of degree d, and so $\bar{A}$ becomes a graded $\bar{E}$-module. Moreover the definition of the filtration on E shows that $\bar{A}$ is a faithful $\bar{E}$-module. Thus we have an injective homomorphism of graded k-algebras

(2) $\bar{E} \underset{\alpha}{\longrightarrow} gr\,End_{k-mod}(\bar{A}).$

Let $H = E^{\times}$, filtered by $H_d = (1 + E_d)^{\times}$. It follows from (13.4) and the injection (2) above that we have:

(3) $\begin{cases}
(a) & (H_d, H_e) \leq H_{d+e} \ \forall d, e \geq 0. \\
(b) & \text{We have injective group homomorphisms} \\
    & L_0(H) \to \bar{E}^{\times}_{(0)} \underset{\alpha}{\longrightarrow} Aut_{gr \ k-mod}(\bar{A}) \\
(c) & \text{We have injective homomorphisms of graded Lie} \\
    & \mathbf{Z}\text{-algebras, } L_+(H) \to \bar{E} \underset{\alpha}{\longrightarrow} gr \ End_{k-mod}(\bar{A})
\end{cases}$

Now finally we have $G = Aut^{filt}_{k-alg}(A) \leq H$, and the filtration defined above on G is just $G_d = G \cap H_d$. It follows that $(G_d, G_e) \leq G_{d+e} \ \forall d, e \geq 0$, and $L_d(G) \to L_d(H)$ is injective. The injection $L_0(G) \to Aut_{gr \ k-mod}(\bar{A})$ obtained from (3)(b) assigns to $g_{(0)} \in L_0(G)$ the induced automorphism of $\bar{A}$, which, like g, is an algebra automorphism. Thus we have an injective homomorphism $L_0(G) \rightarrow$ $Aut_{gr\ k-alg}(\bar{A})$.

It remains to show that the injection $L_+(G) \rightarrow gr End_{k-mod}(\bar{A})$ actually maps $L_+(G)$ into $gr Der_{k-alg}(\bar{A})$. For $g \in G_d$ $(d \geq 1), g_{(d)}$ is mapped to $D = (g-1)_{(d)}$: $\bar{A} \rightarrow \bar{A}$. To show that this is a derivation, consider $a \in A_n$ and $b \in A_m$. Then

$$D(a_{(n)} \cdot b_{(n)}) = (g-1)_{(d)}((ab)_{(n+m)})$$

$$= (g(ab) - ab)_{(n+m+d)}$$

$$= (g(a)g(b) - ab)_{(n+m+d)}$$

$$= [(g(a) - a)g(b) + a(g(b) - b)]_{(n+m+d)}$$

$$= D(a_{(n)})g(b)_{(m)} + a_{(n)}D(b_{(m)})$$

$$= D(a_{(n)})b_{(m)} + a_{(n)}D(b_{(m)})$$

since, with $d \geq 1, g(b) \equiv b\ mod\ A_{m+1}$.

This concludes the proof of (15.2).

(15.3) COROLLARY. Let $\bar{k}_{(0)}$ = the image of k in $\bar{A}_{(0)}$, so that $\bar{A}$ is a graded $\bar{k}_{(0)}$-algebra. Assume that:

(i) For all $d \geq 0, \bar{A}_{(d)}$ is embeddable in a finitely generated free $\bar{k}_{(0)}$-module.

(ii) There is a $d_0 \geq 0$ such that

$$\bar{A}_{(\leq d_0)} := \bigoplus_{d \leq d_0} \bar{A}_{(d)}$$

is a finitely generated projective $\bar{k}_{(0)}$-module, and every $\bar{k}_{(0)}$-algebra automorphism or derivation of $\bar{A}$ is determined by its restriction to $\bar{A}_{(\leq d_0)}$.

Then the filtration $(G_d)_{d \geq 0}$ on $G = Aut^{filt}_{k-alg}(A)$ given in (15.2) is a $\bar{k}_{(0)}$-linear-central filtration.

Proof. Hypothesis (ii) gives an embedding $Aut_{gr\ \bar{k}_{(0)}-alg}(\bar{A})$ into $Aut_{\bar{k}_{(0)}-mod}$ $(\bar{A}_{(\leq d_0)})$. Since $\bar{A}_{(\leq d_0)}$ is finitely generated and projective, its automorphism group embeds in some $GL_{N_0}(\bar{k}_{(0)})$ (cf. proof of (13.5)).

Similarly, the $\bar{k}_{(0)}$-module of $\bar{k}_{(0)}$-derivations of degree e in embeds, by (ii), into

$$\bigoplus_{d \leq d_0} Hom_{\bar{k}_{(0)} - mod}(\bar{A}_{(d)}, \bar{A}_{(d+e)})$$

If $\bar{A}_{(d)} \oplus Q_d \cong \bar{k}_{(0)}^{N_d}$ for $d \leq d_0$, and $\bar{A}_{(d+e)} \leq \bar{k}_{(0)}^{N_{d+e}}$ then $Hom_{\bar{k}_{(0)} - mod}$ $(\bar{A}_{(d)}, \bar{A}_{(d+e)})$ embeds into $Hom_{\bar{k}_{(0)} - mod}(\bar{k}_{(0)}^{N_d}, \bar{k}_{(0)}^{N_{d+e}}) \cong \bar{k}_{(0)}^{N_d \cdot N_{d+e}}$.

Thus, as required, we have $L_d(G)$ embedded in $\bar{k}_{(0)}^N$, where $N = \sum_{d \leq d_0} N_d \cdot N_{d+e}$.

(15.4) **EXAMPLE.** Let $k^{[N]} = k[X_1, \ldots, X_N]$, a polynomial k-algebra in N variables. Let M denote the ideal generated by $X_1, \ldots, X_N$, and filter $k^{[N]}$ by $(M^d)_{d \geq 0}$; the associated graded algebra is just $k^{[N]}$ again, graded with each X_i of degree 1. We put

$$GA_N(k) = Aut_{k-alg}(k^{[N]})$$

and

$$GA_N^0(k) = Aut_{k-alg}^{filt}(k^{[N]}) = \text{the stabilizer of } M.$$

The k-linear-central-filtration on $GA_N^0(k)$ is given by

$$GA_N^d(k) = Ker(GA_N^0(k) \rightarrow Aut_{k-alg}(k^{[N]}/M^{d+1}))$$

Then we obtain an injective group homomorphism (in fact isomorphism)

$$\delta_0 : L_0(GA_N^0(k)) \rightarrow Aut_k(M/M^2) \cong GL_N(k)$$

and an injective homomorphism of graded Lie **Z**-algebras,

$$\delta_+ : L_+(GA_N^0(k)) \rightarrow \underline{ga}_N(k) = grDer_{k-alg}(k^{[N]}).$$

In fact it is shown in [Pit], Prop. (3.9), that $Im(\delta_+)$ consists of all derivations D whose divergence $Div(D)$ is nilpotent. Here, if we put $D_i = \frac{\partial}{\partial X_i}$ and write $D = f_1 D_1 + \ldots + f_N D_N$ then $Div(D) = D_1 f_1 + \ldots + D_N f_N \in k^{[N]}$. Thus, if k is an integral domain (or even just reduced) then $Im(\delta_+)$ is the degree ≥ 1 part of the Lie algebra $\underline{sa}_N(k)$ of divergence zero derivations.

In case $k = \mathbf{Z}$, this separating $\mathbf{Z}$-linear-central filtration $(GA_N^d(\mathbf{Z}))_{d \geq 0}$ gives a good candidate on which to test the Tits alternative. (Cf. (6.2).)

16. Actions on character varieties.

(16.1) Character varieties. As in (14.1), let F be a finitely generated group with representation variety $R_N(F, \mathbf{C}) = Hom(F, GL_N(\mathbf{C}))$, and "character variety"

(1) $$X_N(F, \mathbf{C}) = \text{"}GL_N(\mathbf{C}) \backslash\backslash R_N(F, \mathbf{C}),\text{"}$$

where $GL_N(\mathbf{C})$ acts via conjugation. This translates into an action on $\mathbf{C}_N(F)$, and the affine algebra of $X_N(F, \mathbf{C})$ is the algebra

(2) $$\mathbf{C}_N'(F) = \mathbf{C}_N(F)^{GL_N(\mathbf{C})}$$

of invariants. For $\gamma \in F$ we have

(3) $$\chi(\gamma) := Trace(\rho_{N,F}(\gamma)) \in \mathbf{Z}_N(F) \subset \mathbf{C}_N(F),$$

and these $\chi(\gamma)$ belong to $\mathbf{C}_N'(\Gamma)$, and finitely many of them generate $\mathbf{C}_N'(\Gamma)$ as a $\mathbf{C}$-algebra.

(16.2) The actions of $G = Aut(F)$ and $H = Out(F)$. The natural action of G on $R_N(\Gamma, \mathbf{C})$ commutes with the $GL_N(\mathbf{C})$-action; the resulting action of G on $X_N(F, \mathbf{C})$ factors through H. Thus we obtain $\mathbf{C}$-algebra actions of G on $\mathbf{C}_N(F)$ and H on $\mathbf{C}_N'(F)$. These actions preserve the prime ideals P_0 and $P_0' = P_0 \cap \mathbf{C}_N'(F)$, respectively, where P_0 is the prime of the trivial representation. Hence G and H preserve the filtrations $(P_0^d)_{d \geq 0}$ and $(P_0'^d)_{d \geq 0}$, respectively. Let $\overline{\mathbf{C}_N(F)}$ and $\overline{\mathbf{C}_N'(F)}$ denote the corresponding associated graded algebras, which are generated by their degree 1 components over $\mathbf{C}$ in degree 0.

(16.3) The filtrations on G and H. The above actions correspond to homomorphisms from G and H into the filtration preserving automorphism groups of the corresponding algebras. Pulling back the filtrations on the latter automorphism groups defined in (15.1), we obtain, by (15.2) and (15.3), $\mathbf{C}$-linear-central filtrations $(G_d)_{d\geq 0}$ and $(H_d)_{d\geq 0}$, with the following properties.

$$
(1) \quad
\begin{cases}
(a) & \text{There is an injective group homomorphism} \\
    & \qquad L_0(G) \to Aut_{\mathbf{C}}(P_0/P_0^2) \\
    & \\
(b) & \text{There is an injective graded Lie } \mathbf{Z}\text{-algebra} \\
    & \text{homomorphism } L_+(G) \to gr\,Der_{\mathbf{C}-alg}(\overline{\mathbf{C}_N(F)}).
\end{cases}
$$

$$
(2) \quad
\begin{cases}
(a) & \text{There is an injective group homomorphism} \\
    & \qquad L_0(H) \to Aut_{\mathbf{C}}(P_0'/P_0'^2) \\
    & \\
(b) & \text{There is an injective graded Lie } \mathbf{Z}\text{-algebra} \\
    & \text{homomorphism } L_+(H) \to gr\,Der_{\mathbf{C}-alg}(\overline{\mathbf{C}_N'(F)}).
\end{cases}
$$

$$
(3) \quad
\begin{array}{c}
\text{If } G \text{ acts faithfully on } R_N(F,\mathbf{C}) \text{ and} \\
\bigcap_d P_0^d = \{0\} \text{ then } \bigcap_d G_d = \{1\}.
\end{array}
$$

$$
(4) \quad
\begin{array}{c}
\text{If } H \text{ acts faithfully on } X_N(F,\mathbf{C}) \text{ and} \\
\bigcap_d P_0'^d = \{0\} \text{ then } \bigcap_d H_d = \{1\}.
\end{array}
$$

References

[AS] R. Alperin and P. Shalen, "Linear groups of finite cohomological dimension," *Invent. Math.* **66** (1982) 89-98.

[AS1] M. Asada and M. Kaneko, "On the automorphism group of some pro-ℓ-fundamental groups," *Adv. Studies in Pure Math.* **12** (1987), *Galois Representations and Arithmetic Algebraic Geometry*, pp. 137-159.

[As2] M. Asada, "Two properties of the filtration of the outer automorphism groups of certain groups", (Preprint, 1993).

[Aus] L. Auslander, "On a problem of Philip Hall," *Annals of Math.* **86** (1967) 112-116.

[BH] M. Bestrina and M. Handel, "Train tracks and automorphisms of free groups", *Ann. Math.* **135** (1992) 1-15.

[BL] H. Bass and A. Lubotzky, "Automorphisms of groups and of schemes of finite type," *Israel Jour. Math.* **44** (1983) 1-22.

[GB1] G. Baumslag, "Residual nilpotence and relations in free groups," *Jour. Algebra* **2** (1965) 271-282.

[GB2] G. Baumslag, "On generalized free products," *Math. Zeit.* **78** (1962) 423-438.

[BLM] J.S. Birman, A. Lubotzky, and J. McCarthy, "Abelian and solvable subgroups of the mapping class group," *Duke Math. Jour.* **50** (1983) 1107-1120.

[Bl] N. Blackburn, "Conjugacy in nilpotent groups," *Proc. AMS* **16** (1965) 143-148.

[NB1] N. Bourbaki, *Algebra I,* Chs 1-3, Hermann, Paris (1974).

[NB2] N. Bourbaki, *Lie groups and Lie algebras I,* Chs 1-3, Hermann, Paris (1975).

[NB3] N. Bourbaki, *Commutative algebra,* Hermann, Paris (1972).

[Br] K. Brown, *Cohomology of groups,* Graduate Texts in Mathematics, Springer-Verlag (1982).

[CV] M. Culler and K. Vogtmann, "Moduli of graphs and automorphisms of free groups," *Invent. Math.* **84** (1986) 91-119.

[FP] E. Formanek and C. Procesi, "The automorphism group of a free group is not linear", *Jour. Algebra* **149** (1992) 494-499.

[HBG] H.B. Griffiths, "A covering-space approach to residual properties of groups," *Michigan Math Jour.* **14** (1967) 335-348.

[EKG] Edna K. Grossman, "On the residual finiteness of certain mapping class groups," *Jour. London Math. Soc.* **9** (1974) 160-164.

[G] K. Gruenberg, *Cohomological topics in group theory,* LNM 143 (1970).

[PH] Philip Hall, *Nilpotent groups,* Queen Mary College Mathematics Notes, London (1979).

[Lab1] J. Labute, "On the descending central series of groups with a single defining relation," *Jour. Algebra* **14** (1970) 16-23.

[Lab2] J. Labute, "The determination of the Lie algebra associated to the lower central series of a groups," *Trans. Amer. Math. Soc.* **288** (1985) 39-67.

[Laz] M. Lazard, "Sur les groupes nilpotents et les anneaux de Lie," *Ann. École Normale Supér.* **71** (1954) 101-190.

[LS] R.C.L. Lyndon and P.E. Schupp, *Combinatorial Group Theory*, Springer-Verlag, Berlin-Heidelberg-New York (1977).

[MS] A.M. Macbeath and D. Singerman, "Spaces of subgroups and Teichmuller space," *Proc. London Math. Soc.* **31** (1975) 211-256.

[Mag1] W. Magnus, "Beziehungen zwischen gruppen und Idealen in einem speziellen Ring," *Math. Ann.* **111** (1935) 259-280.

[Mag2] W. Magnus, "Über Beziehungen Zwischen hoheren Kommutatoren," *Jour. Crelle* **177** (1937) 105-115.

[Mag3] W. Magnus, "Rings of Fricke Characters and automorphisms of free groups," *Math. Zeit.* **170** (1980) 91-103.

[MKS] W. Magnus, A Karass, and D. Solitar, *Combinatorial Group Theory*, Interscience Publ., John Wiley (1966).

[Mal] A.I. Malcev, "On certain classes of infinite solvable groups," *Math. Sbornik* **28** (1951) 567-588.

[Mc] John McCarthy, "A Tits-alternative for subgroups of surface mapping class groups," *Trans. AMS* **291** (1985) 583-612.

[McCo1] J. McCool, "A faithful polynomial representation of $Out(F_3)$," *Math. Proc. Comb. Phil. Soc.* **106** (1989) 207-213.

[McCo2] J. McCool, "Virtually-residually-p automorphism groups of group rings, " *Jour. Algebra,* **130** (1990) 106-112.

[Pit] M. Pittaluga, "On the automorphism group of a polynomial algebra," thesis, Columbia University (1984).

[RB-K] A. Rapinchuk and V. Benyash-Krivetz, "Geometric representation theory for fundamental groups of compact orientable surfaces," (preprint).

[Seg] Daniel Segal, *Polycyclic Groups*, Cambridge University Press (1983).

[St] P.F. Stebe, "Conjugacy separability of certain Fuchsian groups," *Trans. AMS* **163** (1972) 173-188.

[T] J. Tits, "Free subgroups in linear groups," *J. Algebra* **20** (1972) 250-270.

[W] B.A.F. Wehrfritz, *Infinite linear groups*, Ergeb. der Math. und ihrer grenzgebiet, **76**, Springer-Verlag (1973).

Department of Mathematics
Columbia University
New York, NY 10027
hb@math.columbia.edu

Department of Mathematics
Hebrew University
Jerusalem, Israel
alexlub@hujivms.huji.ac.il

Contemporary Mathematics
Volume 169, 1994

MUSINGS ON MAGNUS

Gilbert Baumslag

ABSTRACT. The object of this paper is to describe a simple method for proving that certain groups are residually torsion-free nilpotent, to describe some new parafree groups and to raise some new problems in honour of the memory of Wilhelm Magnus.

1. Introduction

I first heard of Wilhelm Magnus in 1956, when I was attending some lectures by B.H. Neumann on amalgamated products. At some point during the course of these lectures, Neumann remarked that Magnus was the first mathematician to recognize the value of amalgamated products and had shown just how effective a tool they were, in his work on groups defined by a single relation.

I was working, at that time, on an universal algebra variation of free groups, involving groups with unique roots, which I called $\mathcal{D}$-groups [1]. Consequently, in an attempt to find analogues of various theorems about free groups for $\mathcal{D}$-groups, I found myself reading a beautiful paper of Magnus, in which he proved the residual torsion-free nilpotence of free groups.

Much of my talk today will be concerned with these two topics, one-relator groups and residual nilpotence.

Let me begin by reminding you of some of the definitions involved.

Let $\mathcal{P}$ be a property of groups.

Definition. We say that a group G is *residually* a $\mathcal{P}$-group if for each $g \in G$, $g \neq 1$, there exists a normal subgroup N of G, such that $g \notin N$ and G/N has $\mathcal{P}$.

The properties that I will be mainly concerned with here are *freeness, nilpotence* and *torsion-free nilpotence*. I will make use of the usual commutator notation. Thus if H and K are subgroups of a group G then

$$[H, K] = gp(h^{-1}k^{-1}hk \mid h \in H, k \in K)$$

is the subgroup generated by all the commutators $[h, k] = h^{-1}k^{-1}hk$. The *lower central series of* G is defined to be the series

$$G = \gamma_1(G) \geq \gamma_2(G) \geq \cdots \geq \gamma_n(G) \geq \ldots,$$

where $\gamma_{n+1}(G) = [\gamma_n(G), G]$. G is termed *nilpotent* if $\gamma_{c+1}(G) = 1$ for some c.

I am now in a position to formulate the theorem of Magnus [12] that I alluded to before.

1991 *Mathematics Subject Classification.* Primary 54C40, 14E20; Secondary 46E25, 20C20.
Key words and phrases. Residually torsion-free nilpotent groups, one-relator groups, $\mathcal{D}$-groups.
The author was supported in part by NSF Grant #9103098

Theorem 1. *Free groups are residually torsion-free nilpotent.*

The basic idea involved in the proof of Theorem 1 is beautifully simple. Magnus concocts a faithful representation of a given free group F in the group of units of a carefully chosen ring R with 1. Each of the elements $f \in F$ takes the form $1 + \phi$, where ϕ lies in an ideal R^+ of R. R^+ carries with it the structure of a metric space designed so as to ensure that if $f \in \gamma_n(F)$, then $d(\phi, 0) \leq 1/n$, where here $d(\phi, 0)$ denotes the distance between ϕ and 0. This suffices to ensure that

$$\bigcap_{n=1}^{\infty} \gamma_n(F) = 1.$$

The sketch of the proof, below, amplifies these remarks.

Proof. Let F be free on $x_1, \ldots, x_q$. Consider the ring R of power series in the non-commuting indeterminates $\xi_1, \ldots, \xi_q$ with rational coefficients. Each element $r \in R$ can be thought of as an infinite sum which takes the form

$$r = r_0 + r_1 + \cdots + r_n + \ldots,$$

where $r_0 \in \mathbb{Q}$ and the *homogeneous component* r_n of r of degree $n > 0$ is a finite sum of integral multiples $c\xi_{i_1} \ldots \xi_{i_n}$ of *monomials* $\xi_{i_1} \ldots \xi_{i_n}$ of degree n, $c \in \mathbb{Q}, i_j \in \{1, \ldots, q\}$. Here $\xi_{i_1} \ldots \xi_{i_n} = \xi_{j_1} \ldots \xi_{j_m}$ only if $n = m$ and $i_r = j_r$ for each $r = 1, \ldots, n$. Each of the elements $a_i = 1 + \xi_i$ is invertible in R, with inverse $a_i^{-1} = 1 - \xi_i + \xi_i^2 - \xi_i^3 + \ldots$. It then turns out that the elements $a_1, \ldots, a_q$ freely generate a free subgroup of rank n, of the multiplicative group of units of R. We define R^+ to be the ideal of R consisting of those elements $r \in R$ such that $r_0 = 0$. We define a metric d on R^+ by setting $d(r, 0) = 0$ if $r = 0$ and, if $r \neq 0$,

$$d(r, 0) = 1/n$$

where n is the degree of the first non-zero homogeneous component of r. Magnus proves that if $f = 1 + \phi \in \gamma_n(F)$, then all of the homogeneous components $f_1 = f_2 = \cdots = f_{n-1} = 0$, i.e., $d(\phi, 0) \leq 1/n$. It is not hard then to deduce that F is residually torsion-free nilpotent.

One of the many consequences of this theorem of Magnus is the following characterisation of finitely generated free groups, which Magnus proved a few years later in [13].

Theorem 2. *If the group G can be be generated by q elements, where q is finite, and if $G/\gamma_n(G) \cong F/\gamma_n(F)$, where F is a free group of rank q, then $G \cong F$.*

Theorem 2 will play a role here in due course.

2. ONE-RELATOR GROUPS

My next encounter with Magnus came some years later. After spending a post-doctoral year in Manchester, as a Special Lecturer, I came to Princeton in 1959 as an instructor. Some time towards the end of that year, Trueman MacHenry, a doctoral student of Magnus whom I had met in Manchester, came to visit me in Princeton. He was accompanied by Bruce Chandler, another of Magnus' doctoral students. MacHenry brought greetings from Magnus and an implicit offer of a job at the Courant Institute. I was very happy at the prospect of working with Magnus, and came up to New York to give a talk early in 1960. This was the first time that I had actually met Magnus and I was very impressed with his quickness and his vast knowledge, not only of mathematics but of almost everything else as well. He must have been amused by my talk, which was to an audience which consisted mainly of analysts, because I talked about an extremely esoteric theorem that Norman Blackburn and I [4] had proved about 9 months earlier, a theorem that only the most special of specialists would have found interesting. Nonetheless, the analysts were apparently convinced by Magnus that I should be offered a position and the offer was made explicit soon after. I immediately accepted the position and came to the Courant in the late summer of 1960. I spent part of that summer in Pasadena, where I met Danny Gorenstein and Roger Lyndon. Roger Lyndon had been working on staggered presentations [9] and asked me a number of questions about one-relator groups while we were in Pasadena. I had not read Magnus' papers on one-relator groups, and so I was forced to spend part of that summer trying to understand Magnus' work.

There are two main theorems that I would like to describe. The first of these is Magnus' celebrated "Freiheitssatz" [10].

Theorem 3. *Let*

$$G = < x_1, \ldots, x_q; r = 1 >$$

be a group defined by a single relator r. If the first and last letters of r are not inverses and if x_1 appears in r, then the subgroup of G generated by $x_2, \ldots, x_q$ is a free group, freely generated by $x_2, \ldots, x_q$.

One of the by-products of the proof of the Freheitssatz was an extradordinary unravelling of the structure of these groups, which allowed Magnus to deduce, in due course, that one-relator groups have solvable word problem [11]:

Theorem 4. *Let G be a group defined by a single relation. Then G has a solvable word problem.*

3. Surface groups

As I remarked earlier, I came to the Courant Institute in the late summer of 1960. Magnus was part of the electromagnetic group of Morris Kline and used to go to a seminar on electromagnetism. I remember one occasion, after tea, when Morris Kline and Magnus and I were in the elevator, in the old hat factory which had become the Courant Institute. Kline was talking to Magnus and addressed him as Bill. I saw Magnus shudder at the prospect of being called Bill. Magnus was much too polite to say anything, but I did not feel at all constrained to be silent. So I said something like this to Kline: "Morris, Wilhelm is a Geheimrat and so he simply cannot be called Bill. Wilhelm is more appropriate". Morris Kline smiled and indicated that this was okay with him. To which Magnus responded by muttering under his breath: "Thank God." He later told me that he had hated being addressed as Bill, and was grateful to me for telling Kline to call him Wilhelm.

Magnus had a large group of Ph.D. students in 1961. They included Karen Fredericks, Bruce Chandler, Seymour Lipschutz and Trueman MacHenry, all of whom worked with Magnus on Combinatorial Group Theory. Martin Greendlinger had already completed his beautiful work on small cancellation theory, but was still around. In addition, Magnus had some other students who worked with him on Hill's equation. There was always a line of students outside his office and I offered to take on some of them to relieve him of some of the burden. Magnus ran a seminar on Combinatorial Group Theory at the time. The seminar was a mixture of pure research and reports by students on papers that they were reading. Shortly after I arrived, the research part of the seminar was broadened and the role of students was reduced essentially to zero. One of Magnus' habits was to propose a number of problems in the seminar. He was always interested in purely algebraic proofs of theorems that had been proved by other means. In particular, he asked late in 1961, whether there was a direct proof of the residual finiteness of the fundamental groups of two-dimensional orientable surfaces. Both Karen Frederick and I began to work on this problem and we both, independently, came up with a solution. Her solution [7] was very closely tied to the actual presentation of these surface groups. I took a somewhat more general approach which yielded somewhat more [2]. The net result was the following

Theorem 5. *Let F be a free group on X, A a free abelian group on Y, f an element of F which is not a proper power in F, and a an element of A which is not a proper power in A. Then the group*

$$G = < X \cup Y; [y, y'] = 1 \ (y, y' \in Y), f = a >$$

is residually free.

In particular, it follows from this theorem that surface groups are residually free and hence residually torsion-free nilpotent (and also residually finite).

The residual nilpotence of one-relator groups was itself a topic of some interest to Magnus. He later put Bruce Chandler to work on this topic. Chandler [6] eventually found an alternative proof of the residual torsion-free nilpotence of surface groups by making use of the ring R that I described at the outset. Some years later I found yet another means of proving the residual nilpotence of surface groups. I never discussed this method with Magnus, and subsequently forgot about it. A few months ago, J. Lewin told me that he had also found a very simple argument to prove the residual nilpotence of surface groups. This prompted me to rethink some of my old ideas and it is to these thoughts that I want to turn next.

4. RESIDUALLY NILPOTENT GROUPS

The fundamental groups of two dimensional orientable surfaces all contain a normal subgroup, which is free and such that the resultant factor group is infinite cyclic. Thus they have the same form as the groups covered by the following theorem.

Theorem 6. *Let G be a finitely generated group. Suppose that G contains a free, normal subgroup N such that G/N is infinite cyclic. If $G/\gamma_2(N)$ is residually torsion-free nilpotent, then so is G.*

This theorem is similar in spirit to a theorem of P. Hall [8], who proved that if G is a group with a normal, nilpotent subgroup N, then G is nilpotent if $G/\gamma_2(N)$ is nilpotent.

Theorem 6 leads to a host of new examples of residually nilpotent groups. In particular, it can be used to give yet another proof of the residual torsion-free nilpotence of surface groups.

There is a further use of Theorem 6 that I want to describe here. To this end, let me recall the following definition.

Definition. A group G is termed *parafree* if G is residually nilpotent and there exists a free group F such that

$$G/\gamma_n(G) \cong F/\gamma_n(F) \; for \; all \; n.$$

Parafree groups, which can be likened to free groups, exist in profusion, see, e.g., [3]. It should be pointed out that it follows from Magnus' Theorem 2 that a non-free, finitely generated parafree group G with the same nilpotent factor groups $G/\gamma_n(G)$ as a free group of rank q cannot be generated by q elements. It is not known how closely a parafree group can resemble a free group. I want to describe next some new, non-free parafree groups, which very closely resemble free groups. The proof that these groups are parafree is an easy application of Theorem 5 (see [6]).

Theorem 7. *Let F be the free group on $s, t, a_1, \ldots, a_q$ and let w be an element of F which involves a_1 and does not involve s. In addition suppose that w lies in the $k - -th$ term $F^{(k)}$ of the derived series of F. Then the one-relator group*

$$G_w =< s, t, a_1, \ldots, a_q; a_1 = w s^{-1} t^{-1} s t >$$

is parafree. Moreover

$$G_w/G_w^{(k)} \cong H/H^{(k)},$$

where H is a free group of rank $q + 1$.

I will sketch the proofs of Theorems 6 and 7 in section **6**.

5. Some problems on $\mathcal{D}$-groups

Let $p_1 = 2$, $p_2 = 3, \ldots$ be the set of all primes in ascending order of magnitude.

Definition. A group G is called a $\mathcal{D}$-group if it admits a set of unary operators

$$\pi_1, \pi_2, \ldots$$

such that for all $g \in G$

$$g^{p_i} \pi_i = g = (g \pi_i)^{p_i}.$$

It is not hard to verify that these $\mathcal{D}$-groups consist precisely of those groups G in which extraction of n-th roots is uniquely possible, for every positive integer n. The class of all $\mathcal{D}$-groups is a *variety of (universal) algebras*. The precise technical description of these terms does not matter. It suffices only to say that in such a variety one has the notion of a free $\mathcal{D}$-group, as well as all of the other notions that one makes use of in group theory.

There are a number of properties of such free $\mathcal{D}$-groups that are similar to properties of free groups. For example one has the following

Theorem 8. *$\mathcal{D}$-subgroups of free $\mathcal{D}$-groups are free.*

I was told about this theorem by Tekla Taylor-Lewin, but I have not been able to locate a reference.

Magnus was fond of these $\mathcal{D}$-groups, and discussed them in his book with Karrass and Solitar [14]. It seems appropriate, therefore, to raise here some new problems about free $\mathcal{D}$-groups, in his memory.

To this end, let F be the free $\mathcal{D}$-group on

$$x_1, \ldots, x_q.$$

It is not hard to see that the mapping

$$x_i \mapsto 1 + \xi_i \ (i = 1, \ldots, q)$$

defines a homomorphism ϕ of F into the group of units of R.

Problem 1. *Is ϕ a monomorphism?*

Problem 2. *Let*

$$G = < x_1, \ldots, x_q; r = 1 > .$$

If extraction of n-th roots in G is unique, whenever such roots exist, can G be embedded in a $\mathcal{D}$-group?

Problem 3. *Suppose that the one-relator group*

$$G = < x_1, \ldots, x_q; r = 1 >$$

can be embedded in a $\mathcal{D}$-group. If H is the one-relator $\mathcal{D}$-group generated, as a $\mathcal{D}$-group, by $x_1, \ldots, x_q$ and defined, as a $\mathcal{D}$-group, by the single relation $r = 1$, is the word problem solvable for H? In general, is the word problem solvable for one-relator $\mathcal{D}$-groups?

Problem 4. *Is there a freiheitssatz for one-relator $\mathcal{D}$-groups?*

Problem 5. *Can free groups be characterised by a length function?*

Problem 6. *Does a free $\mathcal{D}$-group act freely on a Λ-tree, for a suitable choice of Λ?*

6. Proofs

I want to sketch here the proofs of Theorem 6 and Theorem 7.

I would like to begin with the proof of Theorem 6. I will adopt the notation used in the statement of the theorem. Since G/N is infinite cyclic, we can choose an element $t \in G$, such that $G = gp(N, t)$. So t is of infinite order modulo N. Suppose that $g \in G, g \neq 1$. We want to find a normal subgroup K of G such that G/K is torsion-free nilpotent and $g \notin K$. It is clear that it suffices to consider the case where $g \in N$. Since N is free there exists an integer k such that $g \notin \gamma_k(G)$. Notice that $H = N/\gamma_k(N)$ is a torsion-free nilpotent group and so we can form the Mal'cev completion $\overline{H}$ of H (see [15]). This group $\overline{H}$ is a minimal $\mathcal{D}$-group containing H. It is again nilpotent of class $k - 1$ and torsion-free. Moreover if we denote by τ the automorphism that t induces on H, then τ extends uniquely to an automorphism $\overline{\tau}$ of $\overline{H}$. Let $\overline{G}$ be the semidirect product of $\overline{H}$ with the infinite cyclic group generated by an element $\overline{t}$, where $\overline{t}$ acts on $\overline{H}$ as $\overline{\tau}$. Then $M = \overline{H}/\gamma_2(\overline{H})$ is a direct sum of copies of the additive group of $\mathbb{Q}$ (see [1]). Moreover, it is not hard to deduce from the fact that $G/\gamma_2(N)$ is residually torsion-free nilpotent, that $\overline{G}/\gamma_2(\overline{H})$ is also residually torsion-free nilpotent. We now view M as a module over the rational group algebra Γ of the infinite cyclic group on $\overline{t}$. Since Γ is a principal ideal domain and the group G is finitely generated, the Γ-module M is finitely generated and hence a direct sum of cyclic modules. This decomposition of M makes it possible to unravel the action of $\overline{t}$ on $\overline{H}$ and, using the fact that $H = F/\gamma_k(F)$, to deduce that $\overline{H}$ is residually torsion-free nilpotent. It is easy to deduce that G is itself residually torsion-free nilpotent.

There are three steps in the proof of Theorem 7. The first makes use of Magnus' method of unravelling the structure of one-relator groups. This method allows one to prove, easily, that the normal closure N of $s, a_1, \ldots, a_q$ in G_w is free. The second step involves the verification that $G/\gamma_2(N)$ is residually torsion-free nilpotent. So, by Theorem 6, G_w is residually torsion-free nilpotent. The other properties of G_w follow directly from the form of the defining relation of G_w. The final step in the proof of Theorem 7 is the verification that G_w is not free. This is accomplished by invoking an algorithm of J.H.C. Whitehead [15].

I have only talked here about a few of Magnus' theorems. All of his work ([5]) is filled with beautiful, new ideas, giving joy to all of us. Wilhelm Magnus is sorely missed, but his work will be with us always.

REFERENCES

1. Gilbert Baumslag, *Some aspects of groups with unique roots*, Acta Mathematica **104** (1960), 217–303.
2. ———, *On generalised free products*, Math. Zeitschrift **78** (1962), 423–438.
3. ———, *More groups that are just about free*, Bull. Amer. Math. Soc. **72** (1968), 752–754.
4. Gilbert Baumslag and Norman Blackburn, *Groups with cyclic upper central factors*, Proc. London Math. Soc. **(3) 10** (1960), 531–544.
5. Gilbert Baumslag and Bruce Chandler (eds.), *Wilhelm Magnus Collected Papers*, Springer Verlag, New York, Heidelberg, Berlin, 1984.
6. B. Chandler, *The representation of a generalized free product in an associative ring*, Comm. Pure Appl. Math. **21** (1968), 271–288.
7. K.N. Frederick, *The Hopfian property for a class of fundamental groups*, Comm. Pure Appl. Math. **16** (1963), 1–8.
8. P. Hall, *Some sufficient conditions for a group to be nilpotent*, Illinois J. Math. **2** (1958), 787–801.
9. R.C. Lyndon, *Groups with parametric exponents*, Trans. American Math. Soc. **96** (1960), 445–457.
10. W. Magnus, *Über diskontinuierliche Gruppen mit einer definierenden Relation (Der Freiheitssatz)*, J. Reine Angew. Math. **163** (1930), 141–165.
11. ———, *Das Identitätsproblem für Gruppen mit einer definierenden Relation*, Math. Ann. **106** (1932), 295–307.
12. ———, *Beziehungen zwischen Gruppen und Idealen in einem speziellen Ring*, Math. Ann. **111** (1935), 259–280.
13. ———, *Über freie Faktorgruppen und freie Untergruppen gegebener Gruppen*, Monatsh. Math. **47** (1939), 307–313.
14. W. Magnus, A. Karrass and D. Solitar, *Combinatorial group theory*, Wiley, New York, 1966.

15. A.I. Mal'cev, *On a class of homogeneous spaces*, Izvestiya Akad. Nauk SSSR Ser. Mat. **13** (1949), 9–32; English transl. in Amer. Math. Soc. Transl. (1) **9** (1962), 276–307.
16. J.H.C. Whitehead, *On equivalent sets of elements in a free group*, Ann. of Math. **37** (1936), 782–800.

DEPARTMENT OF MATHEMATICS, CITY COLLEGE OF NEW YORK, NEW YORK, N.Y. 10031

Contemporary Mathematics
Volume **169**, 1994

The Monodromy Group of
a Transcendental Function

KEVIN BERRY AND MARVIN TRETKOFF

ABSTRACT. The paper represents the invited lecture "Riemann surfaces, differential equations and combinatorial group theory" presented by the second author at The Mathematical Legacy of Wilhelm Magnus meeting at Polytechnic University, May 1-3, 1992. The research reported on herein is part of the Ph.D. dissertation of the first author, supervised by the second author, at the Stevens Institute of Technology (1991).

Introduction

Wilhelm Magnus was a remarkable person. I had the good fortune to know him for almost thirty years, beginning with my second year at New York University. With the passage of time, Wilhelm's message has grown clearer because he taught by example. In mathematics, this is clearly reflected by the program of this meeting. Here, I would like to add a few remarks about Wilhelm's mathematical personality that may not be apparent to those who only met him through his publications or secondary sources. First, Wilhelm loved number theory. This was clearly reflected in private conversation and in the classes he taught. Some hint of his feelings for the subject can be gleaned from the *Mathemtical Recollections* he wrote for his collected works [**4**]. He notes that he contributed to the subject as one of the editors of Hilbert's number-theoretic work and that he published a paper on the class number of quadratic forms. I believe that he was rather proud of that paper, and he mentioned it to me on numerous occasions. Next, I want to mention Wilhelm's long friendship with C. L. Siegel, one of the great mathematicians of this century. Siegel was one of Wilhelm's mathematical heros, and through him we were afforded personal glimpses of a giant in our subject. In my case, this led to participation in the translation of Siegel's *Topics*

1991 *Mathematics Subject Classification*. Primary 33E99, 30B40; Secondary 20B99.

in Complex Function Theory [7]. From those volumes I learned about the monodromy group of an analytic function. This topic interested Wilhelm, and it is the subject of one of his last publications.

The monodromy group is an algebraic analogue of the Riemann surface of a multi-valued analytic function; it gives a global description of its branching properties. I shall turn to its precise definition shortly, and I shall adopt the viewpoint taken by Siegel [7] that it is a group of permutations of a countable set, the branches of the function at a regular point. First, however, I wish to note that if the function is a solution of a homogeneous linear ordinary differential equation with single-valued coefficients, then the monodromy group admits a faithful representation as a group of complex matrices whose degree equals the order of the equation. Thus, we are led to the *monodromy group of the differential equation*. Wilhelm investigated the group of Hill's equation (see [4, pp. 645-660]), and he encouraged K. Kuiken to investigate the groups of the hypergeometric equation and Heun's equation. In the case of differential equations of the Fuchsian class, the Zariski closure of the monodromy group is the differential Galois group of the equation. We may therefore apply the differential Galois theory, invented by Picard and Vessiot and perfected by E. Kolchin, to the study of differential equations in the same way Galois studied algebraic equations. I have discussed these matters in [8], so I will not repeat them here.

I had many discussions with Wilhelm about the monodromy group as a tool for translating results in group theory to function theory. In fact, we even discussed the combinatorial problems that I shall come to later, so I feel this is an appropriate time to present them. The motivation for these studies is the following somewhat vague question. From the viewpoint of their branching behavior, which transcendental functions should be considered the simplest after $\log(z)$? This question leads to a search for certain two-generator groups of permutations of the integers. In particular, we seek permutations with specific cycle structures and, after showing that these exist, we investigate the algebraic structure of the groups they generate.

Notation and Terminology:

$\mathbb{Z}$ = integers

Σ = the group of all one-to-one mappings of $\mathbb{Z}$ onto itself. We will refer to Σ as the *infinite symmetric group*; its elements will be called *permutations*.

τ denotes the permutation given by $\tau(n) = n + 1$, $n \in \mathbb{Z}$.

A permutation σ will be said to act *transitively* on $\mathbb{Z}$ if the cyclic group generated by σ acts transitively on $\mathbb{Z}$. Clearly, σ acts transitively if and only if it is conjugate to τ. In this case, we may also use the notation $(\ldots\ a_{n-1}a_n a_{n+1}\ \ldots)$ for σ when $\sigma(a_n) = a_{n+1}$, $n \in \mathbb{Z}$.

$\sigma\tau$ denotes τ *followed by* σ.

1. Analytical Aspects

We begin with $w = \log(z) = \log|z| + i\arg(z)$, $z = x + iy$. Locally, this formula defines an analytic function for all values of z except $z = 0$ and $z = \infty$, where the function is undefined. Observe that as z traverses a simple closed loop Γ, for example the unit circle, counterclockwise about the origin, the value of $\log(z)$ increases by $2\pi i$. Therefore, $\log(z)$ has infinitely many values for $z \neq 0, \infty$. Traditionally, we view this "multi-valued function" as a single-valued function on a Riemann surface, $\mathcal{R}$, that is constructed from countably many copies, $\widehat{\mathbb{C}}_n$, $n \in \mathbb{Z}$, of the Riemann sphere $\widehat{\mathbb{C}}$ with the points corresponding to $z = 0$ and $z = \infty$ deleted from each of them. These copies of $\widehat{\mathbb{C}}$ are called the "sheets" of $\mathcal{R}$, and they are joined together so that if a point z_0 on $\widehat{\mathbb{C}}_n$ traverses a copy of Γ counterclockwise, it will end up at the point on $\widehat{\mathbb{C}}_{n+1}$ corresponding to z_0. This leads to the classical picture of $\mathcal{R}$ as an infinite sheeted covering of $\widehat{\mathbb{C}}$ with a single "infinite spiral ramp" above a neighborhood of $z = 0$. Combinatorially, we have associated to the point $z = 0$ the permutation $\tau: n \mapsto n + 1$, $n \in \mathbb{Z}$. Applying the monodromy theorem, we see that traversing Γ clockwise about $z = 0$ has the same effect on $\log(z)$ as traversing a simple closed loop counterclockwise about $z = \infty$. Therefore, associated with $z = \infty$ is the permutation of the sheets of $\mathcal{R}$ given by $\tau^{-1}: n \mapsto n - 1$, $n \in \mathbb{Z}$. Thus, we see that $\mathcal{R}$ has a single infinite spiral ramp above a neighborhood of $z = \infty$. Because $\log(z)$ is locally well-defined for all values $z = z_0 \neq 0, \infty$, we see that $\mathcal{R}$ is unramified above each of these points. In other words, if we select a disk D on $\widehat{\mathbb{C}}$ with center z_0 and with a sufficiently small radius, then the portion of $\mathcal{R}$ corresponding to D consists of a countable infinitude of pairwise disjoint copies, D_n, $n \in \mathbb{Z}$, of D. We say that each D_n lies *above* D and is mapped one-to-one onto it by *projection*. We now turn to a few function-theoretic properties of $\log(z)$.

First, we note that as z approaches 0 or ∞, $|\log(z)|$ approaches ∞. Were it not for the multi-valued nature of $\log(z)$, we might say that it has a pole at $z = 0$ and at $z = \infty$. Of course, each branch of $\log(z)$ approaches ∞ as z approaches zero in any sector whose vertex is $z = 0$. The same remark applies to the behavior of $\log(z)$ as z approaches ∞. Next, we prove the well-known fact that $\log(z)$ is a transcendental function. Namely, by definition, an algebraic function $w = w(z)$ is given by the solutions to a polynomial equation $P(z, w) = 0$, where the coefficients of the polynomial $P(z, w)$ are complex numbers. Thus, for each value $z = z_0$ of the independent variable, $w = w(z_0)$ attains a finite set of values. Since $w = \log(z)$ takes on infinitely many values when $z_0 \neq 0, \infty$, it must be a transcendental function. Of course, this reasoning also proves that any analytic function that has infinitely many branches at a point $z = z_0$ is transcendental. Finally, we note that $w = \log(z)$ satisfies the second order homogeneous linear differential equation $zw'' + w' = 0$. This equation has singularities at $z = 0$ and $z = \infty$. The reader can easily check that these two points are regular singular points, so the differential equation is of *Fuchsian class* (see, for example, [3] and [6] for the definition of these terms). Of course, this is also a reflection of the fact that each solution to the differential equation approaches a limit as z approaches a singular point in any sector with vertex at that singularity.

In many ways, $\log(z)$ is the simplest transcendental function. For example, it has the fewest singularities of all analytic functions that approach definite limits as z approaches each of its singular points. Indeed, it is a classical fact that an entire transcendental function has an essential singularity at $z = \infty$. From this viewpoint, the exponential function and the elementary trigonometric functions are more complicated than $\log(z)$. On the other hand, a finitely many valued analytic function whose branches all approach definite limits as z approaches each of its singular points must be an algebraic function (see, for example [7]). We note, too, that the Riemann surface of $\log(z)$ has the fewest possible branch points and the simplest ramification structure. Indeed, the sphere with a single point deleted is simply connected, so it follows from the monodromy theorem that a multi-valued analytic function must have more than one branch point. Consequently, we must consider functions with three singular points in our search for the "next simplest" transcendental function beyond $\log(z)$.

Applying a preliminary fractional linear transformation to the independent variable, we may suppose that the desired function has its singularities at $z = 0, 1, \infty$. A classical source of such functions is the *hypergeometric equation*

$$z(1 - z)w'' + (c - (1 + a + b)z)w' - abw = 0.$$

This is the most general differential equation of Fuchsian class with three singularities. It has been the subject of intense investigation and application, starting with Euler, Kummer and Gauss and continuing to the present day (see, for example, [1], [3], and [6] for detailed discussions and further references). Here, we merely wish to note that different types of solutions arise as the parameters a, b, c appearing in the coefficients vary. For certain values, all the solutions of the hypergeometric equation are algebraic functions (see, for example, [1] for an account of this classical topic, initiated by H. A. Schwarz, and its connection to current research). On the other hand, if we set $a = b = \frac{1}{2}$ and $c = 1$, then all the solutions of the hypergeometric equation are transcendental functions. Moreover, in this case, the inverse, $\tau = \tau(\lambda)$, of the celebrated Legendre elliptic modular function $\lambda = \lambda(\tau)$ is the quotient of two linearly independent solutions to the hypergeometric equation. Here, we are following tradition by writing τ and λ in place of the variables z and w, respectively. The function $\lambda(\tau)$ is automorphic with respect to the principal congruence subgroup of level two in the modular group and it maps the upper half of the τ-plane conformally onto the universal covering of the λ-sphere with the points $\lambda = 0, 1, \infty$ deleted. Viewed as a branched covering of the λ-sphere, this universal covering has infinitely many infinite spiral ramps above a neighborhood of each of the singularities $\lambda = 0, 1, \infty$. Moreover, this is the Riemann surface of the multi-valued function $\tau(\lambda)$; that is, $\tau(\lambda)$ is not single-valued on a Riemann surface that lies between the thrice punctured λ-sphere and its universal covering. Thus, we see that $\tau(\lambda)$ is a transcendental function with three singular points, but its branching behavior is far more complicated than that of $\log(z)$.

Our goal is to prove the existence of transcendental functions whose Riemann surfaces have a single infinite spiral ramp above their three singular points $z = 0, 1, \infty$. It transpires that there are many, essentially distinct, such functions. Unfortunately, we have not been able to determine a concrete representation for any of them as a power series, an integral, or a solution of a differential equation. This is a result of our method of proof. Namely, we first show that there are Riemann surfaces with the desired combinatorial properties, and then we prove that there are multi-valued functions on the z-sphere that define these Riemann surfaces. The second step has already been established in [2] and [9], where different ramification properties were required. At the heart of the proof is the fact that non-compact Riemann surfaces are Stein manifolds. This result is proved using sophisticated analytic techniques that do not appear to be constructive, so we do not obtain an explicit description of our function. Moreover, these analytic methods do not provide any information about the behavior of our functions as z approaches the singular points. In the future, we hope to investigate this matter and the question of whether our functions satisfy homogeneous linear ordinary differential equations. Now, we turn to the combinatorial aspects of our problem.

Certain combinatorial data, nowadays called a *Hurwitz system*, determine the structure of a Riemann surface as a branched covering of the sphere. This data consists of the following:

(1) branch points $z = b_1, \ldots, z = b_{t+1}$, where $t \geq 1$;
(2) the number n of sheets;
(3) permutations $\pi_1, \ldots, \pi_{t+1}$ of n symbols subject to the condition
$$\pi_1 \pi_2 \cdots \pi_{t+1} = 1;$$
(4) a base point $z_0 \neq b_1, \ldots, b_{t+1}$;
(5) a system of *cuts* $\ell_1, \ldots, \ell_{t+1}$ joining z_0 to the branch points. The ℓ_j are simple paths that meet pairwise only at z_0 and meet sufficiently small circles centered at z_0 in the indicated order when those circles are traversed counterclockwise.

Since the notion of a Hurwitz system has been discussed in detail in [8] and elsewhere, here we restrict ourselves to a few remarks that are essential to an understanding of our investigation.

First, we note that the letter n can denote the cardinality of the integers as well as any positive integer. In algebro-geometric studies and in [8], the number of sheets is supposed to be finite. However, in the present paper, *the number of sheets will be countably infinite* and the letter n will *not* be reserved for the number of sheets. Next, we note that the requirement that $t \geq 1$ stems from the simple connectivity of the sphere with a single point deleted. The topology of the sphere also imposes the product relation appearing in (3). As for conditions (4) and (5), they are used to impose an equivalence relation on triples of data given by (1), (2) and (3). This relation is required because the choices of base point and cuts given by (4) and (5) are auxiliary to the construction of a Riemann surface from countably many copies of the sphere.

Naturally, the existence and uniqueness of a Riemann surface with prescribed Hurwitz system must be discussed. The existence of such a Riemann surface is proved in [7] in case the number of sheets is finite; the same proof applies when there are infinitely many sheets. However, as we have already noted, conditions (4) and (5) are auxiliary, and by varying the choices they describe, we obtain equivalent Riemann surfaces. To describe the situation more precisely, we let G denote the group generated by the permutations $\pi_1, \ldots, \pi_t$ occurring in (3), and we call G the *monodromy group* of the Riemann surface. Of course, G is a subgroup of the group Σ of all permutations of the integers. It turns out that changing the data (4) and (5) is the same as replacing G by a conjugate subgroup of Σ. Moreover, these conjugacy classes of subgroups of Σ correspond biuniquely to equivalence classes of covering spaces of the sphere with the branch points $z = b_1, \ldots, b_{t+1}$ deleted from it (see, for example, [5] for a discussion of topologically equivalent covering spaces). Finally, we note that the Riemann surface described by a Hurwitz system is connected if and only if its monodromy group acts transitively. This is proved in [7] for the case of a finite sheeted surface; the proof applies in case there are infinitely many sheets, too.

The transcendental functions we are seeking have Riemann surfaces that admit the following description:

(1) the branch points are $z = 0, 1, \infty$;
(2) the number of sheets is countably infinite;
(3) the permutations are $\pi_1 = \tau: n \mapsto n + 1$, $n \in \mathbb{Z}$, $\pi_2 = \alpha\tau\alpha^{-1}$ and $\pi_3 = \beta\tau\beta^{-1}$, where α and β belong to Σ.

Of course, (3) is merely the combinatorial expression of the requirement that there be a single infinite spiral ramp above a neighborhood of each of the branch points. From the viewpoint of branching behavior, two such multi-valued functions will be considered equivalent if their Riemann surfaces have conjugate monodromy groups. We now see that the existence of transcendental functions of the desired type depends on the purely combinatorial problem: *are there permutations of the integers satisfying condition (3)?* We will shortly give examples that answer this question affirmatively. Moreover, at least two of our examples generate non-isomorphic monodromy groups, so the corresponding transcendental functions will have essentially distinct branching behavior even though both of them have Riemann surfaces with a single infinite spiral ramp above $z = 0, 1$ and ∞. In one of these examples, the monodromy group is a free group of rank two. Since the monodromy group of $\log(z)$ is free of rank one, our function may be viewed as possibly the next simplest transcendental function.

From the function-theoretic viewpoint, our investigation yields the following

THEOREM 1.

(a) *There is a family of transcendental functions each of whose singularities occur at $z = 0, 1, \infty$ and whose Riemann surface possesses a single
infinite spiral ramp above a neighborhood of each singularity.*

(b) *at least one function in the family has a free group of rank two as its
monodromy group.*

(c) *at least one of the functions occurring in (b) is transcendentally transcendental. That is, it does not satisfy an algebraic differential equation,
$P(z, w, w', \ldots, w^{(n)}) = 0$, where P is a polynomial with complex coefficients.*

Here it is appropriate to note that part (c) is proved by applying Theorem 1 of
[2]. That result states that any finitely generated infinite subgroup of Σ is the
monodromy group of some transcendentally transcendental function. Of course,
it is possible that there are functions that satisfy conditions (a) and (b) but not
condition (c). Therefore, we ask the following

> **Question**: Is there a solution of a hypergeometric differential
> equation that satisfies condition (a) or conditions (a) and (b)?

2. Combinatorial Aspects

We now turn to the combinatorial aspects of our problem, beginning with

THEOREM 2. *There is a permutation $\sigma \in \Sigma$ such that σ and $\sigma\tau$ both act
transitively on $\mathbb{Z}$.*

Of course, the theorem will be proved if we exhibit a single permutation σ
with the indicated properties. Here, we will give a countable family of such
examples. The first of these is very simple, but it has been of limited use in our
subsequent investigations.

EXAMPLE I. If $\sigma = (\cdots 8, -8, 6, -6, 4, -4, 2, -2, 0, 1, -1, 3, -3, \cdots)$, it is obvious that σ acts transitively on $\mathbb{Z}$. Moreover, we see that

$$\tau\sigma = (\cdots 5, -4, 3, -2, 1, 0, 2, -1, 4, -3, 6, -5, 8, \cdots)$$

and

$$\sigma\tau = (\cdots 6, -7, 4, -5, 2, -3, 0, -1, 1, -2, 3, -4, 5 \cdots).$$

Clearly, $\sigma\tau$ and $\tau\sigma$ also act transitively on $\mathbb{Z}$.

EXAMPLE II. Let $n = 2j + 1$, $j \geq 1$, and set

$$a_1^{(k)} = nk + 1, \ldots, a_n^{(k)} = n(k + 1), \ k \geq 0$$

and

$$a_1^{(-k)} = 1 - kn, \dots, a_n^{(-k)} = (1 - k)n, \ k \geq 1.$$

Now, let $\sigma = \sigma_n$ be the permutation of the integers given by Diagram 1. It is easy to see that $\tau\sigma$ is the permutation given by Diagram 2 and that it acts transitively on $\mathbb{Z}$. Next, we note that $\sigma\tau$ also acts transitively on $\mathbb{Z}$. It is illustrated by Diagram 3.

EXAMPLE III. Let $n = 2j$, $j \geq 3$, and let $a_j^{(k)}$ be defined as in Example II. We then let σ denote the permutation of the integers given by Diagram 4. We leave it to the reader to verify σ, $\sigma\tau$ and $\tau\sigma$ act transitively on $\mathbb{Z}$.

REMARKS.

(1) By a simple construction, the examples can be modified to yield continuously many such permutations. However, the subgroups of the infinite symmetric group that are obtained from them are conjugate to the subgroups provided by I, II and III.

(2) We have also constructed permutations σ and ρ of $\mathbb{Z}$ with the property that τ, σ, ρ and $\rho\sigma\tau$ all act transitively on $\mathbb{Z}$.

(3) Example I is due to F. Haimo and M. Tretkoff and, independently, to Joel Brenner.

Now, let G denote the subgroup of Σ generated by τ and σ, where σ is one of the permutations constructed in our examples.

THEOREM 3. *The group G is **not** free. In particular, the following relations hold:*

(a) *If σ is given by Example I, then*

$$\begin{aligned}
(1) && ((\tau\sigma)^{2n}(\sigma\tau)^{2n})^{2n+1} &= 1, && n = 1, 2, \dots \\
(2) && ((\sigma\tau)^{2n}(\tau\sigma)^{2n})^{2n+1} &= 1, && n = 1, 2, \dots \\
(3) && (\tau^{2n}\sigma^{2n}\tau^{-2n}\sigma^{-2n})^2 &= 1, && n = 1, 2, \dots \\
(4) && (\tau\sigma^{-2n}\tau^{-1}\sigma^{-2n})^{2n+1} &= 1, && n = 1, 2, \dots \\
(5) && (\tau^{-2}\sigma^{-2}\tau\sigma^2\tau\sigma^4)^2 &= 1.
\end{aligned}$$

(b) *If σ is given by Example II, with $n = 5$, then we have:*

$$\begin{aligned}
(1) && ((\sigma^{-14}\tau^{10})(\tau^{-5}\sigma^7)^2)^{10} &= 1 \\
(2) && ((\sigma^{-21}\tau^{15})(\tau^{-5}\sigma^7)^3)^{24} &= 1 \\
(3) && (\sigma^{-7}\tau^5\sigma^7\tau^{-5})^{10} &= 1 \\
(4) && (\sigma^{-14}\tau^{10}\sigma^{14}\tau^{-10})^{24} &= 1 \\
(5) && (\tau^{-10}(\sigma\tau)^6(\tau\sigma)^6)^{80} &= 1.
\end{aligned}$$

$a_1^{(0)}$

$\uparrow$

$a_n^{(k-1)}$ $a_n^{(k-2)}$ $a_n^{(0)}$ $a_2^{(0)}$ $a_2^{(1)}$ $\cdots$ $a_2^{(k)}$ $\cdots$

$\uparrow$ $\uparrow$ $\uparrow$ $\uparrow$ $\uparrow$ $\uparrow$

$a_n^{(-k)}$ $a_n^{(-k+1)}$ $\cdots$ $a_n^{(-1)}$ $\vdots$ $\vdots$ $\vdots$

$\uparrow$ $\uparrow$ $\cdots$ $\uparrow$ $\uparrow$ $\uparrow$ $\uparrow$ $\uparrow$

$a_2^{(-k)}$ $\vdots$ $\vdots$ $a_j^{(0)} \to a_{j+3}^{(0)}$ $a_j^{(1)}$ $\cdots$ $a_j^{(k)}$ $\cdots$

$\uparrow$ $\uparrow$ $\uparrow$ $\uparrow$ $\uparrow$ $\uparrow$

$a_1^{(-k)}$ $a_1^{(-k+1)}$ $a_1^{(-1)}$ $a_{j+2}^{(0)}$ $a_{j+2}^{(1)}$ $\cdots$ $a_{j+2}^{(k)}$ $\cdots$

$\uparrow$ $\uparrow$ $\uparrow$ $\uparrow$ $\uparrow$ $\uparrow$

$a_1^{(k)}$ $a_1^{(k-1)}$ $\cdots$ $a_1^{(1)}$ $a_{j+1}^{(0)}$ $a_{j+1}^{(1)}$ $\cdots$ $a_{j+1}^{(k)}$ $\cdots$

$\vdots$ $\uparrow$ $\uparrow$ $\uparrow$ $\uparrow$ $\uparrow$

Diagram 1

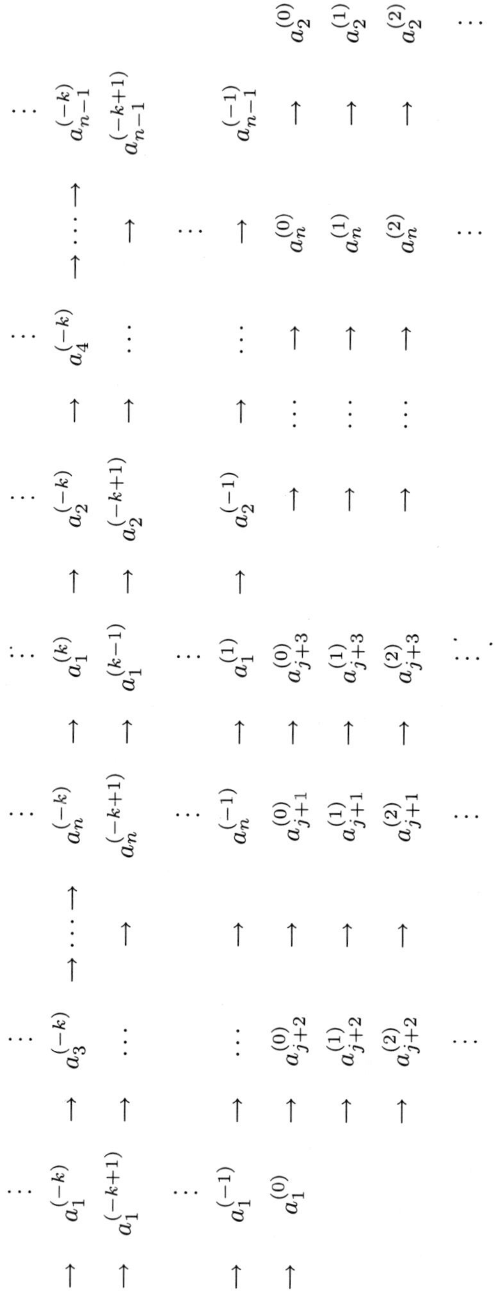

$$a_2^{(0)} \quad a_2^{(1)} \quad a_2^{(2)} \quad \cdots$$

$$\cdots \quad a_{n-1}^{(-k)} \quad a_{n-1}^{(-k+1)} \quad a_{n-1}^{(-1)} \quad \uparrow \quad \uparrow \quad \uparrow$$

$$\uparrow \cdots \uparrow \quad \uparrow \quad \cdots \quad \uparrow \quad a_n^{(0)} \quad a_n^{(1)} \quad a_n^{(2)} \quad \cdots$$

$$\cdots \quad a_4^{(-k)} \quad \vdots \qquad \vdots \quad \uparrow \quad \uparrow \quad \uparrow$$

$$\uparrow \quad \uparrow \qquad \uparrow \quad \vdots \quad \vdots \quad \vdots$$

$$\cdots \quad a_2^{(-k)} \quad a_2^{(-k+1)} \quad a_2^{(-1)} \quad \uparrow \quad \uparrow \quad \uparrow$$

$$\uparrow \quad \uparrow \qquad \uparrow$$

$$\cdots \quad a_1^{(k)} \quad a_1^{(k-1)} \quad \cdots \quad a_1^{(1)} \quad a_{j+3}^{(0)} \quad a_{j+3}^{(1)} \quad a_{j+3}^{(2)} \quad \cdots$$

$$\uparrow \quad \uparrow \qquad \uparrow \quad \uparrow \quad \uparrow \quad \uparrow$$

$$\cdots \quad a_n^{(-k)} \quad a_n^{(-k+1)} \quad \cdots \quad a_n^{(-1)} \quad a_{j+1}^{(0)} \quad a_{j+1}^{(1)} \quad a_{j+1}^{(2)} \quad \cdots$$

$$\uparrow \cdots \uparrow \quad \uparrow \qquad \uparrow \quad \uparrow \quad \uparrow \quad \uparrow$$

$$\cdots \quad a_3^{(-k)} \quad \vdots \qquad \vdots \quad a_{j+2}^{(0)} \quad a_{j+2}^{(1)} \quad a_{j+2}^{(2)} \quad \cdots$$

$$\uparrow \quad \uparrow \qquad \uparrow \quad \uparrow \quad \uparrow \quad \uparrow$$

$$\cdots \quad a_1^{(-k)} \quad a_1^{(-k+1)} \quad \cdots \quad a_1^{(-1)} \quad a_1^{(0)}$$

$$\uparrow \quad \uparrow \qquad \uparrow \quad \uparrow$$

Diagram 2

$$\cdots \quad a_n^{(-k)} \quad a_n^{(-k+1)} \quad \cdots \quad a_n^{(-1)} \quad a_1^{(0)} \quad a_1^{(1)} \quad \cdots \quad a_1^{(k)} \quad \cdots \quad \cdots \quad \cdots$$

$$\cdots \quad a_3^{(-k)} \quad a_3^{(-k+1)} \quad \cdots \quad a_3^{(-1)}$$

$$\cdots \quad a_1^{(-k)} \quad a_1^{(-k+1)} \quad \cdots \quad a_1^{(-1)} \quad a_{j+3}^{(0)} \quad a_{j+3}^{(1)} \quad \cdots \quad a_{j+3}^{(k)} \quad \cdots \quad \cdots \quad \cdots$$

$$\cdots \quad a_n^{(k-1)} \quad a_n^{(k-2)} \quad \cdots \quad a_n^{(0)} \quad a_{j-1}^{(0)} \quad a_{j-1}^{(1)} \quad \cdots \quad a_{j-1}^{(k)} \quad \cdots \quad \cdots \quad \cdots$$

$$\cdots \quad a_{n-1}^{(-k)} \quad a_{n-1}^{(-k+1)} \quad \cdots \quad a_{n-1}^{(-1)} \quad a_{j+2}^{(0)} \quad a_{j+2}^{(1)} \quad \cdots \quad a_{j+2}^{(k)} \quad \cdots \quad \cdots \quad \cdots$$

$$\cdots \quad a_4^{(-k)} \quad a_4^{(-k+1)} \quad \cdots \quad a_j^{(0)} \quad a_j^{(1)} \quad \cdots \quad a_j^{(k)} \quad \cdots \quad \cdots \quad \cdots$$

$$\cdots \quad a_2^{(-k)} \quad a_2^{(-k+1)} \quad \cdots \quad a_2^{(-1)} \quad a_{j+1}^{(0)} \quad a_{j+1}^{(1)} \quad \cdots \quad a_{j+1}^{(k)} \quad a_{j+1}^{(k+1)} \quad \cdots \quad \cdots$$

Diagram 3

$$
\begin{array}{ccccccccccc}
a_j^{(-k)} & \uparrow & a_{j+1}^{(-k)} & \uparrow & a_{j-1}^{(-k)} & \cdots & a_1^{(-k)} & \uparrow & a_n^{(-k)} \\
a_j^{(k-1)} & \uparrow & a_{j+1}^{(k-1)} & \uparrow & a_{j-1}^{(k-1)} & \cdots & a_{n-1}^{(k-1)} & \uparrow & a_1^{(k-1)} \\
a_j^{(-k+1)} & \uparrow & & & & & & & \\
\cdots & & \vdots & & & & & & \\
a_n^{(-1)} & \uparrow & a_j^{(0)} & \uparrow & a_{j+1}^{(0)} & \cdots & a_n^{(0)} & & \\
a_n^{(1)} & \uparrow & a_n^{(2)} & & a_n^{(k)} & & & & \\
\end{array}
$$

Diagram 4

In addition, there is an integer-valued function $k = k(n)$ such that

$$((\sigma^{-1}\tau)^{-3}((\sigma^{-1}\tau)\tau^{2n+1}(\tau^{-1}\sigma)\tau^{-(2n+1)}(\sigma^{-1}\tau))^3)^k = 1$$

if σ is given by Example II with $n = 2j + 1$, $j \geq 1$.

(c)   *There is an integer-valued function $k = k(n)$ such that*

$$(\tau^{-2n}\sigma^{-2n+1}\tau^{2n}\sigma^{2n-1})^k = 1$$

if σ is given by Example III with $n = 2j$, $j \geq 3$.

Unfortunately, the proof of Theorem 3 is a tedious direct verification that must be omitted here. This work is eased somewhat by the introduction of diagrams, essentially coset diagrams, as we shall now indicate.

Let F be the free group of rank two with the free generators x and y. Of course, there is an epimorphism $f: F \to G$ given by $f(x) = \tau$ and $f(y) = \sigma$. Let E denote the subgroup of F consisting of elements w whose images $f(w)$ in G are permutations that fix the integer $n = 0$. It is well-known that we obtain a permutation representation of F by letting its elements act on the collection of cosets of E in F by left multipliction. Clearly, the permutations defined by elements of E fix the coset $1 \cdot E$, and the kernel of the permutation representation is the intersection of all the conjugates of E in F. Moreover, since G acts transitively on $\mathbb{Z}$, this kernel is also the kernel of f. We may therefore identify the collection of cosets with the set of integers, and we may suppose that the coset $1 \cdot E$ corresponds to $n = 0$. Then, the Schreier coset diagram Γ for E in F with respect to the generators x and y is a graph whose vertices are labelled by the integers and whose edges are partitioned into two disjoint sets S and T, where for each vertex, n, there is a unique directed edge in S joining n to $\sigma(n)$ and there is a unique directed edge in T joining n to $\tau(n) = n + 1$. If σ is the permutation defined in Example I, then Γ may be embedded in the plane as illustrated in Figure 1. There, the set T is represented by the horizontal line, and the remaining arcs represent the set S.

Clearly, T is a maximal subtree of Γ, and we may use it to determine the fundamental group of Γ based at $n = 0$. This group is freely generated by all irreducible loops in Γ that begin and end at $n = 0$, so it is freely generated by loops in one-to-one correspondence with the edges belonging to S. Namely, if n is an arbitrary integer, we form a loop by starting at 0 and traversing n edges in T until we reach vertex n. Then, we traverse the edge in S joining n to $\sigma(n)$ and we return to 0 along an irreducible path in T. In terms of permutations, we see that

$$\tau^{-\sigma(n)}\sigma\tau^n(0) = 0, \quad n \in \mathbb{Z}.$$

Alternatively, the set

$$x^{-\sigma(n)}yx^n, \quad n \in \mathbb{Z}$$

freely generates E. This is because Γ may be viewed as a covering of the space Δ consisting of two circles C_1 and C_2 that are joined at a single point P. A

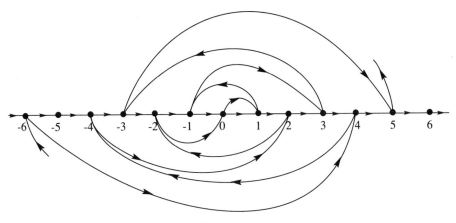

FIGURE 1

projection mapping from Γ to Δ is obtained if we first map each edge belonging
to S bijectively onto C_1, except that its end points are sent to P. The edges of T
are mapped to C_2 in the same way. Then, the general theory of the fundamental
group and covering spaces (see [**5**, chapter 6, especially Figures 6.1 and 6.2])
allows us to conclude that the projection mapping induces an injection of E into
the fundamental group of Δ and that this latter group is free of rank two.

Now, if A, B, C and D denote successively the positive even, positive odd,
negative even and negative odd integers, we see that, with the exception of a
finite set of values, σ and τ induce the following permutations:

$$s = \begin{pmatrix} A & B & C & D \\ C & D & A & B \end{pmatrix}$$

and

$$t = \begin{pmatrix} A & B & C & D \\ B & A & D & C \end{pmatrix}.$$

Clearly, the words given by (1)–(5) in part (a) of Theorem 3 induce the identity permutation on the four symbols A, B, C, D. The outermost exponents
occurring in these relators are required in order to deal with the finite set of
exceptional values. Parts (b) and (c) of Theorem 3 are proved the same way,
but the details are more complicated.

We wish to record the following

COROLLARY 1. *Let G be the group generated by τ and the permutation σ in
Example I. Denoting the images of σ and τ in the abelianization of G by $\bar{\sigma}$ and
$\bar{\tau}$ respectively, we have*

$$(\bar{\sigma})^4 = 1 \quad and \quad (\bar{\tau})^4 = 1.$$

PROOF. It is immediate from relation (5) that $\bar{\sigma}^8 = 1$. Moreover, setting $n = 1$ in relation (4), we also have $(\bar{\sigma})^{12} = 1$, so $(\bar{\sigma})^4 = 1$. Now, if we successively let $n = 1$ and $n = 2$ in relation (1), then we obtain $(\bar{\tau})^{12} = 1$ and $(\bar{\tau})^{40} = 1$. Thus, $(\bar{\tau})^4 = 1$ and the Corollary is proved.

REMARK. The relations occurring in Theorem 3 *may not* represent a complete set of defining relations for the group. The algebraic properties of these groups is the subject of an ongoing Ph.D. thesis by Michael Miniere at the Stevens Institute of Technology.

We conclude by stating the combinatorial result that yields part (b) of Theorem 1. Unfortunately, the proof is too complicated to present here, but we want to make two comments about it. First, our proof is constructive, although we have not obtained a simple formula describing σ. Second, we note that the permutation we construct is obtained by a complicated sequence of modifications of Example II with $n = 5$. Thus far, we have not been able to use Example I for this purpose.

THEOREM 4. *There is a permutation σ of the integers such that*

(1) *σ and $\sigma\tau$ act transitively on $\mathbb{Z}$*

and

(2) *the group G generated by σ and τ is a free group of rank two.*

Finally, we note that F, the free group of rank two, is isomorphic to the fundamental group of $\widehat{\mathbb{C}}$ with the points $z = 0, 1, \infty$ deleted from it. Therefore, F may be viewed as the principal congruence subgroup of level two in the modular group. From this viewpoint, E is the fundamental group of the Riemann surface of the function defined by part (b) of Theorem 1. It is therefore an *infinitely generated free subgroup of the modular group*, so we are led to the following

Question: Can we describe the group E arithmetically?

REFERENCES

1. F. Baldassarri and B. Dwork, *On second order linear differential equations with algebraic solutions*, Am. J. Math. **101** (1979, No. 1), 42–76.
2. F. Haimo, M. F. Singer, and M. Tretkoff, *Remarks on analytic continuation*, Bull. London Math. Soc. **12** (1980), 9–12.
3. E. Hille, *Ordinary differential equations in the complex domain*, Wiley-Interscience, New York, 1971.
4. W. Magnus, *Collected Papers*, Springer-Verlag, 1984.
5. W. Massey, *Algebraic Topology, An Introduction*, Gradute Texts in Mathematics, vol. 56, Springer-Verlag, 1977.
6. E. G. C. Poole, *Introduction to the theory of liner differential equations*, reprint, Dover, New York, 1960.
7. C. L. Siegel, *Topics in Complex Function Theory,*; vol. I, (1969); vol. II, (1971); Vol. III, (1973), Wiley-Interscience, New York.
8. C. L. Tretkoff and M. D. Tretkoff, *Combinatorial Group Theory, Riemann Surfaces and Differential Equations*, Contemporary Mathematics **33** (1984), 467–519.

9. M. Tretkoff, *A new type of transcendental function*, Bull. London Math. Soc. **4** (1972), 167–170.

KEVIN BERRY, DEPARTMENT OF MATHEMATICS, STEVENS INSTITUTE OF TECHNOLOGY, HOBOKEN, NJ 07030

MARVIN TRETKOFF, DEPARTMENT OF MATHEMATICS, STEVENS INSTITUTE OF TECHNOLOGY, HOBOKEN, NJ 07030

Current address: Marvin Tretkoff, 43 Brook Drive East, Princeton, NJ 08540.

Contemporary Mathematics
Volume **169**, 1994

Finite-Dimensional Representations of Artin's Braid Group.

J.S. Birman *   D.D. Long †   J.A. Moody

Dedicated to the memory of Wilhelm Magnus

1   Some linear representations.

This paper attempts a brief survey of what is currently known about an aspect of the linear representation theory of Artin's braid group. We shall freely use standard notation from the theory of braid groups, see [16], [1] or other papers in this volume.

A few years ago a survey such as this would have been short. Only one really interesting linear representation was known, the so-called *Burau representation*, see [1]. What "interesting" should mean is open to interpretation, but it seems reasonable to focus on those representations which have infinite image in a non-trivial way. As an example of what we wish to exclude, recall that a theorem of Baumslag implies that braid groups are residually finite and so have many linear representations with finite image. Moreover, any linear representation ρ may be adjusted by defining $\rho^*(\sigma) = \rho(\sigma).t^{\alpha(\sigma)}$ where $\alpha : B_n \to \mathbf{Z}$ is the abelianisation map and t is an indeterminate. This representation has infinite image, but could hardly be regarded as more interesting than the finite image representation. This is not to say that the finite representations of the braid group are uninteresting, (indeed they are probably very interesting), but rather that the investigation of finite representations is not a direction which we wish to pursue here.

The Burau representation admits many definitions, each in its own way giving some insight. The definition which is usually regarded as classical comes via free differential calculus. Fix the generators $x_1, \ldots x_n$ for a free group of rank n, and define a derivation on the group ring $\mathbf{C}[F_n]$ as follows. On the elements of the free group we define the derivation inductively by:

(i) $\frac{\partial(x_j)}{\partial x_i} = \delta_{ij}$

(ii) $\frac{\partial(x_j.w)}{\partial x_i} = \delta_{ij} + x_j.\frac{\partial(w)}{\partial x_i}$ *where* $w \in F_n$.

We extend the derivation to the whole group ring by linearity. Then we may

1991 *Mathematics Subject Classification.* 20F36, 20F29.

*Supported in part by NSF Grant DMS-92-06584

†Supported in part by the A. P. Sloan and NSF Grant DMS-900-1062

define the matrix corresponding to the n-braid σ by:

$$\beta_n(\sigma) = (\alpha \frac{\partial(\sigma(x_j))}{\partial x_i})$$

Here α is the abelianisation map of the previous paragraph extended to the group ring by linearity. This is a special case of a construction due to Magnus, see [1]. One finds that this representation is actually reducible and splits into an irreducible representation of dimension $n-1$ and a one dimensional representation which is trivial. The $n-1$ dimensional representation is what is usually referred to as the *reduced Burau representation*, we shall denote it by β_n^r.

An alternative description of these representations comes from the consideration of covering spaces. If we set D_n to be the n-punctured disc, choosing the basepoint on the boundary, we may form the fundamental group $\pi_1(D_n)$ and identify this in the obvious way with F_n. There is a homomorphism r : $\pi_1(D_n) \to \mathbf{Z}$ defined by sending each of the generators to the generator t of $\mathbf{Z}$. This homomorphism defines a covering $p : \tilde{D}_n \to D_n$. We consider the homology group $H_1(\tilde{D}_n)$ as a module over $\Lambda = \mathbf{Z}[t, t^{-1}]$ where it becomes finitely generated and free of rank $n-1$. As is well known, there is a description of the braid group as orientation preserving homeomorphisms of D_n where two such homeomorphisms are regarded as equivalent if they are isotopic relative to the boundary. Given this description, we see that the group B_n acts as on $\tilde{D}_n$ and hence on the group $H_1(\tilde{D}_n)$. One computes easily that this action is via module homomorphisms and we obtain a representation of the braid group into $GL(n-1, \Lambda)$; this is the reduced Burau representation.

For later use we also observe that one may also obtain the unreduced representation by consideration of relative homology groups. This is done in the following way. Let p_0 be the basepoint in D_n and let $\tilde{p}_0$ denote the full preimage of this point in $\tilde{D}_n$. Then the (unreduced) Burau representation arises as the action of the braid group on $H_1(\tilde{D}_n, \tilde{p}_0)$; this is a Λ-module which is free of rank n. The natural sequence $H_1(\tilde{D}_n) \to H_1(\tilde{D}_n, \tilde{p}_0) \to H_1(\tilde{p}_0)$ gives the splitting of Burau into trivial and unreduced parts. This point of view is clearly reminiscent of the covering space description of the Alexander polynomial of a knot or link. The fact that there is a concrete connection is between the Burau representation and the Alexander polynomial is well established: If α is a braid whose closure in the 3-sphere is the knot $\hat{\alpha}$, then apart from a normalisation factor, the Alexander polynomial of $\hat{\alpha}$ is given by $det(\beta_n^r(\alpha) - Id)$.

A good mathematical idea usually has many different interpretations, and yet another way of looking at the Burau representation is given in [13]. As is well known, the braid group B_n is a subgroup of $Aut(F_n)$. Starting with this fact, let $R = R(F_n, SU(2, \mathbf{C}))$ be the representation variety of F_n, topologized by the compact open topology. Fix once and for all some generating set $x_1, x_2, \ldots, x_n$ for F_n. Using this basis we see that since F_n is free any representation determines an n-tuple of matrices and any n-tuple of matrices determines a representation. Since the group $SU(2, \mathbf{C})$ is homeomorphic to the 3-sphere S^3, we may thus identify the representation space R with with $\mathcal{R} = S^3 \times S^3 \times \ldots \times S^3$. The action of an element of B_n as an automorphism of F_n then induces a diffeomorphism of $\mathcal{R}$, and the natural map $B_n \to Diff(\mathcal{R})$ is an injective group homomorphism. It turns out that there is a circle of fixed points. Parametrizing the circle by $e^{2\pi it}$, one finds that the induced action on the tangent space to $\mathcal{R}$ at a fixed point gives a linear representation of B_n, and since there is a one-parameter family of fixed points one obtains in this way a representation of B_n which contains the Burau representation.

A whole new family of representations was discovered by Jones in [9]. The construction given is much more mysterious and comes from the theory of Hecke algebras. (See [3]). One considers the $\mathbf{C}$-algebra with a 1 which is generated by $g_1, \ldots, g_{n-1}$ and has relators:

(i) $g_i g_{i+1} g_i = g_{i+1} g_i g_{i+1}$ $1 \le i \le n-1$
(ii) $g_i g_j = g_j g_i$ $|i-j| > 1$
(iii) $g_i^2 = (q-1)g_i + q$

Denote this algebra $H_n(q)$; one way to view this algebra is as a deformation of the complex group algebra of the symmetric group Σ_n, which occurs in this setting as $H_n(1)$. We can summarise most of the salient propeties in the following theorem, the first proof of which is essentially due to Tits [3]:

Theorem 1.1 *(i) The algebra $H_n(q)$ has complex dimension $= n!$ for generic q.*
(ii) For q sufficiently close to 1, $H_n(q)$ is semisimple.
(iii) The simple $H_n(q)$ modules are in one to one correspondence with Young diagrams and their decomposition rules and dimensions are the same as for Σ_n.

We may define the *Jones representation* of B_n by mapping $\sigma_i \to g_i$ and then using the left regular representation. The theorem implies that the Jones representation is completely reducible and that the irreducible subrepresentations correspond to the Young diagrams for the representations of the symmetric group, Σ_n. It is convenient to use the terminology of Young diagrams, so we recall this briefly: The ordinary irreducible representations of the symmetric group Σ_n are parametrised by sequences of integers $n_1 \ge n_2 \ge \ldots \ge n_k$ with $\Sigma n_i = n$. Such a sequence is a *Young diagram* and will be annotated $(n_1, \ldots, n_k)$ with the convention that m^b is the sequence consisting of b consecutive appearances of m. Having (arbitrarily) decided which of (1^n) and (n) is the trivial representation and which corresponds to the signature homomorphism, then the diagram determines the representation. The only property we will use is the restriction rule which describes how the representation of Σ_n with Young diagram $\mathcal{Y}$ breaks up when considered as a representation of Σ_{n-1}; here the rule is the most natural one for which one might hope. Consider all possible Young diagrams obtained from $\mathcal{Y}$ by decreasing one of the n_j's by 1; this describes, with multiplicities, the representation of Σ_{n-1}. In principle this gives an inductive description of the representation corresponding to any Young diagram (given the convention above) although it is not very practical and direct methods exist.

The Jones representation is a generalization of the Burau representation in the sense that one of its irreducible summands is reduced Burau; we shall choose things so that it corresponds to Young diagram $(n-1, 1)$. There are two simple ways to pick out summands which generalise the Burau summand. One is to consider the exterior powers. This is classical and is essentially what controls the Alexander module. The other collection of summands is all those of the shape $(n - m, m)$ where $m \leq n/2$. This is the *Temperley-Lieb algebra* and is the source of the original one-variable Jones polynomial, which appears as a normalised weighted trace function on this algebra. The two-variable polynomial is constructed as a certain trace on the whole Jones representation.

In [2] this process is actually reversed, and starting from the Kauffman polynomial of a knot, an algebra $C_n(l, m)$ is constructed which yields another family of braid group representations, this time with two parameters. The algebra $C_n(l, m)$ has dimension $1.3.5\ldots\ldots(2n - 1)$, is semisimple and has quotients isomorphic to $H_n(q)$.

The representations of B_n in $H_n(q)$ and $C_n(l, m)$ are but two special cases of finite-dimensional matrix representations of B_n which support a "Markov trace", and so give rise to polynomial invariants of knots and links. The description of the trace functions and the associated link invariants goes beyond the scope of this review, however it seems appropriate to describe the "method of R-matrices" which constructs them all. Let E be the ring of Laurent polynomials over the integers in a single variable $\sqrt{q}$, let $m \geq 1$ be an integer, and let V be a free E-module of rank m. For each $n \geq 1$ let $V^{\otimes n}$ denote the n-fold tensor product $V \otimes_E \ldots \otimes_E V$. Choose a basis $v_1, \ldots, v_m$ for V, and choose a corresponding basis $\{v_{i_1} \otimes \ldots \otimes v_{i_n}\}; 1 \leq i_1, \ldots, i_n \leq m\}$ for $V^{\otimes n}$. An E-linear isomorphism f of $V^{\otimes n}$ may then be represented by an m^n-dimensional matrix $(f_{i_1 \ldots i_n}^{j_1 \ldots j_n})$ over E, where the i_k's (resp. j_k's) are row (resp. column) indices.

The family of representations of B_n which we wish to describe have a very special form. They are completely determined by the choice of the integer m and an E-linear isomorphism $R : V^{\otimes 2} \to V^{\otimes 2}$ (the so-called *R-matrix*) with matrix $[R_{i_1 i_2}^{j_1 j_1}]$ as above. Let I_V denote the identity map on the vector space V. The representation $\rho(R) : B_n \to GL_{m^n}(E)$ which we seek is defined by

$$\rho(R) : \sigma_i \to I_V \otimes \ldots \otimes I_V \otimes R \otimes I_V \otimes \ldots \otimes I_V$$

where R acts on the ith and $(i+1)^{st}$ copies of V in $V^{\otimes n}$. Thus, if we know how R acts on $V^{\otimes 2}$ we know $\rho(R)$ for every natural number n.

What properties must R satisfy for $\rho(R)$ to be a representation? The first thing to notice is that if $|i - j| \geq 2$, the non-trivial parts of $\rho(\sigma_i)$ and $\rho(\sigma_j)$ will not interfere with one-another, so the first braid relation $\sigma_i \sigma_j = \sigma_j \sigma_i$ if $|i - j| \geq 2$ is satisfied by construction, independently of the choice of R. As for the second braid relation $\sigma_i \sigma_{i+1} \sigma_i = \sigma_{i+1} \sigma_i \sigma_{i+1}$, it is clear that we only need to look at the actions of $R \otimes I_V$ and $I_V \otimes R$ on $V^{\otimes 3}$. If

$$(R \otimes I_V)(I_V \otimes R)(R \otimes I_V) = (I_V \otimes R)(R \otimes I_V)(I_V \otimes R)$$

then the second braid relation, and therefore both braid relations, will be satisfied. This equation is the clue to the construction. It is known as the *quantum Yang-Baxter* (QYB) equation. It may be thought of as a combinatorial restriction on the entries in the matrix R. It turns out that the theory of quantum groups has lead to an effective classification of solutions to the QYB equation, and so to the construction of all possible R-matrix representations of B_n. For more on this subject, and for explicit examples, see [21] and [8].

2   Linearity and Effectiveness.

Given that linear representations exist, a natural question is whether the group B_n is actually a *linear group* that is to say, can it be *faithfully* represented as a group of matrices. The work of several authors, notably McCarthy [18] and Ivanov [7] has shown that many properties which a linear group must have are shared by braid groups. It is also known that if a faithful representation exists, there is an irreducible faithful representation. (See [5] or [12])

For $n = 2, 3$ it is known that the groups B_n are linear, the first case being a triviality and the second case proved by Magnus and Peluso in [17], where it is shown that the Burau representation is faithful for $n = 3$. To date all other cases of this question remain open. The case $n = 4$ is especially intriguing as it was shown in [5] that the linearity of B_4 is equivalent to the linearity of the group $Aut(F_2)$. Even more, Formanek and Processi have proved in ([6] that this is the only possible case when $Aut(F_n)$ could be linear. Structurally this group is simpler than the other braid groups and this means that special reductions are possible in this case. For example, B_4 contains a normal subgroup which is free of rank 2, and so it follows from [12] that a linear representation of B_4 is faithful if and only if the image of this free group is free of rank 2. This is a generalisation of the famous pair of matrices contained in [1] the freeness of which is shown to give a necessary and sufficient condition for the faithfulness of the Burau representation of B_4.

One can be less ambitious and ask: "When is the Burau representation faithful" and until 1990, this remained open despite work by many authors. Almost all cases are now covered by:

Theorem 2.1 ([19] & [15]) *The Burau representation is not faithful for $n \geq 6$.*

The is the combination of two results, the inital breakthrough of [19], where a slightly different point of view of the covering space description of the Burau representation was employed to show that β_k is not faithful for $n \geq 10$, together with a sharpening of this result in [15], which was used to show that β_k is not faithful for $n \geq 6$. The cases $n = 4, 5$ remain unresolved. This is despite the fact that [15] gives a criterion which is necessary and sufficient to determine faithfulness. We briefly describe the method and then recast the idea in a manner so that it can be generalised. Let ξ_j be the arc shown in Figure 1. Then we define a map

$$\int \omega_j : H_1(\tilde{D}_n, \tilde{p}_0) \to \Lambda$$

by requiring that if α represents some homology class in $H_1(\tilde{D}_n, \tilde{p}_0)$, we set

$$\int_\alpha \omega_j = \sum_{k \in Z} t^k.(\alpha, t^k \xi_j)$$

where $(\alpha, t^k \xi_j)$ is the algebraic intersection number of the arcs in $\tilde{D}_n$. With suitable conventions we may extend $\int \omega_j$ to D_n and one can compute the action of the Burau matrix of a braid by consideration of these integrals. Roughly speaking, $\int_\alpha \omega_j$ can be considered as an obstruction to isotoping the arc α off the arc

ξ_j. The results of 2.1 are formulated in a way which gives elements in the kernel of the Burau representation if there are simple arcs for which certain of these obstructions vanish. The disadvantage with this method is that as it stands it is uniquely suited to the geometric description given for the Burau representation and to provide insight into representations such as the Jones representation it is necessary to have some more geometric description of them.

We now describe a construction which generalises the above and produces new representations of the braid group. It uses an idea which occurs both in the work of Magnus and in the work of Lawrence [11]: given a representation of the free group F_n and certain compatibility conditions one may construct a representation of B_n. The idea seems to be general enough to construct all the summands in the Temperley-Lieb algebra. Conjecturally it could construct all linear representations of the groups B_n, but this remains open. It is studied in detail in [14]. In order to describe the idea behind the construction of [14] we recall the notion of homology or cohomology of a space with coefficients in a flat vector bundle. Suppose that X is a manifold and that we are given a representation $\rho : \pi_1(X) \rightarrow GL(V)$. This enables us to define a flat vector bundle E_ρ: Let $\tilde{X}$ be the universal covering of X. The group $\pi_1(X)$ acts on $\tilde{X} \times V$ by $g.(\tilde{x}, \mathbf{v}) = (g.\tilde{x}, \rho(g).\mathbf{v})$. Then E_ρ is the quotient of $\tilde{X} \times V$ by this action. We now form the cohomology groups of 1-forms with coefficients in E_ρ, denoting these by $H^1(X; \rho)$ or $H_c^1(X; \rho)$ for compactly supported cochains. Relative versions also exist, but we shall omit discussion here. In order to get an action of the braid groups, recall that we have a natural inclusion of B_n as a subgroup of $Aut(F_n)$ so that there is a canonical way of forming a split extension $F_n \rtimes B_n$. It turns out that in order to get an action on the twisted cohomology group what is required is exactly a representation of this split extension:

Theorem 2.2 *Given a representation $\rho : F_n \rtimes B_n \rightarrow GL(V)$ we may construct another representation $\rho_s^+ : B_n \rightarrow H_c^1(D_n; \rho)$ where s is another parameter.*

This works in exactly the way one might expect. The representation restricted to the free factor gives rise to the local system on the punctured disc and thus the twisted cohomology group and the compatibility condition provided by the split extension structure gives the braid group action.

Various comments are in order concerning Theorem 2.2. The first is that although the theorem is stated abstractly, there is a concrete recipe which enables one to write down the description of ρ_s^+ given ρ. The second is that at first sight, it might seem that this theorem is of limited usefulness, since it requires a representation of the more complicated group, namely the split extension, but in fact the algebraic structure of the braid group is very well-understood, and so it has been known for some time that the group B_{n+1} contains subgroups isomorphic to $F_n \rtimes B_n$. Thus we deduce:

Theorem 2.3 *Given a representation $\rho : B_{n+1} \rightarrow GL(V)$, we may construct a representation $\rho_s^+ : B_n \rightarrow H_c^1(D_n; \rho)$.*

The theorem shows that given a k parameter representation of the braid group, the construction yields a $k + 1$ parameter representation, apparently in a nontrivial way. For example, if one starts with a (zero parameter) trivial representation of $F_n \rtimes B_n$, the theorem produces the Burau representation. However the role of this extra parameter is not purely to add extra complication - it also adds extra structure. For there is a natural notion of what it should mean for a representation of a braid group to be unitary (See [20], for example) and the results of Deligne-Mostow and Kohno imply:

Theorem 2.4 *In the above notation, if ρ is unitary, then for generic values of s, so is ρ_s^+.*

Moreover, we may iterate this construction and it is possible to identify a certain sequence of local systems with the construction of [11], where it is proved that such a procedure suffices to produce, as composition factors, the simple algebras in the Jones representation corresponding to the Temperley-Lieb algebra. Thus we have:

Theorem 2.5 [11] *Iteration of the augmenting construction, beginning with the trivial representation will eventually yield all summands of the Temperley-Lieb algebra.*

Our approach has the obvious advantage that it is extremely geometric and one can write down criteria for faithfulness or otherwise of the representations so produced in terms of cup and cap products in the twisted cohomology groups; this is the promised generalisation of the cohomology classes $\int \omega_j$. It leads to the notion of an *effective local system*. Roughly, effectiveness amounts to asking whether geometric intersections are detected by the algebra coming from Poincaré duality in the local system. If the pairing is effective this easily implies that the representation is faithful. The converse need not be true, however and in general it seems possible that the local system and braid group representation provided by $\rho : F_n \rtimes B_n \to GL(V)$ could be (respectively) noneffective and nonfaithful, but piece together to give a faithful representation ρ_s^+. However it is shown in [14] that this can happen only finitely often. The exact result requires the following notation.

Suppose that $\rho_n : F_n \rtimes B_n \to GL(V_n)$ is a sequence of representations with the property that $V_1 \subset V_2 \subset \ldots$ and that if ρ_n is restricted to B_{n-1} this is the representation ρ_{n-1}. If all the ρ_n are faithful representations of B_n, there is nothing more to do so we suppose that this is not the case and we set r to be the smallest number for which ρ_r is not faithful when restricted to the braid group factor. We take s to be the smallest number so that the local system coming from ρ_s restricted to the free factor has noneffective intersection pairing. Finally, we define r^+ to be the largest number so that $\rho_{r^+}^+$ is a faithful representation of B_{r^+-1}. The result is:

Theorem 2.6

$$s \le r^+ \le s + 2r - 2$$

For example, consider the sequence of one dimensional representations $\tau_n : F_n \rtimes B_n \to GL(\mathbf{C})$ where τ_n is induced from the representation of B_{n+1} where the generator σ_i is multiplication by the complex number t. In this case, $r = 3$ and although the exact value of s is not known, the results of [19] and [15] imply that $3 < s \leq 6$. The representation augments to Burau, and we deduce from 2.6 that the Burau representation has range of faithfulness given by $s \leq r^+ \leq s + 4$. In this case the information is weaker than what is already known.

In the case when $n \geq 4$ establishing whether a local system is effective or not seems to be a hard problem. Indeed, it seems possible that there are no effective local systems at all. Currently, even knowing this was true does not seem to suffice to show that the Temperley-Lieb representation of the braid group is eventually non-faithful; what this would show would be that for each fixed m, the representations $(n - m, m)$ become nonfaithful as n tends to infinity.

This problem highlights one of the difficulties of dealing "locally" with objects of the nature of the Jones representation which could be essentially global. For as pointed out above, by Theorem 2.2 of [12] one cannot make a faithful representation of B_n by piecing together sums of nonfaithful representations. This means that if one wants to deal with faithfulness questions, a summand by summand examination of say, the Jones representation is possible. However unless the information obtained is very detailed, this may not suffice to establish faithfulness of the whole representation. For there seems to be no particular reason for thinking that the Burau representation is alone in being "stably nonfaithful" amongst the summands of the Jones representation. But as n becomes large, the number of "types" of summands increases very rapidly, so that every time a particular type of summand becomes nonfaithful, some other type present is still sufficiently complicated so as to be faithful. Even to prove results about the Temperley-Lieb algebra, it would be necessary to sharpen Theorem 2.5 to show that eventually one obtains a representation of the braid group which contains all the two row representations simultaneously.

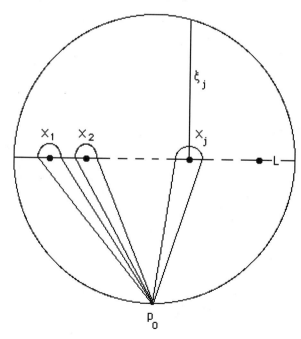

Figure 1

References

[1] J. S. Birman. *Braids, Links and Mapping Class Groups* Annals of Math. Studies, No. 82. Princeton University Press.

[2] J. S. Birman and H. Wenzl. *Braids, link polynomials and a new algebra* Trans. A.M.S. vol. 313, (1989), pp. 249 - 273.

[3] N. Bourbaki. *Groupes et Algèbres de Lie* IV, V VI, Masson, Paris (1982).

[4] P. Deligne and G. D. Mostow. *Monodromy of hypergeometric functions and non-lattice integral monodromy.* Publ. I.H.E.S., No. 63 (1986), pp. 5 - 90.

[5] J. Dyer, E. Formanek and E. Grossman. *On the linearity of the automorphism groups of free groups.* Arch. Math. vol. 38 (1982), pp. 404 - 409.

[6] E. Formanek and C. Procesi. *The automorphism group of a free group is not linear.* Preprint.

[7] N. Ivanov. *Algebraic properties of the Teichmuller modular group.* Soviet Math. Doklady, vol.29, No 2 (1984), pp. 288-291.

[8] M. Jimbo. *Quantum R-matrix for the generalized Toda system.* Comm. Math. Physics, vol.102 (1986), pp. 537-547.

[9] V. F. R. Jones. *Hecke algebra representations f braid groups and link polynomials.* Ann. of Math. vol. 126 (1987), pp. 335 - 388.

[10] T. Kohno. *Linear representations of braid groups and classical Yang-Baxter equations.* Braids : Contemp. Math. vol. 78 (1986).

[11] R. Lawrence. *Homological representations of the Hecke algebra.* Comm. in Math. Phys. vol. 135 (1990), pp. 141 - 192.

[12] D.D. Long. *A note on the normal subgroups of mapping class groups* Math. Proc. Camb. Phil. Soc. vol. 99 (1986), pp. 295 - 303.

[13] D.D. Long, *On the linear representations of the braid group* Trans. AMS Vol. 311, No. 2 (1989), pp. 535-560.

[14] D.D. Long. *Constructing representations of braid groups* preprint, University of California Santa Barbara, 1993.

[15] D. D. Long and M. Paton. *The Burau representation is not faithful for n = 6*. To appear in Topology.

[16] W. Magnus, A. Karrass and D. Solitar. *Combinatorial Group Theory* John Wiley and Sons 1966.

[17] W. Magnus and A. Peluso. *On a theorem of V. I. Arnold*. Comm. in Pure and Applied Math. XXII, pp. 683 - 692.

[18] J. McCarthy *A Tits alternative for subgroups of mapping class groups* Trans. A.M.S. vol. 291 (1985), pp.583-612.

[19] J. Moody. *The faithfulness question for the Burau representation*. To appear in Proc. A.M.S.

[20] C. Squier. *The Burau representation is unitary*. Proc. A.M.S. vol. 90 (1984), pp. 199 - 202.

[21] V. Turaev. *The Yang-Baxter equation and invariants of links*. Invent. Math. vol. 92 (1988), 527-553.

Department of Mathematics
Columbia University
New York, New York 10027

Department of Mathematics
University of California
Santa Barbara, California 93106

Department of Mathematics,
Warwick University,
Coventry CV4 7AL, England

Contemporary Mathematics
Volume **169**, 1994

SQUARING RECTANGLES: THE FINITE
RIEMANN MAPPING THEOREM

J. W. CANNON, W. J. FLOYD AND W. R. PARRY

ABSTRACT. The classical Riemann mapping theorem asserts that any topological quadrilateral in the complex plane can be mapped conformally onto a rectangle. The finite Riemann mapping theorem asserts that any topological quadrilateral tiled by finitely many 2-cells can be mapped with minimal combinatorial distortion onto a rectangle tiled by squares. We prove the finite Riemann mapping theorem, discuss its connections with the classical theories of conformal mapping and electric circuits, and develop algorithms for calculating the finite Riemann mapping.

Table of contents.

0. Introduction.

1. Squared rectangles.

1.1. Scientific American.

1.2. Resistive circuits and the laws of Ohm and Kirchhoff.

1.3. The solution of the resistive circuit problem.

1.4. Planar circuits and rectangles tiled by rectangles.

1.5. Max Dehn's contribution.

2. Optimal weight functions.

2.1. The classical Riemann mapping theorem.

2.2. The finite problem and the existence and uniqueness of its solution.

2.3. The geometry of an optimal weight function: general results.

2.4. The geometry of an optimal weight function: special results for quadrilaterals and rings.

1991 *Mathematics Subject Classification*. Primary 20F32, 30C62, 31A15; Secondary 30F10, 30F40, 57N10.

This research was supported in part by The Geometry Center, University of Minnesota, an STC funded by NSF, DOE, and Minnesota Technology, Inc. The first author was supported in part by NSF Research Grant No. DM-8902071.

2.4.1. Further definitions.

2.4.2. The correspondence.

2.4.3. Level curves and the relationship between skinny flows and fat cuts.

2.4.4. The relationship between fat flows and skinny cuts for tilings.

2.4.5. The relationship between the four moduli.

3. The finite Riemann mapping theorem.

3.1. Examples.

3.2. The Kirchhoff inequalities.

4. Algorithms which calculate the finite Riemann mapping.

4.1. The minimal path algorithm.

4.2. Proofs of the claims about the minimal path algorithm.

4.3. The efficiency of the minimal path algorithm.

4.4. Flow diagrams and a hybrid algorithm for the finite Riemann mapping problem.

4.5. An example of the effectiveness of the hybrid algorithm.

5. Optimal weight functions for 2-layer valence 3 tilings of quadrilaterals.

5.1. Basic definitions and properties.

5.2. Unique prime factorization theorem.

5.3. Prime tilings.

5.4. Preparations for factorization algorithms.

5.5. The left factor algorithm.

5.6. The joint ratio algorithm.

6. Approximating combinatorial moduli.

6.1. Skinny approximation.

6.2. The averaging trick.

6.3. Examples.

7. Variable negative curvature versus constant curvature groups.

0. Introduction

We dedicate this paper to Wilhelm Magnus with admiration and appreciation. This paper has mild ties with Magnus through the work of his thesis advisor, Max Dehn, who proved long ago that, for any rectangle which can be tiled by finitely many squares, the ratio of height to width is rational. This paper is directed in the long term to the recognition of discrete groups of isometries in hyperbolic 3-space, and the subject of discrete groups was one of Magnus' favorites.

Riemann, in formulating his famous Riemann mapping theorem, surely relied on the physics of electrical networks and conducting metal plates for motivation. In turn, Riemann's theorem and its many proofs can be used as starting points for a number of beautiful finite approximations which have combinatorial, geometric, and physical interpretations. These **finite** Riemann mapping theorems are the subject of very intense interest. One of them is the subject of Cannon's paper [8]. Another is of central importance to this paper. The paper is partly expository, especially in the beginning. However, in the later sections, we develop material important to our program of recognizing groups of hyperbolic motions combinatorially.

1. Squared rectangles

This section is purely expository. Our aim is to show a correspondence between rectangles tiled by rectangles and connected planar resistive circuits with one battery.

1.1. Scientific American. We begin our story with a puzzle whose solution is recounted in Chapter 17 of The 2nd Scientific American Book of Mathematical Puzzles and Diversions by Martin Gardner [13]: *Can a square be subdivided into smaller squares of which no two are alike?* Gardner's story teller is William T. Tutte, renowned geometer and combinatorialist at the University of Toronto. Tutte, as a student at Trinity College, Cambridge, during the years 1936-38 pursued an answer to this question with three other students, C. A. B. Smith, A. H. Stone, and R. L. Brooks. "Stone was intrigued by a statement in Dudeney's Canterbury Puzzles [12] which seemed to imply that it is impossible to cut up a **square** into unequal smaller squares." Soon the four friends "were spending much time constructing, and arguing about, dissections of rectangles into squares." They used a mixture of geometric and algebraic methods to create a large catalog of squared rectangles. "In the next stage of the research [they] abandoned experiment in favor of theory." The students "tried to represent squared rectangles by diagrams of different kinds." Smith suggested a diagram which showed that the problem could be reinterpreted as a part of the theory of electrical networks.

We shall now explain Smith's diagram. Consider as an example the rectangle of Figure 1.1.1 which is tiled by smaller rectangles. Each maximal horizontal line segment of the diagram is to be represented by a node of a simple resistive circuit. The top and the bottom of the rectangle represent nodes of the circuit which are to be connected to the terminals of a battery. Each small rectangle is to be represented by an edge of the circuit consisting of a single resistor. This resistor joins the nodes (maximal horizontal line segments) in which the top and bottom of the corresponding rectangle lie. We must assign values to the voltage of the battery and the resistance of the resistors. We interpret height as voltage drop and width as current flow. Hence the height of the large rectangle is assigned as the voltage of the battery. We apply Ohm's law which gives resistance as the quotient of voltage by current to determine the resistance of each resistor as the ratio of height to width in the corresponding small rectangle. We observe that a square in the tiling corresponds to a resistance of 1.

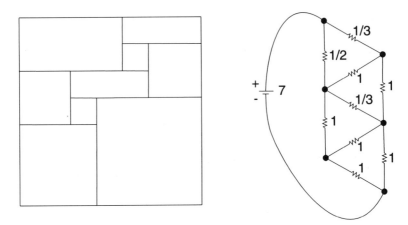

Figure 1.1.1.

One half of Smith's observation is that every rectangle tiled by rectangles gives rise to a circuit in which the sizes of both the large rectangle and its subrectangles are determined by the solution to the corresponding circuit laws. The second half of Smith's observation is that every connected planar resistive circuit with one battery gives rise to a rectangle tiled by rectangles. We shall outline proofs of these observations in subsequent sections.

Our four students used Smith's observation and classical circuit theory to find, after some considerable work, solutions to the problem with which this section began. We shall leave the reader to read their work in the literature. We conclude our discussion of their work by mentioning one related recreational puzzle. A rectangle tiled by rectangles can be cut into its underlying tiles. Many tiled rectangles can be reassembled in only one way into a rectangle. This reassembly constitutes an interesting jigsaw puzzle problem.

We express appreciation to Geoffrey Mess who, from his remarkable memory, was able to refer us to the original paper [4] by these four students in the Duke Mathematical Journal of 1940.

1.2. Resistive circuits and the laws of Ohm and Kirchhoff. Consider a finite directed and connected graph Γ, one edge of which, denoted e, represents a battery of voltage v, and each other edge of which, denoted generically by f, represents a resistor of resistance $r(f)$ and conductance $c(f) = 1/r(f) > 0$. We let $i(f)$ denote the current through the directed edge f and $v(f)$ denote the voltage or electrical potential across f. The flow of electricity in this circuit is governed by laws which we call **circuit laws**, laws attributed specifically to the names of Ohm and Kirchhoff.

Ohm's law. For every edge f in Γ with $f \neq e$,

$$i(f) = c(f)v(f).$$

Note that the current $i(f)$ and voltage $v(f)$ may be positive, negative, or zero, while the conductance $c(f)$, for f not the battery edge, is always positive.

Kirchhoff's voltage law. Let $P = E_1 \cdots E_k$ be a closed edge path in Γ for which the directions of the edges E_j do not necessarily agree with the orientation of P determined by the ordering of the edges $E_1, \ldots, E_k$. Then

$$(1)_P \qquad \pm v(E_1) \pm \cdots \pm v(E_k) = 0,$$

where the sign before $v(E_j)$ in the left side of $(1)_P$ is $+$ if the direction of E_j agrees with the orientation of P and $-$ otherwise. If one of these edges E_j is the directed battery edge e of Γ, then $v(E_j) = v$.

Kirchhoff's current law. If x is a vertex or node of the graph Γ distinct from the two end points of e, if $E_1, \cdots, E_k$ are the directed edges of Γ ending at x, and if $F_1, \cdots, F_l$ are the directed edges of Γ emanating from x, then

$$i(E_1) + \cdots + i(E_k) = i(F_1) + \cdots + i(F_l).$$

Equivalently, by Ohm's law,

$$(2)_x \qquad c(E_1)v(E_1) + \cdots + c(E_k)v(E_k) = c(F_1)v(F_1) + \cdots + c(F_l)v(F_l).$$

1.3. The solution of the resistive circuit problem.

Theorem 1.3.1. *There are unique values of the variables $v(f)$ and $i(f)$ satisfying the circuit laws.*

Proof. This marvelous and well-known result offers a perfect illustration of a fascinating fact of mathematical life described by Henri Poincaré in his wonderful book, The Value of Science [17], pp. 15-17.

> "It is impossible to study the works of the great mathematicians ... without noticing and distinguishing two ... entirely different kinds of minds. The one sort are above all preoccupied with logic .... The other sort are guided by intuition.

> "Look at [the intuition of] Professor Klein: ... He replaces his Riemann surface by a metallic surface whose electrical conductivity varies according to certain laws. He connects two of its points with the two poles of a battery. The current, says he, must pass, and the distribution of this current on the surface will define a function whose singularities will be precisely those called for ...."

Klein's argument applies almost verbatim to our situation and supplies an intuitive proof of Theorem 1.3.1.

But the analytical proof is almost as succinct and fascinating:

By means of Ohm's law we may remove the variables $i(f)$ from consideration and consider only the two sets of Kirchhoff's laws, linear equations in the variables $v(f)$. There are infinitely many closed loops P in any nontrivial circuit so that the equations $(1)_P$ are potentially infinite in number, but there are finite subsystems equivalent to the whole which we might obtain as follows. Pick a maximal tree T in Γ which contains as one of its edges the battery edge e. For each edge f of $\Gamma \backslash T$, consider the unique closed edge path $P = P(f, T)$ in $T \cup \{f\}$ which has f as one of its directed edges and is oriented in the direction of f. Retain the corresponding equation $(1)_P$.

Theorem 1.3.2. *The system $(1) \cup (2)$, with (1) reduced to a basis as indicated in the previous paragraph, is a nonsingular linear system.*

Proof of Theorem 1.3.2. The first step is to note that the reduced system has the same number of equations as unknowns. We leave the proof to the reader: use the Euler characteristic of the connected graph Γ and remember that there is one voltage variable for each nonbattery edge, one vertex equation for each node not attached to the battery, and one loop equation for each edge omitted by a maximal tree.

The second and final step is to prove that the corresponding homogeneous linear system obtained by setting the battery voltage to 0 has a unique solution. Stephen DiPippo suggested completeing the proof by examining the corresponding homogeneous linear system. It follows therefrom that the matrix of the system is nonsingular. Suppose given a solution to the homogeneous system. The Kirchhoff loop equations show that we may assign a well-defined potential to each node of the circuit as follows. Assign the two battery nodes the potential 0. For any other node x, choose an edge path Q from a battery node to x. Sum the values $\pm v(f)$ over the directed edges f of Q using the sign convention of $(1)_P$. Assign the result to x as its potential. The Kirchhoff vertex equations then show that this potential function cannot obtain a maximum or minimum other than 0. It follows that $v(f) = 0$ for every edge f of the circuit. $\Diamond$ Theorems 1.3.1 and 1.3.2 $\Diamond$

We end our discussion of existence and uniqueness for solutions of the circuit equations by noting a result that is not as well-known among geometric topologists as it deserves to be. We have proved that the reduced linear system described above is nonsingular. Therefore the matrix has nonzero determinant. What is that determinant?

Theorem 1.3.3. *The determinant of the reduced linear system of circuit equations is nonzero. It has one nonzero term for each maximal tree in the circuit which contains the battery edge. Each nonzero term has the same sign. The absolute value of the term corresponding to a given maximal tree is the product of the conductances of the nonbattery edges of the tree. In particular, if each of* *the resistances is 1, then the absolute value of the determinant is the number of maximal trees in the circuit containing the battery edge.*

REMARK. The proof of Theorem 1.3.3 is a delightful exercise of which we would not deprive the reader.

1.4. Planar circuits and rectangles tiled by rectangles. We assume the same setting as that of the previous sections except that we assume in addition that the circuit has a planar diagram.

Theorem 1.4.1. *If the circuit Γ is planar, and if the voltages $v(f)$ and currents $i(f)$ are given by the unique solution to the circuit laws, then rectangular tiles of height $|v(f)|$ and width $|i(f)|$ can be assembled so as to tile a rectangle of height $|v|$, where v is the battery voltage, and width $|i|$, where i is the total current exiting or entering a battery node. The ratio $|v/i|$ may be interpreted as the total resistance of the circuit between the battery nodes. The Smith diagram of this tiled rectangle corresponds to the original circuit modified by removing those edges through which no current flows.*

Proof. We must first establish a number of simplifying assumptions, notations, and conventions.

We simplify matters by considering only the case where the voltage v and the conductances $c(f)$ are rational. The general case follows from a simple limiting argument. Alternatively, where we begin later on in the rational case to set out a finite grid, we could in the general case use transverse foliations in place of that grid.

In the rational case, Cramer's rule shows that all of the voltages $v(f)$ and currents $i(f)$ are rational as well. Multiplying by the least common multiple of the denominators, we find that we may assume all of the constants and variables involved are in fact integers.

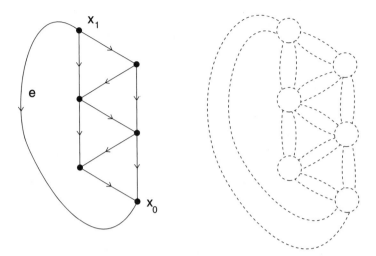

Figure 1.4.2.

We take a regular neighborhood N of the graph Γ in the plane and decompose N into small disks or 0-handles $D(x)$ surrounding the nodes x of the circuit and small strips or 1-handles $H(f)$ covering the uncovered portion of the edges f and joining the disks corresponding to the end points of f. See Figure 1.4.2.

We assign to one end point x_0 of the battery edge e the potential 0 and to the other end point x_1 the potential v. We may assume that $v < 0$ so that current flows from the vertex x_1 through the circuit to the vertex x_0. We proceed then to assign to each node x a potential $p(x)$ as in the proof of the existence and uniqueness theorem above. Actually it is customary to replace this potential function $p(x)$ by its negative $q(x) = -p(x)$ so that current flows along all nonbattery edges from vertices of higher potential to those of lower potential. We will follow this custom. The associated formula is the following. We assume that every nonbattery edge f is directed from its initial end point x of higher potential to its terminal end point y of lower potential. Then

$$v(f) = q(x) - q(y) = p(y) - p(x)$$

and

$$i(f) = c(f)v(f).$$

We obtain particular unity in our picture if we think of a current i flowing through the battery edge **from vertex x_0 to vertex x_1 from lower potential to higher potential.** Only along the battery edge does current flow from lower to higher potential. **The effect of this convention is that the Kirchhoff**

current node law is then satisfied at every node of the network and not just at the nonbattery nodes.

With all of these preliminary matters out of the way, we are ready to prove the theorem. The idea is to impose a finite "rectangular" grid on the neighborhood N in the following way.

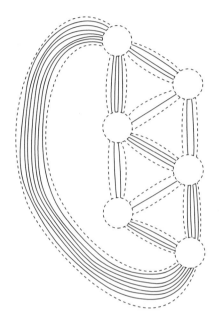

Figure 1.4.3.

Suppose that f is an edge and $H(f)$ the corresponding 1-handle. Construct $i(f)$ parallel directed segments through the 1-handle $H(f)$ joining one of the end 0-handles to the other. See Figure 1.4.3. The direction should coincide with the direction of the current. Call these segments **current curves**. Kirchhoff's current node law implies that, in each 0-handle, the end points of the current curves can be matched in pairs, incoming curves to outgoing curves. A simple induction shows that this matching can be realized in such a way that, if matched curves are joined across the 0-handle by directed arcs, the result is a family of disjoint oriented simple closed **current curves** in the plane, each curve actually lying in the neighborhood N. See Figure 1.4.4.

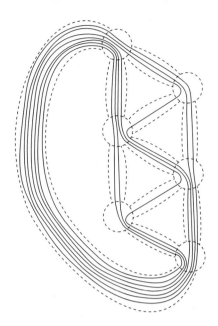

Figure 1.4.4.

We construct a second family of curves called **voltage** or **equipotential** curves as follows. Let f be as above. Assume f oriented from higher potential to lower potential. If f happens to be the battery edge, and in no other case, this orientation will oppose the current flow along the edge f. Construct $v(f)$ parallel segments across the 1-handle $H(f)$ "perpendicular" to the directed edge f, directed in such a way that any ordered pair of (intersecting) segments, (segment, f) forms a local right-handed coordinate system in the plane. Kirchhoff's voltage loop law implies that around the boundary curve of any complementary domain of the neighborhood N the end points of the voltage curves can be matched in pairs, incoming curves to outgoing curves. As before, this matching can be realized in such a way that, if matched curves are joined across the complementary domains by disjoint directed arcs, the result is a family of disjoint oriented simple closed **voltage curves** in the plane, each curve crossing N only in the 1-handles of N. See Figure 1.4.5.

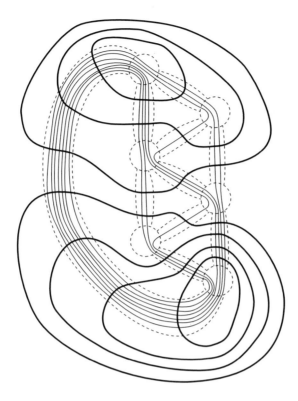

Figure 1.4.5.

It remains to analyze our two families of curves and their intersections. The key observations are these: (1) if we calculate the intersection number $sc(J, K)$ of an oriented voltage curve J with an oriented current curve K, then each intersection outside the battery handle contributes 1 to the intersection number and each intersection inside the battery handle contributes -1; (2) $sc(J, K) = 0$.

CLAIM 1. Each voltage curve and each current curve crosses the battery handle exactly once. For if one missed the battery handle, then all of its intersections with curves of the opposite kind would have the same orientation and the intersection numbers could not be 0. Likewise, it cannot cross the battery handle more than once without having a nonzero intersection number with an arc different from ±1.

CLAIM 2. Each voltage curve hits each current curve exactly once outside of the battery handle. Indeed, all of the intersections outside the boundary handle have the same sign and, taken together, must exactly cancel algebraically the one intersection inside the battery handle.

The truth of the theorem now becomes apparent. The voltage curves are naturally ordered by the way they cross the voltage handle. The current curves are naturally ordered by the way they cross the voltage handle. Therefore the

two families define rectilinear, orthogonal, integer coordinates which naturally turn the entire circuit into a geometric rectangle while turning each 1-handle into a subrectangle of size $i(f) \times v(f)$. The curves viewed in their entirety form a grid like graph paper which gives global integral coordinates.
◊ Theorem 1.4.1 ◊

1.5. Max Dehn's contribution.

Theorem 1.5.1 (Dehn). *Let R denote a rectangle that can be tiled by squares. Then, if H is the height of the rectangle and W its width, the ratio H/W is rational.*

Proof. Scale R so that $H = 1$. Let Γ denote the Smith diagram associated with a tiling of R by squares. Then the voltage of the battery and all of the conductances are 1. Consequently all of the constants in the reduced circuit equations are integers. Since the voltages of the solution are given by Cramer's rule, they are all rational. Since the currents are expressed in terms of the voltages, the conductances, and Ohm's law, they too are rational. The width W of the rectangle is the sum of the currents entering or exiting a battery terminal; hence, it too is rational.
◊ Dehn's Theorem ◊

2. Optimal weight functions

We assume now that we are given a combinatorial tiling of a topological quadrilateral by topological disks. We consider the problem of introducing a circuit whose solution would turn this tiling into a tiling of a rectangle by squares. We would consider the solution to such a circuit as giving a finite Riemann mapping defined on the tiled topological quadrilateral. There are problems in this task. First, a solution is incompatible with the given topology since, in a tiling of a rectangle by rectangles, at most four tiles can come together at a point while, in a combinatorial tiling, any number of tiles can come together at a point. Second, the only natural graph associated with the combinatorial tiling is the dual graph; and in that graph the tiles correspond to nodes and not to edges. Third, while we could mimic the idea of the Smith diagram so as to put the tiles on the edges of the diagram, in a combinatorial tiling there is no well-defined notion of "maximal horizontal edge" in the diagram so that any number of circuits might conceivably correspond to our combinatorial tiling. In view of these difficulties, it is amazing that this "impossible" problem essentially has a solution and that the solution is unique up to scaling. Historically, the variational formula giving its solution was recognized before it was understood that the formula had the geometric content necessary to solve the problem. Cannon [8] had been exploring various finite approximations to the classical Riemann mapping theorem with the aim of extracting from combinatorial tilings and their subdivisions inherent analytic information. Parry discovered that the formula Cannon chose as a working tool had the geometric properties required by the problem. Subsequently and independently, Schramm [18] and Robertson [unpublished], who were aware of Cannon's formula, discovered the same geometric interpretation.

2.1. The classical Riemann mapping theorem.

We use the classical Riemann mapping theorem as a motivation for Cannon's formula. One version of the classical theorem is the following.

Theorem (Riemann). *Suppose given a quadrilateral, that is, a topological disk in the complex plane with four distinguished points on its boundary. Then there is a conformal mapping taking the interior of the quadrilateral onto the interior of a rectangle in such a way that the induced boundary map takes the four distinguished boundary points to the four corners of the rectangle.*

We have always been amazed, in view of the topological difficulties involved, that this theorem is true, let alone that Riemann could have conjectured and proved it. What may have happened is this. Riemann, like Klein in the passage quoted from Poincaré, may have considered the quadrilateral as a metallic conducting plate with battery terminals connected to its "top" and "bottom." "The current must pass," as Klein is supposed to have said. The current flow lines, connecting top to bottom, would have filled the quadrilateral from side to side, one line through each point of the quadrilateral. Equipotential lines, connecting side to side, would likewise have filled the quadrilateral from top to bottom. The pair of families would meet one another orthogonally and give rectilinear flat coordinates for the quadrilateral.

There are three obvious parameters associated with Riemann's theorem, namely the voltage or height H of the image rectangle, the current or width W of the image rectangle, and the total resistance H/W. It is a remarkable fact that this resistance can be realized as a conformal invariant of either the quadrilateral or its conformal equivalent, the image rectangle. In the terminology of complex variables, this total resistance is called the analytic conformal modulus of the quadrilateral. Note that, if $A = HW$ denotes the area of the image rectangle, then

$$H/W = \frac{H^2}{A} = \frac{A}{W^2}.$$

There is a wonderful trick for creating conformal invariants. (See Ahlfors [1].) For a fixed Riemannian surface, one simply assigns a number to each metric conformally equivalent to the given one and then takes either the supremum or the infimum of those numbers over all of the metrics.

The resistance or modulus H/W is precisely such an invariant. (See Lehto and Virtanen [16], Chapter 1.) It may be realized as follows. Begin with the standard Euclidean metric $|dz|$ on the quadrilateral with its accompanying area form $dA = dx \cdot dy$. With each positive metric multiplier $\rho = \rho(z) > 0$ associate the conformally equivalent metric $\rho|dz|$ with its associated area form $\rho^2 dA$. The new metric and area form define a new area $A(\rho)$, a new height $H(\rho)$, and a new width $W(\rho)$ for the quadrilateral which give respectively the area, the minimal distance between the ends, and the minimal distance between the sides of the quadrilateral with respect to the new Riemannian metric $\rho \cdot |dz|$ and the new area form $\rho^2 dx \cdot dy$. Then we have

$$H/W = \sup_\rho \frac{H(\rho)^2}{A(\rho)} = \inf_\rho \frac{A(\rho)}{W(\rho)^2}.$$

Furthermore, both the supremum and infimum are realized by that positive multiplier function ρ which turns the quadrilateral into a rectangle. That is, the optimal function ρ is the absolute value of the derivative of the Riemann mapping. This circumstance plays a central role in the finite theorem where the lack of local coordinates makes the definition of derivative, let alone its use, difficult.

2.2. The finite problem and the existence and uniqueness of its solution.
We are dealing with a combinatorial situation in which no Riemannian structure is given. We wish to use a finite version of the Riemann mapping theorem as a step toward imposing a Riemannian structure. We attempt to copy the classical procedure presented above. As setting, we have a quadrilateral combinatorially tiled by topological disks. We use the tiling to define an approximate metric and approximate area for subsets of the quadrilateral. We simply define both the length and area of a subset to be the number of tiles that intersect it. That is, we assume that each tile has length and area equal to 1 so that it behaves as if it were a unit square. It is then analogous to the classical case if we make a "conformal" change of approximate metric by changing the length of the tile to ρ and the area of the tile to ρ^2. The number ρ may be an arbitrary nonnegative function of the tiles not identically equal to 0. The ρ-length and ρ-area of a subset are then simply the sums of the tile lengths and areas for tiles intersecting that subset. We obtain thereby heights, widths, and areas $H(\rho)$, $W(\rho)$, and $A(\rho)$ by the same formulas used in the classical case. Varying ρ over all possibilities, we obtain two approximate conformal moduli,

$$M_{\sup} = \sup_\rho \frac{H(\rho)^2}{A(\rho)} \quad \text{and} \quad m_{\inf} = \inf_\rho \frac{A(\rho)}{W(\rho)^2}.$$

In contrast to the classical case, it is in general not true that these two moduli are the same. See line 2.4.5.2 (where $m_f = m_{inf}$ and $M_f = M_{sup}$) and Example 6.2.2.

A metric multiplier ρ is said to be an **optimal weight function** if it realizes the supremum in the definition of $M_{\sup}$.

Theorem 2.2.1. *There is a unique optimal weight function ρ such that $A(\rho) = 1$. All other optimal weight functions are scalar multiples of this one.*

REMARK AND DEFINITIONS. In developing the properties of optimal weight functions, we will be well-served by a general setting in linear algebra. We shall prove a number of things in more generality than is required by the study of tiled quadrilaterals. This will lead to some slight changes in notation and terminology.

We therefore fix a positive integer n, which we view as the number of tiles in our tiling. We let P denote a nonempty finite subset of $\mathbf{N}^n \setminus \{0\}$, where $\mathbf{N}$ is the set of nonnegative integers. Elements of P will be called **path vectors**. A **weight vector** is a vector $w = (w_1, \ldots, w_n) \in \mathbf{R}^n \setminus \{0\}$ with $w_i \geq 0$ for $i = 1$, $\ldots$, n. Let $\langle, \rangle$ be the standard inner product on $\mathbf{R}^n$. The **length** of a path vector p relative to the weight vector w is by definition $\langle p, w \rangle$. Given a weight vector w, define **height** H_w relative to w to be the minimum length of a path vector in P relative to w. A path vector p such that $\langle p, w \rangle = H_w$, namely, a path vector whose length relative to w is minimal will be called a **w-minimal path vector**. Define **area** A_w by $A_w = \langle w, w \rangle$. Define **modulus** M by

$$M = \sup_w \frac{H_w^2}{A_w},$$

where the supremum varies over all weight vectors. Define an **optimal weight vector** to be a weight vector w with $M = H_w^2 / A_w$.

We are now ready for the proof of the existence and uniqueness of optimal weight functions. It is clear that to prove Theorem 2.2.1 it suffices to prove Theorem 2.2.2.

Theorem 2.2.2. *There is a unique optimal weight vector w with $A_w = 1$. All other optimal weight vectors are scalar multiples of this one.*

Proof. Scaling w does not change the value of H_w^2 / A_w, and so in defining M the weight vector w may be restricted to the $(n-1)$-sphere S^{n-1}. Moreover, the function which maps a weight vector w in S^{n-1} to H_w^2 is the minimum of a finite number of continuous functions, and so it is also continuous. Thus it is easy to see that M is a positive real number attained for some weight vectors. This argument proves existence of optimal weight vectors.

The proof of uniqueness is a convexity argument. Let $w_0 \neq w_1$ be weight vectors in S^{n-1} with $H_{w_0} \leq H_{w_1}$. Let $w = (1-t)w_0 + tw_1$ for some real number t with $0 < t < 1$. If p is a path vector, then

$$\langle p, w \rangle = (1-t)\langle p, w_0 \rangle + t\langle p, w_1 \rangle \geq (1-t)H_{w_0} + tH_{w_1} \geq H_{w_0}.$$

Since $0 < \|w\| < 1$, $(1/\|w\|)w \in S^{n-1}$ and $\langle p, (1/\|w\|)w \rangle > H_{w_0}$. This argument easily establishes uniqueness.
$\Diamond$ Theorems 2.2.1 and 2.2.2 $\Diamond$

Denote by w_M the unique optimal weight vector in S^{n-1}.

2.3. The geometry of an optimal weight function: general results.

We next present a number of general results in the above vector space setting. Each result will appear in a numbered line followed by an explanation or proof.

2.3.1. Optimal weight vectors are in some sense generalized spherical circumcenters.

Following is an explanation of this statement. Let $x_1, \ldots, x_k$ be points in S^{n-1}. Set

$$r = \inf_{y \in S^{n-1}} \max_i \|x_i - y\|.$$

Then there exists at least one point x in S^{n-1} such that $\|x_i - x\| \leq r$ for $i = 1, \ldots, k$. Such a point x is called a spherical circumcenter of the set $\{x_1, \ldots, x_k\}$. Since

$$\|x_i - y\|^2 = \langle x_i - y, x_i - y \rangle = \langle x_i, x_i \rangle - 2\langle x_i, y \rangle + \langle y, y \rangle = 2 - 2\langle x_i, y \rangle,$$

maximizing $\|x_i - y\|$ is the same as minimizing $\langle x_i, y \rangle$. Thus

$$1 - \frac{1}{2}r^2 = \sup_{y \in S^{n-1}} \min_i \langle x_i, y \rangle,$$

and a spherical circumcenter of the set $\{x_1, \ldots, x_k\}$ is a point x in S^{n-1} such that $\langle x_i, x \rangle \geq 1 - \frac{1}{2}r^2$ for $i = 1, \ldots, k$. Now suppose that all of the path vectors $p_1, \ldots, p_k$ in P have the same length. In this special case the optimal weight vector in S^{n-1} is the spherical circumcenter of the set $\{x_1, \ldots, x_k\}$, where $x_i = \frac{1}{\|p_i\|} p_i$ for $i = 1, \ldots, k$. Thus in general it might be said that optimal weight vectors are generalized spherical circumcenters.

2.3.2. The optimal weight vector w_M lies in the cone in $\mathbf{R}^n$ spanned by the w_M-minimal path vectors.

Proof. Suppose line 2.3.2 is false. Then there exists a hyperplane V in $\mathbf{R}^n$ which separates w_M from the cone spanned by the w_M-minimal path vectors. Let v be the projection of w_M onto V, and let $u = w_M - v$, so that $u + v = w_M$ and $\langle u, v \rangle = 0$. Because w_M and the w_M-minimal path vectors lie on opposite sides of V, $\langle u, p \rangle < 0$ for every w_M-minimal path vector p. Thus $v - u \in S^{n-1}$ and $\langle v - u, p \rangle > \sqrt{M}$ for every w_M-minimal path vector p. By choosing u to be sufficiently small, equivalently, by choosing V to be sufficiently close to w_M, it follows that $\langle v - u, p \rangle > \sqrt{M}$ for every path vector p. This is a contradiction.
◇ 2.3.2 ◇

2.3.3. Suppose $w_M = (w_1, \ldots, w_n)$. Then for every $i = 1, \ldots, n$ either $w_i = 0$ or there exists a w_M-minimal path vector $p = (p_1, \ldots, p_n)$ such that $p_i \neq 0$.

Proof. This follows immediately from line 2.3.2.
◇ 2.3.3 ◇

REMARK. Line 2.3.3 corresponds to Proposition 4.1.2 in Cannon [8]. It is interesting to compare their proofs.

2.3.4. Let w be a weight vector, and let $p_1, \ldots, p_k$ be path vectors whose lengths relative to w are minimal, namely, $\langle p_i, w \rangle = H_w$ for $i = 1, \ldots, k$. If there exist real numbers $a_1, \ldots, a_k$ such that

$$w = \sum_{i=1}^{k} a_i p_i,$$

then

$$A_w = H_w \sum_{i=1}^{k} a_i.$$

Proof. This can be seen from the following.

$$A_w = \langle w, w \rangle = \left\langle \sum_{i=1}^{k} a_i p_i, w \right\rangle = \sum_{i=1}^{k} a_i \langle p_i, w \rangle = H_w \sum_{i=1}^{k} a_i.$$

$\diamond$ 2.3.4 $\diamond$

2.3.5. Suppose that $w_M = \sum_{i=1}^{k} a_i p_i$, where $a_i, \ldots, a_k$ are real numbers and $p_1, \ldots, p_k$ are path vectors whose lengths relative to w_M are minimal. Then $\sum_{i=1}^{k} a_i = \frac{1}{\sqrt{M}}$.

Proof. This is a special case of line 2.3.4.
$\diamond$ 2.3.5 $\diamond$

Line 2.3.2 states that w_M lies in the cone spanned by its minimal path vectors. The next result shows that this property characterizes w_M.

2.3.6. Let w be a weight vector in S^{n-1} such that w lies in the cone spanned by the path vectors whose lengths relative to w are minimal. Then w is the optimal weight vector w_M.

Proof. Let $p_1, \ldots, p_k$ be the path vectors whose lengths relative to w are minimal, and let $a_1, \ldots, a_k$ be nonnegative real numbers such that $w = \sum_{i=1}^{k} a_i p_i$. Then

$$\langle w, w_M \rangle = \left\langle \sum_{i=1}^{k} a_i p_i, w_M \right\rangle = \sum_{i=1}^{k} a_i \langle p_i, w_M \rangle \geq \sum_{i=1}^{k} a_i \sqrt{M} = \frac{\sqrt{M}}{H_w},$$

the last equation coming from line 2.3.4. Since $\sqrt{M} \geq H_w$, $\frac{\sqrt{M}}{H_w} \geq 1$. Thus w and w_M are vectors in S^{n-1} whose inner product is at least 1, and so $w = w_M$.
$\diamond$ 2.3.6 $\diamond$

The next result is well known, but we include its proof for completeness.

2.3.7. Let Q be a finite set of vectors in $\mathbf{R}^n$, and for every $p \in Q$ let b_p be a nonnegative real number so that the vector $x = \sum_{p \in Q} b_p p$ is nonzero. Then there exists a linearly independent subset S of Q and for every $p \in S$ a nonnegative real number a_p such that $x = \sum_{p \in S} a_p p$.

Proof. The proof of line 2.3.7 will proceed by induction on the order $|Q|$ of Q. Because $x \neq 0$, $|Q| \neq 0$ and line 2.3.7 is clear if $|Q| = 1$. Suppose that $|Q| > 1$ and that Q is linearly dependent. Then $\sum_{p \in Q} c_p p = 0$ for some real numbers c_p which are not all 0. It may be assumed that $c_p < 0$ for some p. Then

$$x = \sum_{p \in Q} (b_p + t c_p) p$$

for every nonnegative real number t. Because $c_p < 0$ for some p, there exists a nonnegative value of t for which one of the coefficients $b_p + t c_p$ is 0. Let t_0 be the smallest value of t for which one of these coefficients is 0. Then

$$x = \sum_{p \in Q} (b_p + t_0 c_p) p,$$

which expresses x in the desired form with fewer than $|Q|$ nonzero coefficients.
$\Diamond$ 2.3.7 $\Diamond$

The next result sharpens line 2.3.2.

2.3.8. A scalar multiple of the optimal weight vector w_M is a linear combination of w_M-minimal path vectors in which the coefficients are nonnegative integers. In particular, some optimal weight vector is a vector of integers.

Proof. To prove line 2.3.8, apply line 2.3.2: w_M is a nonnegative linear combination of path vectors $p_1, \ldots, p_k$ with $\langle p_i, w_M \rangle = \sqrt{M}$ for $i = 1, \ldots, k$. By line 2.3.7 it may furthermore be assumed that $p_1, \ldots, p_k$ are linearly independent. Then $\frac{1}{\sqrt{M}} w_M$ is the unique vector v in the subspace generated by $p_1, \ldots, p_k$ with $\langle p_i, v \rangle = 1$. In terms of matrices, this means that the invertible matrix, whose entries are the integers $\langle p_i, p_j \rangle$, multiplies the column vector v to the column vector whose entries are all 1. It follows that $\frac{1}{\sqrt{M}} w_M$ is a nonnegative rational linear combination of $p_1, \ldots, p_k$. This easily proves line 2.3.8.
$\Diamond$ 2.3.8 $\Diamond$

We define a **fundamental family of paths** to be a k-tuple $(p_1, \ldots, p_k)$ of w_M-minimal path vectors $p_1, \ldots, p_k$ such that w_M is a scalar multiple of $\sum_{i=1}^k p_i$. Line 2.3.8 shows that a fundamental family of paths always exists.

We define the **reduced integral optimal weight vector** to be the optimal weight vector given by line 2.3.8 whose coordinates are integers with greatest common divisor 1.

2.4. The geometry of an optimal weight function: special results for quadrilaterals and rings. The results of Section 2.3, though introduced in the setting of tiled quadrilaterals, are completely valid in an abstract vector space setting. The results of this section make use of the plane geometry of quadrilaterals and rings. Furthermore, in developing a duality important for the study of the moduli of quadrilaterals and rings, we need to introduce an additional pair of moduli. Thus our treatment requires a substantial number of new definitions.

2.4.1. Further definitions. We define a **quadrilateral** to be a closed topological disk in the plane $\mathbf{C}$ with four distinguished points on its boundary, and we define a **ring** to be a closed topological annulus in the plane $\mathbf{C}$. Unless stated otherwise, X will denote a quadrilateral or ring in Section 2.4. A **shingling** of X is a finite set of compact connected subsets of X, called **shingles**, which cover X. Let S be a shingling of X. A **path** is a nonempty subset of S whose union is connected. A **fat path** is the set of all of the shingles in S which meet a given topological path in X. A **skinny path** is a nonempty subset of S consisting of distinct shingles which can be ordered as s_1, ..., s_m such that $s_{i-1} \cap s_i \neq \emptyset$ for $i = 2$, ..., m. A fat path is **closed** if it has an underlying topological path which is closed, and a skinny path as above is **closed** if $s_1 \cap s_m \neq \emptyset$. It is easy to see that fat paths and skinny paths are paths.

If X is a quadrilateral, then one of its four distinguished boundary segments X_1 is called the **top** of X, and if X is a ring, then one of its boundary components X_1 is called the **top** of X. The opposite boundary segment or component X_0 is called the **bottom** of X. A **flow** is a path which meets both X_1 and X_0. A **cut** is a path which separates X_1 from X_0 (there does not exist a topological path from X_1 to X_0 which misses all of the shingles in the given path). For brevity it will be said that a cut **separates the ends** of X. A **fat flow** is a fat path which has an underlying topological path joining the ends of X. A **fat cut** is a fat path which has an underlying topological path separating the ends of X. A **skinny flow** is a skinny path whose shingles can be ordered so that one of its extreme shingles meets X_1 and the other extreme shingle meets X_0. If X is a quadrilateral, then a **skinny cut** is analogous to a skinny flow. If X is a ring, then a **skinny cut** is a closed skinny path which separates the ends of X.

Before we continue with our definitions, it is useful to note the following topological fact which we leave to the reader as an exercise.

2.4.1.1. Every flow contains a subflow which is a skinny flow. Every cut contains a subcut which is a skinny cut.

A **weight function** on S is a function which assigns a nonnegative real number to every shingle in S such that some shingle in S has positive weight. Let w be a weight function on S. The **length** of a path relative to w is the sum of the w-weights of the shingles in that path. A **minimal skinny flow** relative to w is a skinny flow of minimal w-length which does not contain a proper skinny subflow. A **minimal skinny cut** relative to w is a skinny cut of minimal w-length which does not contain a proper skinny subcut. Line 2.4.1.1 implies that minimizing skinny flows and cuts is the same as minimizing flows and cuts. The **skinny height** of X relative to w is the length $H_{w,s}$ of a minimal skinny flow relative to w. The **skinny circumference** of X relative to w is the length $C_{w,s}$ of a minimal skinny cut relative to w. The **area** of X relative to w is the sum A_w of the squares of all w-weights. Define the **skinny cut modulus** and **skinny flow modulus** of X relative to S by

$$m_s = \inf_w \frac{A_w}{C_{w,s}^2}, \quad \text{and} \quad M_s = \sup_w \frac{H_{w,s}^2}{A_w},$$

where the infimum and supremum vary over all weight functions of the fixed shingling S. We have analogous **minimal fat flows** whose lengths are $H_{w,f}$ and **minimal fat cuts** whose lengths are $C_{w,f}$. Define the **fat cut modulus** and **fat flow modulus** of X relative to S by

$$m_f = \inf_w \frac{A_w}{C_{w,f}^2}, \quad \text{and} \quad M_f = \sup_w \frac{H_{w,f}^2}{A_w}.$$

2.4.2. The correspondence. We now point out the natural correspondence between the vector space setting introduced in Section 2.2 and the setting of this section. Suppose that S consists of distinct shingles $s_1, \ldots, s_n$. Then every path p corresponds to a path vector $\bar{p}$ in $\mathbf{N}^n \setminus \{0\}$ as follows. If the shingle s_i belongs to p, then the i-th component of $\bar{p}$ is 1. Otherwise the i-th component of $\bar{p}$ is 0. Likewise, every weight function w corresponds to a weight vector $\bar{w}$ so that the i-th component of $\bar{w}$ is $w(s_i)$. The problem of optimizing skinny flows in the present setting corresponds to the optimization problem in the vector space setting in which the set of path vectors P consists of the path vectors of skinny flows. The situation is the same for fat flows. The correspondence also holds for cuts although the modulus in the vector space setting is the reciprocal of the present modulus. The results of the vector space setting apply to each of the four cases under consideration. We thus have four types of **optimal weight functions**. To simplify notation we will identify p with $\bar{p}$ and w with $\bar{w}$.

2.4.3. Level curves and the relationship between skinny flows and fat cuts.

Consider the optimization problem for skinny flows. According to Theorem 2.2.2 there exists a unique optimal weight function w_{M_s} such that the area $A_{w_{M_s}}$ of X relative to w_{M_s} is 1 and

$$M_s = \frac{H_{w_{M_s},s}^2}{A_{w_{M_s}}} = H_{w_{M_s},s}^2.$$

Just as immediately after line 2.3.8, there is a **reduced integral optimal weight function** w, which is an optimal weight function whose values are integers with greatest common divisor 1. From here until line 2.4.3.4 the weight function w will be fixed. Let H denote the skinny height of X relative to w.

Let h be an integer with $1 \leq h \leq H$, and define a **level curve** L_h, which is a set of shingles of S, as follows. First, for a shingle s to belong to L_h, it is necessary that $w(s) \neq 0$. Now, given a shingle s in S with $w(s) \neq 0$, line 2.3.3 implies that there exists a minimal skinny flow f containing s. Suppose that $f = \{s_1, \ldots, s_m\}$, where s_1 meets X_1, s_m meets X_0 and $s_{i-1} \cap s_i \neq \emptyset$ for $i = 2$, $\ldots$, m. Also suppose that $s = s_i$. Then s belongs to L_h if the length of the path $\{s_1, \ldots, s_{i-1}\}$ is less than h and the length of the path $\{s_1, \ldots, s_i\}$ is at least h. To see that this is well-defined, suppose that $g = \{t_1, \ldots, t_k\}$ is also a minimal skinny flow from X_1 to X_0 with $s = t_j$ for some j. The minimality of f and g implies that the length of $\{s_1, \ldots, s_{i-1}\}$ equals the length of $\{t_1, \ldots, t_{j-1}\}$, for otherwise either the skinny flow $\{s_1, \ldots, s_{i-1}, t_j, \ldots, t_k\}$ or the skinny flow $\{t_1, \ldots, t_{j-1}, s_i, \ldots, s_m\}$ is shorter than f and g. Thus L_h is well-defined. Line 2.4.3.1 is clear from the definition.

2.4.3.1. Every shingle s in S belongs to exactly $w(s)$ consecutive level curves and every minimal skinny flow contains exactly one shingle in every level curve.

2.4.3.2. Every level curve is a fat cut. Moreover, there exist smooth arcs or simple closed curves $\alpha_1, \ldots, \alpha_H$ (depending on whether X is a quadrilateral or a ring) in $X \setminus (X_0 \cup X_1)$ which separate the ends of X such that α_h is contained in the connected component of $X \setminus \alpha_{h-1}$ which contains X_0 for $h = 2, \ldots, H$ and the set of shingles which meet α_h is L_h for $h = 1, \ldots, H$.

Proof. The curves $\alpha_1, \ldots, \alpha_H$ will be defined inductively. To aid this definition, another curve α_0 will be defined to begin the induction. Simply let α_0 be the top of X. No claims such as those in line 2.4.3.2 are being made about α_0.

Having defined α_0, suppose that curves $\alpha_1, \ldots, \alpha_{h-1}$ as in line 2.4.3.2 are defined for some h from 1 to H, and define α_h as follows. Let the subset K_1 of X be the union of i) α_{h-1}, ii) the connected component of $X \setminus \alpha_{h-1}$ which contains X_1 (the empty set for $h = 1$) and iii) every shingle s in S for which there exists a skinny path $\{s_1, \ldots, s_m\}$ of length less than h such that s_1 meets X_1 and $s_m = s$. Let the subset K_0 of X be the union of i) X_0 and ii) every shingle s in S such that $s \cap K_1 = \emptyset$. Then K_1 and K_0 are compact subsets of X, K_1 contains X_1, K_0 contains X_0 and it is easy to see that $K_1 \cap K_0 = \emptyset$. Now a standard theorem of plane topology implies that there exists a smooth arc or simple closed curve α_h in X which misses $K_1 \cup K_0$ and separates the ends of X.

In this paragraph it will be shown that the set of shingles of S which meet α_h is L_h. First of all it is easy to see that $K_1 \cup K_0$ contains every shingle with weight 0, so α_h misses every shingle with weight 0. Now suppose that s is a shingle in S with $w(s) \neq 0$. If the level curves to which s belongs have indices less than h, then there exists a skinny path $\{s_1, \ldots, s_m\}$ of length less than h such that s_1 meets X_1 and $s_m = s$. Thus $s \subset K_1$, and α_h misses s. Now suppose that the level curves to which s belongs have indices greater than h. Let f be a minimal skinny flow containing s. If $h > 1$, then by induction α_{h-1} meets exactly one shingle t in f and $\{t\} = f \cap L_{h-1}$. If $h = 1$, then it is easy to see that α_0 meets exactly one shingle t in f and t is the first shingle in f. Hence s

occurs after t in f, and s is contained in the connected component of $X \setminus \alpha_{h-1}$ which contains X_0. It easily follows that $s \cap K_1 = \emptyset$. Thus $s \subset K_0$, and α_h misses s. The argument above shows that every shingle which α_h meets is in L_h. For the opposite inclusion, let $s \in L_h$. Then there exists a minimal skinny flow f containing s. It is easy to see that α_h meets some shingle t in f. Hence $t \in L_h$, and so $t = s$ by line 2.4.3.1. This proves that L_h is precisely the set of shingles which meet α_h.

$\lozenge$ 2.4.3.2 $\lozenge$

The following results will be proven next.

2.4.3.3. Every level curve is a minimal fat cut.

2.4.3.4. $\sum_{h=1}^{H} L_h = w$

2.4.3.5. $w_{M_s} = w_{m_f}$

2.4.3.6. $M_s = m_f$

To begin these proofs, let c be any fat cut. Then there exists a topological path α that separates the ends of X such that c is the set of shingles in S which meet α. It is easy to see that α meets some shingle in every skinny flow f. Thus c and f have at least one shingle in common, and so $\langle c, f \rangle \geq 1$. According to line 2.3.2 there exist minimal skinny flows $f_1, \ldots, f_k$ and nonnegative real numbers $a_1, \ldots, a_k$ such that $w_{M_s} = \sum_{i=1}^{k} a_i f_i$. Thus

2.4.3.7.

$$\langle c, w_{M_s} \rangle = \left\langle c, \sum_{i=1}^{k} a_i f_i \right\rangle = \sum_{i=1}^{k} a_i \langle c, f_i \rangle \geq \sum_{i=1}^{k} a_i = \frac{1}{\sqrt{M_s}},$$

where the last equality comes from line 2.3.5. If c is replaced by a level curve, then line 2.4.3.1 implies that the inequality in line 2.4.3.7 is an equality. This proves line 2.4.3.3. Line 2.4.3.1 implies line 2.4.3.4. Lines 2.3.6 and 2.4.3.4 easily imply line 2.4.3.5. Now line 2.4.3.7 with c replaced by a level curve gives line 2.4.3.6.

$\lozenge$ 2.4.3.3–2.4.3.6 $\lozenge$

We say that two fat cuts are **parallel** if they have underlying topological paths which are disjoint. There is an obvious analogous notion for fat flows, and there are obvious generalizations to families of parallel fat cuts or flows. If $(p_1, \ldots, p_k)$ is a family of parallel fat cuts or flows, then a **family of parallel underlying topological paths** for $(p_1, \ldots, p_k)$ consists of disjoint topological paths $\alpha_1, \ldots, \alpha_k$ such that α_i is an underlying topological path for p_i for $i = 1, \ldots, k$.

Lines 2.4.3.2 and 2.4.3.4 show that every shingling of X has a fundamental family of parallel fat cuts. It follows that if X is a quadrilateral, then every shingling of X has a fundamental family of parallel fat flows.

2.4.4. The relationship between fat flows and skinny cuts for tilings.

A **tiling** of X is a shingling T of X whose shingles, called **tiles**, are the closed 2-cells in a finite cellular decomposition of X. A **vertex** or **edge** of T is a vertex or edge of the corresponding cellular decomposition of X. Fix a tiling T of X.

2.4.4.1. Every minimal fat flow and minimal fat cut consists of all of the tiles that meet a topological path α which either joins or separates the ends of X and misses every vertex of T.

Proof. Simply observe that if the path α meets a vertex, then it can be slightly deformed to miss that vertex so that the resulting fat path is a subpath of the original one.
◊ 2.4.4.1 ◊

It easily follows from line 2.4.4.1 that every minimal fat flow and minimal fat cut for a tiling is a skinny flow or a skinny cut. However, such a skinny path need not be a minimal skinny flow or a minimal skinny cut.

Now consider the optimization problem for fat flows for the tiling T. There exists a reduced integral optimal weight function w for this case just as for the case of skinny flows in Section 2.4.3. Let H denote the fat height of X relative to w. Because every minimal fat flow is a skinny flow, it is possible to define level curves just as for the case of skinny flows in Section 2.4.3. The following analog of line 2.4.3.1 clearly holds.

2.4.4.2. Every tile t in T belongs to exactly $w(t)$ consecutive level curves, and every minimal fat flow contains exactly one tile in every level curve.

2.4.4.3. Every level curve is a skinny cut. Moreover, there exist piecewise smooth arcs or simple closed curves $\alpha_1, \ldots, \alpha_H$ (depending on whether X is a quadrilateral or a ring) in X which separate the ends of X such that α_h is contained in the closure of the connected component of $X \setminus \alpha_{h-1}$ which contains X_0 for $h = 2, \ldots, H$ and the tiles in L_h cover α_h for $h = 1, \ldots, H$. The curves α_h meet every tile t in either the empty set, a set of vertices of t or an arc which meets the boundary of t only at its end points, and they meet each other and the ends of X only at vertices.

Proof. Let h be an integer with $1 \le h \le H$. It will first be shown that L_h separates the ends of X, namely, there does not exist a topological path α joining the ends of X which misses all of the tiles in L_h. Suppose that such an α exists. Then as in the proof of line 2.4.4.1, α can be deformed so that the fat flow which it determines is a skinny flow $f = \{t_1, \ldots, t_m\}$. By assumption, $f \cap L_h = \emptyset$. Note that either $w(t_1) = 0$ or $t_1 \in L_1$. Now let i be the largest index such that $t_i \in L_1 \cup \cdots \cup L_{h-1}$ if there is such an index, and if not let $i = 1$. Let j be the smallest index such that $j > i$ and $w(t_j) \ne 0$ if there is such an index, and if not let $j = m$. Then $w(t_k) = 0$ for every index k such that $i < k < j$. Now construct a fat flow g by concatenating i) a minimal fat path from X_1 to t_i, ii) $\{t_{i+1}, \dots , t_{j-1}\}$ and iii) a minimal fat path from t_j to X_0. It follows that the length of g is less than H, which is impossible. Thus L_h separates the ends of X.

Since L_h separates the ends of X, some connected component of the union of the tiles in L_h separates the ends of X. It will next be shown that the union of the tiles in L_h is connected. Choose a connected component of the union of the tiles in L_h which separates the ends of X, and suppose that L_h contains a tile t not in this connected component. Let f be a minimal fat flow containing t. Then f meets the chosen connected component which separates the ends of X. But then f contains at least two tiles in L_h, contrary to line 2.4.4.2. Thus the union of the tiles in L_h is connected, and so L_h is a cut.

It will be convenient to prove the following results next before completing the proof of line 2.4.4.3.

2.4.4.4. Every level curve is a minimal skinny cut.

2.4.4.5. $\sum_{h=1}^{H} L_h = w$.

2.4.4.6. $w_{M_f} = w_{m_s}$.

2.4.4.7. $M_f = m_s$.

These statements will be proven next by following the proofs of lines 2.4.3.3 through 2.4.3.6. Let c be any cut. It is easy to see that c meets every fat flow f. Thus $\langle c, f \rangle \geq 1$. According to line 2.3.2 there exist minimal fat flows $f_1, \dots, f_k$ and nonnegative real numbers $a_1, \dots, a_k$ such that $w_{M_f} = \sum_{i=1}^{k} a_i f_i$. Thus

2.4.4.8.

$$\langle c, w_{M_f} \rangle = \langle c, \sum_{i=1}^{k} a_i f_i \rangle = \sum_{i=1}^{k} a_i \langle c, f_i \rangle \geq \sum_{i=1}^{k} a_i = \frac{1}{\sqrt{M_f}}.$$

If c is replaced by a level curve, then line 2.4.4.2 implies that the inequality in line 2.4.4.8 is an equality. Thus every level curve is a minimal cut. Now line 2.4.1.1 shows that every level curve is a minimal skinny cut. This proves line 2.4.4.4 and the first statement in line 2.4.4.3. Line 2.4.4.2 implies line 2.4.4.5. Lines 2.3.6 and 2.4.4.5 imply line 2.4.4.6. Now line 2.4.4.8 with c replaced by a level curve gives line 2.4.4.7.
◇ 2.4.4.4–2.4.4.7 ◇

It only remains to construct the curves $\alpha_1, \dots, \alpha_H$ in line 2.4.4.3. A somewhat brief description of such a construction will now be given. Construct α_1 as follows. Suppose that L_1 is given as a skinny cut by the ordered tiles $t_1, \dots, t_m$. If X is a quadrilateral, define points $x_0, \dots, x_m$ in X as follows. It may be assumed that t_1 meets the left side of X and t_m meets the right side of X. If the intersection of t_1 with the left side of X contains an edge, let x_0 be an interior point of that edge, and otherwise let x_0 be a vertex in the intersection of t_1 with the left side of X. In the same way define x_m in the intersection of t_m and the right side of X. Likewise for $i = 1, \ldots, m-1$, if $t_i \cap t_{i+1}$ contains an edge, let x_i be an interior point in that edge, and otherwise let x_i be a vertex in $t_i \cap t_{i+1}$. Join x_{i-1} to x_i by a smooth arc whose interior lies in the interior of t_i for $i = 1, \ldots, m$, and let α_1 be the concatenation of these arcs. If X is a ring, then define α_1 analogously.

Having defined α_1, suppose that $\alpha_1, \ldots, \alpha_{h-1}$ are defined for $h \geq 2$. It is a straightforward matter to define α_h in the same way as α_1, taking care to stay strictly between X_0 and α_{h-1} except at vertices. It is possible to do this because no tile in L_h lies between α_{h-1} and X_1.

◇ 2.4.4.3 ◇

It is a straightforward matter to define what it means for a family of skinny flows or cuts to be **parallel**. It is also straightforward to define the notion of a **family of parallel underlying arcs** or **simple closed curves** for a family of parallel skinny flows or cuts of a tiling.

2.4.4.9. Let T be a tiling of a quadrilateral or ring X. Then T has a fundamental family of parallel paths of all four types: fat cuts, skinny cuts, fat flows and skinny flows.

Proof. Lines 2.4.3.2 and 2.4.3.4 show that T has a fundamental family of parallel fat cuts. Lines 2.4.4.3 and 2.4.4.5 show that T has a fundamental family of parallel skinny cuts.

So let $(f_1, \ldots, f_k)$ be either a fundamental family of fat flows or a fundamental family of skinny flows of T. Let $(\alpha_1, \ldots, \alpha_k)$ be a family of underlying piecewise smooth arcs for $(f_1, \ldots, f_k)$. In the case of fat flows, we apply line 2.4.4.1 and assume that no arc α_i contains a vertex of T. Thus the flows $f_1, \ldots, f_k$ are skinny even if they are fat.

It is easy to see that we may make the following assumptions. The set of points in X which are contained in more than one of the arcs α_i is finite. Every arc α_i meets every tile t in T in either the empty set, a set of vertices of t or an arc which meets the boundary of t only at its end points.

The argument will now proceed by induction on the number n of triples (x, α_i, α_j) such that x is a point in the interior of X and the intersection number of the arcs α_i and α_j at x is not 0. If $n = 0$, then $f_1, \ldots, f_k$ are parallel, as desired. So suppose that $n > 0$ and that line 2.4.4.9 is true for smaller values of n.

Suppose that (x, α_i, α_j) is a triple such that x is a point in the interior of X and the intersection number of the arcs α_i and α_j at x is not 0. Define new arcs α_i' and α_j' so that α_i' consists of the segment of α_j preceding x followed by the segment of α_i following x and α_j' consists of the segment of α_i preceding x followed by the segment of α_j following x. It is possible to define flows f_i' and f_j' analogously because f_i and f_j are skinny. Just as when defining level curves, it follows that f_i' and f_j' have the same length as f_i and f_j. The result is a new fundamental family of flows and a new family of underlying piecewise smooth arcs, and it is not difficult to see that this operation reduces n. Thus $f_1, \ldots, f_k$ can be inductively transformed to a fundamental family of parallel flows, which proves line 2.4.4.9.

$\diamond$ 2.4.4.9 $\diamond$

EXAMPLE 2.4.4.10. Line 2.4.4.9 shows for every tiling of a quadrilateral or ring that there exists a fundamental family of parallel paths of all four types. The shingling S of a quadrilateral X below indicates that this result does not generalize to skinny flows of shinglings. This shingling has 8 shingles, one of which is shaded just to help identify it. The 5 heavy horizontal line segments joining the sides of X are underlying arcs for a fundamental family of fat cuts. They determine the reduced integral optimal weight function of S for fat cuts and skinny flows. The 4 heavy arcs joining the top and the bottom of X are what might be called underlying arcs for a fundamental family of skinny flows of S. These skinny flows are the only minimal skinny flows of S for this optimal weight function.

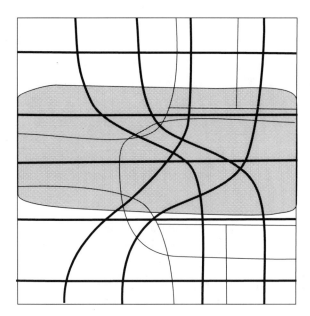

2.4.5. The relationship between the four moduli.

2.4.5.1. For every shingling S of a quadrilateral or ring X,

$$M_s = m_f \leq m_s \leq M_f.$$

Proof. The equality is given by line 2.4.3.6. To prove the first inequality, let c be a minimal fat cut relative to the weight function w_{m_s}. By line 2.4.1.1 c contains a subcut b which is a skinny cut. Thus

$$m_f \leq \frac{1}{\langle c, w_{m_s} \rangle^2} \leq \frac{1}{\langle b, w_{m_s} \rangle^2} \leq \frac{1}{C^2_{w_{m_s}, s}} = m_s,$$

which proves the first inequality. For the second inequality, let c be a minimal skinny cut relative to the weight function w_{M_f}. Although line 2.4.4.8 occurs in the context of tilings, it is also valid for general shinglings. Using line 2.4.4.8 it can easily be seen that

$$m_s \leq \frac{1}{\langle c, w_{M_f} \rangle^2} \leq M_f.$$

◇ 2.4.5.1 ◇

2.4.5.2. For every shingling S of a quadrilateral X,

$$M_s = m_f \leq m_s = M_f.$$

Proof. This follows easily from line 2.4.5.1.
◇ 2.4.5.2 ◇

2.4.5.3. For every tiling T of a quadrilateral or ring X,

$$M_s = m_f \leq m_s = M_f.$$

Proof. This result follows from lines 2.4.5.1 and 2.4.4.7.
◇ 2.4.5.3 ◇

Lines 2.4.5.2 and 2.4.5.3 show that $m_s = M_f$ unless the shingling S is not a tiling and X is a ring. Following is a shingling of a ring for which $m_s < M_f$.

Example 2.4.5.4. Consider the shingling S of a ring X given below. This shingling has 8 shingles, one of which is shaded just to help identify it. The 5 heavy arcs joining the top and bottom of X are underlying arcs for a fundamental family of parallel fat flows of S. They determine the reduced integral optimal weight function w of S for fat flows. Hence $M_f = \frac{H^2_{w,f}}{A_w} = \frac{16}{20} = \frac{4}{5}$. However, there is just one skinny cut which is minimal for w, and it has w-length 5. Hence $\frac{A_w}{C^2_{w,s}} = \frac{20}{25} = \frac{4}{5}$. Since w is not in the cone spanned by its minimal skinny cuts, w is not an optimal weight function for skinny cuts of S by line 2.3.2. Thus $m_s < \frac{A_w}{C^2_{w,s}} = M_f$.

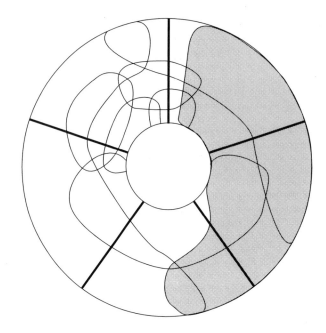

3. The finite Riemann mapping theorem.

We consider a quadrilateral or ring X and a tiling T of X. We optimize fat flows and skinny cuts so as to realize the equality $M_f = m_s$. The analysis of the previous sections shows that the reduced integral optimal weight function can be simultaneously realized as a sum of a fundamental family of parallel fat flows and a fundamental family of parallel skinny cuts. These minimal fat flows have underlying disjoint arcs joining the top and bottom of X, and these minimal skinny cuts have underlying arcs that are disjoint except where they pass through vertices of the tiling T, each arc separating the ends of X. Finally, each arc underlying a flow intersects each arc underlying a cut in precisely one point which is in the interior of some tile. As in Section 1.4, these arcs form a grid which can be pictured as graph paper that indicates how X can be realized as a rectangle or right circular cylinder tiled by squares (squares, since the number of flows passing through the interior of a tile in T is the same as the number of cuts passing through the same interior). Thus we have the following remarkable theorem.

Theorem 3.0.1. (Finite Riemann Mapping Theorem). *Every tiled quadrilateral corresponds uniquely under fat flow optimization to a squared rectangle. Under fat cut optimization it corresponds uniquely to another squared rectangle. Analogous results hold for tiled rings.*

Since squared rectangles solve the resistive circuit problems given by their Smith diagrams (Section 1), tiled quadrilaterals have a natural correspondence with planar circuits with edges representing resistors of unit resistance. There are two natural questions given by this correspondence.

Question 3.0.2. Of the many natural circuits that one could conceivably associate with a tiled quadrilateral, how do the ones associated with fat flow and fat cut optimization differ from the others?

Question 3.0.3. How can we find the optimal circuits associated with a tiled quadrilateral?

We shall answer Question 3.0.2 in Section 3.1. The answer is essentially this: our optimization problems give circuits whose squared rectangles are as like the original tiling combinatorially as possible. Question 3.0.3 is an important question whose answer relates intimately to the problem of calculating the solution to the finite Riemann mapping problem explicitly. We shall discuss algorithms designed to solve this mapping problem in Section 4. Obtaining the appropriate circuits will be discussed particularly in Section 4.4.

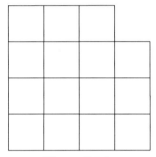

Figure 3.1.1.

3.1. Examples. In Figure 3.1.1 we give a sample tiled quadrilateral: a 4 × 4 square tiled by unit squares from which the upper right hand corner square has been removed. We consider the two upper right hand "indented" edges as the "top" of our quadrilateral and the left side and bottom of the square as the "bottom" of our quadrilateral. In Figure 3.1.2 we interpret certain collections of edges as potential "maximal horizontal segments" by means of which we form three associated Smith circuit diagrams. The Smith diagrams appear in Figure 3.1.3, and the associated squared rectangles appear in Figure 3.1.4.

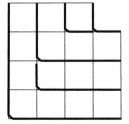 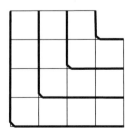 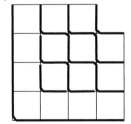

Figure 3.1.2.

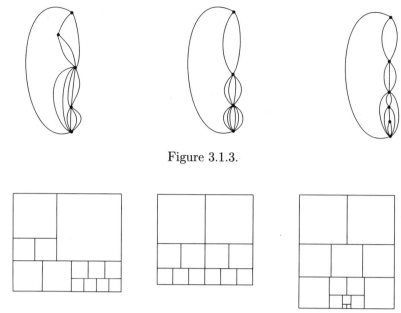

Figure 3.1.3.

Figure 3.1.4.

In each of Figures 3.1.2, 3.1.3, and 3.1.4 the last of the three parallel figures represents the appropriate diagram for optimization of fat flows and skinny cuts. Note that the first two of the three parallel figures, while in some sense geometrically reasonable, do more geometric damage to the original combinatorics in the process of squaring than does the optimization squaring. The damage is easily quantifiable if we refer to the graph paper grid of the squaring developed in Section 1.4 given by the collections of current curves and voltage curves. If these curves are developed in the original tiling, then in the optimization pattern the curves can actually be realized as paths which travel from one tile into an adjacent tile. The fat flows cross edges from one tile to the next. The skinny cuts cross edges or vertices from one tile to the next. In the other diagrams the curves may jump from one tile to a nonadjacent tile. That is, **the optimization curves of the finite Riemann mapping theorem respect, in as far as is possible, the combinatorics of the original tiling.** There is a converse which will be studied in Section 6.1: **if the current and voltage curves, when developed in the original tiling, always pass from tile to adjacent tile, then the resulting modulus or resistance of the circuit (height/width) lies between** m_f **and** M_f.

In summary, the optimization conditions of the finite Riemann mapping theorem delineate a family of circuits related to an arbitrary tiled quadrilateral which, when solved, square a rectangle in a way which is combinatorially optimal or nearly optimal.

3.2. The Kirchhoff inequalities. We have demonstrated that the finite Riemann mapping theorem squares a rectangle, hence is associated with the solution of a finite resistive circuit with one battery. The difficulty already noted is that it is not at all clear from a tiled quadrilateral what the appropriate circuit should be. Nevertheless, we can formulate a family of equations and inequalities analogous to the Kirchhoff equations whose solution solves the finite Riemann mapping problem. We call these equations and inequalities the **Kirchhoff inequalities**. Unfortunately, there is a certain degree of circularity in the formulation which makes the inequalities suitable for *checking* a potential solution but not readily applicable for *finding* a potential solution.

The formulation views the dual graph of the tiling as a generalized electrical circuit in which resistances appear not on the edges but at the nodes. Such an interpretation was developed in discussions with Peter Doyle and his students, including Oded Schramm, at Princeton during the year 1988-89. Schramm, in his study of the finite Riemann mapping theorem, has made some effort to make the analogy precise.

We add to the tiling two new tiles t_0 and t_1 representing the bottom and top edge of the quadrilateral, respectively. We consider the undirected graph dual to the tiling. The tiles represent vertices of the same name. Pairs of tiles meeting along an edge represent edges of the dual graph.

The variables. If s and t are tiles meeting along an edge, then the pairs (s,t) and (t,s) represent an edge of the dual graph. We assign to this edge a variable $i(s,t) = -i(t,s)$ which by convention represents current flow from tile s to tile t along edge (s,t). If $i(s,t) > 0$, then we think of this current as flowing *out* of the tile s and *into* the tile t. The **weight** of a tile t is denoted by $w(t)$ and is the sum of the currents flowing *into* the tile t. Note that $w(t) \geq 0$. Our aim is to solve the finite Riemann mapping problem by using the nonnegative numbers $w(t)$, $t \neq t_0, t_1$, as our weight function. The Kirchhoff inequalities state necessary and sufficient conditions on these weights in order that they optimize fat flows from t_0 to t_1.

The vertex equations. For every vertex s other than t_0 and t_1, the sum of the currents flowing into s equals the sum of the currents flowing out of s.

The loop equations. Let $s_0, \ldots, s_k$ denote a sequence of tiles forming the vertices of a closed path or loop in our dual graph. Assume in addition that each edge (s_0, s_1), (s_1, s_2), $\ldots$, (s_k, s_0) carries nonzero current. We say that the loop is *extreme* at vertex s_i if the two currents $i(s_{i-1}, s_i)$ and $i(s_i, s_{i+1})$ have opposite signs. The loop is *rising* at vertex s_i if the two currents just mentioned are positive, and *falling* at vertex s_i if the two currents are negative. The loop equation associated with our given loop requires that we add the weights of the vertices at which the loop is rising and subtract the weights of the vertices at which the loop is falling. The sum should then be 0.

The loop inequalities. We need to accommodate the possibility that some edges may carry no current ($i(s,t) = 0$). We define the length of a fat path to be the sum of the weights of its vertices. We require that at least one minimal path from t_0 to t_1 be strictly rising in the sense of the previous paragraph at each vertex other than t_0 and t_1. This requirement is equivalent to a whole family of inequalities that imply that a rising path cannot be shortened by inserting splices across edges which carry no current.

Theorem 3.2.1. *Suppose that there exist currents $i(s,t)$, not all 0, satisfying the Kirchhoff inequalities. Then the associated weights $w(s)$, $s \neq t_0, t_1$, form an optimal weight function w which optimizes fat flows from t_0 to t_1.*

Proof. Form $|i(s,t)|$ current lines parallel to the edge (s,t) of the dual graph in the direction of current flow. The vertex equations imply that, except at the extreme vertices, t_0 and t_1, the current lines may be extended through the vertices. By the definition of the weight function w, we see that w is the sum of the resulting paths given by the current lines. Each current line is a path which is rising at each of its vertices with the possible exception of t_0 and t_1.

The loop equations imply that no current path can be closed; for otherwise, since the path is rising at each vertex, the weight sum around it would be positive. Hence each current line must be an arc joining t_0 and t_1. The Kirchhoff inequalities guarantee the existence of a strictly rising minimal path from t_0 to t_1. If some current path rises from t_1 to t_0, then the concatenation of this and a path rising from t_0 to t_1 is a closed loop rising at every vertex other than t_0 and t_1, again contradicting the associated loop equation. Hence all current flows from t_0 to t_1. Furthermore, the loop equations imply that all have the same length.

Since the current paths are w-minimal flows from t_0 to t_1, w is a sum of its minimal flows. By line 2.3.6 w is an optimal weight function for fat flows. ◇ Theorem 3.2.1 ◇

4. Algorithms which calculate the finite Riemann mapping.

Although there does not seem to be a simple system of linear equations whose solution gives an optimal weight function for the finite Riemann mapping problem, nevertheless there are finite algorithms for calculating an optimal weight function. We give a beautiful algorithm which is slow and a hybrid algorithm that is substantially faster.

4.1. The minimal path algorithm. This algorithm involves the construction of a sequence of weight vectors. Line 4.1.1 is a description of this construction. We return to the vector space setting introduced in Section 2.2.

4.1.1. For every nonempty subset Q of the set P of paths fix a vector x_Q in $\mathbf{N}^n \setminus \{0\}$ which lies in the cone spanned by the path vectors in Q. Inductively define a sequence of weight vectors $w_1, w_2, w_3, \ldots$ in $\mathbf{N}^n$ as follows. Let w_1 be any weight vector in $\mathbf{N}^n$. Now suppose that w_ν is defined for some integer $\nu \geq 1$. Let Q_ν be the set of w_ν-minimal path vectors in P, and set $w_{\nu+1} = w_\nu + x_{Q_\nu}$.

The following three statements make successively stronger statements about the sequence w_ν. The first shows that the sequence, normalized to have norm 1, converges to the optimal weight vector of norm 1. The third is a key fact which will allow us to extract an exact solution to the mapping problem from the sequence. The remainder of Section 4.1 will be devoted to a description of the minimal path algorithm. We prove lines 4.1.2, 4.1.3 and 4.1.4 in Section 4.2.

4.1.2. The sequence of weight vectors $\frac{1}{||w_1||}w_1$, $\frac{1}{||w_2||}w_2$, $\frac{1}{||w_3||}w_3$, $\ldots$ in S^{n-1} converges to the optimal weight vector w_M.

4.1.3. The sequence of subsets $Q_1, Q_2, Q_3, \ldots$ of P is eventually periodic; that is, there exist positive integers N and μ such that if $\nu \geq N$, then $Q_{\mu+\nu} = Q_\nu$.

4.1.4. In the notation of line 4.1.3, if $\nu \geq N$, then $w_{\mu+\nu} - w_\nu$ is an optimal weight vector.

Here is how an exact algorithm can be extracted from line 4.1.4.

4.1.5. For every subset Q of P choose x_Q to be the sum of all of the path vectors in Q. Choose $w_1 = (1, 1, 1, \ldots)$.

Lines 4.1.1 and 4.1.5 determine a sequence $w_1, w_2, w_3, \ldots$ of weight vectors in $\mathbf{N}^n$.

At this point one encounters the problem of determining when one has an optimal weight vector as in line 4.1.4. Here is one approach to this problem.

4.1.6. At step number ν compute w_ν, record w_ν and record $\sum_{p \in Q_\nu} a_p$, where $x_{Q_\nu} = \sum_{p \in Q_\nu} a_p p$. Then let the integer ν' vary from 1 to $\nu - 1$. Choose such a ν', and set $w = w_\nu - w_{\nu'}$.

If x_Q is chosen as in line 4.1.5, then every a_p in line 4.1.6 equals 1.

This paragraph describes a test, which if satisfied, implies that w is an optimal weight vector. For this, compute H_w and A_w. There exist positive real numbers $b_1, \ldots, b_k$ and path vectors $p_1, \ldots, p_k$ such that $w = \sum_{i=1}^{k} b_i p_i$. The b_i's and the p_i's are easily gotten from the a_p's and the p's which occur in line 4.1.6. Hence

4.1.7.

$$A_w = \langle w, w \rangle = \left\langle \sum_{i=1}^{k} b_i p_i, w \right\rangle = \sum_{i=1}^{k} b_i \langle p_i, w \rangle \geq H_w \sum_{i=1}^{k} b_i.$$

Furthermore, equality holds throughout line 4.1.7 if and only if p_i is a w-minimal path vector for every i. Thus if $A_w = H_w \sum_{i=1}^{k} b_i$, then w lies in the cone spanned by its minimal path vectors, and so $\frac{1}{||w||}w = w_M$ by line 2.3.6. So, to test whether or not w is an optimal weight vector,

4.1.8. compute A_w, H_w and $\sum_{i=1}^{k} b_i$; if $A_w = H_w \sum_{i=1}^{k} b_i$, then w is an optimal weight vector.

It remains to show that there exist positive integers $\nu' < \nu$ such that equality holds throughout line 4.1.7. For this it is easy to see that there exists a neighborhood U of w_M in S^{n-1} such that if p is a path vector whose length is minimal relative to some weight vector in U, then the length of p is minimal relative to w_M. Combining this with line 4.1.2 shows that if ν' is sufficiently large, then the length of every path vector in $Q_{\nu'}, Q_{\nu'+1}, Q_{\nu'+2}, \ldots$ is minimal relative to w_M. This and line 4.1.4 show that there exist ν' and ν such that w is a scalar multiple of w_M and $Q_{\nu'}, Q_{\nu'+1}, Q_{\nu'+2}, \ldots$ consist of path vectors whose lengths are minimal relative to w_M. For such integers ν' and ν, equality holds throughout line 4.1.7. Thus the algorithm eventually stops.

In conclusion, to compute w_M, compute weight vectors $w_1, w_2, w_3, \ldots$ and their differences as described in lines 4.1.1, 4.1.5 and 4.1.6. Line 4.1.8 describes a test which eventually identifies a difference of the form $w_\nu - w_{\nu'}$ as an optimal weight vector. At this point the algorithm stops.

This paragraph deals with a slight modification of the algorithm described in the previous paragraph which improves computing efficiency. Instead of allowing ν' to be any integer from 1 to $\nu - 1$, take ν' to be the largest power of 2 less than ν. In other words, first take $\nu' = 1$ and $\nu = 2$, then $\nu' = 2$ and $\nu = 3, 4$, then $\nu' = 4$ and $\nu = 5, 6, 7, 8$ and so on. In the case where ν' is any integer from 1 to $\nu - 1$, the number of times that the computation in line 4.1.8 must be performed grows quadratically in $\mu + N$, where N and μ are as in line 4.1.3. In the case where ν' is taken to be a power of 2, the number of times that the computation in line 4.1.8 must be performed grows linearly in $\mu + N$.

4.2. Proofs of the claims about the minimal path algorithm. This section is concerned with proving lines 4.1.2, 4.1.3, and 4.1.4. The first thing to be done here is to present a construction of sequences of weight vectors which is more general than the construction given in line 4.1.1.

Fix two positive real numbers $K < L$. For every nonempty subset Q of P, let $C_Q(K, L)$ denote the set of vectors x in $\mathbf{R}^n$ which lie in the cone spanned by the path vectors in Q such that $K \leq ||x|| \leq L$. Inductively define a sequence of weight vectors $w_1, w_2, w_3, \ldots$ in $\mathbf{R}^n$ as follows. Let w_1 be any weight vector in $\mathbf{R}^n$. Now suppose that w_ν is defined for some integer $\nu \geq 1$. Let Q be the set of w_ν-minimal path vectors in P. Choose a vector x_ν in $C_Q(K, L)$, and set $w_{\nu+1} = w_\nu + x_\nu$.

This construction is clearly more general than the one given in line 4.1.1. Differences between this construction and the one in line 4.1.1 are that the previous weight vectors w_ν are weight vectors of integers and the previous w_ν uniquely determines the previous $w_{\nu+1}$. Thus to prove line 4.1.2, it suffices to prove line 4.2.1 for the present construction.

4.2.1. The sequence of weight vectors $\frac{1}{||w_1||} w_1, \frac{1}{||w_2||} w_2, \frac{1}{||w_3||} w_3, \ldots$ in S^{n-1} converges to the optimal weight vector w_M.

The proof of line 4.2.1 will now begin. Let V denote the linear hyperplane in

$\mathbf{R}^n$ which is orthogonal to w_M. Let $\pi : \mathbf{R}^n \to V$ denote the canonical projection. Thus for every x in $\mathbf{R}^n$, $x = \pi(x) + \langle x, w_M \rangle w_M$.

The assertion in line 4.2.1 is equivalent to saying that if C is a cone in $\mathbf{R}^n$ which is a neighborhood of w_M, then C contains all but finitely many of the vectors w_1, w_2, w_3, ... Because the coordinates of the vectors x_1, x_2, x_3, ... are nonnegative and $\|x_\nu\| \geq K$ for every ν, the lengths of w_1, w_2, w_3, ... increase monotonically to ∞. Combining the last two sentences easily shows the following.

4.2.2. To prove line 4.2.1 it suffices to prove that the sequence of vectors $\pi(w_1)$, $\pi(w_2)$, $\pi(w_3)$, ... is bounded.

In other words, if the vectors w_1, w_2, w_3, ... are contained in a tubular neighborhood of the ray spanned by w_M, then every cone which is a neighborhood of w_M contains all but finitely many of them.

This leads to an investigation of the projection π. The first result in this direction is the following.

4.2.3. If w is a weight vector and p is a w-minimal path vector, then

$$\langle \pi(p), \pi(w) \rangle \leq 0.$$

To begin the proof of line 4.2.3, let p_1, ..., p_k be the w_M-minimal path vectors. By line 2.3.2 there exist nonnegative real numbers a_1, ..., a_k such that $w_M = \sum_{i=1}^k a_i p_i$. Hence

$$0 = \pi(w_M) = \pi\left(\sum_{i=1}^k a_i p_i\right) = \sum_{i=1}^k a_i \pi(p_i),$$

that is, 0 is in the convex hull of the vectors $\pi(p_1)$, ..., $\pi(p_k)$. This easily implies that if w is any weight vector, then $\langle \pi(p_i), \pi(w) \rangle \leq 0$ for some i. Thus for this value of i

4.2.4.
$$\begin{aligned}
H_w &\leq \langle p_i, w \rangle \\
&= \langle \pi(p_i) + \langle p_i, w_M \rangle w_M, \pi(w) + \langle w, w_M \rangle w_M \rangle \\
&= \langle \pi(p_i), \pi(w) \rangle + \langle p_i, w_M \rangle \langle w, w_M \rangle \\
&\leq H_{w_M} \langle w, w_M \rangle.
\end{aligned}$$

On the other hand, if p is a w-minimal path vector, then

4.2.5.
$$\begin{aligned}
H_w &= \langle p, w \rangle \\
&= \langle \pi(p), \pi(w) \rangle + \langle p, w_M \rangle \langle w, w_M \rangle \\
&\geq \langle \pi(p), \pi(w) \rangle + H_{w_M} \langle w, w_M \rangle.
\end{aligned}$$

Combining lines 4.2.4 and 4.2.5 gives

$$\langle \pi(p), \pi(w) \rangle + H_{w_M} \langle w, w_M \rangle \leq H_{w_M} \langle w, w_M \rangle,$$

and so

$$\langle \pi(p), \pi(w) \rangle \leq 0.$$

This proves line 4.2.3.
$\diamond$ 4.2.3 $\diamond$

Now let ν be a positive integer, and consider how $\pi(w_{\nu+1})$ is gotten from $\pi(w_\nu)$. Let Q be the set of w_ν-minimal path vectors. By definition $w_{\nu+1} = w_\nu + x_\nu$, where x_ν is an element of $C_Q(K, L)$. In particular, $\pi(w_{\nu+1}) = \pi(w_\nu) + \pi(x_\nu)$. Furthermore, since $||x_\nu|| \leq L$, it is easy to see that $||\pi(x_\nu)|| \leq L$, and line 4.2.3 shows that $\pi(x_\nu)$ lies in the cone spanned by the vectors in $\pi(P)$ whose inner products with $\pi(w_\nu)$ are nonpositive.

The discussion from line 4.2.2 to here shows that proving line 4.2.1 reduces to proving line 4.2.6, which follows shortly.

Notation will change from here to line 4.2.18, which is the end of the proof of line 4.2.6. The essential difference is that the projection π is omitted. The integer $n - 1$ becomes m.

To prepare for the statement of line 4.2.6, first fix some Euclidean space V of dimension $m \geq 1$. Let P be a finite set of vectors in V with 0 in its convex hull. Fix a positive real number L, and for every nonempty subset Q of P, let $C_Q(L)$ denote the set of vectors x in V which lie in the cone spanned by the vectors in Q such that $||x|| \leq L$. Inductively define a sequence of vectors $w_1, w_2, w_3, \ldots$ in V as follows. Let w_1 be any vector in V. Now suppose that w_ν is defined for some integer $\nu \geq 1$. Let Q be the set of vectors p in P such that $\langle p, w_\nu \rangle \leq 0$. Choose a vector x_ν in $C_Q(L)$, and set $w_{\nu+1} = w_\nu + x_\nu$.

4.2.6. The sequence of vectors $w_1, w_2, w_3, \ldots$ in V is bounded.

It is easy to see that proving line 4.2.6 in turn reduces to proving line 4.2.7. The following definitions are needed to state line 4.2.7.

Let $\lambda_0, \ldots, \lambda_m$ be real numbers such that

$$0 = \lambda_m < \cdots < \lambda_2 < \lambda_1 < \lambda_0 = 1.$$

For every subset Q of P let V_Q be the subspace of V spanned by Q, let $d_Q = \dim(V_Q)$ and for every positive real number r let

$$N_Q(r) = \{x + y \in V | x \in V_Q, y \in V_Q^\perp, ||y|| \leq \lambda_{d_Q} r\}.$$

The set $N_Q(r)$ consists of all vectors in V having distance at most $\lambda_{d_Q} r$ from V_Q. Let

$$B(r) = \cap_{Q \subset P} N_Q(r).$$

It is possible for Q to be the empty set, and $N_\emptyset(r)$ is the closed ball of radius r in V centered at 0. Thus $B(r)$ is contained in this closed ball of radius r.

4.2.7. There exists a choice of $\lambda_1, \ldots, \lambda_{m-1}$ and a positive real number R all of which depend only on P and L such that the following holds. If $w \in B(r)$ for some real number $r \geq R$ and Q is the set of vectors p in P such that $\langle p, w \rangle \leq 0$, then $w + x \in B(r)$ for every vector x in $C_Q(L)$.

To begin the proof of line 4.2.7 observe that $N_Q(r)$ is convex for every Q and r. Thus $B(r)$ is convex for every r. Moreover, since $N_Q(r) = rN_Q(1)$ for every Q and r, $B(r) = rB(1)$ for every r.

Now suppose that $w \in B(r)$ and $x \in C_Q(L)$ as in line 4.2.7. Then $x = \sum_{p \in Q} a_p p$ for some nonnegative real numbers a_p, and $\|x\| \leq L$. Since line 4.2.7 is obvious if $x = 0$, it may be assumed that $x \neq 0$. Line 2.3.7 shows that we may assume that the set of vectors $S = \{p \in Q : a_p \neq 0\}$ is linearly independent. Hence there exists a linear functional $f : V \to \mathbf{R}$ such that $f(p) = 1$ for every p in S. Clearly, $f(x) = \sum_{p \in Q} a_p$. Because $C_Q(L)$ is compact, $f(x)$ is bounded by a bound which depends only on Q and L. This and the fact that P is a finite set imply the following.

4.2.8. There exists a positive real number Λ which depends only on P and L such that it is possible to express x as $x = \sum_{p \in Q} a_p p$, where the a_p's are nonnegative real numbers and $(\sum_{p \in Q} a_p)\|q\| \leq \Lambda$ for every q in Q.

Now suppose that x is expressed as in line 4.2.8. Let $a = \sum_{p \in Q} a_p$, and let $b_p = a_p/a$ for every p in Q. Then $x = \sum_{p \in Q} b_p(ap)$, $\sum_{p \in Q} b_p = 1$ and $\|ap\| \leq \Lambda$ for every p in Q. In particular, this expresses x as a convex combination of the vectors ap.

Suppose that it can be shown that $w + ap \in B(r)$ for every p in Q. Then because $B(r)$ is convex, it also contains

$$\sum_{p \in Q} b_p(w + ap) = \sum_{p \in Q} b_p w + \sum_{p \in Q} b_p(ap) = w + x.$$

Thus to prove line 4.2.7 it suffices to prove the following.

4.2.9. There exists a choice of $\lambda_1, \ldots, \lambda_{m-1}$ and a positive real number R all of which depend only on P and L such that the following holds. If $w \in B(r)$ for some real number $r \geq R$ and p is a vector in P such that $\langle p, w \rangle \leq 0$, then $w + ap \in B(r)$ for every nonnegative real number a such that $\|ap\| \leq \Lambda$.

The next thing to do is to choose the numbers $\lambda_1, \ldots, \lambda_{m-1}$. To prepare for this the following definition will be made. Given a subset Q of P and a positive real number r, let

$$D_Q(r) = V_Q^\perp \cap (\cap_{Q \subset V_{Q'}} N_{Q'}(r)).$$

The choice of $\lambda_1, \ldots, \lambda_{m-1}$ will be made so that the following holds.

4.2.10. If Q is a subset of P consisting of exactly one nonzero vector and Q' is a subset of P for which $V_{Q'}$ does not contain Q, then $N_{Q'}(1)$ is a neighborhood of $D_Q(1)$.

The numbers $\lambda_1, \ldots, \lambda_{m-1}$ will be chosen so that line 4.2.10 holds for all such subsets Q' by means of a backward induction on $d_{Q'}$. The case $d_{Q'} = m$ is vacuously true with no restriction on $\lambda_1, \ldots, \lambda_{m-1}$. Now suppose that $d_{Q'} = k - 1$ for some integer k with $1 \leq k \leq m$ and that $\lambda_k, \ldots, \lambda_{m-1}$ have been chosen to satisfy line 4.2.10 for subsets of the form Q' which generate subspaces of V having dimensions larger than $d_{Q'}$.

Set $Q'' = Q' \cup Q$. Since $V_{Q'}$ does not contain Q and Q contains exactly one nonzero vector, $d_{Q''} = k$. Let ϵ be a positive real number such that $\epsilon^2 + \epsilon < 1 - \lambda_k^2$. Lines 4.2.11 and 4.2.12 will be needed to choose λ_{k-1}.

4.2.11. If $x \in D_Q(1)$ and the distance from x to $V_{Q''}^{\perp}$ is at most ϵ, then $\|x\| < \sqrt{1 - \epsilon}$.

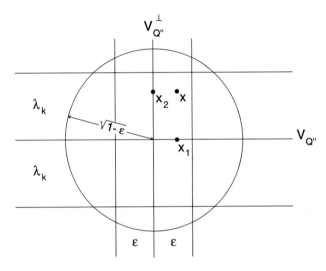

The proof of line 4.2.11 follows easily from the fact that if $x \in D_Q(1)$, then $x \in N_{Q''}(1)$, and so $x = x_1 + x_2$ with $x_1 \in V_{Q''}$, $x_2 \in V_{Q''}^{\perp}$, $\|x_1\| \leq \epsilon$ and $\|x_2\| \leq \lambda_k$. Thus

$$\|x\|^2 \leq \epsilon^2 + \lambda_k^2 < (1 - \epsilon - \lambda_k^2) + \lambda_k^2 = 1 - \epsilon.$$

This proves line 4.2.11.
◇ 4.2.11 ◇

The next thing to be proven is line 4.2.12.

4.2.12. There exists a positive real number $\delta < 1$ for which the following holds. For every choice of Q, Q', and Q'' as immediately before line 4.2.11, if x is a vector in $V_Q^{\perp}$ whose distance from $V_{Q''}^{\perp}$ is greater than ϵ, then the distance from x to $V_{Q'}^{\perp}$ is greater than δ.

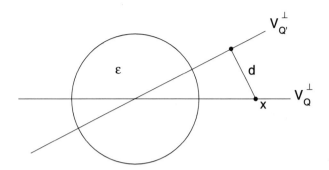

To prove line 4.2.12 first choose Q, Q', and Q'' as immediately before line 4.2.11. Choose sets of vectors S_1, S_2, and S_3 in V such that S_1 is a basis of $V_{Q''}^{\perp}$, $S_1 \cup S_2$ is a basis of $V_Q^{\perp}$ and $S_1 \cup S_3$ is a basis of $V_{Q'}^{\perp}$. It is easy to see that S_3 contains just one vector, and $S_3 \not\subset V_Q^{\perp}$ because $Q \not\subset V_{Q'}$. Thus $S_1 \cup S_2 \cup S_3$ is a basis of V.

Now choose a linear isomorphism $T : V \to \mathbf{R}^m$ such that the sets $T(S_1)$, $T(S_2)$, $T(S_3)$ are mutually orthogonal. Then $T(x) = x_1 + x_2$, where x_i is in the subspace of $\mathbf{R}^m$ generated by $T(S_i)$ for $i = 1, 2$. Because T is a linear isomorphism, there exists a positive real number ϵ' independent of x such that the distance from $T(x)$ to $T(V_{Q''}^{\perp})$ is greater than ϵ'. In other words, $\|x_2\| > \epsilon'$. But the distance from $T(x)$ to $T(V_Q^{\perp})$ is also $\|x_2\|$. Thus for the same reason that ϵ' exists there exists a positive real number δ' independent of x such that the distance from x to $V_Q^{\perp}$ is greater than δ'. This proves line 4.2.12 because there are only finitely many possibilities for Q, Q', and Q''.
◇ 4.2.12 ◇

Now that lines 4.2.11 and 4.2.12 are established, choose λ_{k-1} so that $\sqrt{1-\epsilon} \leq \lambda_{k-1}$ and $\sqrt{1-\delta^2} \leq \lambda_{k-1}$. To prove that $N_{Q'}(1)$ is a neighborhood of $D_Q(1)$, choose $x \in D_Q(1)$. If the distance from x to $V_{Q''}^{\perp}$ is at most ϵ, then $\|x\| < \sqrt{1-\epsilon} \leq \lambda_{k-1}$ by line 4.2.11. Thus $N_{Q'}(1)$ is a neighborhood of x because $N_{Q'}(1)$ contains the closed ball in V of radius λ_{k-1} centered at 0. On the other hand, if the distance from x to $V_{Q''}^{\perp}$ is greater than ϵ, then the distance from x to $V_Q^{\perp}$ is greater than δ by line 4.2.12. Since $\|x\| \leq 1$, this implies that the length of the projection of x to $V_{Q'}^{\perp}$ is less than $\sqrt{1-\delta^2}$. Since $\sqrt{1-\delta^2} \leq \lambda_{k-1}$, it follows that $N_{Q'}(1)$ is a neighborhood of x. The proof of line 4.2.10 is now complete.
◇ 4.2.10 ◇

Let Q continue to denote a subset of P which consists of exactly one nonzero vector. Line 4.2.10 implies that

4.2.13. $\cap_{Q \not\subset V_{Q'}} N_{Q'}(1)$ is a neighborhood of $D_Q(1)$.

Thus

4.2.14. $D_Q(1) = V_Q^{\perp} \cap B(1)$.

On the other hand it is easy to see that

4.2.15.
$$\cap_{Q \subset V_{Q'}} N_{Q'}(1) = V_Q \perp D_Q(1).$$

(The notation $A = B \perp C$ means that A is the direct product of B and C and that B and C are orthogonal to one another.)

Now let $\pi : V \to V_Q^{\perp}$ denote the canonical projection. Suppose that $x \in B(1)$. Then $x \in \cap_{Q \subset V_{Q'}} N_{Q'}(1)$. Hence $\pi(x) \in D_Q(1)$ by line 4.2.15. Now line 4.2.14 implies that $\pi(B(1)) \subset B(1)$, and so scaling by r yields

4.2.16. $\pi(B(r)) \subset B(r)$ for every positive real number r.

Since $\cap_{Q \not\subset V_{Q'}} N_{Q'}(1)$ is a neighborhood of $D_Q(1)$ by line 4.2.13, lines 4.2.14 and 4.2.15 show that $B(1)$ is locally a product near $V_Q^{\perp}$. More precisely, there exists a positive real number ϵ such that $\{x \in V_Q \| \|x\| \le \epsilon\} \perp D_Q(1)$ is the closed ϵ-neighborhood of $V_Q^{\perp} \cap B(1)$ in $B(1)$. This proves the following.

4.2.17. There exists a positive real number R such that $\{x \in V_Q \| \|x\| \le \Lambda\} \perp D_Q(r)$ is the closed Λ-neighborhood of $V_Q^{\perp} \cap B(r)$ in $B(r)$ for every real number $r \ge R$.

The preparations have finally been made to prove line 4.2.9. The numbers $\lambda_1, \ldots, \lambda_{m-1}$ and R have been chosen to depend only on P and L. Let w, p, and a be given as in line 4.2.9. Let $Q = \{p\}$.

First suppose that the distance from w to $V_Q^{\perp}$ is at least Λ. Because $\langle p, w \rangle \le 0$, the vectors w and p are on opposite sides of $V_Q^{\perp}$, so that $w + ap$ is closer to $V_Q^{\perp}$ than w. By line 4.2.16, $\pi(w) \in B(r)$. Hence the line segment from w to $\pi(w)$ is contained in $B(r)$ because $B(r)$ is convex. Thus $w + ap \in B(r)$ if the distance from w to $V_Q^{\perp}$ is at least Λ.

Now suppose that the distance from w to $V_Q^{\perp}$ is at most Λ. As before, either w and p are on opposite sides of $V_Q^{\perp}$ or $w \in V_Q^{\perp}$. In this case the local product structure of $B(r)$ described in line 4.2.17 easily shows that $w + ap \in B(r)$.

4.2.18. This completes the proof of line 4.2.9 and therefore the proofs of lines 4.2.7, 4.2.6, 4.2.1 and 4.1.2.
◇ ◇

Line 4.1.3 will be proven now. As after line 4.2.1 let V denote the linear hyperplane in $\mathbf{R}^n$ which is orthogonal to w_M. Let $\pi : \mathbf{R}^n \to V$ denote the canonical projection. Then $x = \pi(x) + \langle x, w_M \rangle w_M$ for every x in $\mathbf{R}^n$, and so

4.2.19.
$$\pi(x) = x - \frac{\langle x, aw_M \rangle}{a^2}(aw_M)$$

for every vector x in $\mathbf{R}^n$, where a is a positive real number such that aw_M is a vector of integers. Such an a exists by line 2.3.8. Since a is the length of a vector of integers, $a^2 \in \mathbf{Z}$.

According to line 4.1.2 the sequence of weight vectors $\frac{1}{\|w_1\|}w_1$, $\frac{1}{\|w_2\|}w_2$, $\frac{1}{\|w_3\|}w_3$, ... in S^{n-1} converges to w_M. It is easy to see that given any weight vector w in S^{n-1} there exists a neighborhood U of w in S^{n-1} such that if p is a path vector whose length is minimal relative to some weight vector in U, then the length of p is minimal relative to w. Thus taking $w = w_M$ shows that there exists a positive integer N_0 such that if ν is an integer with $\nu \geq N_0$, then every w_ν-minimal path vector is a w_M-minimal path vector. In other words, if $\nu \geq N_0$, then Q_ν consists of w_M-minimal path vectors.

Now let p be a w_M-minimal path vector. Then

$$p = \pi(p) + \langle p, w_M \rangle w_M = \pi(p) + H_{w_M} w_M.$$

Thus

$$\langle p, w_\nu \rangle = \langle \pi(p) + H_{w_M} w_M, \pi(w_\nu) + \langle w_\nu, w_M \rangle w_M \rangle$$
$$= \langle \pi(p), \pi(w_\nu) \rangle + H_{w_M} \langle w_\nu, w_M \rangle.$$

This shows that $\langle p, w_\nu \rangle$ is minimal if and only if $\langle \pi(p), \pi(w_\nu) \rangle$ is minimal.

The argument above shows for $\nu \geq N_0$ that $\pi(w_{\nu+1})$ is obtained from $\pi(w_\nu)$ as follows. Replace the sets P and $\pi(P)$ of all path vectors and their projections to V by the sets of all w_M-minimal path vectors and their projections to V. Let $\pi(Q_\nu)$ be the projection to V of the set Q_ν of w_M-minimal weight vectors p for which $\langle \pi(p), \pi(w_\nu) \rangle$ is minimal. Then $\pi(w_{\nu+1}) = \pi(w_\nu) + \pi(x_{Q_\nu})$, where, of course, $\pi(x_{Q_\nu})$ is a fixed vector in the cone spanned by $\pi(Q_\nu)$.

This construction of $\pi(w_{N_0})$, $\pi(w_{N_0+1})$, $\pi(w_{N_0+2})$, ... is a special case of the construction immediately preceding line 4.2.6. Thus not only is the sequence $\pi(w_{N_0})$, $\pi(w_{N_0+1})$, $\pi(w_{N_0+2})$, ... bounded as given by line 4.2.6, but Q_ν and hence $\pi(w_{\nu+1})$ are uniquely determined by $\pi(w_\nu)$ for $\nu \geq N_0$.

Since the coordinates of w_1, w_2, w_3, ... are integers, line 4.2.19 shows that the coordinates of $\pi(w_1)$, $\pi(w_2)$, $\pi(w_3)$, ... are rational numbers whose denominators are bounded. Combining the last statement with the previous paragraph shows that there exist positive integers N and μ such that $N \geq N_0$ and $\pi(w_{\mu+N}) = \pi(w_N)$. Because $\pi(w_N)$ uniquely determines $\pi(w_{N+1})$, $\pi(w_{\mu+N+1}) = \pi(w_{N+1})$. A straightforward induction argument now gives line 4.1.3.
$\diamondsuit$ 4.1.3 $\diamondsuit$

To prove line 4.1.4 let $w = w_{\mu+\nu} - w_\nu$. Then $w_{m\mu+\nu} = w_\nu + mw$ for every positive integer m. This implies that the sequence $\frac{1}{\|w_1\|}w_1$, $\frac{1}{\|w_2\|}w_2$, $\frac{1}{\|w_3\|}w_3$, ... converges to $\frac{1}{\|w\|}w$. Thus line 4.1.2 implies that w is a scalar multiple of w_M.
$\diamondsuit$ 4.1.4 $\diamondsuit$

4.3. The efficiency of the minimal path algorithm. The minimal path algorithm depends upon finding a cycle in the path additions of the algorithm. We next construct tilings which are juxtapositions of tilings such that the cycle length of the juxtaposition is the least common multiple of the cycle lengths of the component tilings. For this construction we first consider quadrilaterals which are actually rectangles. Such a rectangle R is divided into two halves by a vertical line segment. We tile R so that the left half of R is a single tile and the right half of R is tiled arbitrarily. It is easy to see that when the minimal path algorithm for either skinny flows or fat flows is applied to such a tiling of a rectangle R that at step i the weight of the left half of R is i and the height of R is i. Now consider a side-to-side juxtaposition of several congruent rectangles tiled in this way. It is easy to see that the i-th step of the minimal path algorithm for such a juxtaposition is gotten by combining the i-th steps of the minimal path algorithm for the component tilings. It follows that the cycle length of the juxtaposition is the least common multiple of the cycle lengths of the component tilings. As a result, the minimal path algorithm tends to run very slowly for such juxtapositions unless component cycle lengths have small least common multiple. As a consequence, we will develop a hybrid algorithm in the next section which does not suffer from this defect. We shall give an example in Section 4.5.

4.4. Flow diagrams and a hybrid algorithm for the finite Riemann mapping problem. In this section we present a hybrid algorithm for computing optimal weight functions for fat flows of tilings of quadrilaterals. Our hybrid algorithm depends on using the main construction of the minimal path algorithm to form a guess as to the appropriate circuit diagram to associate with a tiled quadrilateral. We then solve the circuit problem and check whether the answer is indeed a solution to the finite Riemann mapping problem. If not, we iterate the process to make a better guess for the circuit. In general, the hybrid algorithm works many times faster than the pure minimal path algorithm and avoids the least common multiple problem of the minimal path algorithm. It is convenient to work not with the circuit of the appropriate Smith diagram but rather directly with the generalized circuit described in Section 3.2 on the Kirchhoff inequalities. The guessing process is designed to find the edges along which positive current will flow in the solution of the finite Riemann mapping problem as described in Section 3.2. We deal with a subgraph Γ' of the dual graph Γ of the tiling as in Section 3.2. We did not direct the edges of Γ, but we will direct the edges of Γ'. Here is the hybrid algorithm.

(1) Choose strictly increasing sequences $u_1, u_2, u_3, \ldots$ and $v_1, v_2, v_3, \ldots$ of positive integers such that the differences $u_j - v_j$ are positive and approach ∞ as j approaches ∞. The integer i will index iterations of the hybrid algorithm. First set $i = 1$.

(2) Construct the sequence of weight functions $w_1, w_2, w_3, \ldots$ given by the minimal path algorithm and keep track of all the edges appearing in the minimal paths added during the steps which form w_{j+1} from w_j for $j = v_i, \ldots, u_i - 1$. In order to direct these edges, we assume that each of the minimal paths is oriented

from the vertex t_0 corresponding to the bottom edge of the tiled quadrilateral to the vertex t_1 representing the top edge of the tiled quadrilateral. The orientations of these minimal paths unambiguously determine directions for all such edges. The union of these directed edges forms a subgraph Γ' of the graph Γ whose vertices correspond to some subset of the tiles of the tiling and whose edges correspond to some subset of ordered pairs of tiles intersecting along a common edge. This graph Γ' is necessarily connected, connecting t_0 to t_1.

(3) Associate a *current* variable $i(s,t)$ with each directed edge (s,t) of the graph Γ'. Require that this variable only be allowed to take on positive values: $i(s,t) > 0$ if (s,t) is a directed edge associated with the directed graph Γ'. Associate with each vertex tile s of Γ' a vertex equation as in Section 3.2. Choose a basis for the first homology of Γ' and associate with each of the basis loops a loop equation as in Section 3.2. Ignore for the time being the third condition raised by the Kirchhoff inequalities, the only one of the three conditions in which actual inequalities arise. The result is a system of equations in the current variables.

(4) Solve the system of equations derived in step (3). The expected solution space is one dimensional. If that is the case, and if there is a positive solution in this one-dimensional space, proceed to step 5. Otherwise return to step (2) with i incremented by 1.

(5) With the positive variables determined, it is possible to define the weight $w(s)$ of each vertex tile s of Γ': add up the value of the currents flowing into that tile. Assign the weight 0 to each tile not in the subgraph Γ'. Find the length of the w-minimal path from t_0 to t_1 in the full dual graph Γ. If this length agrees with the (equal) lengths of the paths used in forming the directed graph Γ', then w is the desired optimal weight function by Theorem 3.2.1. Otherwise return to step (2) with i incremented by 1.

A combination of line 4.1.4 and Theorem 3.2.1 guarantees that the algorithm finds an optimal weight function. Line 4.1.4 guarantees that step (2) will eventually supply the directed edges through which the solution to the finite Riemann mapping problem will send positive current. A comparison with the appropriate Smith diagram shows that we are one equation short of the number needed to determine the currents uniquely. What is missing is the voltage of the battery edge which changes the solution by a single scale factor. Finally, if there is a positive solution, then Theorem 3.2.1 tells how to check to see whether the solution is correct as in step (5).

4.5. An example of the effectiveness of the hybrid algorithm.

We first tested the effectiveness of the hybrid algorithm on what we call a **corner**. As with the tiling in Section 3.1 we removed one tile from the upper right corner of a large square to form a small "top", with left side and bottom serving as "bottom". On a 9×9 large square the cyclic algorithm ran ove rnight without producing the optimal weight function. We made a hand implementation of the hybrid algorithm by running the cyclic algorithm a thousand cycles, observing the edges crossed by the minimal paths, and sending the resulting equations to a linear algebra package. The total computer time involved was something under four seconds. From the numbers alone it is easy to see that the minimal path algorithm must run through at least 6,476,565 cycles in order to find the optimal weight function, and with our actual implementation had to run at least 19,439,490 cycles. Here is the tiling with optimal weight function.

6476565	6476565	6476565	6479830	7775796	9719745	12959660	19439490	
4748175	4748175	4748175	5185170	6479830	7775796	9719745	12959660	19439490
3379050	3379050	3379050	4107375	5185170	6479830	7775796	9719745	12959660
2247750	2247750	2417850	3223800	4107375	5185170	6479830	7775796	9719745
1385100	1385100	1725300	2417850	3223800	4107375	5185170	6479830	7775796
765450	801900	1202850	1725300	2417850	3223800	4107375	5185170	6479830
328050	437400	801900	1202850	1725300	2417850	3379050	4748175	6476565
109350	218700	437400	801900	1385100	2247750	3379050	4748175	6476565
0	109350	328050	765450	1385100	2247750	3379050	4748175	6476565

5. Optimal weight functions for 2-layer valence 3 tilings of quadrilaterals.

Optimal weight functions for tiled quadrilaterals are in general very complicated. Even simple tilings supply tantalizingly difficult problems. We consider here a special case which can be used as a student puzzle.

Consider congruent rectangles R_i, $i = 0, 1$, and divide R_i into rectangular tiles by vertical line segments joining the top and bottom edges of R_i. Now stack R_1 on top of R_0 and assume that none of the vertical segments subdividing R_0 meets a vertical segment subdividing R_1. The result is a 2-**layer valence** 3 **tiling** T of the quadrilateral $R = R_0 \cup R_1$.

Example 5.0.1. Here is a sample 2-layer valence 3 tiling.

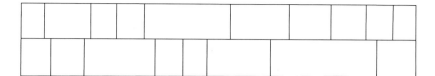

Our goal is to understand optimal weight functions for T relative to fat flows joining the top and bottom of R. We shall see that the tiling T has a natural and unique factorization into **prime** 2-layer valence 3 tilings, that the prime tilings can be completely understood in simple terms, that prime tilings can be multiplied in a simple way, and that the prime decomposition of a 2-layer valence 3 tiling can be found by efficient algorithms. We challenge the reader to carry out the analysis without hints.

We found that the minimal path algorithm was enough to get students started on the project. In just such a way, an undergraduate student at BYU by the name of Geo Meyer helped our research by discovering some of the first properties of the solution.

5.1. Basic definitions and properties.

Combinatorially equivalent tilings are said to be **equivalent**. The equivalence class of a tiling T is denoted by $[T]$.

The **weight** of a tile in the tiling T will be taken here to be the value of the reduced integral optimal weight function of T.

The intersection of R_0 with R_1 is called the **midline** of R. The tiles in R_0 form the **lower layer** of T, and the tiles in R_1 form the **upper layer**. We refer as needed to the left side, right side, top and bottom of R as the **left side, right side, top**, and **bottom** of T.

If u is a union of tiles in a layer of T and v is a union of tiles in the other layer of T, then u **dominates** v if $u \cap m \supset v \cap m$, where m is the midline of R.

Let T be a 2-layer valence 3 tiling of a quadrilateral R. A **right joint** of T is a union $j = e_1 \cup e_2 \cup e_3$ of three edges e_1, e_2, e_3 of T (viewing T as giving a cellular decomposition of R) such that i) e_2 is contained in the interior of the midline of T, ii) e_1 joins the top of T to the left vertex of e_2, and iii) e_3 joins the bottom of T to the right vertex of e_2. The right joint j appears in R as follows.

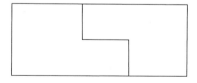

The right joint j separates R into two quadrilaterals Q_1 and Q_2, where j is the right side of Q_1 and the left side of Q_2. Similarly, j determines two subtilings T_1 and T_2 of T, where T_1 and T_2 are 2-layer valence 3 tilings of Q_1

and Q_2, respectively. The tiling T is called the **join** of T_1 and T_2. In general the expression "join of T_1 and T_2" means that T_1 is left of the joint and T_2 is right of the joint. If the optimal weight function for T induces on T_1 and T_2 scalar multiples of their optimal weight functions, then T is called the **(right) product** of T_1 and T_2. In this case j is called a **right break**.

There are analogous notions of **left joint, left join, (left) product** and **left break**.

If a tiling T is the product of tilings T_1 and T_2 as above, then T_1 and T_2 are called **factors** of T. It will be convenient to say that every tiling is a factor of itself. A **prime tiling** is a 2-layer valence 3 tiling of a quadrilateral which has no proper factors.

Example 5.1.1. Here is an example of a 2-layer valence 3 tiling of a quadrilateral which is the (right) product of two prime tilings. The right break between the factors is highlighted, and the weight of every tile appears in that tile.

3	3	3		8	
	9			4	4

The next proposition provides a simple criterion to determine when a join is a product.

Proposition 5.1.2. *Let T be a 2-layer valence 3 tiling of a quadrilateral which is the join of two subtilings T_1 and T_2. Using the reduced integral optimal weight function for T_1, let the rightmost upper, respectively lower, tile of T_1 have weight a, respectively b. Using the reduced integral optimal weight function for T_2, let the leftmost upper, respectively lower, tile of T_2 have weight c, respectively d. If the join is a right join, then it is a product if and only if $\frac{a}{b} \leq \frac{c}{d}$. If the join is a left join, then it is a product if and only if $\frac{a}{b} \geq \frac{c}{d}$.*

Proof. Suppose that T is the right product of T_1 and T_2. Furthermore suppose that the reduced integral optimal weight functions for T_1 and T_2 are scaled by factors r and s to give the reduced integral optimal weight function for T.

	ra	sc	
	rb	sd	

This situation leads to the following.

$$ra + rb = sc + sd \leq sc + rb$$

if and only if

$$\frac{s}{r} = \frac{a+b}{c+d} \text{ and } a+b \leq \frac{s}{r}c + b$$

if and only if

$$\frac{s}{r} = \frac{a+b}{c+d} \text{ and } a+b \leq \frac{a+b}{c+d}c + b$$

if and only if

$$\frac{s}{r} = \frac{a+b}{c+d} \text{ and } \frac{a}{b} \leq \frac{c}{d}.$$

Thus $\frac{a}{b} \leq \frac{c}{d}$.

Conversely, if $\frac{a}{b} \leq \frac{c}{d}$, and r, s are chosen so that $\frac{s}{r} = \frac{a+b}{c+d}$, then the calculations just completed easily show that minimal fat flows for T_1 and T_2 are minimal fat flows for T. Hence the weight function on T induced from the weight functions on T_1 and T_2 lies in the cone spanned by its minimal fat flows by line 2.3.2, and so it is an optimal weight function by line 2.3.6. Thus the right join of T_1 and T_2 is a product.

Analogous arguments prove the corresponding assertions for left joins. ◊ Proposition 5.1.2 ◊

It is clear that the notions of left join, right join, left product, and right product extend to the set of equivalence classes of 2-layer valence 3 tilings of quadrilaterals.

5.2. Unique prime factorization theorem.

Following is a unique prime factorization theorem for 2-layer valence 3 tilings of quadrilaterals.

Theorem 5.2.1. *Every 2-layer valence 3 tiling of a quadrilateral can be uniquely expressed as a product of prime tilings.*

Proof. Let T be a 2-layer valence 3 tiling of a quadrilateral. Let w be the reduced integral optimal weight function of T. Let $F = (f_1, \ldots, f_k)$ be a fundamental family of parallel fat flows whose sum is w in order from left to right. Let $s_1, \ldots, s_m$ be the upper tiles of T in order from left to right. Let $t_1, \ldots, t_n$ be the lower tiles of T in order from left to right. Let the weight of s_1 be ra, and let the weight of t_1 be rb, where a, b, r are positive integers with $GCD(a, b) = 1$.

Line 5.2.2 will be proved next.

5.2.2. If i is an integer with $1 \leq i < b$ and $i \leq m$, then $i + 1 \leq m$ and there exists an integer h with $1 \leq h < k$ such that f_h contains s_i, f_{h+1} contains s_{i+1}, and the lower tiles in f_h and f_{h+1} are equal.

It suffices to prove line 5.2.2 under the assumption that it is true for smaller values of i, using the fact that it is vacuously true for $i = 0$. Let $f_1, \ldots, f_h$ be the flows in F which contain one of $s_1, \ldots, s_i$, and let $t_1, \ldots, t_j$ be the lower tiles which are dominated by $s_1 \cup \cdots \cup s_i$. Because line 5.2.2 is true for smaller values of i and every flow in F has length $ra + rb$, it is not difficult to see that $s_1, \ldots, s_i$ all have weight ra and $t_1, \ldots, t_j$ all have weight rb.

In this paragraph we prove that $s_1, \ldots, s_i$ are not the upper tiles of a factor of T. For this first note that $h = rai$. Because $i < b$ and $GCD(a, b) = 1$, it is impossible for rb to divide h. On the other hand, it is easy to see that $rbj \le h \le rb(j + 1)$. Hence $rbj < h < rb(j + 1)$. This proves that $s_1, \ldots, s_i$ are not the upper tiles of a factor of T.

Since $s_1, \ldots, s_i$ are not the upper tiles of a factor of T, it follows that t_{j+1} exists and f_h contains s_i and t_{j+1}. Moreover, not only do f_{h+1} and s_{i+1} exist, but f_{h+1} contains s_{i+1} and t_{j+1}. This proves line 5.2.2.
◇ 5.2.2 ◇

Since line 5.2.2 is proved, $b \le m$. As was observed above, line 5.2.2 implies that $s_1, \ldots, s_b$ all have weight ra. Now let $f_1, \ldots, f_h$ be the flows in F which contain one of $s_1, \ldots, s_b$. It follows that $h = rab$. From this it is easy to see that $t_1, \ldots, t_a$ exist, they all have weight rb, and $f_1, \ldots, f_h$ are exactly the flows in F which contain $t_1, \ldots, t_a$. Let P be the subtiling of T which consists of $s_1, \ldots, s_b, t_1, \ldots, t_a$. Then P is a factor of T. It is proved in the next-to-last paragraph that P is prime. The prime tiling P is uniquely determined by T because the optimal weight function is unique.

Here is what the argument above proves. Given a 2-layer valence 3 tiling of a quadrilateral, there exists one and only one prime tiling which is a left factor of T. An obvious induction argument completes the proof of Theorem 5.2.1.
◇ Theorem 5.2.1 ◇

5.3. Prime tilings.

The proof of Theorem 5.2.1 yields information about prime tilings. In particular, there is a map from the set of prime tilings to the set of positive rational numbers. The prime tiling P maps to the rational number $\frac{a}{b}$, where P has b upper tiles each with weight a and a lower tiles each with weight b. The number $\frac{a}{b}$ will be called the **weight ratio** of the prime tiling P. In general a fraction $\frac{a}{b}$ will be called the **reduced weight ratio** of a prime tiling P if a, b are relatively prime positive integers and $\frac{a}{b}$ is the weight ratio of P.

It is clear that prime tilings are equivalent if and only if they have equal weight ratios. Thus the weight ratio map induces an injective map from the set of equivalence classes of prime tilings to the set of positive rational numbers.

Prime tilings with arbitrarily prescribed weight ratios will be constructed in the following paragraphs. In particular, this construction shows that the weight ratio map from the set of equivalence classes of prime tilings to the set of positive rational numbers is bijective.

Let a and b be positive integers with $GCD(a, b) = 1$. A prime tiling with weight ratio $\frac{a}{b}$ will now be constructed assuming that $a \geq b$. Let $x_1, \ldots, x_b$ be integers such that $x_i \equiv ia \mod b$ and $0 \leq x_i < b$ for $i = 1, \ldots, b$. Let $a = bq + r$, where q, r are integers such that $0 \leq r < b$. In particular, $r = x_1$. Construct a 2-layer valence 3 tiling P of a rectangle R as follows. The midline of the tiling is the line segment joining the midpoints of the sides of R. Let the upper layer of P consist of b rectangular tiles $s_1, \ldots, s_b$. Construct the lower layer of P to consist of rectangular tiles none of which dominates an upper tile so that for every integer i with $1 \leq i \leq b$ tile s_i dominates q tiles if $x_i \leq r$ and tile s_i dominates $q - 1$ tiles if $x_i > r$. This describes the tiling P.

To see that P is a prime tiling with weight ratio $\frac{a}{b}$, it suffices to prove that the weight of every upper tile of P is a and the weight of every lower tile in P is b. For this it suffices to construct a family F of fat flows for P such that i) every flow in F consists of one upper tile and one lower tile, ii) every lower tile is contained in b of the flows in F and iii) every upper tile is contained in a of the flows in F. To this end define integers $y_1, \ldots, y_b$ with $0 \leq y_i < b$ such that the following holds.

5.3.1. $x_i + y_i \equiv a \mod b$ for $i = 1, \ldots, b$.

Since $x_{i+1} \equiv x_i + a \mod b$ and $x_{i+1} + y_{i+1} \equiv a \mod b$, $x_i + y_{i+1} \equiv 0 \mod b$ for $i = 1, \ldots, b - 1$, which gives the following.

5.3.2. $x_i + y_{i+1} = b$ for $i = 1, \ldots, b - 1$.

The family F of fat flows will be constructed in this paragraph. Suppose given an upper tile s_i. If $i > 1$, then let t be the leftmost lower tile which meets s_i. If $i < b$, then let t' be the rightmost lower tile which meets s_i.

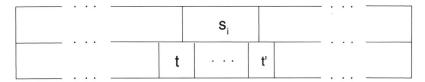

If $i > 1$, then let F include y_i paths consisting of s_i and t. If $i < b$, then let F include x_i paths consisting of s_i and t'. The family F will also include bq or $b(q-1)$ paths consisting of s_i and a tile below s_i depending upon whether $x_i \leq r$ or $x_i > r$.

It will now be verified that F satisfies the necessary conditions. It is clear that condition i) is satisfied. Condition ii) follows easily from line 5.3.2. To verify condition iii), let s_i be an upper tile. If $x_i \leq r$, then line 5.3.1 implies that $x_i + y_i = r$. Hence in this case $a = bq + x_i + y_i$, and so s_i is contained in a flows in F. If $x_i > r$, then 5.3.1 implies that $x_i + y_i = b + r$. Hence in this case $a = b(q - 1) + x_i + y_i$, and again s_i is contained in a flows of F.

This completes the construction of a prime tiling with weight ratio $\frac{a}{b}$ if $\frac{a}{b} \geq 1$. The construction for rational numbers less than 1 is analogous. (Alternatively, the horizontal reflection of a prime tiling with weight ratio $\frac{a}{b}$ is a prime tiling with weight ratio $\frac{b}{a}$. See below for reflections.)

Example 5.3.3. A prime tiling with weight ratio $\frac{17}{7}$ can be constructed as follows. First note that $17 = 7 \cdot 2 + 3$. Next compute the least nonnegative residues of multiples of 3 modulo 7: $3, 6, 2, 5, 1, 4, 0$. Because $6, 5, 4 > 3$, the second, fourth and sixth upper tiles dominate one tile. The other upper tiles dominate two tiles. Thus the following is a prime tiling with weight ratio $\frac{17}{7}$.

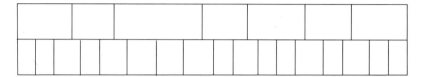

There is a **vertical reflection** (that is, reflection in a vertical line) on the set of equivalence classes of 2-layer valence 3 tilings of quadrilaterals which is defined as follows. Given an equivalence class of 2-layer valence 3 tilings of quadrilaterals, it is possible to choose a representative T which is a tiling of a rectangle. The vertical reflection maps $[T]$ to the equivalence class represented by the tiling gotten by reflecting T through a vertical line parallel to the sides of T. A tiling T will be called **symmetric** if $[T]$ is fixed by vertical reflection. There is a **horizontal reflection** (that is, reflection in a horizontal line) on the set of equivalence classes of 2-layer valence 3 tilings of quadrilaterals which is analogous to vertical reflection.

Proposition 5.3.4. *Prime tilings are symmetric (under reflection in a vertical line).*

Proof. Suppose that P is a prime tiling with reduced weight ratio $\frac{a}{b}$. If Q is a representative of the image of $[P]$ under the vertical reflection, then Q has the property that each of its upper tiles has weight a and each of its lower tiles has weight b. Thus every prime factor of Q has weight ratio $\frac{a}{b}$. This easily proves Proposition 5.3.4.
$\Diamond$ Proposition 5.3.4 $\Diamond$

5.4. Preparations for factorization algorithms.

Given a 2-layer valence 3 tiling T of a quadrilateral, one would like to be able to compute its optimal weight function. The unique prime factorization theorem provides a way to do this: computing the optimal weight function of a prime tiling merely involves counting tiles, and the optimal weight function of T is gotten by scaling the optimal weight functions of its prime factors so that their heights agree. Thus the problem of computing the optimal weight function of T reduces to computing the prime factorization of T. The results of this section will be used to factor 2-layer valence 3 tilings of quadrilaterals.

It is clear for each of the following results in this section that analogous results hold for representatives of the images of the equivalence class of the given tiling under the vertical and horizontal reflections. When the following results are cited, it will be left to the reader to apply reflections if necessary.

Lemma 5.4.1. *Suppose given an upper tile s in a 2-layer valence 3 tiling T of a quadrilateral which dominates n tiles $t_1, \ldots, t_n$ with $n \geq 1$. Let s have weight a, and let $t_1, \ldots, t_n$ have weight b.*

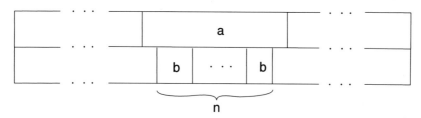

Then $n \leq \frac{a}{b} \leq m$, where m is the number of lower tiles meeting s. Since $m \leq n+2$, $n \leq \frac{a}{b} \leq n+2$. Moreover, $\frac{a}{b} = n$ if and only if s and $t_1, \ldots, t_n$ form a factor of T, and $\frac{a}{b} = n+2$ if and only if there is a lower tile t_0 immediately left of $t_1, \ldots, t_n$ and a lower tile t_{n+1} immediately right of $t_1, \ldots, t_n$ such that s and $t_0, \ldots, t_{n+1}$ form a factor of T.

Proof. Let w be the reduced integral optimal weight function of T. Let F be a fundamental family of parallel fat flows for T whose sum is w. It is clear that every flow in F which contains one of $t_1, \ldots, t_n$ also contains s, and so $nb \leq a$. Thus $n \leq \frac{a}{b}$. Moreover, equality holds if and only if the flows in F which contain one of $t_1, \ldots, t_n$ are exactly the flows which contain s. This occurs if and only if s and $t_1, \ldots, t_n$ form a factor of T. If some flow in F contains s but none of $t_1, \ldots, t_n$, then either t_0 or t_{n+1} exists and its weight is b. Now it is easy to see that $a \leq mb$. Thus $\frac{a}{b} \leq m$. The final assertion of the lemma is now easy to prove.
◊ Lemma 5.4.1 ◊

Lemma 5.4.2. *Suppose given an upper tile s in a 2-layer valence 3 tiling of a quadrilateral which dominates n tiles $t_1, \ldots, t_n$ with $n \geq 0$. Suppose that there exists a lower tile $t_0 \notin \{t_1, \ldots, t_n\}$ which meets the lower left vertex of s. Let s have weight a, and let t_0 have weight b.*

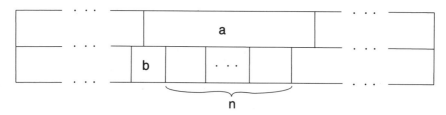

Then $\frac{a}{b} \leq m$, where m is the number of lower tiles meeting s. Since $m \leq n+2$, $\frac{a}{b} \leq n+2$. Moreover, $\frac{a}{b} = n+2$ if and only if there is a lower tile t_{n+1} immediately right of $t_1, \ldots, t_n$ such that s and $t_0, \ldots, t_{n+1}$ form a factor of T.

Proof. It is clear that if a minimal fat flow contains s, then the lower tile in it has weight at most b. As in the proof of Lemma 5.4.1, it easily follows that $a \le mb$. Thus $\frac{a}{b} \le m$. The rest of the lemma can be proved as in Lemma 5.4.1. $\Diamond$ Lemma 5.4.2 $\Diamond$

Lemma 5.4.3. *Suppose given an upper tile in a 2-layer valence 3 tiling of a quadrilateral which dominates $m \ge 0$ tiles. Suppose that immediately to the right of this upper tile is another upper tile which dominates $n \ge m$ tiles. Define points x, y as in the diagram below.*

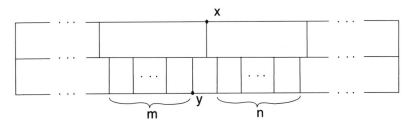

If there is a break from x to y, then the prime tilings on either side of this break are equivalent.

Proof. Suppose that there is a break from x to y. Let the reduced weight ratio of the prime tiling left, respectively right, of this break be $\frac{a}{b}$, respectively $\frac{c}{d}$. Then Lemmas 5.4.1 and 5.4.2 applied to these prime tilings show that

$$\frac{a}{b} \le m + 1 \le n + 1 \le \frac{c}{d}.$$

Since Proposition 5.1.2 states that $\frac{a}{b} \ge \frac{c}{d}$, it follows that the two prime tilings have equal weight ratios, and so the two prime tilings are equivalent. $\Diamond$ Lemma 5.4.3 $\Diamond$

Proposition 5.4.4. *Suppose given an upper tile in a 2-layer valence 3 tiling of a quadrilateral which dominates $m \ge 0$ tiles. Suppose that immediately to the right of this upper tile are $k \ge 0$ upper tiles each of which dominates $m + 1$ tiles. Suppose that immediately to the right of these upper tiles is an upper tile which dominates $n \ge m + 2$ tiles. Define points x, y as in the diagram below.*

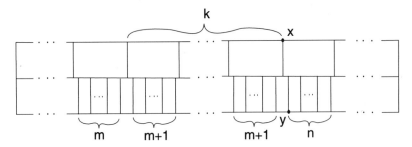

Then the joint from x to y is a break.

Proof. Let the reduced weight ratio of the prime tiling P containing the leftmost of the $k+2$ upper tiles under consideration be $\frac{a}{b}$. Similarly, let the reduced weight ratio of the prime tiling Q containing the rightmost of the $k+2$ upper tiles under consideration be $\frac{c}{d}$. According to Lemmas 5.4.1 and 5.4.2

$$\frac{a}{b} \le m+2 \le n \le \frac{c}{d}.$$

If $\frac{c}{d} = n$, then Lemma 5.4.1 shows that there is a break joining x and y as desired. Hence it may be assumed that $\frac{c}{d} > n$. Thus $P \ne Q$, and so there is a break between P and Q.

Suppose that the joint from x to y is not a break.

In this paragraph it will be shown by contradiction that the break at the left side of Q is not a right break. Suppose that it is a right break. Then Lemma 5.4.1 shows that $\frac{c}{d} \le m+2 \le n$. This contradicts the assumption that $\frac{c}{d} > n$.

Thus the break at the left side of Q is a left break. Now Lemma 5.4.3 shows that Q is equivalent to the prime tiling to its left. This is easily seen to be impossible using Proposition 5.3.4, which states that Q is symmetric. $\Diamond$ Proposition 5.4.4 $\Diamond$

The next proposition can be viewed as an extension of Proposition 5.4.4. It might be said that Proposition 5.4.5 extends Proposition 5.4.4 to the case in which m is negative and n is positive.

Proposition 5.4.5. *Suppose given a lower tile in a 2-layer valence 3 tiling of a quadrilateral which dominates m tiles with $m \ge 1$. Suppose that immediately to the right of this lower tile are $k \ge 0$ lower tiles each of which dominates $m-1$ tiles. Suppose that immediately to the right of these upper tiles is an upper tile which dominates n tiles with $n \ge 1$. Define points x, y as in the diagram below.*

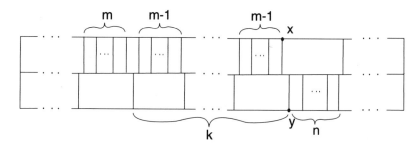

Then the joint from x to y is a break.

Proof. The proof of Proposition 5.4.4 easily extends to prove Proposition 5.4.5. $\Diamond$ Proposition 5.4.5 $\Diamond$

5.5. The left factor algorithm.

An algorithm, called the **left factor algorithm**, for computing prime factorizations of 2-layer valence 3 tilings of quadrilaterals will be described in this section. The left factor algorithm consists of three phases. The first phase is described immediately below.

Let T be a 2-layer valence 3 tiling of a quadrilateral. Using Propositions 5.4.4 and 5.4.5, the first phase of the left factor algorithm expresses T as a product of factors which have a very special form as follows. Suppose that T contains two tiles s, respectively t, which dominate m, respectively n, tiles. Suppose that $n \geq m+2$ if s, t are in the same layer of T, and suppose that $m > 0$, $n > 0$ if s, t are in different layers of T. Further suppose that s, t are closest such tiles. The following two sentences are easy to see under these assumptions. If s, t are in the same layer and there are tiles between s and t in this layer, then $n = m + 2$ and every tile between s and t in this layer dominates $m + 1$ tiles. If s, t are in different layers and there are tiles "between" s and t, then $m = n = 1$ and every tile "between" s and t dominates no tile. Thus under these assumptions either Proposition 5.4.4 or Proposition 5.4.5 provides an easily computable factorization of T. The first phase of the left factor algorithm consists of repeatedly applying Propositions 5.4.4 and 5.4.5 in this manner to obtain a factorization of T for which every factor S has the following properties: i) either no lower tile of S dominates a tile or no upper tile of S dominates a tile; ii) if no lower tile of S dominates a tile, then there exists an integer k such that every upper tile dominates either k or $k + 1$ tiles; iii) if no upper tile of S dominates a tile, then there exists an integer k such that every lower tile dominates either k or $k + 1$ tiles. Thus (using the horizontal reflection) the first phase of the left factor algorithm easily reduces computing the prime factorization of T to computing prime factorizations of tilings S such that no lower tile of S dominates a tile and there exists an integer k for which every upper tile of S dominates either k or $k + 1$ tiles.

A function will be defined in this paragraph which will be used by the second phase of the left factor algorithm to simplify these latter tilings S. Let $\mathcal{T}$ denote the set of equivalence classes of 2-layer valence 3 tilings of quadrilaterals. Let $\mathcal{T}^+$ denote the set of equivalence classes $[T]$ in $\mathcal{T}$ such that every upper tile in T dominates at least one tile but T is not a prime tiling with weight ratio 1, that is, T does not consist of one upper tile and one lower tile. Define the **deletion map** $\delta : \mathcal{T}^+ \to \mathcal{T}$ so that for every equivalence class $[T]$ in $\mathcal{T}^+$ the equivalence class $\delta([T])$ is represented by the tiling gotten from T by deleting one tile below every upper tile.

Example 5.5.1. The equivalence class of the tiling

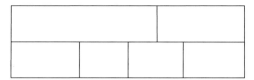

is in $\mathcal{T}^+$, and its image under the deletion map δ is represented by the following.

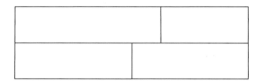

The deletion map is introduced because of the following proposition.

Proposition 5.5.2. *If P is a prime tiling with weight ratio $\frac{a}{b}$ such that $[P]$ lies in $\mathcal{T}^+$, then $\delta([P])$ is the equivalence class of prime tilings with weight ratio $\frac{a}{b} - 1$. Furthermore, given an equivalence class $[T]$ in $\mathcal{T}^+$ such that T is the product of prime tilings $P_1, \ldots, P_n$ in order from left to right, then $[P_1], \ldots, [P_n]$ lie in $\mathcal{T}^+$ and $\delta([T])$ is the product of $\delta([P_1]), \ldots, \delta([P_n])$ in order from left to right.*

Proof. Let P be a prime tiling with reduced weight ratio $\frac{a}{b}$ such that $[P]$ lies in $\mathcal{T}^+$. It is clear that $a \geq b$. Moreover, $a > b$ because the equivalence class of prime tilings with weight ratio 1 is not in $\mathcal{T}^+$. Let $a = bq + r$, where q, r are integers with $0 \leq r < b$. Suppose that $q > 1$. Then since $a - b = b(q - 1) + r$ with $q - 1 \geq 1$, the construction of prime tilings in Section 5.3 shows that $\delta([P])$ is represented by a prime tiling with weight ratio $\frac{a-b}{b} = \frac{a}{b} - 1$. Now suppose that $q = 1$. In this case because every upper tile of P dominates at least one tile, the construction of prime tilings shows that every upper tile of P dominates exactly one tile and that $r = b - 1$. It easily follows that if $q = 1$, then $\delta([P])$ is represented by a prime tiling with weight ratio $\frac{b-1}{b} = \frac{r}{b} = \frac{b+r}{b} - 1 = \frac{a}{b} - 1$. This proves the first assertion of Proposition 5.5.2.

In this paragraph it will be proved that if P is a prime factor of a tiling T such that $[T]$ lies in $\mathcal{T}^+$, then $[P]$ is also in $\mathcal{T}^+$. It is easy to see that every upper tile in P dominates at least one tile. Thus what must be proved is that the weight ratio of P is not 1. Suppose that the weight ratio of P is 1. Then because every upper tile in T dominates at least one tile, P occurs in T in one of the following three forms.

5.5.3.

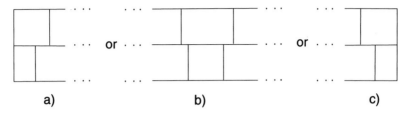

a)   b)   c)

If P occurs in T as in a) of Figure 5.5.3, then Lemma 5.4.3 shows that the prime tiling Q to the right of P is equivalent to P. But Q does not occur in T in one of the three forms given in Figure 5.5.3. Thus P does not occur in T as in a) of Figure 5.5.3. Similar arguments conclude the proof that the weight ratio of P is not 1, and so $[P]$ lies in $\mathcal{T}^+$.

It is now easy to complete the proof of Proposition 5.5.2 by using Proposition 5.1.2 and the fact that δ reduces weight ratios of equivalence classes of prime tilings by 1.
◇ Proposition 5.5.2 ◇

The description of the left factor algorithm now turns to the second phase. Let T be a 2-layer valence 3 tiling of a quadrilateral for which there exists a positive integer k such that every upper tile of T dominates either k or $k+1$ tiles. Assuming that T is not the prime tiling with weight ratio 1, Proposition 5.5.2 shows that computing the prime factorization of T is equivalent to computing the prime factorization of a tiling S which represents $\delta([T])$. Note that Example 5.5.1 shows for $k = 1$ that a lower tile of S might dominate an upper tile. If this occurs, then apply the first phase of the left factor algorithm to S. The second phase of the left factor algorithm replaces T by a tiling which represents the image of $[T]$ under the largest possible iterate of δ and applies the first phase of the algorithm to this tiling.

Thus the factorization of a 2-layer valence 3 tiling T of a quadrilateral easily reduces to the case in which no lower tile of T dominates a tile and every upper tile of T dominates at most one tile. It is interesting that this seems to be the most difficult case. The next lemma gives information about the location of breaks for such tilings T.

Lemma 5.5.4. *Let T be a 2-layer valence 3 tiling of a quadrilateral such that no lower tile of T dominates a tile and every upper tile of T dominates at most one tile. Let b be a left break of T, and let s, respectively t, be the upper tile just left, respectively right, of b. Then s dominates one tile and t dominates no tile. An analogous result holds for right breaks.*

Proof. The proof will proceed by induction on the number k of prime factors of T. If $k = 1$, then the lemma is vacuously true, so suppose that $k > 1$ and that the lemma is true for smaller values of k.

Suppose that s dominates m tiles and t dominates n tiles with $m, n \in \{0, 1\}$. The proof of the induction step will proceed by contradiction. Suppose that $m \le n$. Let P be the prime tiling immediately left of b, and let Q be the prime tiling immediately right of B. Lemma 5.4.3 shows that $[P] = [Q]$. Let P, Q have weight ratio $\frac{a}{b}$. Lemmas 5.4.1 and 5.4.2 show that

$$\frac{a}{b} \le m + 1 \le n + 1 \le \frac{a}{b}.$$

Thus $\frac{a}{b} = m + 1 = n + 1$. This implies that P consists of s, the m tiles which s dominates and one more lower tile which meets the lower left vertex of P. Similarly, Q consists of t, the n tiles which t dominates and one more lower tile which meets s and t.

Now suppose that $m = n = 0$. Then the right side of Q is not the right side of T, for otherwise the lower tile of Q dominates the upper tile of Q, contrary to assumption. Thus the right side of Q is a left break of the tiling S which consists of the tiles between b and the right side of T. The induction hypothesis applies to S and shows that Q dominates one tile. This contradiction proves the induction step in this case.

It remains to consider the case in which $m = n = 1$. This case can be proved just as the case in which $m = n = 0$ by working with the factor of T left of b. $\Diamond$ Lemma 5.5.4 $\Diamond$

Now let T be a 2-layer valence 3 tiling of a quadrilateral such that no lower tile of T dominates a tile and every upper tile of T dominates at most one tile. The third phase of the left factor algorithm, which computes the prime factorization of T will now be given. The left factor algorithm is named for this third and most difficult phase. Call a joint of T a **potential break** if it is not ruled out by Lemma 5.5.4. If T has no potential breaks, then T is prime. Otherwise, let $b_1, \ldots, b_n$ be the potential breaks of T.

Now the left factor algorithm searches for the leftmost prime factor of T as follows. Let T_1 be the tiling which consists of the tiles of T between the left side of T and b_1. The next step of the algorithm is to determine whether or not T_1 is prime. An easy, although possibly inconclusive, test to apply is to test T_1 for symmetry. If T_1 is not symmetric, then T_1 is not prime by Proposition 5.3.4. If T_1 is symmetric, then suppose that T_1 has a lower tiles and b upper tiles. If T_1 is prime, then its weight ratio is $\frac{a}{b}$. It is only slightly more difficult than testing for symmetry to check whether or not T_1 is a prime tiling with weight ratio $\frac{a}{b}$, using the construction of prime tilings given in Section 5.3. If T_1 is not prime, then redefine T_1 to be the tiling which consists of the tiles between the left side of T and b_2 (the right side of T if $n = 1$). Continue in this manner until a prime tiling T_1 is found. This tiling T_1 is a potential leftmost prime factor of T.

If $T_1 = T$, then there is nothing more to do. If $T_1 \neq T$, then find as above a potential leftmost prime factor T_2 of the tiling S which consists of the tiles of T between the right side of T_1 and the right side of T.

Use Proposition 5.1.2 to test whether or not the join of T_1 and T_2 is a product. If the join of T_1 and T_2 is not a product, then continue the search for a suitable T_2, that is, a prime tiling T_2 such that the join of T_1 and T_2 is a product and T_2 consists of the tiles between the left side of S and a potential break of S or the right side of S. If no suitable T_2 is found, then T_1 is not the leftmost prime factor of T, so continue the search for a leftmost prime factor of T.

Suppose that a suitable tiling T_2 is found. If T is the product of T_1 and T_2, then there is nothing more to do. Otherwise, search for a suitable T_3. It is clear that this process continues until the prime factorization of T is found. This completes the description of the left factor algorithm.

Example 5.5.5. Following is an application of the third phase of the left factor algorithm. Let T be the tiling shown below.

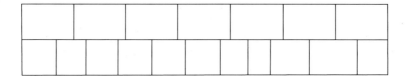

No lower tile of T dominates a tile and every upper tile of T dominates at most one tile. Thus the third phase of the left factor algorithm applies to T.

The potential breaks of T are labeled b_1, b_2, b_3, b_4 below, and the sides of T are labeled b_0, b_5.

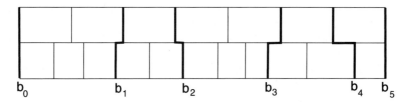

For integers i, j with $0 \le i < j \le 5$, let T_{ij} be the tiling which consists of the tiles in T between b_i and b_j. The table below describes how the left factor algorithm computes the prime factorization of T. Every box is either empty or it contains some T_{ij}. If T_{ij} is not symmetric, then "NS" appears. If T_{ij} is prime, then "P" appears with the weight ratio of T_{ij}. If T_{ij} is prime, but the join of the previous potential prime factor of T and T_{ij} is not a product, then "NP" appears. If T_{ij} is prime and the join of the previous potential prime factor of T and T_{ij} is a product, then move right in the table. Otherwise, move down and left.

$T_{0,1}$ P,3/2	$T_{1,2}$ P,2,NP	
	$T_{1,3}$ P,5/3,NP	
	$T_{1,4}$ P,7/4,NP	
	$T_{1,5}$ NS	
$T_{0,2}$ P,5/3	$T_{2,3}$ P,3/2,NP	
	$T_{2,4}$ P,5/3	$T_{4,5}$ P,1,NP
	$T_{2,5}$ NS	
$T_{0,3}$ P,8/5	$T_{3,4}$ P,2,NP	
	$T_{3,5}$ P,3/2	

Thus T is the product of $T_{0,3}$ and $T_{3,5}$, prime tilings with weight ratios $\frac{8}{5}$ and $\frac{3}{2}$, respectively.

Here is the reduced integral optimal weight function.

40	40	40	40	40	39	39

25	25	25	25	25	25	25	25	26	26	26

5.6. The joint ratio algorithm.

In this section another algorithm, called the joint ratio algorithm, for computing prime factorizations of 2-layer valence 3 tilings of quadrilaterals will be described. Joint ratios are defined next.

Suppose given a 2-layer valence 3 tiling T of a quadrilateral. Define the **right joint ratio** for a given right joint of T to be the number of lower tiles in T left of the joint divided by the number of upper tiles in T left of the joint. Define a right joint ratio for the right side of T as well; this is the number of lower tiles in T divided by the number of upper tiles in T. Define **left joint ratios** in the same way using left joints instead of right joints. The left joint ratio for the right side of T equals the right joint ratio for the right side of T.

Example 5.6.1. The right and left joint ratios for

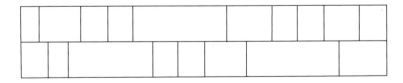

are

$$\frac{1}{1} \quad \frac{3}{4} \quad \frac{6}{5} \quad \frac{7}{8} \quad \frac{8}{10}$$

and

$$\frac{2}{2} \quad \frac{5}{5} \quad \frac{6}{6} \quad \frac{7}{9} \quad \frac{8}{10}.$$

The next theorem is the basis for the joint ratio algorithm.

Theorem 5.6.2. *Suppose given a 2-layer valence 3 tiling T of a quadrilateral.*

1) The minimal right joint ratios of T occur at breaks or the right side of T.

2) The maximal left joint ratios of T occur at breaks or the right side of T.

3) If the joint ratio of the right side of T is both the unique minimal right joint ratio and the unique maximal left joint ratio of T, then T is a prime tiling.

Proof. The proof will begin by establishing some notation. Suppose that T has a right joint j. Let j_1 be a right joint of T left of j. If there is no right joint of T left of j, then let j_1 be the left side of T. Let j_2 be a right joint of T right of j. If there is no right joint of T right of j, then let j_2 be the right side of T. Let u, respectively v, be the number of upper, respectively lower, tiles left of j. Let x, respectively y, be the number of upper, respectively lower, tiles between j and j_1. Let w, respectively z, be the number of upper, respectively lower, tiles between j and j_2.

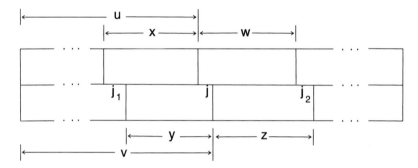

Thus the right joint ratios for j, j_1, j_2 are $\frac{v}{u}, \frac{v-y}{u-x}, \frac{v+z}{u+w}$, except that j_1 has no right joint ratio if it is the left side of T. This leads to line 5.6.3, which is easily verified.

5.6.3.

$$\frac{v}{u} \le \frac{v-y}{u-x} \iff uv - vx \le uv - uy \iff vx \ge uy \iff \frac{v}{u} \ge \frac{y}{x}$$

$$\frac{v}{u} \le \frac{v+z}{u+w} \iff uv + vw \le uv + uz \iff vw \le uz \iff \frac{v}{u} \le \frac{z}{w}$$

Now consider statement 1) of Theorem 5.6.2. Suppose that the right joint ratio of the above joint j is minimal and that j is not a break. This will lead to a contradiction. Choose j_1 and j_2 to be right breaks or sides of T so that there are no right breaks between j_1 and j_2. Because the right joint ratio for j is less than or equal to the right joint ratios for j_1 (if it exists) and j_2, $\frac{v}{u} \le \frac{v-y}{u-x}$ (if meaningful) and $\frac{v}{u} \le \frac{v+z}{u+w}$. From line 5.6.3 it follows that

5.6.4.

$$\frac{v}{u} \ge \frac{y}{x} \text{ and } \frac{v}{u} \le \frac{z}{w}.$$

The ratio $\frac{y}{x}$ will be investigated in this paragraph. Let $P_1, \ldots, P_n$ be the consecutive prime factors of T for which j_1 is the left side of P_1 and j is contained in P_n.

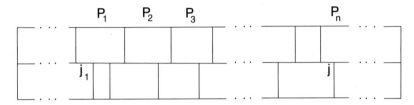

Let the reduced weight ratios of $P_1, \ldots, P_n$ be $\frac{a_1}{b_1}, \ldots, \frac{a_n}{b_n}$. Because the breaks between the P_i's are left breaks, Proposition 5.1.2 shows that

$$\frac{a_1}{b_1} \ge \ldots \ge \frac{a_n}{b_n}.$$

For $i = 1, \ldots, n-1$ let $x_i = b_i$ and $y_i = a_i$. Choose x_n and y_n so that $\sum_{i=1}^{n} x_i = x$ and $\sum_{i=1}^{n} y_i = y$. Then $\frac{y_i}{x_i} = \frac{a_i}{b_i} \ge \frac{a_n}{b_n}$, and so

5.6.5.

$$y_i b_n \ge x_i a_n \quad \text{for} \quad i = 1, \ldots, n-1.$$

Furthermore, because P_n is a prime tiling containing the right joint j having x_n tiles with weights a_n in its upper layer left of j and y_n tiles with weights b_n in its lower layer left of j,

5.6.6.

$$y_n b_n > x_n a_n.$$

Combining lines 5.6.5 and 5.6.6 gives $y b_n > x a_n$, and so

5.6.7.

$$\frac{y}{x} > \frac{a_n}{b_n}.$$

Combining lines 5.6.4 and 5.6.7 gives $\frac{v}{u} > \frac{a_n}{b_n}$. An argument just like that of the previous paragraph gives $\frac{z}{w} < \frac{a_n}{b_n}$, and so $\frac{v}{u} < \frac{a_n}{b_n}$. This contradiction concludes the proof of 1).

Statement 2) can be proved just as statement 1) was proved, or it can be reduced to statement 1) by means of the horizontal reflection.

Now consider statement 3). Suppose that T is not prime. It must be shown that the joint ratio of the right side of T is not both the unique minimal right joint ratio and the unique maximal left joint ratio of T.

First suppose that all of the breaks of T are right breaks. Let $P_1, \ldots, P_n$ be the prime factors of T with reduced weight ratios $\frac{a_1}{b_1}, \ldots, \frac{a_n}{b_n}$.

Arguing as in the paragraph following line 5.6.4, it follows that

$$\frac{a_1}{b_1} \leq \cdots \leq \frac{a_n}{b_n},$$

hence $b_i a_1 \leq a_i b_1$ for $i = 1, \ldots, n$, hence $x a_1 \leq y b_1$, where $x = \sum_{i=1}^{n} b_i$ and $y = \sum_{i=1}^{n} a_i$, and so $\frac{a_1}{b_1} \leq \frac{y}{x}$. This shows that the right joint ratio $\frac{y}{x}$ of the right side of T is at least as large as the right joint ratio $\frac{a_1}{b_1}$ of the break between P_1 and P_2, as desired.

The argument above proves 3) if all of the breaks of T are right breaks. It is now easy to see that 3) is also true if all of the breaks of T are left breaks.

Finally suppose that T has both right breaks and left breaks. By applying the horizontal reflection if necessary, it may be assumed that the leftmost break of T is a right break. The argument of the next-to-last paragraph shows that the right joint ratio of this leftmost break of T is less than or equal to the left joint ratio of the leftmost left break of T. Since the right joint ratios are minimized and the left joint ratios are maximized, if the minimal value equals the maximal value, then these values equal the joint ratios of the leftmost right and left breaks of T. This finishes the proof of 3).
◇ Theorem 5.6.2 ◇

Theorem 5.6.2 provides a **joint ratio algorithm** for computing factorizations of 2-layer valence 3 tilings of quadrilaterals as follows. Compute all right joint ratios of T. All minimal values of these joint ratios occur at breaks or the right side of T. Compute all left joint ratios of T. All maximal values of these joint ratios occur at breaks or the right side of T. If no breaks of T arise from this minimization and maximization, then T is prime. Otherwise, perform analogous minimizations and maximizations with the factors of T determined by all of the breaks found above. This process eventually ends with the prime factorization of T. This completes the description of the joint ratio algorithm.

In practice it is easier to apply the first two phases of the left factor algorithm than the joint ratio algorithm to a general 2-layer valence 3 tiling T of a quadrilateral. In other words, it is best to reserve the joint ratio algorithm for tilings of the sort handled by the third phase of the left factor algorithm. In this case it is only necessary to compute joint ratios of the potential breaks given by Lemma 5.5.4.

Example 5.6.8. Consider the tiling T of Example 5.5.5, given below.

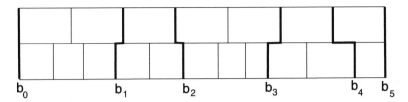

The right joint ratios for b_2, b_4, b_5 are $\frac{5}{3}, \frac{10}{6}, \frac{11}{7}$. The minimum is $\frac{11}{7}$, which yields no breaks. The left joint ratios for b_1, b_3, b_5 are $\frac{3}{2}, \frac{8}{5}, \frac{11}{7}$. The maximum is $\frac{8}{5}$. Thus b_3 is a break.

Now consider $T_{0,3}$, using the notation of Example 5.5.5. The right joint ratios for b_2, b_3 are $\frac{5}{3}, \frac{8}{5}$. The minimum is $\frac{8}{5}$. The left joint ratios for b_1, b_3 are $\frac{3}{2}, \frac{8}{5}$. The maximum is $\frac{8}{5}$. Thus $T_{0,3}$ is prime with weight ratio $\frac{8}{5}$.

Finally, consider $T_{3,5}$. The right joint ratios for b_4, b_5 are $\frac{2}{1}, \frac{3}{2}$. The minimum is $\frac{3}{2}$. The only left joint ratio is $\frac{3}{2}$. Thus $T_{3,5}$ is prime with weight ratio $\frac{3}{2}$.

The weight ratio algorithm shows that T is the product of $T_{0,3}$ and $T_{3,5}$.

6. Approximating combinatorial moduli.

How does one approximate the combinatorial moduli of a shingled ring? In Section 7 we shall explain a connection between combinatorial moduli and the recognition of discrete groups of constant negative curvature in dimension 3. This treatment requires that a whole sequence of combinatorial moduli be approximated.

Recall that if S is a finite collection of sets that covers a ring or quadrilateral R in the plane or 2-sphere, then we have combinatorial moduli $M_{\sup}(R, S)$ and $m_{\inf}(R, S)$. Our aim is to approximate these two numbers.

The conformal or quasiconformal compatibility of an entire sequence of covers is measured by the notion of *conformality*. A sequence $S(1), S(2), \ldots$ of locally finite covers of the 2-sphere S^2 or the plane E^2 with mesh approaching 0 is said to be **conformal** if there is a positive number K having the following properties:

Axiom (1). *Approximate moduli are almost well-defined.* That is, if R is a ring in S^2 (E^2), then there is a positive integer I and a positive number m such that, for all $i \geq I$, the combinatorially defined moduli

$$m_{\inf}(R, S(i)) \quad \text{and} \quad M_{\sup}(R, S(i))$$

both lie in the interval $[m, K \cdot m]$. The number m is called an **approximate modulus** for R.

Axiom (2). *Points of S^2 (E^2) are encircled by small rings of large approximate modulus.* That is, given $p \in S^2$ (E^2), a neighborhood N of p in S^2, and a positive number M, there is a ring R encircling p in N such that any approximate modulus m for R is at least M.

6.1. Skinny approximation.

In this section we obtain estimates on moduli from the existence of skinny paths with special properties.

Given a tiling of a quadrilateral or ring, the finite Riemann mapping theorem shows that there exists a grid consisting of flows and cuts such that the paths in one direction are fat, the paths in the other direction are skinny and the modulus is the numb er of cuts divided by the number of flows. Theorem 6.1.1 states that if we allow the paths in both directions to be skinny, then the number of cuts divided by the number of flows lies between our two moduli.

Theorem 6.1.1. *Let T be a tiling of a quadrilateral or ring X. Let $(f_1, \ldots, f_C)$ be a family of parallel skinny flows of T, and let $(c_1, \ldots, c_H)$ be a family of parallel skinny cuts of T for some positive integers C and H. Let $(\alpha_1, \ldots, \alpha_C)$ be a family of underlying piecewise smooth arcs for $(f_1, \ldots, f_C)$, and let $(\beta_1, \ldots, \beta_H)$ be a family of underlying piecewise smooth arcs or simple closed curves for $(c_1, \ldots, c_H)$. Suppose that every α_i meets every β_j exactly once, and given a tile t in T, the number of α_i's meeting the interior of t equals the number of β_j's meeting the interior of t and every such α_i meets every such β_j in the interior of t. Then*

$$m_f \leq \frac{H}{C} \leq M_f.$$

Proof. Define a weight function w on T as follows. If t is a tile in T, then $w(t)$ is the number of α_i's which meet the interior of t.

We first estimate A_w as follows.

$$A_w = \langle w, w \rangle = \left\langle \sum_{i=1}^{C} f_i, \sum_{j=1}^{H} c_j \right\rangle = \sum_{i=1}^{C}\sum_{j=1}^{H} \langle f_i, c_j \rangle \leq \sum_{i=1}^{C}\sum_{j=1}^{H} 1 = CH.$$

Next let p be a minimal fat cut for w. Then

$$C_{w,f} = \langle p, w \rangle = \left\langle p, \sum_{i=1}^{C} f_i \right\rangle = \sum_{i=1}^{C} \langle p, f_i \rangle \geq \sum_{i=1}^{C} 1 = C.$$

Thus

$$\frac{H}{C} = \frac{CH}{C^2} \geq \frac{A_w}{C_{w,f}^2} \geq m_f,$$

which gives the first inequality of Theorem 6.1.1.

Next let p be a minimal fat flow for w. Then

$$H_{w,f} = \langle p, w \rangle = \left\langle p, \sum_{j=1}^{H} c_j \right\rangle = \sum_{j=1}^{H} \langle p, c_j \rangle \geq \sum_{j=1}^{H} 1 = H.$$

Thus

$$\frac{H}{C} = \frac{H^2}{CH} \leq \frac{H_{w,f}^2}{A_w} \leq M_f,$$

which gives the second inequality of Theorem 6.1.1.

This proves Theorem 6.1.1.

$\Diamond$ Theorem 6.1.1 $\Diamond$

It will be convenient in investigating Axiom (2) of Cannon [8] to allow degenerate rings in which one of the boundary components is a point. That is, a **degenerate ring** is a closed topological disk X together with a distinguished point x_0 in its interior. Suppose S is a shingling of a degenerate ring X with distinguished point x_0. A **fat cut** is a fat path which has an underlying topological path separating the boundary of X and the point x_0. A **skinny flow** is a skinny path whose shingles can be ordered such that one of its extreme shingles contains x_0 and the other extreme shingle meets the boundary of X. With these definitions, one defines the moduli m_f and M_s as for a ring.

Proposition 6.1.2. *Let X be a quadrilateral or ring or degenerate ring, and let T be a tiling of X. Let $w_1, \ldots, w_k$ be weight functions on T with disjoint supports for some positive integer k. Suppose that for each $i \in \{1, \ldots, k\}$, $C_{w_i,f} \geq 1$. Let $A = \sum_{i=1}^{k} A_{w_i}$. Then*

$$m_f \leq \frac{A}{k^2}.$$

Proof. Let $w = \sum_{i=1}^{k} w_i$. Then $A_w = \sum_{i=1}^{k} A_{w_i} = A$. Let p be a minimal fat cut for w. Then $C_{w,f} = \langle p, w \rangle = \langle p, \sum_{i=1}^{k} w_i \rangle \geq k$. Thus $m_f \leq \frac{A_w}{C_{w,f}^2} \leq \frac{A}{k^2}$.

$\Diamond$ Proposition 6.1.2 $\Diamond$

Proposition 6.1.3. *Let X be a quadrilateral or ring or degenerate ring, and let T be a tiling of X. Let $f_1, \ldots, f_k$ be disjoint skinny flows for T for some positive integer k. Suppose that for each $i \in \{1, \ldots, k\}$, f_i contains at most N tiles. Then*

$$m_f \leq \frac{N}{k}.$$

Proof. For each $i \in \{1, \ldots, k\}$, let w_i be the weight function defined by $w_i(t) = 1$ if $t \in f_i$ and $w_i(t) = 0$ otherwise. The proof follows immediately from Proposition 6.1.2.

$\Diamond$ Proposition 6.1.3 $\Diamond$

6.2. The averaging trick. We prove the results of this section using the averaging trick shown to us by Mladen Bestvina which occurs in the proof of Theorem 7.1 in Cannon [8]. In this section we let X denote a quadrilateral or ring.

Let T be a tiling of X. Line 2.4.5.2 shows that $M_s = m_f \leq m_s = M_f$. We wish to know to what extent equality can fail in the inequality. We view T as giving a cellular decomposition of X, and so we can speak of valences of vertices of T. We will see that a bound on the valences of the vertices of T gives a bound on the quotient M_f/m_f.

First consider the simplest case in which every vertex of T has valence 3, except that in case of a quadrilateral the four corner vertices have valence 2. It is easy to see in this case that every skinny path is in fact a fat path (It is here that we use the fact that the four corner vertices have valence 2.) essentially because if two tiles have a vertex in common, then they have an edge in common. It easily follows that optimal weight functions for fat flows are optimal weight functions for skinny flows. This proves the following.

6.2.1. If T is a tiling of a quadrilateral or ring such that every vertex of T has valence 3, except that in case of a quadrilateral the four corner vertices have valence 2, then $M_s = m_f = m_s = M_f$.

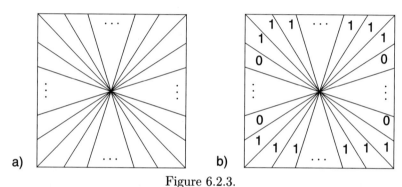

Figure 6.2.3.

Example 6.2.2. This example shows that in general the inequality in line 2.4.5.2 is strict. Let T be the tiling of a square given in a) of Figure 6.2.3. This tiling is invariant under a 90° rotation of the square. Let the valence of the central vertex be $v \geq 8$.

First consider skinny flows for T. It is easy to see that the weight function given in b) of Figure 6.2.3 lies in the cone spanned by its minimal skinny flows. Thus this weight function w is an optimal weight function for skinny flows by line 2.3.6. Hence

$$M_s = \frac{H_{w,s}^2}{A_w} = \frac{4}{\frac{v}{2} + 4} = \frac{8}{v + 8}.$$

We can compute M_f in the same way, but we proceed as follows. Since T is invariant under a 90° rotation, the problem of optimizing skinny cuts for T is equivalent to the problem of optimizing skinny flows for T. It easily follows that $m_s = M_s^{-1}$. Hence line 2.4.5.2 shows that

$$M_s = m_f = \frac{8}{v + 8} < \frac{v + 8}{8} = m_s = M_f.$$

Thus $\frac{M_f}{m_f} = \frac{(v+8)^2}{64}$.

◇ Example 6.2.2 ◇

Although Example 6.2.2 shows that in general the inequality in line 2.4.5.2 is strict, the following result shows that the quotient M_f/m_f in Example 6.2.2 is almost the largest possible.

Theorem 6.2.4. *Let T be a tiling of a quadrilateral or ring X. Let v be a positive integer such that the valence of every vertex in T is at most v. Then*

$$\frac{M_f}{m_f} \le 4v^2.$$

Proof. Let $(p_1, \dots, p_k)$ be a fundamental family of parallel skinny flows for T, which exists by line 2.4.4.9. Let $(\alpha_1, \dots, \alpha_k)$ be a family of parallel underlying arcs for $(p_1, \dots, p_k)$. For every integer i with $1 \le i \le k$ let $\varphi(\alpha_i)$ be the set of all tiles in T which meet α_i, so that $\varphi(\alpha_i)$ is a fat flow for $i = 1, \dots, k$.

Now let w be an optimal weight function for fat flows of T. Let $L_w(p)$ denote the length of a path p relative to w. Then since $H_{w,f} \le L_w(\varphi(\alpha_i))$ for $i = 1, \dots, k$,

$$H_{w,f} \le \frac{1}{k}\sum_{i=1}^{k} L_w(\varphi(\alpha_i)) = \frac{1}{k}\sum_{i=1}^{k}\sum_{t\in\varphi(\alpha_i)} w(t) = \frac{1}{k}\sum_{t\in T} w'(t)w(t),$$

where w' is the weight function on T for which $w'(t)$ is the number of arcs α_i which meet the tile t. This gives the following, where the second inequality comes from the Cauchy-Schwarz inequality.

6.2.5.

$$M_f = \frac{H_{w,f}^2}{A_w} \le \frac{1}{k^2 A_w}\left(\sum_{t\in T} w'(t)w(t)\right)^2 \le \frac{1}{k^2 A_w} A_{w'} A_w = \frac{A_{w'}}{k^2}$$

Our next goal is to estimate the weight function w'. Let $w'' = \sum_{i=1}^{k} p_i$, so that w'' is an optimal weight function for skinny flows of T. For every vertex u in T and tile s in T define $w_u(s)$ as follows. First, $w_u(s) = 0$ if $u \notin s$. If $u \in s$, then $w_u(s)$ is the number of arcs $\alpha_1, \dots, \alpha_k$ which contain u and meet the interior of s.

Now let t be a tile in T. Since the arcs $\alpha_1, \dots, \alpha_k$ are parallel, it is not difficult to see the following. There exist vertices l_t and r_t in t such that if i is an integer for which α_i meets t, then either α_i contains l_t or α_i contains r_t or α_i meets the interior of t. See Figure 6.2.6.

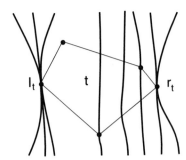

Figure 6.2.6. The tile t, the vertices l_t, r_t and the arcs which meet t.

Define $\psi(t)$ to be the set of tiles in $T \setminus \{t\}$ which contain l_t or r_t. Then,

$$w'(t) \le w''(t) + \sum_{s \in \psi(t)} \sum_{u \in \{l_t, r_t\}} w_u(s).$$

We now use this inequality to estimate $A_{w'}$. Since l_t and r_t have valence at most v, it is easy to see that the number of nonzero terms in the right side of the last inequality of the form $w''(t)$ or $w_u(s)$ is at most $2v$. Thus the Cauchy-Schwarz inequality shows that

$$w'(t)^2 \le 2v \left(w''(t)^2 + \sum_{s \in \psi(t)} \sum_{u \in \{l_t, r_t\}} w_u(s)^2 \right).$$

Hence

$$A_{w'} \le 2v \left(A_{w''} + \sum_{t \in T} \sum_{s \in \psi(t)} \sum_{u \in \{l_t, r_t\}} w_u(s)^2 \right)$$

$$\le 2v \left(A_{w''} + \sum_{s \in T} \sum_{u \in s} \sum_{t \in T \setminus \{s\}, u \in t} w_u(s)^2 \right)$$

$$\le 2v \left(A_{w''} + (v-1) \sum_{s \in T} \sum_{u \in s} w_u(s)^2 \right).$$

Let $s \in T$, and let u be a vertex in s with $w_u(s) \ne 0$. Then one of the arcs α_i meets the interior of s and meets the boundary of s in two points, u and some other point u'. We say that the point in $\{u, u'\}$ which lies closer to the bottom of T along α_i lies in the bottom of s. We say that the point in $\{u, u'\}$ which lies closer to the top of T along α_i lies in the top of s. It is clear that the sum of the terms $w_u(s)$ either over all vertices u in the bottom of s or over all vertices u in the top of s is at most $w''(s)$. Hence the sum of the terms $w_u(s)^2$ either over all vertices u in the bottom of s or over all vertices u in the top of s is at most $w''(s)^2$. Thus

$$A_{w'} \le 2v \left(A_{w''} + (v-1) \sum_{s \in T} 2w''(s)^2 \right) \le 2v \left(A_{w''} + 2(v-1)A_{w''} \right) \le 4v^2 A_{w''}.$$

Combining this estimate for $A_{w'}$ with line 6.2.5 gives that $M_f \le 4v^2 \frac{A_{w''}}{k^2}$. Line 2.3.4 shows that $A_{w''} = kH_{w'',s}$, and so $\frac{A_{w''}}{k^2} = \frac{H^2_{w'',s}}{A_{w''}} = M_s$. Hence $M_f \le 4v^2 M_s = 4v^2 m_f$, the equality coming from line 2.4.3.6. This proves Theorem 6.2.4.

◇ Theorem 6.2.4 ◇

We next prove a theorem which provides a comparison between the moduli of two different shinglings of a quadrilateral or ring.

Theorem 6.2.7. *Let X be a quadrilateral or ring. Let S and T be two shinglings of X such that every shingle in S and T is the closure of its interior. Suppose that there exists a positive integer K such that for every shingle s in S the number of shingles in T which meet the interior of s is at most K. Likewise suppose that there exists a positive integer L such that for every shingle t in T the number of shingles in S which meet the interior of t is at most L. We augment our usual notation for moduli with parentheses and an S or T to indicate the shingling being used, so for example $M_f(S)$ is the fat flow modulus of X relative to S. Then*

$$\frac{1}{KL} M_s(T) \le M_s(S) \le M_f(S) \le KL M_f(T).$$

Proof. We first prove that $M_s(T) \le KL M_s(S)$. To begin, define a function φ from the set of flows of S to the set of flows of T so that if p is a flow of S, then $\varphi(p)$ is the set of all shingles in T which meet the interior of some shingle in p. To see that $\varphi(p)$ is a flow, first note that given a shingle s in S, the shingles in T which meet the interior of s cover s. Hence if p is a flow of S, then the shingles in $\varphi(p)$ cover the union of the shingles in p. It easily follows that if p is a flow of S, then $\varphi(p)$ is a flow of T.

Now let w be an optimal weight function for skinny flows of T. Let $L_w(p)$ denote the length of a path p of T relative to w. Let $p_1, \dots, p_k$ be a fundamental family of skinny flows of S. Since $\varphi(p_i)$ is a flow of T, line 2.4.1.1 shows that $\varphi(p_i)$ contains a skinny flow of T, and so $H_{w,s} \le L_w(\varphi(p_i))$ for $i = 1, \dots, k$. Hence

$$H_{w,s} \le \frac{1}{k} \sum_{i=1}^{k} L_w(\varphi(p_i)) = \frac{1}{k} \sum_{i=1}^{k} \sum_{t \in \varphi(p_i)} w(t) = \frac{1}{k} \sum_{t \in T} w'(t) w(t),$$

where w' is the weight function on T for which $w'(t)$ is the number of flows $\varphi(p_i)$ containing the shingle t. This gives the following, where the second inequality comes from the Cauchy-Schwarz inequality.

6.2.8.

$$M_s(T) = \frac{H^2_{w,s}}{A_w} \le \frac{1}{k^2 A_w} \left(\sum_{t \in T} w'(t) w(t) \right)^2 \le \frac{1}{k^2 A_w} A_{w'} A_w = \frac{A_{w'}}{k^2}$$

Our next goal is to estimate $A_{w'}$. Let $w'' = \sum_{i=1}^{k} p_i$, so that w'' is an optimal weight function for skinny flows of S. Given a shingle s in S, let $\tau(s)$ be the set of shingles in T which meet the interior of s. Likewise, given a shingle t in T, let $\sigma(t)$ be the set of shingles in S which meet the interior of t. Then $w'(t) \leq \sum_{s \in \sigma(t)} w''(s)$. Since $|\sigma(t)| \leq L$, the Cauchy-Schwarz inequality gives that $w'(t)^2 \leq L \sum_{s \in \sigma(t)} w''(s)^2$. Hence

$$A_{w'} = \sum_{t \in T} w'(t)^2 \leq L \sum_{t \in T} \sum_{s \in \sigma(t)} w''(s)^2$$

$$= L \sum_{s \in S} \sum_{t \in \tau(s)} w''(s)^2 \leq KL \sum_{s \in S} w''(s)^2 = KLA_{w''}.$$

This and line 6.2.8 give that $M_s(T) \leq KL\frac{A_{w''}}{k^2}$. Line 2.3.4 shows that $A_{w''} = kH_{w'',s}$, and so $\frac{A_{w''}}{k^2} = \frac{H_{w'',s}^2}{A_{w''}} = M_s(S)$. Thus $M_s(T) \leq KLM_s(S)$.

The inequality $M_f(S) \leq KLM_f(T)$ can be proved in essentially the same way. Here we take $(p_1, \ldots, p_k)$ to be a fundamental family of fat flows of T. Let $(\alpha_1, \ldots, \alpha_k)$ be a family of underlying arcs for $(p_1, \ldots, p_k)$. Define $\varphi(\alpha_i)$ to be the set of shingles in S which meet α_i, so that $\varphi(\alpha_i)$ is a fat flow of S for $i = 1, \ldots, k$. It is easy to see that the above argument shows in this case that $M_f(S) \leq KLM_f(T)$.

Finally, the inequality $M_s(S) \leq M_f(S)$ comes from line 2.4.5.1.

This proves Theorem 6.2.7.
◊ Theorem 6.2.7 ◊

6.3. Examples.

6.3.1. Barycentric subdivision.

Let T_0 be any triangulation of the Euclidean plane, viewed as a tiling. Recursively define a sequence $\{T_i\}$ by defining T_i to be the second barycentric subdivision $B^2(T_{i-1})$ of T_{i-1} for $i > 0$.

Theorem 6.3.1.1. *The sequence T_0, T_1, $\ldots$ is not conformal. In fact, it fails to satisfy either of Axioms (1) and (2).*

Before proving Theorem 6.3.1.1, we make some preliminary estimates.

Let x_0 be a vertex of T_0, let R_0 be the closed star of x_0, let e be an edge of T_0 and let R_1 be the closed star of e. See Figure 6.3.1.2. Let n be the number of tiles in R_0. We consider R_0 as a degenerate ring with distinguished point x_0. We estimate the moduli of R_0 and R_1 with respect to the sequence $\{T_i\}$.

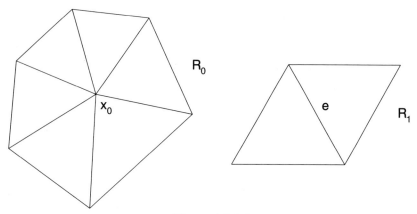

Figure 6.3.1.2.

First use constant weights $w_i \equiv 1$ on the tilings $T_i | R_1$ of R_1 to estimate the suprema $M_f(R_1, T_i)$ and $M_s(R_1, T_i)$ from below and the infima $m_f(R_1, T_i)$ and $m_s(R_1, T_i)$ from above. It is straightforward to show that $\lim_{i \to \infty} M_f(R_1, T_i) = \lim_{i \to \infty} M_s(R_1, T_i) = 0$ and $\lim_{i \to \infty} m_f(R_1, T_i) = \lim_{i \to \infty} m_s(R_1, T_i) = \infty$. We leave the proofs to the reader: the point is that areas grow by powers of 36 while lengths, both fat and skinny, grow by approximately powers of 4. Hence one learns absolutely nothing asymptotically from constant weights.

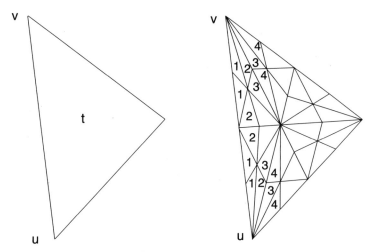

Figure 6.3.1.3.

We next estimate moduli by using skinny paths. The key observation is the following. Let t be a tile of T_i for some i, and let u, v be distinct vertices of t. Then there are 4 disjoint skinny paths $\alpha_1, \ldots, \alpha_4$ in $B^2(t)$, labeled 1, 2, 3 and 4 in Figure 6.3.1.3, such that for $j \in \{1, \ldots, 4\}$ α_j has 4 tiles and they can be ordered such that one of the extreme tiles contains u and the other extreme tile contains v.

6.3.1.4. For every nonnegative integer i, $m_f(R_0, T_i) \leq \frac{1}{n}$.

Proof. There are n disjoint skinny flows of the tiling $T_0|R_0$ of the degenerate ring R_0, each consisting of one tile. It follows inductively from the above observation that if $i \geq 0$, there are $n4^i$ disjoint skinny flows $\alpha_1, \ldots, \alpha_{n4^i}$ in the tiling $T_i|R_0$ of R_0, and each α_j, $j \in \{1, \ldots, n4^i\}$, contains 4^i tiles. It follows from Proposition 6.1.3 that if $i \geq 0$, $m_f(R_0, T_i) \leq \frac{4^i}{n4^i} = \frac{1}{n}$.
$\diamond$ 6.3.1.4 $\diamond$

6.3.1.5.
$$\lim_{i \to \infty} m_f(R_1, T_i) = 0.$$

Proof. First consider the tiling $T_1|R_1$ of the quadrilateral R_1. See Figure 6.3.1.6. For each $j \in \{1, \ldots, 8\}$, let $w_{1,j}$ be the weight function defined on $T_1|R_1$ by $w_{1,j}(t) = 0$ if t is not labeled j, $w_{1,j}(t) = 1/2$ if t is colored grey and is labeled j, and $w_{1,j}(t) = 1$ if t is colored white and is labeled j. Then $w_{1,1}, \ldots, w_{1,8}$ have disjoint supports, and for $j \in \{1, \ldots, 8\}$ $C_{w_{1,j},f} = 1$. Furthermore, $A_{w_{1,j}} = 4$ if $j \in \{1, \ldots, 6\}$ and $A_{w_{1,7}} = A_{w_{1,8}} = 3$. By Proposition 6.1.2, $m_f(R_1, T_1) \leq \frac{6 \cdot 4 + 2 \cdot 3}{8^2} = \frac{15}{32}$. Note that for each $j \in \{1, 3, 5, 7\}$, the tiles in the support of $w_{1,j} + w_{i,j+1}$ can be grouped in blocks of 4 tiles of equal weights as in Figure 6.3.1.7.

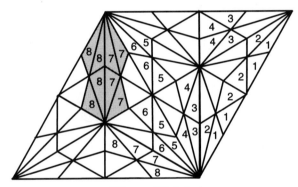

Figure 6.3.1.6.

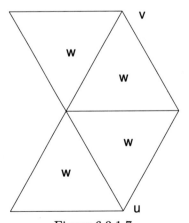

Figure 6.3.1.7.

We show inductively that if $i \in \mathbf{N}$, there are 2^{2i+1} disjointly supported weight functions $w_{i,1}, \ldots, w_{i,2^{2i+1}}$ on $T_i | R_1$ such that $\sum_{j=1}^{2^{2i+1}} A_{w_{i,j}} = 2(15)^i$ and for $j \in \{1, \ldots, 2^{2i+1}\}$, $C_{w_{i,j},f} = 1$, the tiles in the support of $w_{i,j} + w_{i,j+1}$ for j odd can be grouped in fundamental blocks of 4 tiles of equal weights as in Figure 6.3.1.7, and if two fundamental blocks for $w_{i,j} + w_{i,j+1}$ have nonempty intersection the intersection contains one of the vertices labeled as in Figure 6.3.1.7.

Figure 6.3.1.8.

We have already shown this for $i = 1$, so suppose that the above is true for some positive integer i. Let j be an integer such that $2j + 1 \in \{1, \ldots, 2^{2i+1}\}$. Define weight functions $w_{i+1,8j+k}$, $k \in \{1, \ldots, 8\}$, as follows. Let F be a fundamental block in the support of $w_{i,2j+1} + w_{i,2j+2}$ with weight w as shown in Figure 6.3.1.7. Let t be a tile in $B^2(F)$. First suppose that F contains two tiles in the support of $w_{i,2j+1}$ and two tiles in the support of $w_{i,2j+2}$. See Figure 6.3.1.8. In this case $w_{i+1,8j+k}(t) = 0$ if t is not labeled k, $w_{i+1,8j+k}(t) = w/2$ if t is colored grey and is labeled k, and $w_{i+1,8j+k}(t) = w$ if t is colored white and is labeled k. Now suppose that all four tiles in F are in the support of $w_{i,2j+1}$. In this case $w_{i+1,8j+k}(t) = 0$ if $k \in \{5, 6, 7, 8\}$, and if $k \in \{1, 2, 3, 4\}$, then $w_{i+1,8j+k}(t) = 0$ if t is not labeled k or $k + 4$, $w_{i+1,8j+k}(t) = w/2$ if t is colored grey and is labeled $k + 4$, and $w_{i+1,8j+k}(t) = w$ if t is colored white and is labeled k or $k + 4$. Finally suppose that all four tiles in F are in the support of $w_{i,2j+2}$. In this case $w_{i+1,8j+k}(t) - 0$ if $k \in \{1, 2, 3, 4\}$, and if $k \in \{5, 6, 7, 8\}$, then $w_{i+1,8j+k}(t) = 0$ if t is not labeled k or $k - 4$, $w_{i+1,8j+k}(t) = w/2$ if t is colored grey and is labeled k, and $w_{i+1,8j+k}(t) = w$ if t is colored white and is labeled k or $k-4$. Then $C_{w_{i+1,j},f} = 1$ for every j, and the weights in the support of $w_{i+1,j} + w_{i+1,j+1}$ for j odd can be grouped in fundamental blocks of equal weight as in Figure 6.3.1.7. The sum of the $w_{i,j}$-areas of the fundamental block F is $4w^2$, and the sum of the $w_{i+1,j}$-areas of F is $14 \cdot 4 \cdot w^2 + 4 \cdot 4 \cdot \frac{w^2}{4} = 60w^2$. Hence $\sum_{j=1}^{2^{2i+3}} A_{w_{i+1,j}} = 2(15)^{i+1}$. This completes the proof of the induction step.

By Proposition 6.1.2, if $i \in \mathbf{N}$ then $m_f(R_1, T_i) \leq \frac{2(15)^i}{(2^{2i+1})^2} = \frac{1}{2}\left(\frac{15}{16}\right)^i$. Hence $\lim_{i\to\infty} m_f(R_1, T_i) = 0$.
$\Diamond$ 6.3.1.5 $\Diamond$

6.3.1.9.

$$\lim_{i\to\infty} m_f(R_0, T_i) = 0.$$

Proof. The proof is very similar to the proof of line 6.3.1.5. The details are left to the reader.
$\Diamond$ 6.3.1.9 $\Diamond$

Proof of Theorem 6.3.1.1. If $\{T_i\}$ satisfies the first axiom of conformality, then there is a positive real number K such that if R is a ring, then there is a positive real number m such that for i sufficiently large, $m_f(R, T_i)$ and $M_f(R, T_i)$ both lie in the interval $[m, Km]$. Suppose R is any ring in R_0 that is a union of tiles in one of the tilings in $\{T_i\}$, and that separates x_0 and the boundary of R_0. The proofs of lines 6.3.1.5 and 6.3.1.9 were based on constructing disjointly supported weight functions on $T_i|R_0$ and then using Proposition 6.1.2. Since the hypotheses of Proposition 6.1.2 are also satisfied by the restrictions of these weight functions to $T_i|R$, the same argument shows that $\lim_{i\to\infty} m_f(R, T_i) = 0$ and hence it is impossible for $m_f(R, T_i)$ to lie in a K-interval $[m, Km]$ for i sufficiently large. Hence $\{T_i\}$ does not satisfy the first axiom of conformality.

The sequence $\{T_i\}$ fails the first axiom in a stronger fashion. Let $R = R_0 \setminus \text{star}(x_0, B^2(R_0))$. Then $\lim_{i\to\infty} m_f(R, T_i) = 0$ as shown above. There are four parallel skinny cuts in $T_1|R$ constructed by using the labeling of Figure 6.3.1.3 in each of the n tiles of $T_0|R_0$. An adaptation of the arguments used in proving lines 6.3.1.5 and 6.3.1.9 shows that $\lim_{i\to\infty} M_f(R, T_i) = \infty$, and hence for any real number K, $m_f(R, T_i)$ and $M_f(R, T_i)$ do not lie in a K-interval if i is sufficiently large.

If $\{T_i\}$ satisfies the second axiom of conformality, then given a positive real number M, there is a ring R in R_0 separating x_0 and the boundary of R_0 with $m_f(R, T_i) \geq M$ for i sufficiently large. Let j be an integer large enough so that $R \subset R' = R_0 \setminus \text{star}(x_0, B^{2j}(R_0))$. It is not difficult to see that $m_f(R, T_i) \leq m_f(R', T_i)$ for every i. Thus it is impossible that $m_f(R, T_i) \geq M$ for i sufficiently large because we have shown that $\lim_{i\to\infty} m_f(R', T_i) = 0$.
$\Diamond$ Theorem 6.3.1.1 $\Diamond$

6.3.2. Square tilings.

Let T_0 denote the tiling of the plane by squares of unit edge with vertices on the integer lattice. Let T_{i+1} denote the tiling formed by subdividing each square of T_i into four squares of equal size.

Theorem 6.3.2.1. *The sequence T_0, T_1, ... of square tilings of the plane is conformal.*

Proof. We outline two proofs, one geometric and the other combinatorial.

The geometric proof simply quotes a theorem of Cannon [8]. The square tilings of the plane are obviously "almost round" in the sense of Section 7 of [8]; hence the sequence of square tilings gives combinatorial moduli comparable to the standard analytic moduli by Theorem 7.1 of [8]. Hence the sequence is conformal as claimed.

The combinatorial proof will not be complete since we prove Axiom 1 only for special rings. However, the proof we give illustrates important techniques and can be extended by fussy effort to a complete proof. The proof of Axiom 1 will be modeled on the proof of Theorem 7.3 of [8]. The proof of Axiom 2 will use the bounded valence theorem, Theorem 6.2.4.

Axiom 1 for special rings. Let R be a ring that is a union of tiles of T_0, no one of which intersects both ends of R. Let $M_i = M_{\sup}(R, T_i)$ and $m_i = m_{\inf}(R, T_i)$. We claim that there is a constant $K > 0$, independent of i and of R, such that

6.3.2.2.

$$\frac{1}{K} m_0 \le m_i \le M_i \le K M_0.$$

The bounded valence theorem, Theorem 6.2.4, and line 2.4.5.1 imply the existence of a constant $L > 0$, independent of i and R, such that

6.3.2.3.

$$m_i \le M_i \le L m_i.$$

Hence m_i and M_i lie in the interval $[m, K^2 L \cdot m]$, where $m = (1/K) m_0$, i is arbitrary and K and L are independent of R, so that Axiom 1 of conformality is satisfied for these special rings R.

In order to establish the claim, we take weight functions σ and τ realizing M_0 as $H^2(R, \sigma)/A(R, \sigma)$ and m_0 as $A(R, \tau)/C^2(R, \tau)$. We define weight functions σ_i and τ_i on $T_i|R$ as follows. If t' is a tile of $T_i|R$, define

$$\sigma_i(t') = \max\{\sigma(t) \mid t \in T_0|R \text{ and } t' \subset \text{Nbd}(t, 1/2)\},$$

where $\text{Nbd}(X, \epsilon)$ denotes the closed Euclidean ϵ-neighborhood of X in the plane, with the sup norm. Define τ_i by a similar equation which uses τ instead of σ.

An obvious calculation shows that $A(R, \rho_i) \leq 2^{2i+2} A(R, \rho)$ for $\rho = \sigma$ and for $\rho = \tau$. The real point of the choice of the weight functions σ_i and τ_i shows up in the following length estimate. Let α be a path joining the ends of R and let β be a path separating the ends of R, with $L_{\sigma_i}(\alpha) = H(R, \sigma_i)$ and $L_{\tau_i}(\beta) = C(R, \tau_i)$. If α intersects a tile $t \in T_0 | R$, then because t does not intersect both ends of R, α crosses the annulus between t and $E^2 \setminus \mathrm{Nbd}(t, 1/2)$. The intersection with this annulus contributes at least $2^{i-1} \cdot \sigma(t)$ to the σ_i-length of α. Every tile weight in this contribution might be counted with respect to as many as three tiles t. We conclude that

$$H(R, \sigma_i) = L_{\sigma_i}(\alpha) \geq \frac{1}{3} \cdot 2^{i-1} \cdot L_\sigma(\alpha) \geq \frac{1}{3} \cdot 2^{i-1} \cdot H(R, \sigma).$$

Similarly,

$$C(R, \tau_i) = L_{\tau_i}(\beta) \geq \frac{1}{3} \cdot 2^{i-1} \cdot L_\tau(\beta) \geq \frac{1}{3} \cdot 2^{i-1} \cdot C(R, \tau).$$

We conclude that

$$M_i \geq \frac{H^2(R, \sigma_i)}{A(R, \sigma_i)} \geq \frac{(1/9)2^{2i-2} H^2(R, \sigma)}{2^{2i+2} A(R, \sigma)} = \frac{1}{144} M_0.$$

Similarly

$$m_i \leq \frac{A(R, \tau_i)}{C^2(R, \tau_i)} \leq \frac{2^{2i+2} A(R, \tau)}{(1/9)2^{2i-2} C^2(R, \tau)} = 144 m_0.$$

We use 6.3.2.3 to deduce that

$$m_0 \leq M_0 \leq 144 M_i \leq 144 L m_i$$

and

$$M_0 \geq m_0 \geq \frac{1}{144} m_i \geq \frac{1}{144L} M_i.$$

These latter two inequalities yield 6.3.2.2 provided that $K \geq 144L$.

Axiom 2. Pick a tiling T_i. Let R be a ring formed by taking a square with n^2 tiles from T_i and deleting a concentric square with $(m-1)^2$ tiles, $m - 1 < n$. We estimate the modulus of R as follows. Divide the tiles of $T_i | R$ into layers, one tile thick, circling R. Successive layers, from inside out, will have

$$4m, \ 4m + 8, \ 4m + 16, \ \ldots, \ 4n - 4$$

tiles. Define a weight function τ on T_i so that the tiles t in the layer having $4k$ tiles each have the weight $\tau(t) = 1/4k$. Then

$$L m_{\inf}(R, T_i) \geq M_{\sup}(R, T_i) \geq \frac{H^2(R, \tau)}{A(R, \tau)}.$$

$$H(R, \tau) = \frac{1}{4m} + \frac{1}{4m + 8} + \cdots + \frac{1}{4n - 4} = A(R, \tau).$$

We deduce that

$$\frac{H^2(R,\tau)}{A(R,\tau)} = \frac{1}{4m} + \frac{1}{4m+8} + \cdots + \frac{1}{4n-4}$$

$$\geq \int_m^n \frac{dx}{8x}$$

$$= \frac{1}{8} \ln(n/m),$$

so that $m_{\inf}(R, T_i)$ is large for n/m large.

We leave as an exercise the proof, which we do not need, that

$$m_{\inf}(R, T_i) = A(R, \tau)/C^2(R, \tau).$$

7. Variable negative curvature versus constant curvature groups.

Thurston's geometrization conjecture [21] would require that a 3-manifold admitting a Riemannian metric of variable negative curvature also admit a metric of constant negative curvature. The finite Riemann mapping theorem of Section 3 and the combinatorial Riemann mapping theorem of Cannon [8] have potential application to the finding of a constant curvature metric. This expository section explains the connection.

We assume that G is a discrete group that is **negatively curved** (in the large). That is, G is **word hyperbolic** in the sense of Gromov (see [15], [7], [14], [11], [2], [3]). In order to apply the mapping theorems, we need to assume that G has as its space $S_\infty(G)$ at infinity the 2-sphere S^2 (see, for example, [15] or [20]), and we need to extract from G finite coverings of this 2-sphere. Both the space at infinity and the requisite finite coverings are defined in terms of a finitely generated Cayley group graph Γ for G (see [5], or [6], or any of a number of other places for a description of Cayley graphs). Points of $S_\infty(G)$ are equivalence classes $[R]$ of geodesic rays $R : [0, \infty) \to \Gamma$, rays R and S being equivalent if $d(R(t), S(t))$ is bounded for $t \in [0, \infty)$. Both the topology at infinity and the requisite coverings of the space at infinity are defined in terms of what we call **combinatorial half spaces** and **combinatorial disks at infinity** (see [7] for motivations and discussion). Fix a base vertex O of Γ. Let $R : [0, \infty) \to \Gamma$ be a geodesic ray such that $R(0) = O$. Let n be a positive integer. The **half space** $H(R, n)$ is defined by the equation

$$H(R, n) = \{ x \in \Gamma | d(x, R([n, \infty))) \leq d(x, R([0, n))) \},$$

and the **combinatorial disk** $D(R, n)$ is defined by the equation

$$D(R, n) = \{ [R'] \mid R'(0) = O \text{ and } \lim_{t \to \infty} d(R'(t), \Gamma \setminus H(R, n)) = \infty \}.$$

For n a positive integer, we define $O(n)$ to be the collection of all combinatorial disks of the form $D(R, n)$ where R is a geodesic ray with initial point O.

Eric Swenson [20] has proved a number of important facts about these sets. A few of them are summarized in the following theorem.

Theorem 7.1. *[Swenson, 20]. Let G be a negatively curved group, Γ a finitely-generated Cayley graph for G, and O a base vertex for Γ.*

(1) The disks $D(R, n)$ based at O form a basis for a compact metric topology on $S_\infty(G)$, and this topology is equivalent with Gromov's topology.

(2) For each positive integer n, the set $O(n)$ of disks $D(R, n)$ based at O forms a finite open cover of $S_\infty(G)$; and, as n varies, these coverings $O(n)$ have uniformly bounded degree. Furthermore, the mesh of these coverings goes to 0 as $n \to \infty$. Consequently, $\dim(S_\infty(G))$ is finite.

(3) The coverings $O(n+1)$ can be derived from the coverings $O(n)$ by a linearly recursive rule which can be interpreted as subdivision.

Conjecture. If $S_\infty(G)$ is a k-sphere for some k, then each open disk $D(R, n)$ is contractible, or at least has trivial homology.

We say that a discrete group G has **constant negative curvature** (in the large), or that G is **hyperbolic**, if G acts cocompactly, properly discontinuously, and isometrically on some hyperbolic space H^n. Cannon and Cooper [9] have characterized groups of constant negative curvature in terms of the geometry of a finitely-generated Cayley group graph Γ:

Theorem 7.2. *[9]. A discrete group G has constant negative curvature (in the large) in dimension 3 if and only if a finitely-generated Cayley group graph Γ for G is quasi-isometric with H^3.*

Cannon and Cooper are under the impression that their proof works also in higher dimensions. In the cocompact case stated in Theorem 7.2, this follows from Sullivan [19] or Tukia [22].

Cannon and Swenson have characterized 3-dimensional discrete groups of constant negative curvature in terms of the sequence $O(1)$, $O(2)$, ... of covers defined above.

Theorem 7.3. *[10]. A discrete group G acts cocompactly, properly discontinuously, and isometrically on H^3 if and only if the following conditions are satisfied:*

(1) The group G is negatively curved (in the large); its space $S_\infty(G)$ at infinity is the 2-sphere.

(2) The sequence $O(1)$, $O(2)$, ... of covers defined above is conformal in the sense of Section 6.

The proof involves a good deal of the geometry of negatively-curved spaces from Swenson's thesis [20], the combinatorial Riemann mapping theorem [Cannon, 8], and the Sullivan-Tukia result [Sullivan, 19 (see, in particular, p. 468)] on uniformly quasiconformal group actions on S^2. The Cannon-Swenson characterization, Theorem 7.3, reduces the 3-dimensional problem broached by Theorem 7.2 to a (difficult) 2-dimensional problem, namely the approximation of combinatorial moduli. Much remains to be done.

REFERENCES

[1] Ahlfors, L. V., *Conformal Invariants: Topics in Geometric Function Theory*, McGraw-Hill, New York, 1973.

[2] Alonso, J. M., Brady, T., Cooper, D., Ferlini, V., Lustig, M., Mihalik, M., Shapiro, M., Short, H., *Notes on word hyperbolic groups*, Group Theory from a Geometrical Viewpoint: 21 March - 6 April 1990, ICTP, Trieste, Italy (E. Ghys, A. Haefliger and A. Verjovsky, eds.), World Scientific, Singapore, 1991, pp. 3-63.

[3] Bowditch, B. H., *Notes on Gromov's hyperbolicity criterion for path-metric spaces*, Group Theory from a Geometrical Viewpoint: 21 March - 6 April 1990, ICTP, Trieste, Italy (E. Ghys, A. Haefliger and A. Verjovsky, eds.), World Scientific, Singapore, 1991, pp. 64-167.

[4] Brooks, R. L., Smith, C. A. B., Stone, A. H., Tutte, W. T., *The dissection of squares into squares*, Duke Math. J. **7** (1940), 312–340.

[5] Cannon, J. W., *The combinatorial structure of cocompact discrete hyperbolic groups*, Geometriae Dedicata **16** (1984), 123–148.

[6] Cannon, J. W., *Almost convex groups*, Geometriae Dedicata **22** (1987), 197–210.

[7] Cannon, J. W., *The theory of negatively curved spaces and groups*, Ergodic Theory, Symbolic Dynamics and Hyperbolic Spaces (T. Bedford, M. Keane and C. Series, eds.), Oxford University Press, Oxford, 1991, pp. 315–369.

[8] Cannon, J. W., *The combinatorial Riemann mapping theorem*, submitted to Acta Mathematica.

[9] Cannon J. W., Cooper, D., *A characterization of compact hyperbolic and finite-volume groups in dimension three*, to appear in Trans. Amer. Math. Soc..

[10] Cannon, J. W., Swenson, E. L., *Recognizing constant curvature discrete groups*, in preparation.

[11] Coornaert, M., Delzant, T., Papadopoulos, A., *Géométrie et théorie des groupes*, Springer Lecture Notes Series 1441, Springer-Verlag, New York Berlin Heidelberg, 1990.

[12] Dudeney, H. E., *The Canterbury Puzzles*, (This is a reprint of the 1907 edition.), Dover, New York, 1958.

[13] Gardner, M., *The Second Scientific American Book of Mathematical Puzzles and Diversions*, University of Chicago Press, Chicago, 1987.

[14] E. Ghys and P. de la Harpe (eds.), *Sur les Groupes Hyperboliques d'aprés Mikhael Gromov*, Progress in Mathematics. Vol 83, Birkhaüser, Zurich, 1990.

[15] Gromov, M., *Hyperbolic groups*, Essays in Group Theory (S. M. Gersten, ed.), Springer-Verlag, New York Berlin Heidelberg, 1987, pp. 75–264.

[16] Lehto, O., Virtanen, K. I., *Quasiconformal Mappings in the Plane*, Springer-Verlag, New York, 1973.

[17] Poincaré, H., *The Value of Science*, Dover, New York, 1958.

[18] Schramm, O., *Square tilings with prescribed combinatorics*, preprint.

[19] Sullivan, D., *On the ergodic theory at infinity of an arbitrary discrete group of hyperbolic motions*, Riemann Surfaces and Related Topics: Proceedings of the 1978 Stony Brook Conference (I. Kra and B. Maskit, eds.), Princeton University Press, Princeton, 1981, pp. 465–496.

[20] Swenson, E. L., *Negatively curved groups and related topics*, Dissertation, BYU (1993).

[21] Thurston, W. P., *Three dimensional manifolds, Kleinian groups and hyperbolic geometry*, Bull. Amer. Math. Soc. **6, # 3** (1982), 357–381.

[22] Tukia, P., *On quasiconformal groups*, J. Analyse Math. **46** (1986), 318–346.

DEPARTMENT OF MATHEMATICS, BYU, PROVO, UT 84602, U.S.A.

DEPARTMENT OF MATHEMATICS, VPI& SU, BLACKSBURG, VA 24061-0123, U.S.A.

DEPARTMENT OF MATHEMATICS, EMU, YPSILANTI, MI 48197, U.S.A.

Contemporary Mathematics
Volume 169, 1994

THE FREIHEITSSATZ AND ITS EXTENSIONS

BENJAMIN FINE AND GERHARD ROSENBERGER

In memory of Wilhelm Magnus

CONTENTS

1. Introduction
2. The FHS and Amalgams
3. The Classical Freiheitssatz and Magnus' Method
4. Small Cancellation Products and Diagrams
5. One-Relator Products
6. One-Relator Products of Cyclics
7. Multi-Relator Versions of the Freiheitssatz

1. INTRODUCTION

The theory of one-relator groups has always been of central importance in combinatorial group theory. The reasons for this importance are varied and arise from geometric and complex function theoretic as well as algebraic considerations. See the article by G. Baumslag [4] for a complete discussion of this. The cornerstone

Mathematics 1991 Subject Classification Code: Primary 20E05,20E06,20E07 Secondary 20F32,20F06 : Keywords : One-relator Group, Freiheitssatz, one-relator product,small cancellation groups

This is a final version of this paper.

of one-relator group theory is the Freiheitssatz or Independence Theorem originally proposed by M. Dehn and originally proved by W. Magnus [55]. Magnus' method of proof initiated the use of amalgam constructions in the study of infinite discrete groups.

The Freiheitssatz as proven by Magnus says the following: Let $G = <x_1, \ldots, x_n; R>$ be a one-relator group and suppose that the relator R is cyclically reduced in the free group on $x_1, \ldots, x_n$ and that R involves all the generators. Then the subgroup of G generated by any proper subset of the generators is free on that subset.

In coarser language the theorem says that if G is as above, then given $x_1, \ldots, x_{n-1}$ the only relations involving them are the trivial ones. Equivalently the only relations on $x_1, \ldots, x_{n-1}$ are the "obvious ones" from the presentation — in the case of a one-relator group — trivial ones.

The Freiheitssatz can be considered as a non-commutative analog of some more transparent results in commutative algebraic structures. For example suppose F is a field and $V = F^n$, n-dimensional linear space over F. If W is a subspace of V defined by

$$W = \{(x_1, \ldots, x_n) \in V; \sum_{i=1}^{n} m_i x_i = 0$$

with $m_i \in F$ and $m_i \neq 0$ for $i = 1, \ldots, n\}$,

then W has dimension $n - 1$.

For abelian groups consider F to be free abelian on $<w_1, \ldots, w_n>$. Let G be the quotient of F modulo the relation $\sum_{i=1}^{n} m_i w_i$ with $m_i \in \mathbb{Z}$ and $m_i \neq 0$ for $i = 1, \ldots, n$. Then the subgroup of G generated by $w_1, \ldots, w_{n-1}$ is free abelian on these generators.

Finally suppose that we have an irreducible algebraic equation in n complex variables in which all the variables appear. This then cannot be used to derive any irreducible algebraic equation in which not all of these variables appear.

The purpose of this paper is to survey both the classical Freiheitssatz of Magnus and its extensions to classes of groups beyond one-relator groups. To do this we give a general definition of a Freiheitssatz which we will abbreviate FHS. Our outline is then as follows: Magnus' method relies heavily on properties of amalgams, that is, free products with amalgamations and HNN groups, so in section 2 we review these constructions. From our point of view

the importance of these constructions is that they satisfy a FHS
relative to their factors. Both Magnus' method and methods which
mimic this classical approach involve embedding an overgroup G
in an amalgam so that the proposed "FHS factor" (see Section
2) is an amalgam factor. In Section 3 we go over the classical
Freiheitssatz and Magnus' method. Dehn's original suggestion was
geometric so we also point out a topological translation of Magnus'
proof. Some of these topological ideas have evolved into "pictures"
over groups which are important in the study of the Freheitssatz
for one-relator products. The use of pictures in handling the Freiheitssatz was motivated originally by small cancellation products.
We discuss these in Section 4. In Section 5 we look at one-relator
products which are a natural generalization of one-relator groups.
Specifically a one-relator product of a family $\{A_i\}$ of groups is a
quotient $G = (*A_i)/N(R)$ where R is a single element in $*A_i$, the
free product of the A_i, which is cyclically reduced and of syllable
length at least 2 and $N(R)$ is its normal closure. The groups A_i
are called the *factors* of G and R is the *relator*. If $R = S^m$ where
S is a cyclically reduced word of syllable length at least 2 in the
free product of the A_i, then R is a *proper power*. A FHS for a
one-relator product gives conditions for the factors to inject into
the product. There are two general approaches to looking at a FHS
for a one-relator product. The first is to look for conditions on the
factors while the second is to look at conditions on the relator. We
examine both in Section 5 relying heavily on work of J. Howie ([41]
through [47]). In much of the work on one-relator products the
factors are assumed to be torsion-free. In Section 6 we remove this
restriction and handle one-relator products of cyclics. This class of
groups is of interest in that it is the natural algebraic generalization
of Fuchsian groups. The techniques for a FHS in this case depend
on linear representations of these groups. Finally in Section 7 we
survey certain extensions of the FHS to multi-relator groups. As
we will see the techniques used in these extensions borrow from
all the previous work — the algebraic version of Magnus' method,
geometric analysis and representation theory.

2. THE FHS AND AMALGAMS

Let X, Y be disjoint sets of generators and suppose that the
group A has the presentation $A = \; < X; \mathrm{Rel}(X) >$ and that the
group G has the presentation $G = \; < X, Y; \mathrm{Rel}\,(X), \mathrm{Rel}(X, Y) >$.
Then we say that G satisfies a FHS {relative to A} if $< X >_G = A$.

In other words the subgroup of G generated by X is isomorphic to A. As above, in coarser language this says that the complete set of relations on X in G is the "obvious" one from the presentation of G. An alternative way to look at this is that A injects into G under the obvious map taking X to X. In this language Magnus' original FHS can be phrased as a one-relator group which satisfies a FHS relative to the free group on any proper subset of the generators.

Magnus' method involves the use of amalgam constructions — free products with amalgamation and HNN groups. These constructions are the primary decomposition methods for infinite discrete groups and we briefly review their definitions.

Suppose A and B are groups with respective presentations $A = < X; R >$, $B = < Y; S >$. Suppose $v_1, \ldots, v_n$ are elements of A (given by X-words) and $w_1, \ldots, w_n$ are elements of B (given by Y-words), and suppose further that $v_i \to w_i$, $i = 1, \ldots, n$ defines an isomorphism of the subgroup H of A generated by $\{v_i\}$ with the subgroup K of B generated by $\{w_i\}$. Then the group G with the presentation $G = < X, Y; R, S, v_i w_i^{-1}, i = 1, \ldots, n >$ is the free product of A and B amalgamating H to K. We denote this by $A *_H B$. A and B are called the *factors* of G. The properties of amalgamated free products can be found in Lyndon and Schupp [54] or in Magnus, Karrass and Solitar [60].

Next suppose B is a group with presentation $B = <$ gens B; rels $B >$ and suppose $\{L_i\}$, for i in some index set I, is a collection of subgroups of B. Suppose further for each i there is an isomorphism $f_i : L_i \to f_i(L_i)$ of L_i onto another subgroup $f_i(L_i)$ of B. Let G be the group with the presentation $G = <$ gens $B, t_i, i \in I$; rels $B, t_i a t_i^{-1} = f_i(a), a \in L_i >$. G is then called an *HNN group* or an *HNN extension* with *base group* B, *stable letters* t_i (or equivalently *free part generated by* $\{t_i\}$) and *associated subgroups* $\{L_i, f_i(L_i)\}$. Again the properties of HNN groups can be found in [54].

For our purposes what is most important about these constructions is that the elements of the resulting groups have *normal forms*. This in turn implies that the factors A, B for a free product with amalgamation and the base group B in an HNN group inject into the group or equivalently that an amalgam satisfies a FHS relative to its factors.

Theorem 2.1 (Normal Form Theorems). *[54](1) Let $G = A *_H B$, let $\{a_i\}$ be a set of right coset representatives for A mod H and let $\{b_i\}$ be a set of right coset representatives for B mod K.*

Then every $g \in G$ has a unique representation as

$$ha_{i1}b_{i1} \cdots \cdots \cdots a_{ij}b_{ij}$$

where $h \in H$ and $a_{i1}b_{i1} \cdots a_{ij}b_{ij}h$ is a reduced word in the free product $A * B$. The unique representation above is called a *normal form* or *reduced sequence* for the corresponding element of G. In particular the normal form for an element of a factor is a reduced word in that factor.

(2) Suppose G is an HNN group with base B and associated subgroups $\{A_i, f_i(A_i)\}$ and suppose we choose a fixed set of right coset representatives for $\{A_i\}$ and $\{f_i(A_i)\}$ in B. Then a *normal form* in G is a sequence

$$g_0 t_{i_1}^{e_1} g_1 t_{i_2}^{e_2} \cdots \cdots \cdots t_{i_k}^{e_k} g_k$$

where $g_0, g_1, \ldots, g_k$ are elements of B. g_0 can be an arbitrary element of B while if $e_j = -1$, then g_j is a coset representative for A_i in G and if $e_j = 1$, then g_j is a coset representative for $f_i(A_i)$ in G. Finally there is no subsequence of the form $t^e \cdot 1 \cdot t^{-e}$. Every element w in G has a unique representation as a normal form.

Corollary. *(1) Let $G = A *_H B$. Then A and B inject into G under the obvious map. Equivalently a free product with amalgamation satisfies a FHS relative to its factors.*

(2) Let G be an HNN group with base B. Then B injects into G under the obvious map. Equivalently an HNN group satisfies a FHS relative to its base.

Because of the relationship between the FHS and amalgam constructions we say that if G satisfies a FHS relative to A, A is a *FHS factor* of G.

The basic idea in Magnus' method for the classical FHS and certain extensions of it is to embed the group into an amalgam in such a way that the proposed FHS factor embeds into an amalgam factor which in turn contains the proposed FHS factor as a FHS factor. The result can then be obtained by applying the FHS for amalgams.

From Bass-Serre Theory [74] amalgam constructions can also be viewed by considering the actions of groups on simplicial trees. This has led to the following extension — called the *fundamental group of a graph of groups* — of the basic constructions which has been used extensively in work on various extensions of the FHS.

Let X be a connected graph, and suppose that for each vertex v of X there is a group G_v. The groups G_v are called *vertex groups*. Suppose further that for each edge e of X there is a group G_e such that $G_e = G_{e'}$ where e' is the oppositely directed edge and that each G_e injects into its terminal vertex groups. The groups G_e are called *edge groups*. Let **g** stand for the collection of vertex groups, edge groups and injections. Then the pair $(\mathbf{g}, X)$ is called a *graph of groups*. For each e the edge group G_e injects into both G_v and G_w where v, w are its intial and terminal vertices respectively. For each such edge e associate a free generator t_e and a relation $t_e G_v t_e^{-1} = G_w$ and form the HNN extension of the free product $*G_v$ of the vertex groups with stable letters $\{t_e\}$ and subgroup isomorphisms given by the indicated edge group identifications. We denote this HNN extension by $F(\mathbf{g})$. Now let T be a maximal tree in X and impose the additional relations $\{t_e = 1 \text{ for } e \text{ in } T\}$. The resulting group is the *fundamental group of the graph of groups* denoted by $\pi_1(\mathbf{g}, X, T)$. Up to isomorphism this is independent of the maximal tree T. Further if the graph X has only one edge pair with distinct vertices, we simply obtain the free product with amalgamation of the vertex groups.

Theorem 2.2 [71]. *Let G be the fundamental group of a graph of groups. Then any vertex group naturally injects into G. In other words G satisfies a FHS relative to any of its vertex groups.*

A further construction of interest is when the graph is a polygon. A group G is a *polygonal product* if it can be described in the following manner. There is a polygon P. Each vertex v corresponds to a group G_v. Each edge y corresponds to a group G_y which is subgroup of the vertex groups it connects. Adjacent vertices are amalgamated via relations given by identifications along the edges. The group G is then formed by the free product of the G_v modulo the amalagmating edge relations. This is represented in figure 1.

If $n \geq 4$, a polygonal product splits as a free product with amalgamation (splitting G at any subtree of the polygon), and so for $n \geq 4$ the vertex groups inject. For $n = 3$, that is a *triangular product* of groups, the situation is more complicated. It may happen that the whole group may collapse [77], and so clearly in general a triangle of groups does not satisfy a FHS relative to its vertices. However recently for a polygonal product of groups Gersten and Stallings [74] have defined a concept of angle (*Gersten-Stallings angles*) . If the resulting triangle of groups is non-positively curved,

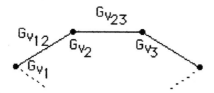

Figure 1. A Polygonal Product.

that is the Gersten-Stallings angle sum is $\leq \pi$ then the vertex groups inject and so in this case a triangular product does satisfy a FHS.

Theorem 2.3 [77]. *Let G be polygonal product with polygon P. Then if either P has 4 or more sides or P is a non-positively curved triangle relative to the Gersten-Stallings angles G satisfies a FHS relative to any vertex group.*

3. THE CLASSICAL FREIHEITSSATZ AND MAGNUS' METHOD

We now present Magnus' method for proving the classical FHS. Actually what we give is an equivalent version which uses the amalgam constructions explicitly. These were implicit in Magnus' original proof but the actual constructions were not yet defined. Magnus' first proof used the concept of a *staggered presentation*, about which we'll say more later (see [54]), and then he proved directly that certain subgroups had the properties of free products with amalgamation, a concept which had been introduced by Schreier [73]. When Magnus realized that he was using Schreier's constructions, he added a footnote to the proof explaining that several of Magnus' technical lemmas could be avoided by the use of Schreier's results (see [15, Chapter II.5]). The proof we outline below is based on McCool and Schupp [62] and Moldavanskii [65].

Theorem 3.1.   (Freiheitssatz). *Let $G = < x_1, \ldots, x_n; R = 1 >$ where R is a cyclically reduced word which involves all the generators. Then the subgroup generated by $x_1, \ldots, x_{n-1}$ is free on these generators.*

Proof. We do an induction on the number of generators n. It is clear for $n = 1$. To avoid using multiple indices, suppose $G =$

$< a, \ldots, c, t; R = 1 >$ where $t = x_n$ renaming the generators without indices. Suppose first that t has exponent sum zero in the relator R. For any generator g, let $g_i = t^i g t^{-i-1}, i \in \mathbb{Z}$. Then the relator R can be expressed as a new word S in the generators $a_i, \ldots, c_i, \ldots$. Let $m =$ minimum of the a-subscripts appearing in S, $M =$ maximum of the a-subscripts appearing in S, $n =$ minimum of the c-subscripts appearing in S, $N =$ maximum of the c-subscripts appearing in S. The word S is then a cyclically reduced word in the generators $a_m, \ldots, a_M, \ldots, c_n, \ldots, c_N$ and further this word S is of shorter length than R. Let H be the one-relator group $< a_m, \ldots, a_M, \ldots, c_n, \ldots, c_N; S = 1 >$. Conjugates of the original $a, \ldots, c$ appear in H and by induction generate a free subgroup, so we must show that H injects into G. Again from the inductive hypothesis, in the group H, the subgroups generated by $\{a_m, \ldots, a_{M-1}, c_n, \ldots, c_{N-1}\}$ and $\{a_{m+1}, \ldots, a_M, c_{n+1}, \ldots, c_N\}$ are free. Therefore the group G^* given by

$$G^* = < t, H; t^{-1} a_m t = a_{m+1}, \ldots, t^{-1} a_{M-1} t$$
$$= a_M, \ldots, t^{-1} c_n t$$
$$= c_{n+1}, \ldots, t^{-1} c_{N-1} t$$
$$= c_N >$$

is an HNN group with base H. But G^* is isomorphic to G, so H injects in G.

Now suppose the exponent sum of t in R is non-zero. Let $\alpha =$ exponent sum of t in R and let $\beta =$ exponent sum of a in R. Let L be $\{a, b, \ldots c\}$ and let K be the one-relator group with the presentation $K = < y, x, b, \ldots c; R(yx^{-\beta}, x^{\alpha}, b, \ldots, c) >$. The map $t \rightarrow yx^{-\beta}$, $a \rightarrow x^{\alpha}$, $b \rightarrow b, \ldots$, $c \rightarrow c$ defines a homomorphism of G into K. Let R_1 be the cyclic reduction of $R(yx^{-\beta}, x^{\alpha}, b, \ldots, c)$. Then the exponent sum of x in R_1 is zero and y occurs in R_1. Then as in the case above K can be expressed as an HNN group with a one-relator group H as the base and stable letter x. Let S be as before the single relator in H. S then has length less than R since all x symbols have been eliminated. By induction the subgroup of K generated by $\{x, b, \ldots, c\}$ is freely generated by this set, and therefore the subgroup generated by $\{x^{\alpha}, b, \ldots, c\}$ is also freely generated by this set. Since $a \rightarrow x^{\alpha}, b \rightarrow b, \ldots, c \rightarrow c$ maps L onto these, L freely generates a subgroup of G. This completes the proof.

Notice that the one-relator group K in the second part of the proof is actually the free product of G with an infinite cyclic group $< x >$ and then identifying a with x^α. By the Freiheitssatz a has infinite order and thus this identification gives a subgroup isomorphism and hence K is the free product with amalgamation of G and $< x >$. It follows that G injects into K. Thus we have actually proven the following theorem of Moldavanskii [65] which in turn implies the Freiheitssatz.

Theorem 3.2. *Every one-relator group whose relator is cyclically reduced and involves at least two generators can be embedded into an HNN group with a single stable letter. The base of this HNN group is a one-relator group whose defining relator is of smaller length than the original relator and whose associated subgroups are free.*

This technique of embedding a one-relator group into an HNN extension with a one-relator base group with shorter relator and then using inductive arguments is what is called *Magnus' Method*. It has been used extensively to prove results about one-relator groups, for example, to show that one-relator groups have solvable word problem.

The theory of HNN groups was not available to Magnus so the original proof is somewhat different. Magnus used the concept of a *staggered presentation*. This arises in the following manner. Let G be a one-relator group with relator R, let f be an epimorphism of G onto an infinite cyclic group (formed as in the proof we gave by finding a generator with exponent sum zero in R) and let $H = \text{Ker } f$. Then from the Reidemeister-Schreier rewriting process H has presentation which is "staggered" in the sense that it has the form $< y_i (i \in \mathbb{Z}), z_j (j \in I); R_i (i \in \mathbb{Z}) >$ for some index set I and that there are constants m and M such that each R_i is a word in $\{y_{i+m}, \ldots, y_{i+M}, z_j (j \in I)\}$ properly involving y_{i+m} and y_{i+M}. Further identifications between staggered layers lead to free products with amalgamation so that the layers inject. In his first proof Magnus proved the necessary embeddings for free products with amalgamation directly, without referring to the work of Schreier. We mention this concept of a staggered presentation because it has been reflected in various geometric arguments concerning one-relator groups which use the concept of a staggered complex (see [41]).

Dehn's original conception of the FHS was geometric. As reported by Magnus [15] Dehn's approach was as follows. Suppose as above that one generator t in the group G has exponent sum zero in the single relator R, and let G_o be the subgroup of G generated by the remaining generators. Dehn visualized the graph of G as a layered structure with each layer a copy of one of the subgroups $t^n G_o t^{-n} = G_n$ for $n \in \mathbb{Z}$. The union of these layers would then form the graph of a normal subgroup N of G with the powers of t as coset representatives. The problem was to prove that each layer is a tree and thus the corresponding subgroup, each isomorphic to G_o, is free.

Several authors have returned to this geometric-topological approach both in reproving the original FHS and in extending it. Below we present a method due to Jim Howie (see [39]) of using the method of *Papakyriapopolous towers* from 3-manifold topology to give a proof of the FHS. This technique is the topological mirror version of the algebraic Magnus technique and therefore is a return to Dehn's ideas. After describing Howie's tower method, we will also introduce a second (different) geometric approach due to Lyndon [53].

A *tower* is a map $g = i_o p_1 i_1 \ldots p_h i_h : K' \to K$ between connected CW-complexes such that each i_j is an inclusion map and each p_j is a covering projection. For the proof of the FHS we restrict the p_j to be infinite cyclic coverings, that is regular, connected coverings with infinite cyclic covering transformation groups. If S is another connected CW-complex, then a *tower lifting* is a commutative triangle, where g is a tower (see Figure 2). It is proper tower lifting if g is not an isomorphism, and it is maximal if f' does not have a proper tower lifting.

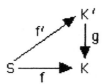

Figure 2. A Tower Lifting.

Notice that Magnus' method corresponds geometrically to a tower. Initially we have a 2-complex with a single 2-cell. This is lifted to an infinite cyclic cover and then restricted to a finite

subcomplex of that cover. Then by an inductive argument we reduce to the case of a subcomplex containing a single 2-cell which
is "simpler" than the original single 2-cell. For the FHS or other
properties of one-relator groups which are proved using Magnus'
method, the above procedure is continued until after finitely many
repetitions we arrive at a situation where there are no more infinite
cyclic covers and the general case can be deduced.

Howie's tower proof of the FHS depends on the following three
lemmas (see [41] for proofs).

Lemma 3.3 [41]. *Let S be a finite CW-complex. Then any simplicial map $f : S \to K$ has a maximal tower lifting unique up to
isomorphism.*

For the other two lemmas we need the concept of a *staggered
2-complex*. This is closely related to the idea of a staggered presentation as discussed earlier. A 2-complex is *staggered* [41] if all
of its 2-cells are attached by cyclically reduced paths of positive
length and if the sets of 1-cells and of 2-cells are equipped with
linear orderings which are compatible in the following sense: when-
ever $\alpha < \beta$ are 2-cells, then $(\min \alpha) < (\min \beta)$ and $(\max \alpha) <
(\max \beta)$. Here $\min \alpha$ (respectively $\max \alpha$) denotes the minimum
(maximum) 1-cell involved in the attaching map for α.

Lemma 3.4 [41]. *Let K be a staggered 2-complex and $g : K \to
K'$ a tower. Then K' is staggered.*

Lemma 3.5 [41]. *Let K' be a staggered 2-complex with at least
one 2-cell such that $H^1(K') = 0$, and suppose that the greatest
2-cell α of K' is not attached by a proper power (in $\pi_1(K'(1))$).
Then K' collapses across α with free edge $\max \alpha$.*

Howie's Tower Proof of the FHS. As before let $G = <
x_1, \ldots, x_n; R = 1 >$ where R is a cyclically reduced word involving
all the generators. We can without loss of generality assume that R
is not a proper power; if $R = S^m$, then replace R by S. Let K be the
2-complex model of the given presentation (see [41]), and let Γ be
the subgraph of its 1-skeleton $K(1)$ consisting of the unique 0-cell
and the 1-cells $x_1, \ldots, x_{n-1}$. It must be shown that $\pi_1(\Gamma) \to \pi_1 K$
is injective. If not, then there exists a map $f : D^2 \to K$ with $f(S^1)\Gamma$
representing a non-trivial element of $\pi_1(\Gamma)$. Suppose f is simplicial, and consider the maximal tower lifting of Figure 3. Notice
K' is staggered and we may assume that $g(\max \alpha) = x_n$ for each
2-cell α of K'. Further K' is finite for otherwise f' would factor

through an inclusion, contradicting maximality, and $H^1(K') = 0$ for otherwise f' would lift over an infinite cyclic covering contradicting maximality. Finally no 2-cell in K' is attached by a proper power. Repeated applications of Lemma 3.5 show that K' collapses to a 1-complex Γ' (necessarily a tree) such that $g(y) = x_n$ for each 1-cell y in $K' - \Gamma'$. In particular $f'(S^1)g^{-1}(\Gamma)\Gamma'$, so $f(S^1)$ is null-homotopic in $g^{-1}(\Gamma)$ and $f(S^1)$ is nullhomotopic in Γ. This is a contradiction so the map must be injective.

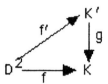

Figure 3. A Tower Lifting.

We mention that this tower approach has also been used by Howie to reprove other results about one-relator groups as well as to derive some entirely new results (see [41] and the references there).

A second geometric-topological approach was due to Lyndon who attempted to translate Magnus' original proof into combinatorial geometry. Lyndon gave a proof based on Van Kampen diagrams and the so-called maximum modulus principle for such diagrams [53]. We will discuss this further in the next section. We note that Lyndon's method was further adapted to one-relator products by H. Short [76].

Another approach that has been employed in proving the FHS is representation theoretic. The basic idea is as follows: Suppose A is a proposed FHS factor of G. Find a representation ρ of G into some linear group such that the restriction of ρ to A is faithful. This will imply that A injects into G. As an example, motivated by Ree-Mendelsohn [68], Baumslag, Morgan and Shalen [5] and P. Shalen [75] the following can be established.

Theorem 3.6 [68],[5],[75],[26]. *(1) Let G be be a one-relator group with torsion that is $G =< x_1, \ldots, x_n; R = 1 >$ where $R =$*

S^m with S cyclically reduced involving all the generators and $m \geq$ 2. Then G admits a representation ρ into $PSL_2(\mathbf{C})$ such that ρ is faithful on the free subgroup generated by $x_1, \ldots, x_{n-1}$, $\rho(x_n)$ has infinite order and $\rho(S)$ has order m.

(2) Let $G = \langle x_1, \ldots, x_n; R = 1 \rangle$ where $R = UV$ with $U = U(x_1, \ldots, x_p)$, $V = V(x_{p+1}, \ldots, x_n)$ non-trivial, cyclically reduced words in the free groups on the respective generators they involve and not proper powers and with $1 \leq p < n$. Then G admits a faithful representation into $PSL_2(\mathbf{C})$.

The groups in part (2) are special cases of *groups of F-type* which are a natural algebraic generalization of Fuchsian groups (see [28]). We also note that Magnus [58] pointed out that not every one-relator group can have a faithful representation in $PSL_2(\mathbf{C})$.

Representation theoretic techniques have proved especially effective in handling one-relator products of cyclics which we will discuss in Section 6.

Linear group techniques were also used to prove the following theorem of Gerstenhaber and Rothaus [35].

Theorem 3.7 [35]. *Let G be a finite group, and let $R = 1$ be an equation over G in the single variable x. If x occurs with non-zero exponent sum in R, then the equation $R = 1$ can be solved in a finite overgroup of G.*

This is relevant to our discussion because of the close tie between the FHS and the *adjunction problem*, that is the *solution of equations over groups*. Basically the adjunction problem is the following. Let G be a group and $W(x_1, \ldots, x_n)$ be an equation (or system of equations) with coefficients in G. This question is whether or not this can be solved in some overgroup of G. In the case of a single equation $W(x) = 1$ in a single unknown x a result of F.Levin [49] shows that there exists a solution over G if x appears in W only with positive exponents.

To see the connection with the FHS consider $G = A* \langle t \rangle /N(R)$ where R is a cyclically reduced word in the free product $A* \langle t \rangle$ of syllable length at least 2. Consider R then as a polynomial $R(t)$ in the variable t with coefficients in A. The equation $R(t) = 1$ has a solution in G if and only if A injects into G, that is, A is a FHS factor of G. More generally Baumslag and Pride [3] have shown that if $\mathbf{X}$ is a class of groups which contains the infinite cyclic group and is closed under free products with finitely many factors, then the existence of a FHS for one-relator products of **X**-groups is equivalent to the fact that any equation is solvable over an **X**-group. By this we mean that if $G = (*H_i)/N(R)$ where the H_i are arbitrary **X**-groups and R is a cyclically reduced word in the free product on the H_i of syllable length at least 2, then each H_i injects into G if and only if any equation is solvable over an **X**-group.

Finally in connection with the adjunction problem we mention the *Kervaire Conjecture* (also called the *Kervaire-Laudenbach Conjecture*). This says that if $G = A* <t> /N(R)$ is trivial, then A is trivial. From the FHS this is clearly true if A is a free group. Recently A.Klyachko [48] proved that the Kervaire conjecture is true whenever A is a torsion-free group. In this paper he also proved a Freiheitssatz. Specifically if A is torsion-free and the exponent sum of t in $R \in A* <t>$ is ± 1, then A injects into $A* <t> /N(R)$. Klyachko's result is a strenghtening of earlier work which showed that the the Kervaire conjecture is true if A is *locally indicable*, that is if every non-trivial, finitely generated subgroup admits an epimorphism onto an infinite cyclic group [43]. We will say more about this in Section 5.

4. SMALL CANCELLATION PRODUCTS AND DIAGRAMS

As a consequence of the FHS and the method developed for its proof, Magnus was able to provide a solution to the word problem for one-relator groups. In fact his proof established the stronger result that if $G = <X; R>$ is a one-relator group and w an arbitrary word in G, then there is an algorithm to determine if w belongs to the subgroup generated by any arbitrary subset of the given generators.

Magnus' work on the word problem was motivated in part by Dehn's solution to the word problem for finitely generated surface groups [20] (see also [15,63]). These groups have one-relator presentations of the form

$$Sg = <x_1, y_1, \ldots, x_g, y_g; R = 1> \text{ where } R = [x_1, y_1] \ldots [x_g, y_g].$$

Dehn proved that in a surface group S_g any word w which represents the identity must contain at least half of the original relator R, that is if $w = 1$ in S_g, then $w = bcd$ where for some cyclic permutation R' of R, $R' = ct$ with $|t| < |c|$ where $|\ |$ represents free group length. It follows then that $w = bt^{-1}d$ and this word representation of w has shorter length than the original. Given an arbitrary w in S_g we can apply this type of reduction process to obtain shorter words. After a finite number of steps we will either arrive at 1 showing that $w = 1$ or at a word that cannot be shortened in which case $w \neq 1$. This procedure solves the word problem for S_g and is known as *Dehn's Algorithm* for a surface group.

In general suppose G has a finite presentation $< X; R >$ (R here is a set of words in X). Let F be the free group on X and N the normal closure in F of R, $N = N_F(R)$ so that $G = F/N$. G, or more precisely the finite presentation $< X; R >$, has a *Dehn Algorithm*, if there exists a finite set of words $D \subset N$ such that any non-empty word w in N can be shortened by applying a relator in D. That is, given any non-empty w in N, w has a factorization $w = ubv$ where there is an element of the form bc in D with $|b| < |c|$. Then applying bc to w we have $w = uc^{-1}v$ where $|uc^{-1}v| < |ubv|$.

By the same argument as in the surface group case the existence of a Dehn Algorithm leads to a solution of the word problem.

In 1947 Tartakovskii [78] using purely algebraic methods showed that certain groups where "not much cancellation is possible in multiplying relators" also satisfied a Dehn Algorithm. Greendlinger [37], Schiek [72] and Britton [10] introduced other "small cancellation conditions" (see below) and also obtained Dehn Algorithms. This work initiated what is called *small cancellation theory*. This theory was placed in a geometric context in the mid 1960's by Lyndon [51]. Dehn's original approach was geometric and relied on an analysis of the tessellation of the hyperbolic plane given by a surface group, and thus Lyndon's geometrization was in a sense a full circle return to Dehn.

Small cancellation theory is pertinent to our discussion of the FHS for two reasons. First of all *small cancellation products*, that is amalgams (free products, free products with amalgamation and HNN groups) modulo relations satisfying certain small cancellation conditions satisfy a FHS relative to the amalgam factors (see below). Second the geometric constructions used in the proofs, Lyndon - Van Kampen diagrams, have been extended and modified for use in proving the FHS for one-relator products.

A complete and readable account of small cancellation theory can be found in Chapter V of Lyndon and Schupp's book [54]. The proofs, both algebraic and geometric, are quite complex. What we will do in this section is define the small cancellation conditions for free groups and then introduce the geometric constructions that go along with them. We then give the extensions to various amalgam constructions and state the relevant FHS. Finally we mention some more recent results on groups satisfying a Dehn Algorithm.

Suppose F is free on a set of generators X. Let R be a *symmetrized* set of words in F. By this we mean that all elements of R are cyclically reduced and for each r in R all cyclically reduced conjugates of both r and r^{-1} are in R. If r_1 and r_2 are distinct elements of R with $r_1 = bc_1$ and $r_2 = bc_2$, then b is called a *piece*. Pieces represent those subwords of elements of R which can be cancelled by multiplying two non-inverse elements of R. The small cancellation hypotheses state that pieces must be relatively small parts of elements of R. We define three small cancellation conditions, the most common being the first which is a metric condition.

The first is a metric condition denoted $C'(\lambda)$ where λ is a positive real number. This condition asserts that if r is an element of R with $r = bc$ and b a piece, then $|b| < \lambda|c|$. If G is a group with a presentation $< X; R >$ where R is symmetrized and satisfies $C'(\lambda)$, then G is called a λ- *group*. So for example, if $\lambda = 1/6$, G is a *sixth group*, etc.

If p is a natural number, the second small cancellation condition is a non-metric one denoted $C(p)$. This asserts that no element of R is a product of fewer than p pieces. Notice that $C'(\lambda)$ implies $C(p)$ for $\lambda \geq 1/(p-1)$.

The final small cancellation condition is also a non-metric condition denoted $T(q)$ for q a natural number. This asserts the following: Suppose $r_1, \ldots, r_h$ with $3 \geq h < q$ are elements of R with no successive pair inverses. Then at least one of the products $r_1 r_2, \ldots, r_{h-1} r_h, r_h r_1$ is reduced without cancellation. $T(q)$ is dual to $C(p)$ for $(1/p) + (1/q) = 1/2$ in a suitable geometric context [54].

Greendlinger [37] proved purely algebraically that sixth-groups satisfy a Dehn Algorithm; Schiek showed the same for fourth groups also satisfying $T(4)$ [69].

Suppose that the group G has a finite presentation $< X; R >$ with R a symmetrized set of words. As before let F be free on X and let N be the normal closure in F of R so that $G = F/N$. If w is a word in G, then $w = 1$ if and only if w as a word in F is a product of conjugates of elements of R, that is, $w = u_1 r_1 u_1^{-1} \cdots u_m r_m u_m^{-1}$ where the u_i are words in F and the r_i are elements of R. The sequence $u_1 r_1 u_1^{-1}, \ldots, u_m r_m u_m^{-1}$ is called an R -*sequence of length* m for w. A *minimal R-sequence* for w is an R-sequence of minimum length. We will associate to any R-sequence for w a connected, simply-connected *diagram* in the Euclidean plane called an

R-diagram. The small cancellation hypotheses will be analyzed by analyzing these diagrams.

A (Lyndon-Van Kampen) *diagram* for a group F consists of a collection M of vertices, oriented edges and regions in the Euclidean plane together with a labelling function assigning to each oriented edge e an element (e) of F. This labelling function must satisfy $(e^{-1}) = (e)^{-1}$ where e^{-1} is the oppositiely oriented edge of e. Further if α is a path in M with $\alpha = e_1 \cdots e_n$, then (α) is defined as $(e_1) \cdots (e_n)$. If D is a region in M, a label of D is an element (α) for any boundary cycle of D. We have the following

Theorem 4.1 [54]. *Let F be a free group and $c_1, \ldots, c_m$ a sequence of non-trivial elements of F. Then there exists a diagram $M = M(c_1, \ldots, c_m)$ over F satsifying the following properties.*

   (i) *If e is an edge of M, $(e) \neq 1$.*
   (ii) *M is connected and simply connected with a distinguished vertex O on the boundary of M. There is a boundary cycle $e_1, \ldots, e_n$ of M (a cycle in M of minimal length which contains all the edges in the boundary of M) beginning at O such that the product $(e_1) \cdots (e_n)$ is reduced without cancellation and $(e_1) \cdots (e_n) = c_1 \cdots c_m$*
   (iii) *If D is a region of M and $e_1, \ldots, e_j$ is a boundary cycle of D, then $(e_1) \cdots (e_j)$ is reduced without cancellation and is a cyclically reduced conjugate of some c_i.*

The following provides a converse to the above theorem and also allows us to relate this to R-sequences.

Theorem 4.2 [54]. *Let M be a connected, simply connected diagram over a group F with regions $D_1, \ldots, D_m$. Let α be a boundary cycle of M beginning at a vertex v_o on the boundary of M and let $w = (\alpha)$. Then there exists boundary labels r_i of D_i and elements u_i of F, $1 \leq i \leq m$ such that $w = u_1 r_1 u_1^{-1} \cdots u_m r_m u_m^{-1}$.*

Now suppose that R is a symmetrized subset of words in a free group F. An *R-diagram* is a diagram M over F such that if ∂ is any boundary cycle of any region D of M, then (∂) is in R. If $G = F/N$ as before, then from the two theorems we obtain the following fact. A word w in F is in N if and only if there exists a connected, simply connected R-diagram M such that the label on the boundary of M is w. Thus connected, simply connected diagrams provide an adequate tool for studying membership in normal subgroups of free groups. The analysis of the small cancellation conditions lies in analyzing the structure of R-diagrams under these conditions. We briefly indicate one result along these lines and then return to our main discussion of the FHS. We mention also that the concept of "pictures" over groups which are used in deriving the FHS for one-relator products are duals to these Van Kampen diagrams. On a historical note , Van Kampen discovered these diagrams in the 1930's [81], and they were apparently not used to any great extent until Lyndon [51] and Weinbaum [79] applied them to small cancellation theory. In doing this Lyndon provided many combinatorial generalizations of regular tessellations of the plane [51,52] , again returning to the ideas of Dehn. In particular Lyndon used Van Kampen diagrams to translate Magnus' original proof of the Freiheitssatz into combinatorial geometry. The main tool employed in this was the *maximum-minimum modulus* principle for these diagrams, which can be described in the following manner. Consider a presentation $G = < X; R >$, and assume that X is a disjoint union of subsets $X\mathcal{T}$. The generators in each are said to be of the same *type*. Assume further that there is an integer-valued function assigning a subscript to each generator in X. If w is a word on X and $\mathcal{T}$ is a type, then $\max\mathcal{T}(w)$ will denote the generator of type $\mathcal{T}$ with maximum subscript which occurs in w (when w involves a generator of type $\mathcal{T}$). Similarly we define $\min\mathcal{T}(w)$. The presentation $< X; R >$ is *staggered* if every relator in R contains at least one generator of each type and the following condition holds for every type $\mathcal{T}$: If $i < j$, then $\max\mathcal{T}(r_i) < max\mathcal{T}(r_j)$ and $\min\mathcal{T}(r_i) < min\mathcal{T}(r_j)$. ($x_i < x_j$ if x_i and x_j are generators of the same type with $i < j$.) Let $< x : r >$ be a staggered presentation and $R*$ the symmetrized set generated by R. Let M be a connected, simply connected reduced $R*$-diagram (see below). For any type $\mathcal{T}$, $\max\mathcal{T}(M)$ (respectively $\min\mathcal{T}(M)$) will denote the generator of type $\mathcal{T}$ with maximum (resp. minimum) subscript which occurs as the label on an edge of M. The maximum-minimum modulus principle then says that for each type $\mathcal{T}$, there are edges in ∂M labelled by $\max\mathcal{T}(M)$ and $\max\mathcal{T}(M)$. Lyndon's proof of the Freiheitssatz not only worked for one-relator groups but also for one-relator products where the factors are subgroups of the additive group of the reals [52,53,54].

A diagram over F is a *reduced diagram* if it satisfies the following additional property. Let D_1 and D_2 be regions of M with an edge $e \subseteq \partial(D_1) \cup \partial(D_2)$. Let $e\delta_1$ and $\delta_2 e^{-1}$ be boundary cycles of D_1 and D_2 respectively. Let $f(\delta_1) = f_1$ and $f(\delta_2) = f_2$. Then one never has $f_2 = f_1^{-1}$. If M is the diagram of a minimal R-sequence, then M is reduced [54].

We have the following:

Theorem 4.3 [54]. *Let R be a symmetrized set of elements of a free group and let M be a reduced R-diagram. Then*

(i) *The minimal length of an R-sequence for w is the number of regions in an R-diagram for a minimal R-sequence.*

(ii) *If R satisfies $C(p)$, then each region D of M such that the intersection of ∂D and ∂M does not contain an edge has $d(D) \geq p$ where $d(D) =$ number of edges in a boundary cycle of D.*

(iii) *If R satisfies $T(q)$, then each interior vertex has $d(v) \geq q$ where $d(v) =$ number of edges having v as an inital vertex.*

To state our FHS for small cancellation products we need to extend the small cancellation hypothesis to symmetrized sets of words over the various amalgam constructions.

First suppose that $F = *F_i$ is a free product of non-trivial groups F_i. A word w in F has *reduced form* $w = uv$ if there is no *cancellation* or *consolidation* in forming the concatenation of the free product normal forms for u and v. This is equivalent to the fact that the last syllable of u is in a different factor than the first syllable of v. w is in *semi-reduced* form if $w = uv$ where there is no cancellation in forming uv but there may be consolidation - the last syllable of u may be in the same factor as the first syllable of v but they don't cancel completely. Suppose R is a symmetrized subset of F. Then b is a *piece* if there exist distinct elements r_1 and r_2 in R with semi-reduced forms $r_1 = bc_1$ and $r_2 = bc_2$.

To extend these concepts to free products with amalgamation and HNN groups we need slightly more general definitions of normal forms than those given previously. Suppose F is the free product of non-trivial groups F_i amalgamated over a common (isomorphically embedded) subgroup A. If $w \neq 1$ is in F, we say that a *normal form* for w consists of any sequence $w = y_1 \cdots y_n$ such that y_i is in a factor of F, successive y_i come from different factors of F and no y_i is in the amalgamated subgroup unless $n = 1$. Under this definition there will be many normal forms for the same element but their lengths will all be the same. If u has normal form $y_1 \cdots y_n$ and v has normal form $x_1 \cdots x_n$, then there is *cancellation* in forming uv if $y_n x_1$ is in the amalgamated subgroup. If $y_n x_1$ is not in the amalgamated subgroup but y_n and x_1 are in the same factor of F, then there is *consolidation*. With these ideas the definition of semi-reduced forms and a piece is the same as for a free product.

Finally suppose F is an HNN group with base H so that $F = <H, t : t^{-1}At = (A) >$. If w is in F, we consider a normal form for w to be any sequence $w = h_o t^{e_1} h_1 \cdots h_n t^{e_n} h_{n+1}$ where no t-reductions are possible. Any two normal forms for w will have the same number of t symbols. If u and v have the respective normal forms $h_o t^{e_1} h_1 \cdots h_n t^{e_n} h_{n+1}$ and $h'_o t^{f_1} h'_1 \cdots h'_m t^{f_n} h'_{n+1}$, then there is *cancellation* in forming uv if either $e_n = -1$, $h_{n+1} h'_o$ is in A and $f_1 = 1$ or if $e_n = 1$, $h_{n+1} h_o'$ is in (A) and $f_1 = -1$. If there is no cancellation in forming uv, we say that uv is in *reduced form*. If R is a symmetrized set in F, then the definition of a piece is as in the previous two cases except here we require the factorizations to be reduced forms.

We now state our metric small cancellation condition for amalgams. If F is either a free product, free product with amalgamation or HNN group and R is a symmetrized set of words in F then the metric condition $C'(\lambda)$ for λ a positive real number is defined as follows:

Condition $C'(\lambda)$. If r is in R with semi-reduced form (reduced form for HNN group) $r = bc$ and b a piece, then $|b| < \lambda|r|$. Also $|r| > 1/\lambda$ for all r in R.

There are similar analogues for $C(p)$ and $T(q)$ but these are not necessary for our FHS result. We define a *small cancellation product* to be a group $G = F/N$ where F is either a free product, free product with amalgamation or HNN group and $N = N(R)$ is the normal closure of a symmetrized set of words in F satisfying a small cancellation condition $C'(\lambda)$ for some λ. Our major result is the following which is actually a summary of several results (see [54]).

Theorem 4.4 (The FHS for Small Cancellation Products) (see[54] and the references there). Let G be a small cancellation product satisfying $C'(\lambda)$ for some $\lambda \leq 1/6$. Then G satisfies a FHS relative to any of the amalgam factors, that is, any factor of the underlying amalgam injects into G under the identity homomorphism.

The proof of these results follows the same general outlines as for small cancellation theory over free groups, that is, the analysis of appropriate diagrams. The approach given in Lyndon and

Schupp[54] is to treat each amalgam construction separately. Recently J. Corson [19] has considered Small Cancellation Theory over graphs of groups. His approach both unifies and extends this classical treatment. He defines *non-spherical* sets of words which are the analogues of symmetrized sets of words satisfying small cancellation conditions. He then proves that if F is graph of groups and N is the normal closure in F of a non-spherical set of words, then the group natural map $F \rightarrow F/N$ embeds the vertex and edge groups of F. Similar results have been done independently by R.M.S. Mahmud [61].

We mention also that Collins and Perraud have proven a FHS for small cancellation products (over a free product) satisfying different small cancellation conditions [18].

We close this section by pointing out the connection between Dehn's algorithm and the *hyperbolic groups* introduced by Gromov [39]. Let G be a finitely generated group with fixed finite generating set X. Let Γ be the Cayley graph of G relative to this generating set X equipped with the word metric. A *geodesic* in the Cayley graph is a path between two points with minimal length relative to the word metric. A geodesic triangle is a triangle with geodesic sides. A geodesic triangle in Γ is δ- *thin* if any point on one side is at a distance less than δ from some point on one of the other two sides. Γ is δ-*hyperbolic* if every geodesic triangle is δ-thin. Finally G is *word-hyperbolic* if G is δ-hyperbolic with respect to some generating set X and some fixed $\delta/ge0$. The tie with Dehn's algorithm is the following remarkable theorem.

Theorem 4.5. *If G is finitely presented, then G is word hyperbolic if and only if G satisfies a Dehn algorithm. ([39,see also 63 and the references there]).*

5. ONE-RELATOR PRODUCTS

A *one-relator product* of a family of groups A_i, i in some index set I, is the quotient $G = A/N(R)$ of their free product $A = *A_i$ by the normal closure $N(R)$ of a single non-trivial word R in the free product. We assume that R is cyclically reduced and of syllable length at least two. The groups A_i are called the *factors* while R is the *relator*. In analogy with the one-relator group case we say R *involves* A_i if R has a non-trivial syllable from A_i. If $R = S^m$ with S a non-trivial cyclically reduced word in the free product and $m \geq 2$, then R is a *proper power*. In this context a one-relator group is just a one-relator product of free groups. Thus one-relator products are a natural way to generalize the vast theory of one-relator groups. We refer to the survey articles by Howie [41,42,46] for a more complete discussion of results on one-relator products beyond the FHS.

A Freiheitssatz holds for a one-relator product G if each factor injects into G via the natural map, that is, each factor is a FHS factor. In general a FHS is not true for a one-relator product. Consider $A = < a >$ and $B = < b >$ to be finite cyclic groups of relatively prime order. Then $G = A*B/N(ab)$ is a trivial group and hence clearly A and B cannot inject. Therefore some restrictions must be imposed. There are two approaches. The first is to impose conditions on the factors while the second is to impose conditions on the relator. We will discuss conditions on the factors first and then conditions on the relator. The most common relator condition is that R is a proper power of suitably high order. Lyndon [53] showed a FHS for one-relator products when the factors are subgroups of the additive group of the real numbers. S. Pride [66] then showed that the FHS holds for one-relator products of locally fully residually free factors while this was extended to locally residually free factors by B. Baumslag and S. Pride [3] . The above results are all examples of *locally indicable factors*. Recall that a group G is *locally indicable* if every non-trivial finitely generated subgroup has the infinite cyclic group as an epimorphic image. Thus the above results are subsumed by the following theorem which is at present the mosr general statement about one-relator products with conditions of the factors.

Theorem 5.1 [2,11,12,43,76]. *The Freiheitssatz holds for one-relator products with locally indicable factors.*

This was proven originally by Brodskii [13]{announced in [11]} using an algebraic method along the lines of Magnus' original proof. J. Howie [42] gave an independent proof using a variation of the tower method. H. Short [76] gave another proof which introduced the concept of "pictures". These pictures are the duals of Lyndon - Van Kampen diagrams for one-relator products. In introducing pictures Short was adapting Lyndon's geometric proof of the FHS to the context of one-relator products of locally indicable groups. B. Baumslag [2] gave an entirely algebraic, self-contained proof which mimiced Magnus' original proof and which was similar to Brodskii's approach. Local indicability of factors seems to be the property which is just strong enough to allow most of the results on one-relator groups to carry over to one-relator products. Again we refer to the articles by J. Howie [41,42].

Theorem 5.1 can be iterated and extended in several ways to obtain the FHS for groups formed by more general constructions than one-relator products. These extensions are generally of the form $(*A_i)/N(Rj)$ where the A_i are locally indicable and the relators R_j are "staggered" [41]. We present two results of this type. The first is a straightforward iteration using a result of Howie while the second, due to M. Edjvet, does not actually fall into this pattern but rather returns to the small cancellation results of the last section.

Theorem 5.2. *Let $A_1, \ldots, A_n$ be locally indicable groups, and for each $i = 1, \ldots, n - 1$, let R_i be a non-trivial word in the free product $G_i = A_1 * A_2 * \cdots * A_{i+1}$ of syllable length at least two and involving A_{i+1}. Assume further that for $i = 1, \ldots, n - 2$, R_i not a proper power. Then each A_i is a FHS factor for $G = (*A_i)/N(R_1, \ldots, R_{n-1})$.*

Proof. Suppose $n = 3$ so that $G = (A * B * C)/N(R_1, R_2)$ with A, B, C locally indicable and R_1 involving A and B not a proper power and R_2 involving C. From a result of Howie [42] a one-relator product of locally indicable factors is locally indicable if the relator is not a proper power. Therefore $G_1 = (A * B)/N(R_1)$ is locally indicable. Now $G = (G_1 * C)/N(R_2)$ is then a one-relator product with locally indicable factors so both G_1 and C inject. Further A and B inject into G_1 and thus into G. The general result then follows easily by induction.

This type of iteration was used also in studying the class of groups of special NEC type by Fine and Rosenberger [33].

M. Edjvet [24] considered the following type of situation which generalizes results of S. Pride [67] on staggered presentations where each relator involves exactly two generators. Let X be a set of generators and $\mathbf{A}$ a class of finite subsets of X. For each $\mathbf{a}$ in $\mathbf{A}$ let $R(\mathbf{a})$ be a set of cyclically reduced words on $\mathbf{a}$ each of which involves all elements of $\mathbf{a}$.

Let $G(\mathbf{A})$ be the group with presentation $G(\mathbf{A}) = < x; r >$ where $r = \cup_{\mathbf{a} \in \mathbf{A}} r(\mathbf{a})$. A *face group* of $G(\mathbf{A})$ is a group $G(\mathbf{a}) = < \mathbf{a}; \cup_{\mathbf{b} \subseteq \mathbf{a}} r(\mathbf{b}) >$.

A Freiheitssatz exists for this situation if $G(\mathbf{B})$ naturally injects into $G(\mathbf{A})$ where X_o is some subset of X and $\mathbf{B}$ is the subset of $\mathbf{A}$ consisting of those members of $\mathbf{A}$ that are subsets of X_o.

To obtain a FHS Edjvet introduces analogs of the standard small cancellation conditions. A *k-factorization* of a word w on $\mathbf{a}$ is a product expression $w = w_1 \cdots w_k$ where each w_i $(1 \leq i \leq k)$ is a word on a proper subset of $\mathbf{a}$. $G(\mathbf{a})$ has *property B_k* if no element of $r(\mathbf{a})$ has a k-factorization. This is analagous to condition $C(k+1)$ of small cancellation theory. (see Section 4)

For S a subset of X define a *d-map* $d_S : 2^X \rightarrow 2^X$ by $d_S(y) = y \cap S$. A *p-gon* (on X) is a subset of 2^X of the form $\{\{x_1, x_2\}, \ldots, \{x_{p-1}, x_p\}, \{x_p, x_1\}\}$ where $x_i \neq x_j$ for $1 \leq i < j \leq p$. $\mathbf{A}$ has property $N(p)$ if no p-element subset of $\mathbf{A}$ can be mapped to a p-gon by some d-map. This property is analagous to property $T(p+1)$.

Finally, let $\bar{r}(\mathbf{a})$ denote the set of words on $\mathbf{a}$ which are equal to 1 in the face group $G(\mathbf{a})$ but are not equal to 1 in the group $< \mathbf{a}; \cup_{\mathbf{b} \subseteq \mathbf{a}} r(\mathbf{b}) >$. Say two members $\mathbf{c}$, $\mathbf{d}$ are *incomparable* if $\mathbf{c} \notin \mathrm{Id}(\mathbf{d})$ and $\mathbf{d} \notin \mathrm{Id}(\mathbf{c})$ where $\mathrm{Id}(\mathbf{a})$ is the set $\{\mathbf{b} \in; , \mathbf{b} \subseteq \mathbf{a}\}$. Then $G(\mathbf{a})$ has *property B_2'* if the following holds: let $\mathbf{a}$, $\mathbf{b}$ be incomparable members of $\mathbf{A}$ and let uu_1u_2, uv_1v_2 be 3-factorizations of elements of $\bar{r}(\mathbf{a})$, $\bar{r}(\mathbf{b})$ respectively; then $u_1^{-1}v_1$ is a word involving either all of $\mathbf{a}$ or all of $\mathbf{b}$.

The following theorem is a FHS for this situation.

Theorem 5.3 [24]. *The Freiheitssatz holds for $G(\mathbf{A})$ if one of the following conditions is satisfied:*

(i) *Each face group $G(\mathbf{a})$ has property B_5.*

(ii) *Each face group $G(\mathbf{a})$ has property B_3, and $\mathbf{A}$ has property $N(3)$.*

(iii) *Each face group $G(\mathbf{a})$ has property B_2 and property B_2', and $\mathbf{A}$ has properties $N(3),N(4),N(5)$.*

As with the analagous FHS results for small cancellation products Edjvet's proof is based on an analysis of diagrams. Locally indicable groups are torsion-free. The following has been a rather standard conjecture about one-relator products.

Conjecture. *The Freiheitssatz holds for one-relator products of torsion-free factors.*

Recently Brodskii and Howie [14] have proved some results on special cases supporting this conjecture. Let $x = (x_1, \ldots, x_n)$ be a sequence of elements in a group G. x_i is isolated in x if no x_j belongs to the cyclic subgroup generated by x_i for $i \neq j$.

Theorem 5.4 [14]. *Let $G = (A*B)/N(W)$ where A and B are torsion-free groups and $W = a_1b_1 \cdots a_kb_k$ is a cyclically reduced word in $A*B$ such that some a_i is isolated in $(a_1, \ldots, a_k)$ and some b_j is isolated in $(b_1, \ldots, b_k)$. Then A and B naturally inject into G.*

If $W = a_1x^{m_1} \cdots a_kx^{m_k}$ is a word in $A* < x >$, then the *sign-index* σ is the number of sign changes in the cyclic sequence $(m_1, \ldots, m_k)$.

Theorem 5.5 [14]. *Let $G = A* < x > /N(W)$ where A is torsion-free and $W = a_1x^{m_1} \ldots a_kx^{m_k}$ is a cyclically reduced word in $A* < x >$ of length at least 2 and sign-index $\sigma \leq 2$. Then A naturally injects into G. (In particular the Kervaire conjecture is satisfied). Further suppose that in addition one of the following holds:*

(i) *The sign-index $\sigma > 0$*
(ii) *$a_1 \cdots a_k$ is in $A \setminus \{1\}$*
(iii) *$\alpha = m_k + m_1a_1 + m_2(a_1a_2) + \ldots + m_{k-1}(a_1 \cdots a_{k-1})$ is not a unit in QG*
(iv) *$k \leq 3$*

Then $< x >$ naturally injects into G . (x has infinite order in G.)

Corollary 5.6 [14]. *Let $G = (A*B)/N(W)$ where A and B are torsion-free groups and $W = a_1b_1 \cdots a_kb_k$ is a cyclically reduced word in $A*B$ such that $a_i \neq 1 \neq b_j$ and $k \leq 3$. Then A and B naturally inject into G.*

Before we leave our discussion of conditions on the factors we mention one final result of Howie and Short [47] which is also related to the adjunction problem. To any finite system of equations $\Sigma : r_1(t_1, \ldots, t_n) = \ldots = r_m(t_1, \ldots, t_n) = 1$ over a group, in variables $\{t_i\}$, we can associate the integer matrix $M(\Sigma) = (\mu_{ij})$ where μ_{ij} is the exponent sum of t_i in the word (equation) r_j. If the rank of $M(\Sigma)$ is equal to the number of equations in Σ, then Σ is called *independent*.

Theorem 5.7 [47]. *Let $G = (A* < t_1, \ldots, t_n >)/N(r_1, \ldots, r_n)$. Then A injects naturally into G under either of the following two conditions:*

(i) *A is locally indicable and the system $(r_1, \ldots, r_n)$ is independent in $\{t_i\}$*
(ii) *A is locally indicable, $n = 1$ and r_1 is not conjugate to an element of A.*

We now turn to condtions on the relator R. The basic results have been concerned with the case where $R = S^m$ with $m \geq 2$, that is, where R is a proper power. Generally the higher the power the more well-behaved the one-relator product. In fact if $R = S^m$ with $m \geq 7$ then almost all results on one-relator groups carry over to the resulting one-relator product and the proofs are relatively simple [22].

Suppose $G = (*A_i)/N(S^m)$ is a one-relator product with relator $R = S^m$ a proper power. If $m/ge7$ then the relator set $\{S^m\}$ satisfies the small cancellation metric condition $C'(1/6)$ so G satisfies the FHS from the results on small cancellation products (Theorem 4.4). If $m = 6$ the FHS was proven using "pictures" (see below) by Gonzalez-Acuna and Short [36]. Independently the FHS for $m = 6$ follows from the results of Collins and Perraud[18] mentioned earlier since in this case S^m satisfies the non-metric small cancellation condition $C(6)$. An analysis of pictures by J. Howie [44] yielded the FHS for $m = 5$ while a much deeper analysis also by J. Howie did the same for $m = 4$ [45] and certain cases when $m = 3$ [22]. (The last result joint with A. Duncan). Thus we have the following.

Theorem 5.8 [22,44,45]. *Suppose $G = (*A_i)/N(S^m)$ is a one-relator product where S is a cyclically reduced word in the free product $(*A_i)$ of syllable length at least 2 and suppose $m/ge4$. Then the FHS holds, that is, each factor A_i naturally injects into G. Further if $m = 3$ and the relator S contains no letters of order 2, then the FHS holds.*

The general case for $m \geq 2$ is still open. However for certain subcases it is known, for example if the factors are locally indicable. In Section 6 we consider further representation theoretic criteria for the FHS to hold when $m \geq 2$.

There are no known examples for which the FHS fails for $m \geq 2$ and therefore we have the following conjecture.

Conjecture. *The FHS holds for any one-relator product $G = (*A_i)/N(R^m)$ where $m/ge2$ and where R is a cyclically reduced word in the free product $(*A_i)$ of syllable length at least 2.*

We close this section with a brief review of "pictures" over one-relator products and how their analysis leads to the FHS. Pictures are essentially the duals of the Lyndon-Van Kampen diagrams introduced in Section 4. They were adapted to the context of one-relator products by H. Short [76]. A full discussion of pictures can be found in [25] while the following description is summarized from [21] and [44,45].

A *picture* Γ of a one-relator product $T = (A * B) \setminus N(R^m)$ on a compact surface Σ consists of:

    (i) A disjoint union of (small) discs $v_1, \ldots, v_n$ in int(Σ) called the *vertices* of Γ.

    (ii) A properly embedded 1-submanifold ξ of $\Sigma_0 = \Sigma \setminus$ int($\cup v_i$). The components of this submanifold are called the *arcs* of Γ.

    (iii) An orientation of $\partial\Sigma_0$ and a *labelling function* that associates to each component of $\partial\Sigma_0 \setminus \xi$ a *label* which is an element of $A \cup B$.

A picture over S^2 is called a *spherical picture*.

The data above is also required to satisfy a number of properties reflecting the fact that it is to represent the one-relator product.

    (a) In any *region* Δ of Γ–component of $\Sigma \setminus \cup v_i \cup \xi$–either all labels belong to an A-region or a B-region accordingly.

    (b) Each arc separates an A-region from a B-region.

    (c) The *vertex label* of any vertex v_i–the word consisting of the labels of ∂v_i read in the direction of orientation from some starting point– is identically equal to R^m in the free monoid on $A \cup B$ up to cyclic permutation.

    (d) Suppose Δ is an orientable region of Γ of genus g with k boundary components. Then each boundary component has a *boundary label* $a_1^{\epsilon_1} \cdots a_t^{\epsilon_t}$ where $a_1, \ldots, a_t$ are the labels of that boundary component in the cyclic order induced from some fixed orientation of Δ and $\epsilon_i = +1$ if the orientation of the segment of $\partial\Sigma_0$ labelled a_i agrees with that of the boundary component of Δ and -1 otherwise. If $\alpha_1, \ldots, \alpha_k$ are the boundary labels of Δ, then the equation $X_1\alpha_1 X_1^{-1} \cdots X_k\alpha_k X_k^{-1}[Y_1, Z_1] \cdots [Y_{g'}, Z_g]$ is solvable for (X_i, Y_j, Z_j) in A if Δ is an A-region or in B if Δ is a B-region.

Under the conditions (a) through (d) above, pictures can be constructed over G. Further there are certain allowable operations on pictures which when applied yield an essentially equivalent picture. These are called *bridge moves* and insertion or deletion of *floating dipoles* or *exceptional pictures*. We refer to [21] or [44] for a precise definition. Two pictures over G on the same surface are *equivalent* if each can be obtained from the other by a sequence of these allowable operations. Pictures can be associated to maps from the surface Σ to a certain space with fundamental group G [21] and equivalent pictures have maps which differ only by a certain homotopy. A picture on Σ over G is *efficient* if it has the least number of vertices in its equivalence class.

For the proof of the FHS the crucial condition on pictures is the following:

Conjecture F. *Suppose $G = (*A_i)/N(R^m)$ is a one-relator product where S is a cyclically reduced word in the free product $(*A_i)$ of syllable length at least 2 and suppose $m/ge2$. Let Γ be an efficient picture on D^2 over G such that at most 3 vertices of Γ are connected to ∂D^2 by arcs. Then every vertex of Γ is connected to ∂D^2 by arcs.*

Theorem 5.9 [21]. *Suppose $G = (*A_i)/N(R^m)$ is a one-relator product where R is a cyclically reduced word in the free product $(*Ai)$ of syllable length at least 2 and suppose $m \geq 2$. If conjecture F holds for pictures over G on the disc, then the FHS holds for G.*

Howie's results for $m = 5$ and $m = 4$ then follow by showing that in these cases conjecture F holds. Howie also gives examples where for $m = 2$ and $m = 3$ conjecture F fails [21]. However, even in these cases the FHS holds.

6. ONE-RELATOR PRODUCTS OF CYCLICS

The results of the last section led to a FHS for a one-relator product when the factors are torsion-free or the relator has order 4 or greater. We now turn our attention to the case where the factors may have torsion. In particular we consider *one-relator products of cyclics*. Such groups have the form

(6.1)
$$G =< a_1, a_2, \dots, a_n; a_i^{e_i} = 1, i = 1, \dots, n, R(a_1, \dots, a_n) = 1 >$$

where $e_i = 0$ or $e_i > 2$ for each i and $R(a_1, \dots, a_n)$ is cyclically reduced in the free product on $a_1, \dots, a_n$. We will be especially interested in the case when $R = S^m$ with $S = S(a_1, \dots, a_n)$ cyclically reduced in the free product on $a_1, \dots, a_n$ and $m/ge2$.

Of independent interest are the 2-generator cases of the above with $R = S^m$ and $m \geq 2$. These groups then have the form

(6.2) $$G =< a, b; a^p = b^t = S^m(a, b) = 1 >$$

and are called the *generalized triangle groups*. Notice that if $S(a, b) = ab$, then we have an ordinary triangle group.

The groups of form (6.1) provide a natural algebraic generalization of Fuchsian groups which are one-relator products of cyclics relative to the *standard Poincare presentation* [54]

$$F =< e_1, \ldots, e_p, h_1, \ldots, h_t, a_1, b_1, \ldots, a_g, b_g; e_i^{m_i} = 1,$$

(6.3) $i = 1, \ldots, p, R = 1 >$

where $R = UV$ with $U = e_1 \cdots e_p h_1 \cdots h_t, V = [a_1, b_1] \cdots [a_g, b_g], p \geq 0, t \geq 0, g \geq 0, p + t + g > 0$, and $m_i \geq 2$ for $i = 1, \ldots, p$. U or V could possibly be trivial.

Further, if in form (6.1) above, the relator $R = UV$ with $U = U(a_1, \ldots, a_p), V = V(a_p, \ldots, a_{p+1})$ non-trivial cyclically reduced words of infinite order in the free product on the respective generators which they involve and $1 \leq p < n$, the resulting group is called a *group of F-type*. It was shown in [28] that groups of F-type exhibit many of the same algebraic and linear properties as Fuchsian groups.

In handling the FHS for one-relator products of cyclics, representation theoretic techniques are used. As mentioned in Section 3 the basic idea is to find a representation of the group G such that this representation is faithful on the proposed FHS factor. This will ensure that the FHS factor naturally injects. For one-relator products of form (6.1) where the relator S^m is a proper power $(m \geq 2)$ such that S involves all the generators, an *essential representation* is a representation $\rho : G \rightarrow \{$Linear Group$\}$ such that $\rho(a_i)$ has order e_i if $e_i/ge2$ and infinite order if $e_i = 0$ and $\rho(S)$ has order m. G. Baumslag, Morgan and Shalen [5] using a technique based on work of Ree and Mendelsohn [68] proved the following.

Theorem 6.1 [5]. *A generalized triangle group (form 6.2 above with m/ge2) admits an essential representation into $PSL_2(\mathbb{C})$. In particular no generalized triangle group is trivial.*

This result was proven independently by S. Boyer [9] who showed the existence of an essential representation into SU(2).

A refinement and extension of the technique in Baumslag, Morgan and Shalen allowed a proof of the following theorem.

Theorem 6.2 [24]. *Let A and B be groups which admit faithful representations in $PSL_2(\mathbb{C})$, and suppose R is a cyclically reduced word in the free product of A and B of length ≥ 2. Suppose $G = (A*B)/N(S^m)$ where $m \geq 2$. Then G admits a representation $\rho : G \to PSL_2(\mathbb{C})$ such that $\rho|_A$ and $\rho|_B$ are faithful and $\rho(S)$ has order m. In particular the maps $A \to G$ and $B \to G$ are injective, that is, the Freiheitssatz holds.*

We outline the proof. Write the relator S as $S = a_1b_1...a_kb_k$ with $a_i \in A$ and $b_i \in B$. Since S is cyclically reduced of length at least 2, we can assume that $a_i \neq 1$ and $b_i \neq 1$.

Choose faithful representations $\rho_A : A \to PSL_2(\mathbb{C})$ and $\rho_B : B \to PSL_2(\mathbb{C})$ such that after suitable conjugation if necessary

$$\rho_A(a_i) = \pm \begin{pmatrix} * & x_i \\ * & * \end{pmatrix} \qquad \text{with } x_i \neq 0$$

$$\rho_B(b_i) = \pm \begin{pmatrix} * & y_i \\ * & * \end{pmatrix} \qquad \text{with } y_i \neq 0 \text{ for all } i = 1, \dots, k$$

For $w \in \mathbb{C}$ let ρ_A^w denote the representation ρ_A conjugated by $\begin{pmatrix} 1 & 0 \\ w & 1 \end{pmatrix}$, that is,

$$\rho_A^w(a) = \begin{pmatrix} 1 & 0 \\ w & 1 \end{pmatrix} \rho_A(a) \begin{pmatrix} 1 & 0 \\ -w & 1 \end{pmatrix}.$$

Considering w as a variable, define

$$T(w) = \text{trace}(\rho_A^w(a_1)\rho_B(b_1) \cdots \rho_A^w(a_k)\rho_B(b_k)).$$

Then $T(w)$ is a complex polynomial in w and because of the choices of ρ_A and ρ_B it is *non-constant* in w. By the fundamental theorem of algebra there is a solution w_0 to the equation $T(w_0) = 2cos(\pi/m)$.

Now define the representation $\rho : G \to PSL_2(\mathbb{C})$ by $\rho \restriction_A = \rho_A^{w_0}$ and $\rho \restriction_B = \rho_B$. The trace of $\rho(S)$ is then $T(w_0) = 2\cos(\pi/m)$ and so $\rho(S)$ has order m. Further $\rho \restriction_A = \rho_A^{w_0}$ is faithful on A and $\rho \restriction_B$ is faithful on B. Thus ρ is the desired representation.

This theorem then extends the class of groups for which the FHS is true when $m \geq 2$. Thus, for example, the FHS holds when the factors are Fuchsian groups, Kleinian groups, free metabelian groups of rank 2 etc. In particular any free product of cyclic groups can be faithfully represented in $PSL_2(\mathbb{C})$ and therefore we have as a corollary.

Corollary 6.3 [26]. *Let G be a one-relator product of cyclics with relator $R = S_m$ with $m \geq 2$. Then G admits an essential representation ρ into $PSL_2(\mathbb{C})$ such that $\rho| < x_1, \ldots, x_{n-1} >$ is faithful. In particular the FHS holds so that the subgroup generated by $< x_1, \ldots, x_{n-1} >$ is the obvious free product of cyclics.*

In a series of papers, Fine and Rosenberger along with Allenby, Howie, Levin, Rohl and Tang (see [27-32] and the references cited there) have shown that one-relator products, with relator a proper power, satisfy many linearity properties. These include, under suitable restrictions, satisfying the Tits Alternative, conjugacy separability, subgroup separability, being virtually torsion free and admitting a decomposition as a free product with amalgamation. Using a result of P. Shalen [75], the representation technique can be extended to groups of F-type.

Theorem 6.4 [28]. *Let G be a group of F-type. If $R = UV$ is the defining relator and neither U nor V is a proper power in the free product of the generators they involve, then G admits a faithful representation into $PSL_2(\mathbb{C})$.*

Using an amalgam decomposition for groups of F-type, we can also obtain a FHS.

Theorem 6.5 [26]. *Let G be a group of F-type then any $(n-2)$ of the given generators generate the obvious free product of cyclics, that is $< x_1, \ldots, x_{n-2} >$ is a FHS factor.*

7. MULTI-RELATOR VERSIONS OF THE FREIHEITSSATZ

We close this survey with a discussion of several additional multi-relator versions of the Freiheitssatz. The first result we consider is due to I. Anshel [1] and applies to a class of two-relator groups. She essentially uses an extension of the Magnus method to graphs of groups. B. Bogley [7], using the concept of "aspherical presentations" (see below), extended Anshel's result to certain groups with arbitrarily many relators. Along these lines we also present a result of Bogley and Pride [8] on a FHS for groups with asperical relative presentations.

We next discuss some extensions of the FHS for one-relator products of cyclics to certain classes of groups related to these. These are called *groups of special NEC-type* and *generalized tetrahedron groups* and were introduced in [33,34,80]. Finally we mention a very strong result of Romanovskii [69] which is not technically a

FHS (relative to our definition) but certainly within the spirit of this survey.

I. Anshel considers two-relator presentations of the form $G = <A, x, y; S = T = 1 >$ where A is a free group on $\{a_1, \ldots, a_n\}$, x, y are two additional generators and S, T are arbitrary relators (cyclically reduced words in the free group on $\{a_1, \ldots, a_n, x, y\}$) which are assumed to be "independent" from $\{x, y\}$. The concept of *independence* is crucial and defined in the following manner. Let F be the free group on $\{a_1, \ldots, a_n, x, y\} = \{A, x, y\}$, and let $U =$ normal closure in F of $\{S, T\}$ so that $G = F/U$. Let N be the normal closure in F of $\{A, S, T\}$, and let $N_G = N/U$. N_G is the normal closure in G of the image of A in G. Let W be a fixed minimal Schreier coset representative system for $F/N\{= G/N_G\}$ (see [1]), and for g in G let $\bar{g}$ denote its coset representative relative to this system. S, T are then termed *independent relative to* $\{x, y\}$ if given a fixed minimal Schreier system for F/N the following conditions hold.

(a)
    (i) The relators are given in the form $S = x_1 c_1 \cdots x_k c_k$, $T = y_1 d_1 \cdots y_k d_k$ where h, $k \geq 2$ and c_i, d_i are non-trivial and in the free group on A and x_i, y_i are non-trivial in the free group on $\{x, y\}$.
    (ii) The coset representatives $\bar{x}_i \neq 1$ for $i = 1, \ldots, k$ and $\bar{y}_j \neq 1$ for $j = 1, \ldots, h$.

(b) Set

$$P = \{\bar{x}_1, \overline{x_1 x_2}, \ldots, \overline{x_1 \cdots x_k} = 1\}$$
$$Q = \{\bar{y}_1, \overline{y_1 y_2}, \ldots, \overline{y_1 \cdots y_h} = 1\}$$
$$R = P \cup Q \subseteq W(\text{The Schrier system}). \text{ Then}$$

    (i) $P \subseteq Q$ and $Q \subseteq P$
    (ii) $R \cap R,^{-1} = \{1\}$ and the map $< R \setminus \{1\}; > \rightarrow F/N$ is monic.

Her main result is then

Theorem 7.1 [1]. *Let* $G = < A, x, y; S, T >$ *as above. If the relators S and T are independent relative to the generators $\{x, y\}$, then A naturally injects into $G \to A$ is a FHS factor.*

Her proof mimics the Magnus approach. She constructs a non-trivial homomorphism from N_G, the normal closure of A in G to the fundamental group of a graph of groups (γ, Γ) where Γ is the complete graph based on G/N_G. The independence condition is used to verify that the (γ, Γ) constructed is actually a graph of groups. The image of $< A_G >$ is then in a vertex group of this graph and the argument is completed by applying the classical Freiheitssatz of Magnus to analyze the vertex groups. She also shows that without the independence criteria it is possible to construct two-relator presentations of the form of the theorem where the FHS fails.

B. Bogley, in attempting to describe the relation module for groups with presentations as in the Anshel paper, found a way to generalize her results to certain clases of groups with arbitrarily many relations. The class of groups for which Bogley's results apply are those with *semi-staggered presentations*. This is defined in the following way. Let A, X be arbitrary disjoint sets of generators, and let $G =< A, X : R >$ where R is a set of cyclically reduced words in the free group F on $\{A, X\}$. Let U be the normal closure in F of R so that $G = F/U$. Let N be the normal closure in G of $\{A, R\}$, and let $H = F \setminus N$. Assume that the relators are given in the form $r = x_1 a_1 \cdots x_n a_n$ where $n \geq 1$ and for $i = 1, \ldots, n$, x_i is a non-empty reduced word in X and a_i is a non-empty reduced word in A. For a relator r in R, set $P_r = \{x_1 N, x_1 x_2 N, \ldots, x_1 \cdots x_n N = N\}$ and set $\Pi = \cup\{P_r : r \in R\} \subseteq H$. If Ω is a subset of H, write $\Omega^\bullet = \Omega \setminus \{N\}$. The presentation $< A, X : R >$ is then *semi-staggered* in A if the following three conditions hold:

    (i) $P_r^\bullet \neq \emptyset$ for all $r \in R$

    (ii) There exist linear orderings on R and $\Pi^\bullet$ such that if r, $s \in R$ and $r < s$, then

        (a) $\min P_r^\bullet < \min P_s^\bullet$ and

        (b) $\max P_r^\bullet < \max P_s^\bullet$.

    (iii) $\Pi^\bullet$ is a basis for a free subgroup of H.

Bogley points out that this is a proper generalization of Anshel's independence condition. The Anshel type presentation $< A, x, y; S, T >$ is semi-staggered in A if and only if S, T are independent relative to x, y. Bogley's version of the FHS is then

Theorem 7.2 [7]. *If $G =< A, X; R >$ is semi-staggered in A then A is a basis for a free subgroup of G. (In our language, the free group on A is a FHS factor of G.)*

The proof depends upon constructing an "aspherical complex" for G (see [7]) and then using the following geometric version of the original Freiheitssatz of Magnus.

Geometric Version of the Classical FHS. Let L be a CW complex with a single two-cell, and suppose that L has a maximal tree that does not contain an open one-cell c^1 of L. If the attaching map for the two-cell of L strictly involves c^1, (that is, is not freely homotopic in the one-skeleton L^1 of L to a map into $L^1 \setminus c^1$) then the inclusion of $L^1 \setminus c^1$ in L^1 induces a monomorphism of fundamental groups.

These ideas were further extended by Bogley and Pride to groups with what they termed *aspherical relative presentations* [8]. If $H = < Y; S >$ and X is a set of generators disjoint from Y, then the group G has *relative presentation* $< Y, X; S, R >$ if R is set of cyclically reduced words in the free product of H and the free group on X and no word in R is in H. G has an aspherical relative presentation if every non-empty spherical picture (in the sense of Section 5) satisfy some further geometric conditions (see [8]). Under these conditions the group H naturally injects into G.

Theorem 7.3 [8]. *Let* $H = < X; S >$, *and let* $G = < X, Y; S, R >$ *be an orientable aspherical relative presentation . Then* H *is a FHS factor of* G.

The results of Bogley and Bogley and Pride can be considered as extensions of the results on small cancellation products. In the work of Collins and Perraud [17] mentioned earlier, spherical pictures are also employed.

We next mention two results motivated by work on extending properties of Fuchsian groups. A *group of special NEC type* (SN-type) is a group of the form

$$
\begin{aligned}
G = < a_1, \ldots, a_n, b_1, \ldots, b_k \,;\, a_1^{e+1} &= \ldots = a_n^{e_n} \\
&= b_1^{f_1} = \ldots = b_k^{f_k} \\
&= R_1^{m_1} = \ldots = R_k^{m_k} \\
&= 1 >
\end{aligned}
$$

(7.1)

where $n \geq 1$, $k \geq 2$, $e_j = 0$ or $e_j \geq 2$ for $i = 1, \ldots, n$, $f_i = 0$ or $f_i \geq 2$ for $i = 1, \ldots, k$, $m_i \geq 2$ for $i = 1, \ldots, k$ and for each $i = 1, \ldots, k$, $R_i = R_i(a_1, \ldots, a_n, b_i)$ is a cyclically reduced word in the free product on $\{a_1, \ldots, a_n, b_i\}$ which involves b_i and at least one of the a_i's and which is not a proper power. In this case we call the $\{R_i\}$ the *relators*.

From the FHS for one-relator products of cyclics these groups can be considered as iterated amalgams of one-relator products of cyclics. Thus we have the following theorem.

Theorem 7.4 [33]. *Let G be a group of special NEC-type (form (7.1)). Then*

(1) $< a_1, \ldots, a_n >_G = < a_1 ; a_1^{e_1} > * \cdots * < a_n ; a_n^{e_n} >$

(2) *for any subset* $\{b_{i_1}, \ldots, b_{i_t}\}$

$$< a_1, \ldots, a_n, b_{i_1}, \ldots, b_{i_t} >_G = < a_1, \ldots, a_n, b_{i_1}, \ldots, b_{i_t} ;$$

$$a_1^{e_1}, \ldots = a_n^{e_n} = b_{i_1}^{f_{i_1}} = \ldots = b_{i_t}^{f_{i_t}} = R_{i_1}^{m_{i_1}} = \ldots = R_{i_t}^{m_{i_t}} = 1 >$$

(3) *If each R_i involves all $\{a_1, \ldots, a_n\}$, then for any proper subset $\{a_{i_1}, \ldots, a_{i_r}\}$*

$$< a_{i_1}, \ldots, a_{i_r}, b_j >_G = < a_{i_1} ; a_{i_1}^{e_{i_1}} > * \cdots * < b_j ; b_j^{f_j} >$$

The theorem itself is a straighforward consequence of the amalgam construction. However what is of more importance is the next result.

Theorem 7.5 [33]. *A group of special NEC-type admits an essential representation into $PSL_2(\mathbb{C})$. (Here, as for one-relator products of cyclics, an essential representation is a representation where the image of every generator and every relator has exactly the correct order.)*

This leads to a whole set of results on linearity properties for these groups [33]. The same type of techniques handle a second class of groups introduced by Vinberg [80] and independently by Fine and Rosenberger [34]. A *generalized tetrahedron group* is a group of the form

$$< a_1, a_2, a_3 : a_1^{e_1} = a_2^{e_2} = a_3^{e_3} =$$

$$R_1^m(a_1, a_2) = R_2^p(a_1, a_3) = R_3^q(a_2, a_3) = 1 >$$

where $e_i = 0$ or $e_i \geq 2$ for $i = 1, 2, 3$; $2 \leq m, p, q$; $R_1(a_1, a_2)$ is a cyclically reduced word in the free product on a_1, a_2 which involves both a_1 and a_2, $R_2(a_1, a_3)$ is a cyclically reduced word in the free product on a_1, a_3 which involves both a_1 and a_3 and $R_3(a_2, a_3)$ is a cyclically reduced word in the free product on a_2, a_3 which involves both a_2 and a_3. Further each R_i, $i = 1, 2, 3$ is not a proper power in the free product on the generators it involves.

Group theoretically this can be shown to have the structure of a triangular product (see Section 2) of the ordinary triangle groups

$$G_1 = < a_1, a_2; a_1^{e_1} = a_2^{e_2} = R_1^m(a_1, a_2) = 1 >$$
$$G_2 = < a_1, a_3; a_1^{e_1} = a_3^{e_3} = R_2^p(a_1, a_3) = 1 >$$
$$G_3 = < a_2, a_3; a_2^{e_2} = a_3^{e_3} = R_3^q(a_2, a_3) = 1 >$$

with edge amalgamations over the finite cyclic subgroups $< a_1 >$, $< a_2 >$, $< a_3 >$. We call G_1, G_2, G_3 the generalized triangle group factors of the generalized tetrahedron group G.

By using essential representations into $PSL_2(\mathbb{C})$ it can be shown that, relative to the Gersten-Stallings angles, this triangle is non-positively curved if $(1/m) + (1/p) + (1/q) \leq 1$. Therefore we obtain:

Theorem 7.6 [34]. *Let G be a generalized tetrahedron group of form (7.2).*

(i) *If $(1/m) + (1/p) + (1/q) \leq 1$, then each generalized triangle group factor is a FHS factor.*

(ii) *If a generalized triangle group factor G_i admits a faithful representation into $PSL_2(\mathbb{C})$, then G_i is a FHS factor.*

Finally we mention the following very strong result of Romanovskii [69].

Theorem 7.7 [69]. *Let $G = < x_1, \ldots, x_n; R_1, \ldots, R_k >$ have deficiency $d > 0$ ($n - k = d > 0$). Then there exist a set of d of the given generators which are a basis for a free group.*

REFERENCES

1. I. Anshel, *On two relator groups*, Topology and Combinatorial Group Theory **1440** (1990), Spriner Lecture Notes in Math, 1–21.

2. B. Baumslag, *Free products of locally indicable groups with a single relator*, Bull. Austral. Math. Soc. **29** (1984), 401–404.

3. B. Baumslag and S.J. Pride, *An extension of the Freiheitssatz*, Math. Proc. Camb. Phil. Soc. v89 (1981), 35–41.

4. G. Baumslag, *A survey of groups with a single relation*, Proceedings of Groups St. Andrews 1985 LMS lecture Notes Series **121** (1986), 30–58.

5. G. Baumslag, J. Morgan and P. Shalen, *Generalized Triangle Groups*, Math. Proc. Camb. Phil. Soc. **102** (1987), 25–31.

6. G. Baumslag, S. Gersten, M. Shapiro and H. Short, *Automatic Groups and Amalgams*, Algorithms and Classification in Combinatorial Group Theory **23** (1992), Springer-Verlag MSRI Publications, 179–195.

7. W.A. Bogley, *An identity theorem for multi-relator groups*, Math. Proc. Camb. Phil. Soc. v109 (1991), 313–321.

8. W.A. Bogley and S.J. Pride, *Aspherical relative presentations*, to appear Proc. Edinburgh Math. Soc..

9. S. Boyer, *On proper powers in free products and Dehn surgery*, J. Pure and Appl. Alg. **51** (1988), 217–229.

10. J.L. Britton, *Solution of the word problem for certain types of groups I,II*, Proc. Glasgow Math Soc. **3** (1957), 68–90.

11. S.D. Brodskii, *Equations over groups and groups with a single defining relation*, Uspekhi Mat. Nauk **35:4** (1980), 183.

12. S.D. Brodskii, *Anomalous products of locally indicable groups*, Algebraicheskie Sistemy (1981), Ivanovo University, 51–77.

13. S.D. Brodskii, *Equations over groups and groups with a single defining relation*, Siberian Math. J. **25** (1984), 231–251.

14. S.D. Brodskii and J.Howie, *One relator products of torsion-free groups*, to appear.

15. B. Chandler and W. Magnus, *The History of Combinatorial Group Theory: A Case Study in the History of Ideas*, Springer-Verlag ,New York , Heidelberg, Berlin, 1982.

16. D. Collins, *Free subgroups of small cancellation groups*, Proc. London Math. Soc. **26** (1973), 193–206.

17. D. Collins and J. Huebschmann, *Spherical diagrams and identities among relations*, Math. Ann. v261 (1983), 155–183.

18. D. Collins and J. Perraud, *Cohomology and finite subgroups of small cancellation quotients of free products*, Math. Proc. Camb. Phil. Soc. **97** (1985), 243–259.

19. J. Corson, *Small cancellation theory over graphs of groups*, to appear.

20. M. Dehn, *Uber unendliche diskontinuerliche Gruppen*, Math. Ann. **69** (1911), 116–144.

21. A.J. Duncan and J. Howie, *One relator products with high-powered relators*, Proceedings of the Geometric Group Theory Symposium, University of Sussex 1991.

22. A.J. Duncan and J. Howie, *Weinbaum's Conjecture on Unique Subwords of Nonperiodic Words*, Proceedings of the American Mathematical Society **115 4** (1992), 947-954.

23. M. Edjvet, *A Magnus theorem for free products of locally indicable groups*, Glasgow Math. J. **31** (1989), 383–387.

24. M. Edjvet, *On a certain class of group presentations*, Math. Proc. Camb. Phil. Soc. **105** (1989), 25–35.

25. R. Fenn, *Techniques of Geometric Topology*, London Math. Soc. Lecture Notes Series **57** (1983).

26. B. Fine, J. Howie and G. Rosenberger, *One-relator quotients and free products of cyclics*, Proc. Amer. Math. Soc. **102** (1988), 1–6.

27. B. Fine and G.Rosenberger, *Complex Representations and One–Relator Products of Cyclics*, Geometry of Group Representations **74** (1987), Contemporary Math., 131–149.

28. B. Fine and G.Rosenberger, *Generalizing Algebraic Properties of Fuchsian Groups*, London Mathematical Soc. Lecture Notes Series 159, vol. 1, 1989, pp. 124–148.

29. B. Fine, F. Levin and G. Rosenberger, *Free Subgroups and Decompositions of One-Relator Products of Cyclics: Part 1: The Tits Alternative*, Arch.

Math. (1988), 97–109.

30. B. Fine, F. Levin and G. Rosenberger, *Free subgroups and decompositions of one-relator products of cyclics; Part 2 : Normal Torsion-Free Subgroups and FPA Decompositions*, J. Indian Math. Soc. **49** (1985), 237–247.

31. B. Fine and G. Rosenberger, *A Note on Generalized Triangle Groups*, Abh. Math. Sem. Univ. Hamburg **56** (1986), 233–244.

32. B. Fine, J. Howie and G. Rosenberger, *Ree-Mendelsohn Pairs in Generalized Triangle Groups*, Comm. in Algebra **17** (1989), 251–258.

33. B. Fine and G. Rosenberger, *On Groups of Special NEC Type*, to appear.

34. B. Fine, F. Levin, F. Roehl and G. Rosenberger, *On the generalized tetrahedron groups*, to appear.

35. M. Gerstenhaber and O. Rothaus, *The solution of sets of equations in groups*, Proc. Nat. Acad. Sci. U.S.A. **48** (1962), 1531–1533.

36. F. Gonzalez-Acuna and H. Short, *Knot Surgery and Primeness*, Math. Proc. Camb. Phil. Soc. **99** (1986), 89–102.

37. M. Greendlinger, *Dehn's Algorithm for the Word Problem*, Comm. Pure and Applied Math. **13** (1960), 67–83.

38. M. Greendlinger, *On Dehn's Algorithm for conjugacy and word problems with applications*, Comm. Pure and Applied Math. **13** (1960), 641–677.

39. M. Gromov, *Hyperbolic Groups*, Essays in Group Theory S. Gersten ed. MSRI Publication 8, Springer-Verlag, 1987.

40. G.A. Gurevich, *On the conjugacy problem for one relator groups*, Dokladi Akad. Nauk SSSr Ser. Math **207** (1972), 18–20.

41. J. Howie, *How to Generalize One-Relator Group Theory*, Annals of Math. Studies **111, Eds. S.Gersten and J. Stallings,** (1987), Pg. 53–78.

42. J. Howie, *On Locally Indicable Groups*, Math. Z. **180** (1982), 445–461.

43. J. Howie, *On pairs of 2-complexes and systems of equations over groups*, J. Reine angew. Math. **324** (1981), 165–174.

44. J. Howie, *The quotient of a free product of groups by a single high-powered relator. I. Pictures. Fifth and higher powers*, Proc. London Math. Soc. **59** (1989), 507–540.

45. J. Howie, *The quotient of a free product of groups by a single high-powered relator. I. Fourth powers*, Proc. London Math. Soc. **61** (1990), 33–62.

46. J. Howie, *One Relator Products of Groups*, Proceedings of Groups St. Andrews 1985 – LMS lecture Notes Series **121** (1986), 216–2206.

47. J. Howie and H. Short, *The band sum problem*, J. London Math. Soc. **31** (1985), 571–576.

48. A. Klyachko, *A Funny Property of Sphere and Equations over Groups*, to appear, preprint Moscow State University.

49. F. Levin, *Solutions of equations over groups*, Bull. Amer. Math. Soc. **68** (1962), 603–604.

50. R.C. Lyndon, *Length Functions in Groups*, Math. Scand. **12** (1963), 209–234.

51. R.C. Lyndon, *On Dehn's Algorithm*, Math. Ann. **166** (1966), 208–228.

52. R.C. Lyndon, *A maximum principle for graphs*, J. Combinatorial Theory **3** (1967), 34–37.

53. R.C. Lyndon, *On the Freiheitssatz*, J. London Math. Soc. **5** (1972), 95–101.

54. R.C. Lyndon and P.E. Schupp, *Combinatorial Group Theory*, Springer-Verlag 1977.

55. W. Magnus, *Uber diskontinuerliche Gruppen mit einer definerenden Relation*, J. Reine u. Angew. Math. **163** (1930), 141–165.

56. W. Magnus, *Untersuchen uber unendliche diskontinuerliche Gruppen*, Math. Ann. **105** (1931), 52–74.

57. W. Magnus, *Das Identitatsproblem fur Gruppen mit einer definerenden Relation*, Math. Ann. **106** (1932), 295–307.

58. W. Magnus, *The uses of 2 by 2 matrices in combinatorial group theory*, Resultate Math. **4** (1981), 171–192.

59. W. Magnus, *Two generator subgroups of $PSL_2(\mathbb{C})$*, Nachr. Akad. Wiss. Gottingen II. Math. Phys. Kl. **7** (1975), 81– 94.

60. W. Magnus, A. Karrass, and D. Solitar,, *Combinatorial Group Theory* –, Wiley 1966, Second Edition, Dover Publications, New York, 1976.

61. R.M.S. Mahmud, *The Word Problem for Small Cancellation Quotients of Groups Acting on Trees*, Proceedings of the Royal Society of Edinburgh **121A** (1992), 361-374.

62. J. McCool and P.E. Schupp, *On one–relator groups and HNN extensions*, J. Austr. Math. Soc. **16,** (1973), 249–256.

63. C.F. Miller III, *Decision Problems for Groups – Survey and Reflections*, Algorithms and Classification in Combinatorial Group Theory (G. Baumslag and C.F. Miller III Springer-Verlag MSRI Publications, eds.), vol. 23, 1992, pp. 1–61.

64. C.F. Miller III and P.E. Schupp, *The geometry of HNN extensions*, Comm. Pure and Appl. Math. **26** (1973), 787–802.

65. D.I. Moldavanskii, *Certain subgroups of groups with one defining relation*, Sibirsk. Math. Zh. **8** (1967), 1370–1384.

66. S.J. Pride, *One relator quotients of free products*, Math. Proc. Camb. Phil. Soc. **88** (1980), 233–243.

67. S.J. Pride, *Groups with presentations in which each defining relator involves exactly two generators*, J. London Math. Soc. 2 **36** (1987), 245–256.

68. R. Ree and N.S. Mendelsohn, *Free subgroups of groups with a single defining relation*, Arch. Math. 1 **9** (1968), 577–580.

69. N.S. Romanovskii, *Free subgroups of finitely presented groups*, Algebra i Logika **16** (1977), 88–97.

70. G. Rosenberger, *Faithful linear representations and residual finiteness of certain one–relator products of cyclics" to appear J. Siberian Math. Soc.*.

71. P.E. Schupp, *A Strengthened Freiheitssatz*, Math. Ann. **221** (1976), 73–80.

72. H. Schiek, *Ahnlichkeitsanalyse von Gruppenrelationen*, Acta Math. **96** (1956), 157–252.

73. O. Schreier, *Die Untergruppen der freien Gruppen*, Abh. Math.Sem. Hamburg Univ. **5** (1927), 161–183.

74. J.P. Serre, *Trees*, Springer-Verlag, 1980.

75. P.Shalen, *Linear representations of certain amalgamated products*, J. Pure and Appl. Algebra **15** (1979), , 187–197.

76. H. Short, *Topological Methods in Group Theory; the adjunction problem*, Ph.D. Thesis, University of Warwick (1984).

77. J.R. Stallings, *Non-positively curved triangles of groups*, Group Theory from a Geometrical Viewpoint – World Scientific 1991 pg. 491–503.

78. V.A. Tartakovskii, *Solution of the word problem for groups with a k reduced basis for k > 6*, Izv. Akad. Nauk SSSR Ser Math. **13** (1949), 483–494.

79. E.R. Van Kampen, *On some lemmas in the theory of groups*, Amer. J. Math. **55** (1933), 268–273.
80. E.B. Vinberg, *On groups defined by periodic relations*, to appear.
81. C.M. Weinbaum, *On relators and diagrams for groups with a single defining relator*, Illinois J. Math. **16** (1972), 308–322.

BENJAMIN FINE, DEPARTMENT OF MATHEMATICS, FAIRFIELD UNIVERSITY, FAIRFIELD, CONNECTICUT 06430, UNITED STATES

GERHARD ROSENBERGER, FACHBEREICH MATHEMATIK UNIVERSITAT, DORTMUND, POSTFACH 50 05 00, 4600 DORTMUND 50, FEDERAL REPUBLIC OF GERMANY

Contemporary Mathematics
Volume 169, 1994

A Rodrigues-Type Formula for the q-Racah Polynomials and Some Related Results

ISMOR FISCHER

ABSTRACT. In [1], Askey and Wilson describe and prove various properties of a very general class of discrete orthogonal polynomials that are basic hypergeometric extensions of Racah polynomials. In this paper we derive some additional results for that class.

1. Introduction

A basic hypergeometric power series has the standard form

$$
{}_r\phi_s\left({a_1\cdots a_r \atop b_1\cdots b_s};q,\,x\right) = \sum_{n=0}^{\infty} \frac{(a_1,\cdots,a_r;\,q)_n}{(b_1,\cdots,b_s,\,q;\,q)_n}\left[(-1)^n q^{\binom{n}{2}}\right]^{1+s-r} x^n.
$$

The base q is a fixed formal parameter, and we write

$$
(a_1,\cdots,a_r;\,q)_n = (a_1;\,q)_n \cdots (a_r;\,q)_n,
$$

where the q-shifted factorial is given by

$$
(a;\,q)_n = \prod_{k=0}^{n-1}(1-aq^k),\quad n = 0,\,1,\,2,\cdots,
$$

with the empty product defined to be 1. We also define $(a;\,q)_\infty$ in the obvious way when $|q| < 1$. Note that the quotient $(a;\,q)_\infty/(aq^n;\,q)_\infty$ formally reduces to the finite product $(a;\,q)_n$, so that the above restriction on q may be removed in this case. Note also that for $n = 0,\,1,\,2,\cdots$

$$
\lim_{q\to 1}\frac{(q^\alpha;\,q)_n}{(1-q)^n} = (\alpha)_n,
$$

1991 Mathematics Subject Classification. Primary 33A65; Secondary 33A30.
This paper is in final form and no version of it will be submitted for publication elsewhere.

where $(\alpha)_n = \prod_{k=0}^{n-1}(\alpha+k)$ is the ordinary shifted factorial, so called since $(1)_n = n!$. The corresponding power series is known as an ordinary hypergeometric series and is denoted by

$$
{}_rF_s \left(\begin{matrix} \alpha_1, \cdots, \alpha_r \\ \beta_1, \cdots, \beta_s \end{matrix} ; x \right) = \sum_{n=0}^{\infty} \frac{(\alpha_1, \cdots, \alpha_r)_n}{(\beta_1, \cdots, \beta_s, 1)_n} x^n ,
$$

with $(\alpha_1, \cdots, \alpha_r)_n = (\alpha_1)_n \cdots (\alpha_r)_n$.

Hypergeometric and basic hypergeometric series are often used to define various "special functions" of mathematical physics, including families of classical orthogonal polynomials and their q-extensions. Some typical examples are Jacobi polynomials and their discrete generalizations the Hahn and dual Hahn polynomials, which are themselves special cases of a self-dual class, the Racah polynomials; [3] – [7].

In [1], Askey and Wilson define a basic hypergeometric extension of this latter class. The q-Racah polynomials are given by the terminating series

(1) $\quad p_n(\mu(x); a, b, c, d; q) = {}_4\phi_3 \left(\begin{matrix} q^{-n}, & abq^{n+1}, & q^{-x}, & cdq^{x+1} \\ aq, & bdq, & cq \end{matrix} ; q, q \right)$

where $\mu(x) = q^{-x} + cdq^{x+1}$. Assume one of aq, bdq, or cq is q^{-N}. Then these polynomials satisfy a discrete orthogonality relation

(2)
$$
I_{m,n}^{(a, b, c, d; q)} =
$$
$$
\sum_{x=0}^{N} p_n(\mu(x); a, b, c, d; q)\, p_m(\mu(x); a, b, c, d; q)\, w(x; a, b, c, d; q) = \delta_{mn} h_n,
$$
$$
0 \le m, n \le N ,
$$

with respect to the weight function

$$
w(x; a, b, c, d; q) = \frac{q^{-x} - cdq^{x+1}}{1 - cdq} \widetilde{w}(x; a, b, c, d; q) ,
$$

where

$$
\widetilde{w}(x; a, b, c, d; q) = \frac{(aq, bdq, cq, cdq; q)_x (ab)^{-x}}{(a^{-1}cdq, b^{-1}cq, dq, q; q)_x} ,
$$

and $h_n = h_n^{(a, b, c, d; q)} \neq 0$ to be determined later.

The q-Racah polynomials are often identified with an equivalent class–the Askey-Wilson polynomials–and have all the "classical" orthogonal polynomials as special or limiting cases; [1], [2].

2. Main Results

In [2], the authors present a Rodrigues-type formula for the Askey-Wilson polynomials with respect to a complex operator. In this paper, we derive a similar formula for the q-Racah class, and examine some consequences of the proof. We first define the real divided difference operator

$$D_q[f(\mu(x);\; a,\; b,\; c,\; d;\; q)] = (1 - cd)(1 - q)\frac{\Delta f(\mu(x);\; a,\; b,\; c,\; d;\; q)}{\Delta \mu(x)}$$

$$= \frac{1 - cd}{q^{-x-1} - cdq^{x+1}}\Delta f(\mu(x);\; a,\; b,\; c,\; d;\; q),$$

where $\Delta F(x) = F(x + 1) - F(x)$ is the first forward difference operator acting on x.

A straightforward computation on (1) shows that

(3) $$D_q p_m(\mu(x);\; a,\; b,\; c,\; d;\; q) = \frac{-q(1 - q^{-m})(1 - abq^{m+1})(1 - cd)}{(1 - aq)(1 - bdq)(1 - cq)} \times$$

$$p_{m-1}(\mu(x);\; aq,\; bq,\; cq,\; d;\; q),$$

or equivalently,

(4) $$p_m(\mu(x);\; a,\; b,\; c,\; d;\; q) = K_m\frac{\Delta p_{m+1}(\mu(x);\; aq^{-1},\; bq^{-1},\; cq^{-1},\; d;\; q)}{q^{-x} - cdq^{x+1}},$$

where $K_m = K_m(a,\; b,\; c,\; d;\; q) = -\frac{(1-a)(1-bd)(1-c)}{(1-q^{-m-1})(1-abq^m)}$. Now plug (4) into (2):

$$I_{m,\, n}^{(a,\, b,\, c,\, d;\, q)} = K'_m \sum_{x=0}^{N} \widetilde{w}(x;\; a,\; b,\; c,\; d;\; q)p_n(\mu(x);\; a,\; b,\; c,\; d;\; q)\times$$

$$\Delta p_{m+1}(\mu(x);\; aq^{-1},\; bq^{-1},\; cq^{-1},\; d;\; q),$$

where $K'_m = \frac{K_m}{1-cdq}$. Next apply the summation by parts formula

$$\sum_{x=0}^{N} u(x)\Delta v(x) = -\sum_{x=0}^{N+1} v(x)\Delta u(x - 1)$$

with $u(-1) = 0$ to obtain

(5)

$$I_{m,\, n}^{(a,\, b,\, c,\, d;\, q)} = -K'_m \sum_{x=0}^{N} \Delta[\widetilde{w}(x - 1;\; a,\; b,\; c,\; d;\; q)p_n(\mu(x - 1);\; a,\; b,\; c,\; d;\; q)]\times$$

$$p_{m+1}(\mu(x);\; aq^{-1},\; bq^{-1},\; cq^{-1},\; d;\; q) = \delta_{mn}h_n\,.$$

To evaluate the first term in the summand, we may use Leibniz's rule:

(6) $$\Delta[f(x)g(x)] = f(x)\Delta g(x) + g(x + 1)\Delta f(x)$$

with $f(x) = \tilde{w}(x - 1; a, b, c, d; q)$ and $g(x) = p_n(\mu(x-1); a, b, c, d; q)$. Now by a tedious calculation,

$$\Delta f(x) = \frac{(aq, bdq, cq, cdq; q)_{x-1}(ab)^{-x}}{(a^{-1}cdq, b^{-1}cq, dq, q; q)_x}(1 - cdq^{2x})s(q^x),$$

where

(7) $s(q^x) = (1 - ab)(1 - q^x)(1 - cdq^x) + (1 - a)(1 - bd)(1 - c)q^x,$

i. e. , a quadratic polynomial in q^x. From (4) it follows that

$$\Delta g(x) = L_n q^{-x}(1 - cdq^{2x})p_{n-1}(\mu(x-1); aq, bq, cq, d; q)$$

where $L_n = L_n(a, b, c, d; q) = \frac{-q(1-q^{-n})(1-abq^{n+1})}{(1-aq)(1-bdq)(1-cq)}$. Thus from (6),

(8)
$$\Delta[f(x)g(x)] = \frac{(aq, bdq, cq, cdq; q)_{x-1}(ab)^{-x}}{(a^{-1}cdq, b^{-1}cq, dq, q; q)_x}(1 - cdq^{2x})q^x\pi_{n+1}(x; a, b, c, d; q),$$

where

(9) $\pi_{n+1}(x; a, b, c, d; q) = q^{-x}s(q^x)p_n(\mu(x); a, b, c, d; q) + L_n \times$

$(1 - q^x)(1 - a^{-1}cdq^x)(1 - b^{-1}cq^x)(1 - dq^x)q^{-2x}p_{n-1}(\mu(x-1); aq, bq, cq, d; q).$
A simple check verifies that π_{n+1} is a polynomial of degree precisely $n + 1$ in both q^x and q^{-x}.

Substituting (8) back into (5), we obtain

(10) $I_{m, n}^{(a, b, c, d; q)} = K_m'' \sum_{x=0}^{N+1} w(x; aq^{-1}, bq^{-1}, cq^{-1}, d; q)\pi_{n+1}(x; a, b, c, d; q)\times$

$$p_{m+1}(\mu(x); aq^{-1}, bq^{-1}, cq^{-1}, d; q),$$

where

(11) $K_m'' = \frac{-K_m'}{(1 - a)(1 - bd)(1 - c)}.$

Thus by orthogonality, π_{n+1} may be expressed as a linear combination of q-Racah polynomials:

(12) $\pi_{n+1}(x; a, b, c, d; q) = \sum_{j=0}^{n+1} c_j p_j(\mu(x); aq^{-1}, bq^{-1}, cq^{-1}, d; q).$

We would now like to show that the coefficients $c_j = 0$, $j \leq n$. This fact may be determined by substituting (12) into (10):

$$I_{m, n}^{(a, b, c, d; q)} = K_m'' \sum_{j=0}^{n+1} c_j \sum_{x=0}^{N+1} p_{m+1}(\mu(x); aq^{-1}, bq^{-1}, cq^{-1}, d; q)\times$$

$$p_j(\mu(x); aq^{-1}, bq^{-1}, cq^{-1}, d; q)w(x; aq^{-1}, bq^{-1}, cq^{-1}, d; q).$$

Now for each $m \leq n-1$, the left-hand side is zero by orthogonality, but the right side is zero unless $j = m+1 \leq n$. Thus

$$0 = c_{m+1} h_{m+1}^{(aq^{-1}, bq^{-1}, cq^{-1}, d; q)}$$

from which the claim follows.

Hence by (12),

(13) $\qquad \pi_{n+1}(x; a, b, c, d; q) = c_{n+1} p_{n+1}(\mu(x); aq^{-1}, bq^{-1}, cq^{-1}, d; q)$

and by (10),

(14) $\qquad\qquad I_{m, n}^{(a, b, c, d; q)} = K_m'' c_{n+1} I_{m+1, n+1}^{(aq^{-1}, bq^{-1}, cq^{-1}, d; q)}$

To determine the value of c_{n+1}, let $x = 0$ in (13) and use (9) and (7):

(15) $\qquad\qquad\qquad c_{n+1} = (1-a)(1-bd)(1-c).$

Combining (13) together with (5) and (10), we see that

$$- K_m' \Delta[\tilde{w}(x-1; a, b, c, d; q) p_n(\mu(x-1); a, b, c, d; q)] =$$
$$K_m'' w(x; aq^{-1}, bq^{-1}, cq^{-1}, d; q) c_{n+1} p_{n+1}(\mu(x); aq^{-1}, bq^{-1}, cq^{-1}, d; q).$$

The constants cancel by virtue of (11) and (15); multiplying both sides by $\frac{1-cd}{q^{-x}-cdq^x}$ yields

(16)
$$D_q[\tilde{w}(x-1; a, b, c, d; q) p_n(\mu(x-1); a, b, c, d; q)] =$$
$$\tilde{w}(x; aq^{-1}, bq^{-1}, cq^{-1}, d; q) p_{n+1}(\mu(x); aq^{-1}, bq^{-1}, cq^{-1}, d; q)$$

Now use (3) on the left-hand side of (16) with appropriate parameter shifts to obtain the *second order difference equation* for p_n:

$$D_q[\tilde{w}(x; aq, bq, cq, d; q) D_q p_n(\mu(x); a, b, c, d; q)] +$$
$$\frac{q(1-q^{-n})(1-abq^{n+1})(1-cd)}{(1-aq)(1-bdq)(1-cq)} \tilde{w}(x+1; a, b, c, d; q) p_n(\mu(x+1); a, b, c, d; q)$$
$$= 0.$$

Iterating relation (16) — appropriately shifted — yields the *general Rodrigues formula* for p_n:

$$D_q^k[\tilde{w}(x-k; aq^k, bq^k, cq^k, d; q) p_n(\mu(x-k); aq^k, bq^k, cq^k, d; q)] =$$
$$\tilde{w}(x; a, b, c, d; q) p_{n+k}(\mu(x); a, b, c, d; q).$$

Finally, combine (2), (11), (14), and (15) when $m = n$:

$$h_n^{(a, b, c, d; q)} = \frac{(1-a)(1-bd)(1-c)}{(1-cdq)(1-q^{-n-1})(1-abq^n)} h_{n+1}^{(aq^{-1}, bq^{-1}, cq^{-1}, d; q)}$$

or equivalently,

$$h_n^{(a,\,b,\,c,\,d;\,q)} = \frac{(1 - cdq^2)(1 - q^{-n})(1 - abq^{n+1})}{(1 - aq)(1 - bdq)(1 - cq)} h_{n-1}^{(aq,\,bq,\,cq,\,d;\,q)}$$

Iterating this relation n times produces

$$(17) \qquad h_n^{(a,\,b,\,c,\,d;\,q)} = \frac{(cdq^2,\,q^{-n},\,abq^{n+1};\,q)_n}{(aq,\,bdq,\,cq;\,q)_n} h_0^{(aq^n,\,bq^n,\,cq^n,\,d;\,q)}.$$

The value of $h_0^{(a,\,b,\,c,\,d;\,q)}$ can be computed by summing a very well-poised $_6\phi_5$; see [1]. The result when combined with (17) is

$$h_n = h_n^{(a,\,b,\,c,\,d;\,q)} = \frac{(1 - abq)(bq,\,ad^{-1}q,\,abc^{-1}q,\,q;\,q)_n(cdq)^n}{(1 - abq^{2n+1})(aq,\,bdq,\,cq,\,abq;\,q)_n} \times$$
$$\frac{(cdq^2,\,a^{-1}b^{-1}c,\,a^{-1}d,\,b^{-1};\,q)_\infty}{(a^{-1}cdq,\,b^{-1}cq,\,dq,\,a^{-1}b^{-1}q^{-1};\,q)_\infty}.$$

The ideas outlined in the derivations above may be applied to subclasses of the q-Racah polynomials. For example, the q-Hahn polynomials

$$Q_n(q^{-x};\,a,\,b,\,N;\,q) = {}_3\phi_2 \left(\begin{matrix} q^{-n}, & abq^{n+1}, & q^{-x} \\ aq, & q^{-N} \end{matrix} ;\,q,\,q \right),$$

which may be obtained by letting $cq = q^{-N}$ and $d = 0$, satisfy a discrete orthogonality relation

$$\sum_{x=0}^{N} Q_m(q^{-x};\,a,\,b,\,N;\,q)Q_n(q^{-x};\,a,\,b,\,N;\,q)\widetilde{w}(x;\,a,\,b,\,N;\,q)q^{-x} = 0, \quad m \neq n$$

where

$$\widetilde{w}(x;\,a,\,b,\,N;\,q) = \frac{(aq,\,q^{-N};\,q)_x(ab)^{-x}}{(b^{-1}q^{-N},\,q;\,q)_x}.$$

This class also satisfies a general Rodrigues formula

$$\widetilde{w}(x;\,a,\,b,\,N;\,q)Q_{n+k}(q^{-x};\,a,\,b,\,N;\,q) =$$

$$\Delta_q^k[\widetilde{w}(x - k;\,aq^k,\,bq^k,\,N - k;\,q)Q_n(q^{-x+k};\,aq^k,\,bq^k,\,N - k;\,q)]$$

with respect to the divided difference operator

$$\Delta_q f(q^{-x}) = (1 - q)\frac{\Delta f(q^{-x})}{\Delta(q^{-x})}$$
$$= q^{x+1}\Delta f(q^{-x}).$$

Acknowledgment
The author wishes to express his sincere thanks to Professor R. A. Askey for his support during the time of this research at the University of Wisconsin – Madison.

REFERENCES

1. R. Askey and J. Wilson, *A Set of Orthogonal Polynomials that Generalize the Racah Coefficients or $6 - j$ Symbols*, SIAM J. Math. Anal. **10** (1979), 1008–1016.

2. _____, *Some Basic Hypergeometric Orthogonal Polynomials that Generalize Jacobi Polynomials*, vol.54, No. 319, Memoirs Amer. Math. Soc., Providence, R. I., (March 1985).

3. A. Erdélyi, ed., *Higher Transcendental Functions*, Vols. I, II, Robert E. Krieger Pub. Co., Malaba, Florida, 1953.

4. G. Gasper and M. Rahman, *Basic Hypergeometric Series*, Cambridge Univ. Press, Cambridge, New York and Melbourne, 1990.

5. S. Karlin and J. L. McGregor, *The Hahn polynomials, formulas and an application*, Scripta Math., **26** (1961), 33–46.

6. A. F. Nikiforov, S. K. Suslov, and V. B. Uvarov, *Classical Orthogonal Polynomials of a Discrete Variable*, Springer-Verlag, Berlin and New York, 1991.

7. A. F. Nikiforov and V. B. Uvarov, *Special Functions of Mathematical Physics* (1984), "Nauka", Moscow (Russian); English transl. (1988), Birkhäuser.

Contemporary Mathematics
Volume **169**, 1994

Air on the Dirac Strings

GEORGE K. FRANCIS AND LOUIS H. KAUFFMAN

ABSTRACT. Wilhelm Magnus used braids for presenting the mapping class group of punctured spheres. He drew pictures of strings moving between two concentric spheres in a manner now known as the Dirac string trick. In this paper we discuss the trick, we show how it leads to a demonstration of the quaternion group, and we expose the quaternionic recipes for programming real-time interactive computer animations of such phenomena.

1. Introduction

This paper is devoted to the context of the well-known orientation entanglement relation of a mobile object connected to a fixed surround by any number of strings. The fundamental instance of this phenomenon is the isotopic triviality of the the strings after a $720°$ rotation of the object about a fixed axis. Dirac's demonstration of this, using a pair of scissors attached by two strings to the legs of a chair on a desk, was motivated by a corresponding property of quantum mechanical systems. The phase of the wave function of an electron changes sign after a continuous rotation of the observer by $360°$.

M.H.A. Newman [N] observed that this phenomenon has a counterpart in the theory of braid groups. For the spherical braid group the center is generated by a braid which may be regarded as a belt of n-strands, with a $360°$ twist. That this braid has order two in the spherical braid group is another instance of the orientation entanglement relation.

Wilhelm Magnus showed how to use braid groups in presenting the mapping class group of punctured spheres and tori in his classical paper on the automorphisms of the fundamental group for these surfaces [M]. Departing from the

1991 *Mathematics Subject Classification*. Primary 57M25, 57S05; Secondary 81G20.

Key words and phrases. low dimensional geometry and topology, quaternions, braid groups, Euclidean rotation group, mapping class group, fundamental group, Dirac string trick, blackboard drawing, real-time interactive computer animation, twisting and writhing ribbons, ambient isotopy, the belt trick, quantum theory.

The second author was supported in part by NSF Grant Nr. DMS-9205277 and the Program for Mathematics and Molecular Biology, UC, Berkeley.

customary abstraction of group theory, Wilhelm drew pictures illustrating his text. An informal picture-story on this is in [F, p.132ff].

In this paper we shall do the same, illustrating the Dirac string trick by means of many drawings and enough mathematics to explain the workings of real-time interactive computer animations (RTICAs) we have made to illustrate this trick. We also explain how an understanding of the trick leads directly to an understanding of the quaternion group by looking at the symmetries of an object with strings or belts attached to it. In customary advertisements the buyer is guaranteed that "there are no strings attached." For our consumers we guarantee that the strings are attached!

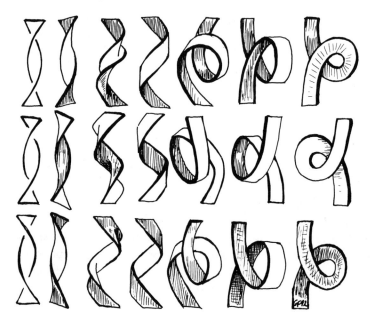

FIGURE 1. SEIFERT'S MOVE OF TWIST THROUGH CURL TO WRITHE. Each row begins with a schematic drawing of a fully twisted ribbon. The first two rows are all right-handed, the third row is left-handed. In the fourth column the ribbon *curls* about an invisible cylinder. A topological ribbon can stretch into the curious seventh position. Such ribbons are said to *writhe* parallel to the plane of the paper.

The paper is organized as follows. Section 2 describes the trick directly in terms of rubber-band topology. There are simple illustrations suitable for an academic topologist to reproduce at the blackboard or on the back of an envelope. The section ends with a demonstration of the quaternions that is based on the belt trick [K2, K3]. Section 3 discusses the mathematics of the quaternions and Hamilton's ingenious way to represent rotations by quaternions. We show how many useful computations in geometrical computer graphics are easily and most naturally derived from the algebraic properties of quaternions. In

particular, we discuss how small rotations about just two orthogonal axes, imple-
mented as graphics hardware primitives and mediated by 2-dimensional (mouse
or track-ball) input devices [H1,H2], suffice to (approximately) control the entire
3-dimensional rotation group. Section 4 returns to the belt trick and explains in
detail the consequences for graphics, geometry and topology of our method of
twisting objects by means of quaternionically formulated self-mappings of space
(ambient isotopies).

The paper concludes with a quaternionic formula parametrizing the belt trick.
The same formula also gives a specific homotopy which takes the square of the
generator of the fundamental group of the rotation group in three dimensions
to the identity. To our knowledge, this is the first publication of an explicit
presentation for this relator in $\pi_1(\mathbb{SO}(3))$.

2. The Trick Itself - In Rubber Band Topology

A belt attached between two concentric spheres and twisted by 720° can be
isotopically deformed to an untwisted belt, keeping the ends of the belt fixed
to the two spheres and restricting the isotopy to the region between the two
spheres.

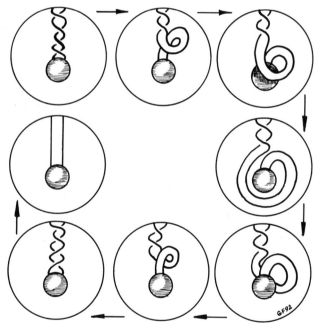

FIGURE 2. TOPOLOGIST'S ILLUSTRATION OF THE BELT TRICK.
A spherical bob hangs inside a larger sphere by a belt with two full,
left-handed twists. The lower one is isotoped to the writhing position.
This is isotopically deformed into its right-handed position. This leads
to a belt with two oppositely oriented twists which cancel.

This is a statement of the belt trick itself. Informally speaking, an *isotopy* is simply a motion of the belt during which it may shrink and stretch, but may neither tear nor pass through itself. The easiest *topological* illustration of the belt trick involves Seifert's [S] exchange of curl for twist shown in Figure 1. A twisted paper ribbon curls naturally into the shapes in the middle of the picture if you relax its ends. A ribbon writhing like the spiral staircase on the right will also twist properly if stretched. The relation of twist and writhe to self-linking was discovered by James White [W].

Seifert's move is fundamental to knot theory and its applications. For example, the phenomenon of DNA supercoiling can be analyzed on the basis of this exchange. Here we employ Seifert's move to draw Dirac's belt trick in a schematic way a topologist might use at the blackboard or on a paper napkin.

In this form we see that the trick can be performed on a belt or on a braid with any number of strands. In braid–form the trick becomes a fundamental relation in the spherical braid group. For an illustrated exposition of this, see [F1, pp. 131ff]. Now this blackboard illustration reveals that the essence of the trick is an exchange of 360° twisted belt for −360° twisted belt that is obtained by passing a curl around the inner concentric sphere. This isotopy is particularly simple, and it makes the belt trick easy to illustrate.

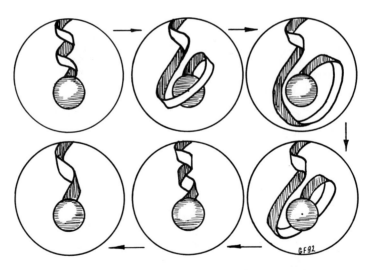

FIGURE 3. THE BELT TRICK IN TERMS OF CURLS.
The belt is initially curled about its longitudinal axis. The lower curl is pulled around the bob so that the two curls can cancel. Compare this to Figure 2, using Figure 1 for picture dictionary.

At this level of topological blackboard illustration, the belt trick may seem amusing but it is not yet apparent that it is actually the tip of an iceberg – an iceberg that contains extraordinary relationships in topology, geometry, algebra and mathematical physics. In order to begin to uncover those relationships, we start by invoking the trick with respect to different spatial axes, and in the process uncover the structure of the quaternion group!

So first let us think of the act of twisting the belt by fixing one end and rotating the other end about some (fixed) axis. A 360° rotation does not bring the belt back to the starting configuration, but a 720° rotation *does* return the belt to the beginning, topologically speaking! In this sense, a 360° rotation is the analog of the number -1, and a 180° rotation is the analog of the complex number $\sqrt{-1}$, since $(\sqrt{-1})^2 = -1$ and $(\sqrt{-1})^4 = 1$. As we shall explain presently, this analogy is more than superficial.

Note how the belt trick already illustrates Sir William Rowan Hamilton's [WRH] brilliant idea that in space there should be a square root of negative one associated with *every* axis of rotation. The 720° rotation can be done about any convenient axis. For example, in Figure 3 the belt trick is shown for rotation about the axis of the belt itself, while in Figure 4, the belt was originally twisted about a transverse axis.

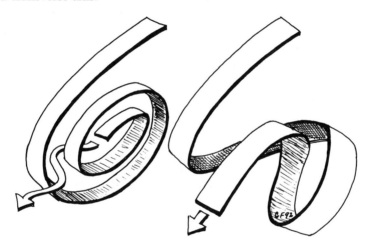

FIGURE 4. DIRAC'S TRICK FOR A PHYSICAL BELT.
A belt curled about a transverse axis, on the left, untwists most surprisingly by pulling it taut from the position on the right.

In Hamilton's system, three orthogonal directions in space are denoted i, j and k, in a right-handed order. See Figure 5. In the quaternions, there is a *fourth* direction denoted by the scalar 1. By Hamilton's principle, one has

$$i^2 = j^2 = k^2 = -1.$$

In fact, one has that $ijk = -1$ and that the algebra formed by $1, i, j, k$ is associative. From these quaternionic rules

$$i^2 = j^2 = k^2 = ijk = -1, \;$$

it follows that

$$ij = k, \quad jk = i, \quad ki = j$$

and that

$$ji = -ji, kj = -jk, ik = -ki.$$

LEMMA 0. *If a, b, c denote real numbers with $a^2 + b^2 + c^2 = 1$, then, among the quaternions,*

$$(ai + bj + ck)^2 = -1.$$

We now have a startling analogy between Hamilton's quaternions and the 180°–belt rotations about axes in three space. Is it more than an analogy? Indeed it is! By composing rotations about different axes and simplifying the belt topologically, we find *exactly* Hamilton's relations. This is illustrated below for $ijk = -1$. We leave other compositions for the reader. This interpretation of the quaternions first appeared in [K1, pp.134f] and [K2, pp.92-97]. In [K3, pp.427-441] it formed the basis of a physical device, the *Quaternion Demonstrator*, which is easily constructed out of materials found on every desk. It also expressed in a simple computer program for the belt trick written in common BASIC. This is the central algorithm in the the real-time interactive computer animation (RTICA) editors used in making the various video-tapes [CFHK, FKS, HKF], the first of which was premiered at the Conference.

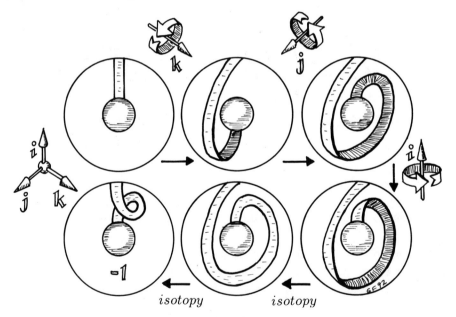

FIGURE 5. THE QUATERNION RELATION $ijk = -1$.
It is accomplished by turning the inner sphere $+180°$ about the axes and dragging the belt along. Compare the last detail with that of Figure 1. Seifert's move verifies that $i^2 = j^2 = k^2 = -1$.

3. The Quaternions

We saw from the belt trick that the fundamental connection between topology and the quaternions takes its origin in their common property of having a $\sqrt{-1}$ for every direction in 3-space. In fact, much of the quaternions' structure derives from this fact alone. For example,

LEMMA 1. *From the rule that $u^2 = -1$ for every unit vector, it follows that*

$$(3.1) \qquad\qquad ij = -ji, \quad ik = -ki, \quad jk = -kj.$$

PROOF. For $u = ai + bj + ck$, with $a^2 + b^2 + c^2 = 1$ and $i^2 = j^2 = k^2 = -1$, we have that

$$u^2 = -(a^2 + b^2 + c^2) + ab(ij + ji) + ac(ik + ki) + bc(jk + ki).$$

Three clever choices of a, b and c complete the demonstration. $\square$

The extra rule $ij = k$ is special to 3–dimensional space. One might suspect that Hamilton manipulated algebra to fit the analogy. But this is not the case. Hamilton invented the algebra of the quaternions to be a precise method for composing rotations in three-dimensional space. Indeed, his student, Peter Guthrie Tait, whose work on knots [T1] is loved by knot theorists, gives the following as his master's favorite definition [T2]:

"From the purely geometrical point of view, a quaternion may be regarded as the quotient of two directed lines in space — or, what comes to the same thing, as the factor, or operator, which changes one directed line into another."

Today we use "vector" for "directed line." Thus, the motion that takes j to k is indeed a rotation about the i axis of 90°. In order to show how Hamilton's remarkable synthesis works, we begin with the unit sphere in $\mathbb{R}^4$:

$$\mathbb{S}^3 = \{g = t + ai + bj + ck : 1 = \|g\| = t^2 + a^2 + b^2 + c^2\}.$$

Thus $g = t + v$ where v is a vector in 3-space. Since $t^2 + \|v\|^2 = 1$ and $v = ru$, where $r = \|v\| = \sqrt{a^2 + b^2 + c^2}$ and $u = v/\|v\|$ is a direction in 3-space, we may write

$$(3.2) \qquad\qquad g = \cos(\alpha) + \sin(\alpha)u = e^{\alpha u}$$

where $\alpha = \arctan(r/t), t = \cos(\alpha)$, and $v = \sin(\alpha)u$. Note that, as expected, DeMoivre's Rule holds:

$$(3.3) \qquad\qquad e^{\alpha u}e^{\beta u} = e^{(\alpha + \beta)u}$$

We may accept this as a definition of the exponential function over the quaternions. However, $u^2 = -1$ implies that the formal power series expansions on both sides match, a fact that would have pleased Euler.

In general, the set of quaternions $\mathbb{H}$ is co-extensive with $\mathbb{R}^4$ but, in view of their structure as a non-commutative field, a more apt definition is

$$\mathbb{H} = \{r + p : r \in \mathbb{R}, p \in \mathbb{R}^3\}$$

where r is a real number and p is a pure vector in 3-space. With addition inherited from $\mathbb{R}^4$ but multiplication defined as above, readers may practice their high-school algebra skills by checking that

$$(3.4) \qquad (r_1 + p_1)(r_2 + p_2) = (r_1 r_2 - p_1 \cdot p_2) + (r_1 p_2 + r_2 p_1) + p_1 \times p_2.$$

Note how this formula expresses the quaternionic product in terms of all four products from the vector calculus of 3 dimensions.

Indeed, we can deduce three formulas from the next figure which are useful for calculating rotations. First, decompose a 3-vector v into components parallel and orthogonal to a given direction u,

$$(3.5) \qquad v = (u \cdot v)u + (u \times v) \times u.$$

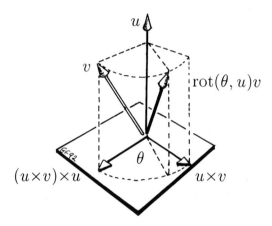

FIGURE 6. DECOMPOSING A VECTOR INTO ITS COMPONENTS PARALLEL AND PERPENDICULAR TO A GIVEN UNIT VECTOR.

Next, reconstruct the rotation of v about u by an angle of θ by applying it to each component:

$$(3.6) \qquad \text{rot}(\theta, u)v = (u \cdot v)u + \cos(\theta)(u \times v) \times u + \sin(\theta)u \times v$$

Finally, substituting (3.5) into (3.6) to remove the double cross product, we arrive at the expression

$$(3.7) \qquad \text{rot}(\theta, u)v = (1 - \cos\theta)u(u \cdot v) + \cos(\theta)v + \sin(\theta)u \times v$$

which readily expands into the classical representation of $\text{rot}(\theta, u)$ as a 3×3 orthogonal matrix. Nevertheless, quaternions will prove to be much more convenient for describing rotations than matrices or vector products, as we shall see presently.

The obvious allusion to the language and notation of complex numbers in the definition of the quaternionic product (3.4) was entirely intentional. Hamilton dubbed r the real part and p the pure part of the quaternion $h = r + p$, and the conjugate, $\bar{h} = r - p$, has the expected property that $h\bar{h} = r^2 + \|p\|^2 = \|h\|^2$. One must however be careful to note that $\overline{h_1 h_2} = \bar{h}_2\ \bar{h}_1$. Thus, for a unit quaternion, $g^{-1} = \bar{g}$, and the quotient of two pure vectors

$$(3.8) \qquad p_1/p_2 = r_1 u_1 (r_2 u_2)^{-1} = \frac{r_1}{r_2} u_1 \bar{u}_2$$

makes Tait's definition above precise. It also conforms to Tait's [T2] description:

"Quaternions (as a mathematical method) is an extension, or improvement, of Cartesian geometry, in which the artifices of co-ordinates, &c., are got rid of, all directions in space being treated on precisely the same terms."

In preparation for presenting Hamilton's fundamental theorem relating the group $\mathbb{G} = \mathbb{S}^3$ of unit quaternions to the group $\mathbb{SO}(3)$ of rotations in $\mathbb{R}^3$, we propose a purely group theoretic lemma.

LEMMA 2. *Write* $\mathcal{R} : \mathbb{G} \to \mathrm{Aut}(\mathbb{H}) \ni \mathcal{R}(g)h = gh\bar{g}$. *We then have*

$$(3.9) \qquad \mathcal{R}(g_1)\mathcal{R}(g_2) = \mathcal{R}(g_1 g_2) = \mathcal{R}(\mathcal{R}(g_1)g_2)\mathcal{R}(g_1)$$

and

$$(3.10) \qquad \mathcal{R}(g)e^{\alpha u} = e^{\alpha \mathcal{R}(g)u}.$$

for $r, \alpha \in \mathbb{R}$ *and* $p, u \in \mathbb{R}^3$.

PROOF. Although the second equality of (3.9) follows from the first, here is a direct verification:

$$g_1 g_2\ h\ \bar{g}_2\bar{g}_1 = g_1 g_2\bar{g}_1\ g_1\ h\ \bar{g}_1\ g_1\bar{g}_2\bar{g}_1$$

(3.10) is an immediate consequence from the observation that

$$\mathcal{R}(g)(r + p) = r + \mathcal{R}(g)p.$$

□

We next draw some conclusions from this lemma. The first deserves to be called a "theorem" since Hamilton was so delighted in its discovery.

THEOREM 1. *The mapping* $\mathbb{S}^3 \to \mathbb{SO}(3) : g \to \mathcal{R}(g)$ *is a 2:1 covering homomorphism that takes the unit quaternion* $e^{\alpha u/2} = c + su$, *where* $c = \cos(\alpha/2)$ *and* $s = \sin(\alpha/2)$, *to the rotation by an angle of* α *about the* u-*axis,*

$$\mathrm{rot}(\alpha, u) = \mathcal{R}(e^{\alpha u/2}).$$

PROOF. We saw that the operator $\mathcal{R}(g)$ preserves pure quaternions, and so $g = c + su$ acts on $\mathbb{R}^3$. The action is linear, and it leaves vectors parallel to the axis u fixed. It thus remains to compute its action on a direction v orthogonal to u. We have

$$(c + su)v(c - su) = c^2 v + 2cs(u \times v) - s^2(u \times v) \times u$$
$$= (c^2 - s^2)v + 2csu \times v$$
$$= \cos(\alpha)v + \sin(\alpha)u \times v,$$

because $u \cdot v = 0$ implies that $v = (u \times v) \times u$ for unit vectors. Since $\{u, v, u \times v\}$ forms a right handed orthogonal frame, the last expression is indeed the rotation of v about u by α radians. The double covering stems from the fact that $\mathcal{R}(-g) = \mathcal{R}(g)$. $\square$

In particular, the composition of two rotations in space is again a rotation, and its magnitude and axis may easily be computed.

COROLLARY 1.
$$\mathrm{rot}(\theta_1, u_1)\,\mathrm{rot}(\theta_2, u_2) = \mathrm{rot}(\theta_3, u_3)$$
where $c_n = \cos(\theta_n)$, $s_n = \sin(\theta_n)$, $n = 1, 2, 3$,

(3.11) $c_3 = c_1 c_2 - s_1 s_2 u_1 \cdot u_2$

and

$$s_3 u_3 = c_2 s_1 u_1 + c_1 s_2 u_2 + s_1 s_2 u_1 \times u_2.$$

$\square$

The next few applications are motivated, in part, from the ubiquitous use of a mouse to manage rotations on contemporary graphics computers, and a desire to limit the number of geometrical primitives which are built into the hardware of the computer. Suppose now that the displacement of the mouse on its horizontal, planar pad is to be interpreted as a rotation of the scene in the computer screen. Though the horizontal position of the mouse is echoed by a cursor on the vertical picture screen, few people have trouble identifying these two planes.

A very common method is to correlate a displacement v of the mouse with an axis u and angle θ of rotation as follows. If u is the clockwise perpendicular to v in the picture plane, then this kind of rotor will give the impression that the viewer is rotating relative to the object. This is useful for simulations in which it is appropriate for the viewer to be flying or swimming about the scene. However, choosing u to be the counter-clockwise perpendicular to v produces the subjective impression of pushing the object about invisible gimbals. Hanson [H1, H2] compared this kind of rotor to manipulating a track-ball device. The utility of the following lemma was discovered by Glenn Chappell while working on [CFHK].

COROLLARY 2. *For any direction u in the x, y-plane, we have that $u = \mathrm{rot}(\alpha, k)i$ for some angle α and*

(3.12) $\mathrm{rot}(\theta, u) = \mathrm{rot}(\alpha, k)\,\mathrm{rot}(\theta, i)\,\mathrm{rot}(-\alpha, k).$

PROOF. Apply $\mathcal{R}$ to the calculation

$$e^{\alpha k/2} e^{\theta i/2} e^{-\alpha k/2} = \mathcal{R}(e^{\alpha k/2}) e^{\theta i/2}$$
$$= e^{\theta \operatorname{rot}(\alpha, k)i/2}$$
$$= e^{\theta u/2}.$$

It would seem that two degrees of freedom (the mouse moving in a plane) are too few to realize every rotation in space. This is far from the truth, as all people who have used a mouse-rotor based on quaternionic principles have experienced. A variety of motions, usually circular, permits one to place the object into every desired position. Consider this factorization of a rotation about the k-axis

COROLLARY 3.

(3.13) $$\operatorname{rot}(\theta, k) = \operatorname{rot}(\pi/2, i) \operatorname{rot}(\theta, j) \operatorname{rot}(-\pi/2, i)$$

PROOF. Apply $\mathcal{R}$ to the calculation

(3.14) $$\mathcal{R}(e^{\pi i/4}) e^{\theta j/2} = e^{\theta \operatorname{rot}(\pi/2, i)j/2} = e^{\theta k/2}.$$

□

We invite the reader to experience the pleasure of quaternionic calculations by proving the following summary.

COROLLARY 4. *For every scalar θ and direction u in $\mathbb{R}^3$, there is a finite sequence $\{\alpha_n, \beta_n\}$ so that*

(3.15) $$\operatorname{rot}(\theta, u) = \prod_n \operatorname{rot}(\alpha_n, i) \operatorname{rot}(\beta_n, j)$$

The final two applications of Hamilton's great theorem are of a more analytical nature and show the special role that that the 1-parameter groups $\operatorname{rot}(\alpha, i)$ and $\operatorname{rot}(\beta, j), \alpha, \beta \in \mathbb{R}$ play in realizing $\mathbb{SO}(3)$ on a graphics computer.

COROLLARY 5. *Let*

$$\alpha i + \beta j = \sqrt{\alpha^2 + \beta^2}\, u = \gamma u$$

be a plane vector. Then

(3.16) $$\lim_{nt \to \mu} (e^{t\alpha i} e^{t\beta j})^n = e^{\gamma \mu u}$$

where, in the limit, $t \to 0$, and $n \to \infty$, but in such a way that $nt \to \mu$, a constant.

HEURISTIC PROOF. The product of the two exponentials is

$$\cos(\alpha t)\cos(\beta t) + i \sin(\alpha t)\cos(\beta t) + j \cos(\alpha t)\sin(\beta t) + k \sin(\alpha t)\sin(\beta t).$$

Expanding the two cosines of the real part, we see that

$$(1 - \alpha^2 t^2/2 + o(3))(1 - \beta^2 t^2/2 + o(3)) = (1 - \gamma^2 t^2/2 + o(3))$$

where $\gamma = \sqrt{\alpha^2 + \beta^2}$. Thus, $\cos(\gamma t)$ is a first order approximation to the real part of the product.

We next approximate the pure part by applying the common substitution of $\cos(\varepsilon) = 1$ and $\sin(\varepsilon) = \varepsilon$ for small enough ε:

$$(\alpha t i + \beta t j) + h.o.t. = (\gamma t)\frac{\alpha i + \beta j}{\gamma} + h.o.t.$$

$$= \sin(\gamma t)u + h.o.t.$$

Thus, ignoring the higher order terms, the unit quaternion $e^{t\alpha i}e^{t\beta j}$ is approximately $e^{\gamma t u}$. Raising both to the n-th power and using the assumption that $e^{\gamma(nt)u}$ converges to $e^{\gamma\mu u}$, so does the left hand side of the limit. $\square$

The final corollary is standard in the theory of Lie groups. We include here a heuristic proof for completeness.

COROLLARY 6.

(3.17)
$$\lim_{nt^2 \to \mu} (e^{-ti}e^{-tj}e^{ti}e^{tj})^n = e^{2\mu k}$$

where the limit is similar to that in the previous proposition.

PROOF. For $c = \cos t$ and $s = \sin t$, compute that the commutator is

$$[e^{ti}, e^{tj}] = (1 - 2s^4) + 2cs^2\sqrt{1 + s^2}\begin{bmatrix} -s/\sqrt{1+s^2} \\ +s/\sqrt{1+s^2} \\ +c/\sqrt{1+s^2} \end{bmatrix} = e^{\sigma\ell}$$

where $\sigma = \arccos(1 - 2s^4)$ and ℓ is the unit vector displayed. Note that this point on $\mathbb{S}^3$ converges nicely to k. The limit $e^{n\sigma\ell} \to e^{2\mu k}$ follows from the approximation, $\sigma \sim 2t^2$, which follows from

$$1 - 2\sin(t)^4 = \cos(\sigma) = 1 - 2\sin(\frac{\sigma}{2})^2.$$

$\square$

Corollary 5 expresses the fact that a quaternionic rotor can be very well approximated by sampling the position of the mouse sufficiently often and interpreting each step as the product of just two small orthogonal rotations. The effectiveness of this procedure was discovered by Chris Hartman during the making of [CFHK]. Corollary 6 says that "cranking" the mouse in sufficiently small circles adequately simulates a rotation in the picture plane. For example, for $t = 0.05$, it takes 314 iterations of the commutator in (3.17) to rotate the object a quarter turn about the z-axis. Since $nt^2 \approx 100\pi/400 = \pi/4$, Theorem 1 and Corollary 6 are confirmed.

All of these phenomena are universally well-known to programmers of real-time interactive computer animations, but have, to our knowledge, never been rigorously derived from first geometrical principles in the coordinate free way envisioned by Hamilton. A more conventional treatment of quaternions, as used in computer graphics, is ably presented in [Sh, BCGH].

We shall devote the next section to a more substantial application of Hamilton's great invention, one that has even greater application to numerical geometry and computer graphics.

4. Returning to the Belt Trick

With the help of these remarks about the quaternions, it is easy to describe specific formulas suitable for computer programming the belt trick. The idea is as follows: *Move the belt by moving the ambient space containing an image of the belt. Create this motion of the ambient space by rotating the spheres concentric to the origin in a smoothly varying way.*

As it will be convenient to work in polar coordinates, we shall extend the usual notation for scalar multiplication of vectors also to sets. For example,

$$\mathbb{R}^3 \setminus \{0\} = (0, \infty)\mathbb{S}^2.$$

Our fundamental isotopy is defined as follows:

DEFINITION. *To every smooth map of a rectangular ribbon to 3-space*

$$(\theta, u)[\rho_0, \rho_1] \times [\tau_0, \tau_1] \to \mathbb{R}^3 \ni (\rho, \tau) \to \theta(\rho)u(\tau)$$

associate the ambient isotopy

(4.1) $\quad \text{twist}(\theta, u) \; : \; \mathbb{R}^3 \times [0, 1] \to \mathbb{R}^3 \ni (x, t) \to \mathcal{R}(e^{\theta(\|x\|)u(t)/2})x$

The rotations themselves can be accomplished by using the quaternionic formalism, by using the more conventional matrix formalisms for rotations, or by an optimized combination of both. Here we describe our methods qualitatively, but everything translates easily into specific formulas. We shall use this abbreviation for the *linear interpolation* between two elements, A, B, in a ring over the reals:

$$\text{lirp}(\rho_0, \rho_1, A, B)\rho = \frac{\rho_1 - \rho}{\rho_1 - \rho_0}A + \frac{\rho - \rho_0}{\rho_1 - \rho_0}B$$

As our first example, we begin with an untwisted belt $\mathbb{B}^2$ obtained as follows. Extrude a short arc $\alpha : \mathbb{R}^1 \to \mathbb{S}^2$ through the north pole of the unit sphere so as to hang between two radii, $0 < \rho_0 < \rho_1$:

(4.2) $\qquad\qquad \mathbb{A}^1 = \alpha([-\omega, +\omega]), \quad \mathbb{B}^2 = [\rho_0, \rho_1]\mathbb{A}^1.$

To twist $\mathbb{B}^2$ one full right-handed turn about its own axis (which we take to be the vertical axis k), apply the diffeomorphism

(4.3) $\qquad x \to \text{twist}(-2\pi\theta, k)(x), \quad \theta = \text{lirp}(\rho_0, \rho_1, 1, 0)\|x\| = \frac{\rho_1 - \|x\|}{\rho_1 - \rho_0}.$

Note that this leaves the ρ_1 sphere fixed while turning the ρ_0 sphere $-360°$ and interpolating the rotations for the intervening spheres. The effect is a belt $\mathbb{B}^2_k$ with a $+360°$ twist. Reversing the sign of 2π in (4.3) reverses the twist in the belt $\mathbb{B}^2_{-k}$.

The simplest belt trick takes the $-360°$ twisted belt $\mathbb{B}^2_{-k}$ and moves it iso-topically into a $360°$ twisted belt $\mathbb{B}^2_k$. This is accomplished by the isotopy

$$(4.4) \qquad x \to \text{twist}(+2\pi\theta(\|x\|), u(t))(x), \quad u(t) = \text{rot}(\pi t, j)k,$$

where $u(t)$ is the roving rotation axis. It is noteworthy here that the outer end $\rho_1 \mathbb{A}^1$ of the belt never moves while the inner end $\rho_0 \mathbb{A}^1$ is rotated one full turn about j.

The *Dirac String Trick* itself may be accomplished by catenating the two above maps and manipulating their parameters. Consider the following two-parameter isotopy

$$(4.5) \quad x \to \text{twist}(2\pi\theta(\|x\|), u(t))(x)\, \text{twist}(s\, 2\pi\theta(\|x\|), k)(x), \quad -1 \le s \le +1,$$

applied to the belt $\mathbb{B}^2$ to produce the family $\mathbb{B}^2(s, t)$ of belts. The reader will find what follows easier to read if we translate (4.5) into quaternionic notation

$$(4.6) \qquad \mathbb{B}^2(s, t) = \mathcal{R}(e^{\theta \pi u(t)} e^{s\ \theta \pi\ k})\mathbb{B}^2$$

When $s = 0$, (4.4) is just (4.3) since the initial diffeomorphism is the identity. Hence $\mathbb{B}^2(0, 0) = \mathbb{B}^2_{-k}$. At $s = -1$, the first map produces $\mathbb{B}^2_{-k}$, which is undone by the isotopy (4.4) at $t = 0$. Hence $\mathbb{B}^2(-1, 0) = \mathbb{B}^2$. As $s \to 1$, this belt has become twisted $-720°$. One might say that the Dirac strings had been "wound up."

Example. For those readers eager to program the Dirac belt trick [K3, p435f] as well as those readers who prefer their abstractions by way of a good example, here are the explicit formulas for a special case of (4.6). Let a belt of unit length have one end at the origin and the other at the north pole of the unit sphere. Suppose it has already been twisted two full turns. Then $s = 0, \rho_0 = 0, \rho_1 = 1$ and $0 \le \theta \le 1$. We abbreviate

$$C_t = \cos(\pi t), S_t = \sin(\pi t),$$

and set

$$(4.7) \qquad \mathcal{D}(\theta, t) := e^{\pi \theta u(t)} e^{\pi \theta k}, \quad u(t) := \text{rot}(\pi t, j)k = iS_t + kC_t.$$

Thus $e^{\pi \theta k}$ represents a $2\pi\theta$ rotation about the vertical axis k, while $e^{\pi \theta u(t)}$ represents a $2\pi\theta$ rotation about $u(t)$, as this axis moves from $u(0) = k$ to $u(1) = -k$.

Here is a useful expansion of this two parameter family of quaternions:

$$
\begin{aligned}
\mathcal{D}(\theta, t) &= (C_\theta + uS_\theta)(C_\theta + kS_\theta) \\
&= C_\theta^2 + (u + k)S_\theta C_\theta + ukS_\theta^2 \\
&= C_\theta^2 + (iS_t + k(C_t + 1))S_\theta C_\theta + (-jS_t - C_t)S_\theta^2 \\
&= (C_\theta^2 - C_t S_\theta^2) + i(S_t S_\theta C_\theta) + j(-S_t S_\theta^2) + k(C_t + 1)S_\theta C_\theta
\end{aligned}
$$

At $t = 0$ we see that

(4.8) $\mathcal{D}(\theta, 0) = e^{\pi \theta k} e^{\pi \theta k} = e^{2\pi \theta k} = 1\ C_{2\theta} + k S_{2\theta}$

is a great circle through the north pole of $\mathbb{S}^3$ over the range of θ, while

(4.9) $\mathcal{D}(\theta, 1) = C_\theta^2 + S_\theta^2 = 1.$

THEOREM 2. $\mathcal{D}(\theta, t)$ *is an explicit parametrization by the unit square of the homotopy from the square of the generator of* $\pi_1(\mathbb{SO}(3))$ *to the identity.*

PROOF. Recall that the elements of the fundamental group $\pi_1(\mathbb{SO}(3))$ of the rotation group of $\mathbb{R}^3$ are homotopy classes of loops in $\mathbb{SO}(3)$, based at the identity transformation on $\mathbb{R}^3$. Such a loop is a one-parameter family of rotations beginning and ending at the identity. However, as a topological space, $\mathbb{SO}(3)$ is homeomorphic to the quotient space of the 3-sphere $\mathbb{S}^3$ upon identification of antipodal pairs. This is the essence of Theorem 1 where we computed this homeomorphism explicitly. Two quaternions of unit length correspond to each rotation, and the representation by inner-automorphism $\mathcal{R}$ of quaternions as rotations exhibits this quotient space structure for $\mathbb{SO}(3)$. From (4.8) it follows that the loop $\mathcal{RD}([0, \frac{1}{2}], 0)$ in $\mathbb{SO}(3)$ consists of the rotations about a single axis from the "un-rotation" to one full turn. Composition in the fundamental group is traditionally multiplicative. Thus the square $\mathcal{RD}([0, 1], 0)$ of this generator is null-homotopic under $\mathcal{RD}([0, 1], [0, 1])$ by (4.9). □

This relationship of the belt trick with the fundamental group of the 3-dimensional rotation group really is fundamental in physics, cf. [BL, Chapter 4, pp. 180–203], [BR], [K3, pp. 427–441], and especially the figure on page 1149 of [MTW, Chapter 41, pp. 1148ff].

It also connects to other topological topics, such as the "J-homomorphism" in the theory of spherical homotopy groups. We shall extend our exposition in the monograph that will accompany a videotape of the real-time interactive computer animations we have made [CFHK].

REFERENCES

[BCGH]   Alan H. Barr, Bena Currin, Steven Gabriel, John F. Hughes, *Smooth interpolation of orientations with angular velocity constraints using quaternions*, Computer Graphics **26,2** (1992), 313–320.

[BL]   L. C. Biedenharn and J. D. Louck, *Angular momentum in quantum physics - Theory and Applications*, Encyclopedia of Mathematics and its Applications, Addison Wesley, Cambridge University Press, 1989.

[BP]   Herbert J. Bernstein and Anthony Phillips, *Bundles and quantum theory*, Scientific American **245,1** (1981).

[BR]   E. P. Battey-Pratt and T. J. Racey, *Geometric models for fundamental particles*, Intern. Jour. Theor. Physics **19,6** (1980).

[CFHK]   Glenn Chappell, George Francis, Chris Hartman, Louis H. Kauffman, *Air on the Dirac Strings,*, 15 min silent videotape, National Center for Supercomputing Applications, University of Illinois, May 31, 1992.

[F]   George K. Francis, *A Topological Picturebook*, Springer Verlag, 1987.

[FKS]   George Francis, Louis H. Kauffman and Dan Sandin, *Air on the Dirac Strings*, 2 min narrated videotape, ACM SIGGRAPH Video Review, Issue 93, 1993.

[WRH]   William Rowan Hamilton, *Elements of Quaternions*, Longmans Green, Green, London, 1866.

[H1]   Andrew Hanson, *The rolling ball: applications of a method for controlling three degrees of freedom using two-dimensional input devices*, Manuscript, dated 7.13.88.

[H2]   _____, *The rolling ball*, Graphics Gems III (David Kirk, ed.), Academic Press, San Diego, 1992.

[HFK]   John Hart, George Francis, Louis H. Kauffman, *Visualizing quaternion rotation*, 23 page ms with 7 drawings, March 1993.

[K1]   Louis H. Kauffman, *Sign and space*, Proc. IASWR Conference (Chris Chapple, ed.), IASWR, Melville Memorial Library, SUNY Stony Brook, New York, 1982.

[K2]   _____, *On Knots*, Ann. Math. Studies 115, Princeton U. Press, 1987.

[K3]   _____, *Knots and Physics*, World Scientific, Singapore, 1991.

[M]   Wilhelm Magnus, *Über Automorphismen von Fundamentalgruppen berandeter Flächen*, Math. Ann. **109** (1934), 617–646.

[MTW]   C. W. Misner, K. S. Thorne, J. A. Wheeler, *Gravitation*, Freeman, San Francisco, 1973.

[N]   M. H. A. Newman, *On a string problem of Dirac*, J. Lond.Math. **17** (1942), 173–177.

[S]   Herbert Seifert, *Über das Geschlecht von Knoten*, Math. Annalen **110** (1934), 571–592.

[Sh]   Ken Shoemake, *Animating rotations with quaternion curves*, Computer Graphics **19(3)** (1985), 245–254.

[T1]   Peter Tait, *On Knots I,II,III*, Scientific Papers, vol. I, Cambr. Univ. Press, London, 1898, pp. 273–347.

[T2]   Peter Tait, *Quaternions*, Encyclopædia Britannica, Eleventh Edition, 1911.

[W]   James White, *Self-linking and the Gauss integral in higher dimensions*, Amer. J. Math. **91** (1969), 693–728.

MATHEMATICS DEPARTMENT, UNIVERSITY OF ILLINOIS, URBANA, ILLINOIS, 61801

E-mail: gfrancis@math.uiuc.edu

MATHEMATICS DEPARTMENT, UNIVERSITY OF ILLINOIS, CHICAGO, ILLINOIS, 60607

E-mail: u10451@uicvm.bitnet

Contemporary Mathematics
Volume **169**, 1994

DOES LYNDON'S LENGTH
FUNCTION IMPLY
THE UNIVERSAL THEORY OF FREE GROUPS?

ANTHONY M. GAGLIONE[1] AND DENNIS SPELLMAN

ABSTRACT. This note shows that every model of the universal theory of the non-Abelian free groups admits a Lyndon length function. The question is then posed as to whether the model class of the universal theory of the non-Abelian free groups is precisely the class of such groups. The authors have subsequently given a negative answer to this question.

Definitions and notation will be those of Bell and Slomson [3] and of Gaglione and Spellman [6], [7] and [8].

Let L be a first-order language with equality. Two L-structures A and B *have the same universal theory* if they satisfy precisely the same universal sentences (and therefore also precisely the same existential sentences) of L. If B is an L-structure, let $\mathrm{Th}(B) \cap (\forall \cup \exists)$ be the set of all universal and existential sentences of L true in B. Evidently the L-structure A has the same universal theory as B if and only if A is a model of $\mathrm{Th}(B) \cap (\forall \cup \exists)$.

If A is a substructure of the L-structure B, then a necessary and sufficient condition that A and B have the same universal theory is that there be a model $^*A$ of $\mathrm{Th}(A) \cap (\forall \cup \exists)$ such that $A \subseteq B \subseteq \,^*A$. This in turn is equivalent to the existence of an index set I and an ultrafilter D on I such that B is embeddable in the ultrapower A^I/D. A different necessary and sufficient condition that an L-structure B and a substructure A have the same universal theory is that A and B satisfy precisely the same primitive sentences of L. (see [3, Ch. 9]).

Let A be a non-trivial, torsion-free, Abelian group. Let $<$ be a strict linear order on A such that for arbitrary $(a, b, c) \in A^3$ we have $a + c < b + c$ whenever $a < b$. Then the ordered pair $\Lambda = (A, <)$ is an *ordered Abelian group*. A non-trivial, torsion-free, Abelian group A is *orderable* provided there is at least one strict linear order $<$ such that $\Lambda = (A, <)$ is an ordered Abelian group.

[1]The research of this author was partially supported by the Identification Systems Branch, Radar Division, Naval Research Laboratory.

1991 Mathematics Subject Classification. Primary $20A15, 20F32, 03C60$. Secondary $03B10, 03C07, 03C20, 20E08$.

It is well-known that every non-trivial, torsion-free, Abelian group is orderable. Nonetheless, we'll present an argument to that effect in this paper. We shall also show the well-known result that this class of Abelian groups is the class having the same universal theory as $\mathbb{Z}$; moreover, we shall show that every model of the universal theory of the non-Abelian free groups admits a Lyndon length function.

We now specify two first-order languages with equality. L_o shall contain a binary operation symbol $\cdot$, a unary operation symbol $^{-1}$ and a constant symbol 1. $\mathcal{L}$ shall contain a binary operation symbol $+$, a unary operation symbol $-$, a constant symbol 0 and a binary relation symbol $<$. L_o shall be the *language of group theory* and $\mathcal{L}$ shall be the *language of ordered Abelian groups*. A primitive sentence of L_o is one of the form

$$\exists \overline{x}(\bigwedge_i p_i(\overline{x}) = P_i(\overline{x})) \bigwedge \bigwedge_j (q_j(\overline{x}) \neq Q_j(\overline{x})))$$

where $\overline{x}$ is a tuple of variables and the $p_i(x), P_i(x), q_j(x)$ and $Q_j(x)$ are terms of L_o.

In case the L_o-structures we are considering are groups, this type of sentence may be simplified to one of the form

$$\exists \overline{x}(\bigwedge_i p_i(\overline{x}) = 1) \bigwedge \bigwedge_i (q_j(\overline{x}) \neq 1))$$

where $\overline{x} = (x_1, \ldots, x_m)$ is a tuple of distinct variables and the $p_i(\overline{x})$ and $q_j(\overline{x})$ are words on $\{x_1, \ldots, x_m\} \cup \{x_1^{-1}, \ldots, x_m^{-1}\}$.

LEMMA 1. *A non-trivial Abelian group A has the same universal theory as $\mathbb{Z}$ if and only if A is torsion-free.*

PROOF. One implication is trivial. Assume A is a non-trivial, torsion-free, Abelian group to show that A is a model of $\mathrm{Th}(\mathbb{Z}) \cap (\forall \cup \exists)$. We write our groups additively here. Let a be a non-zero element of A and put $A_0 = <a> \cong \mathbb{Z}$. It will suffice to show that A_0 satisfies every primitive sentence true in A. To that end, consider the system

$$(*)\begin{cases} \sum_{k=1}^m p_{i,k}x_k = 0, & (1 \leq i \leq I), \\ \sum_{k=1}^m q_{j,k}x_k \neq 0, & (1 \leq j \leq J), \end{cases}$$

of equations and inequations. Suppose (*) has a solution

$$(x_1, \ldots, x_m) = (a_1, \ldots, a_m)$$

in A. It suffices to show that (*) has a solution in A_0. Let B be the subgroup of A generated by $\{a_1, \ldots, a_m\}$. If $B = 0$, then $(x_1, \ldots, x_m) = (0, \ldots, 0)$ is a solution to (*) in A_0. We may therefore assume $B \neq 0$. So B is then a non-trivial, finitely generated, torsion free, Abelian group. Thus, B is free Abelian of some finite rank $r \geq 1$. But it was shown in Gaglione and Spellman [7] that $\mathbb{Z}^r$ and $\mathbb{Z}$ have the same universal theory. Therefore, (*) has a solution in A_0.■

LEMMA 2. *Every non-trivial torsion-free Abelian group A is orderable.*

PROOF. Such a group A has the same universal theory as $\mathbb{Z}$. Therefore A is embedded in an ultrapower $\mathbb{Z}^I/D$ of $\mathbb{Z}$. The order in $\mathbb{Z}$ induces an order in $\mathbb{Z}^I/D$ making $(\mathbb{Z}^I/D, <)$ an ordered Abelian group. The restriction of $<$ to A makes $(A, <)$ an ordered Abelian group.∎

DEFINITION 1 (Lyndon [12]). *Let G be a (multiplicatively written) group. Let $\Lambda = (A, <)$ be an (additively written) ordered Abelian group. Let*

$\lambda : G \to A, g \xmapsto{\lambda} |g|$ *be a function, and let*

$2c : G^2 \to A$ *be defined by* $(g_1, g_2) \xmapsto{2c} |g_1| + |g_2| - |g_1 g_2^{-1}|$. *The the ordered triple (G, Λ, λ) is a* normed *group provided the following six axioms are satisfied:*
$(A0) x \neq 1$ *implies* $|x| < |x^2|$
$(A1) \mid x \mid \geq 0$ *and* $\mid x \mid = 0$ *iff* $x = 1$
$(A2) \mid x^{-1} \mid = \mid x \mid$
$(A3) 2c(x, y) \geq 0$
$(A4) 2c(x, y) > 2c(x, z)$ *implies* $2c(y, z) = 2c(x, z)$
$(CO) 2c(x, y) \equiv 0 \pmod{2A}$
A group G is normable *provided there is at least one ordered Abelian group $\Lambda = (A, <)$ and at least one map $\lambda : G \to A$ such that (G, Λ, λ) is a normed group.*

REMARK. The axioms are not independent. Chiswell [5] has shown that $(A3)$ is a consequence of $(A2), (A4)$ and the following
$(A1') \mid 1 \mid = 0.$

If S is a set of sentences of a first-order language L with equality, let $\mathbb{M}(S)$ be the model class of S.

THEOREM 1. *Let L be a first-order language with equality. Let $\mathcal{X}$ be a class of L-structures. Then the following three properties form a set of necessary and sufficient conditions that $\mathcal{X}$ be of the form $\mathbb{M}(S)$ for at least one set S of sentences of L:*
(i) $\mathcal{X}$ is closed under isomorphism.
(ii) $\mathcal{X}$ is closed under the formation of ultraproducts.
(iii) If $c\mathcal{X}$ is the class of all L-structures not in $\mathcal{X}$, then $c\mathcal{X}$ is closed under the formation of ultrapowers.

Theorem 1 is a deduction of Theorem 3.10, Chapter 7 of [3] without assuming (G.C.H.) using the Keisler-Shelah Theorem (see [3,15]).

Although "the" norm is not generally defined in L_o, norms are internal in the sense that they extend to ultraproducts. It is then straightforward to deduce -

COROLLARY. *The class of all non-Abelian, normable groups is the model class $\mathbb{M}(\Theta)$ of some set Θ of sentences of L_o.*

It is known that the non-Abelian free groups have the same universal theory (see [7]). Thus, if F_2 is free of rank 2 and $\Phi = \mathrm{Th}(F_2) \cap (\forall \cup \exists)$, then every non-Abelian, free group is a model of Φ. It is not difficult to convince oneself that the models of Φ are precisely those non-Abelian groups embeddable in some ultrapower of F_2, since one can easily show that every model of Φ contains a copy of F_2. But if $F_2 = \langle a_1, a_2 \rangle$, then the length function with respect to the free basis $\{a_1, a_2\}$, $\lambda : F_2 \to \mathbb{Z}$ induces a length function $^*\lambda : F_2^I/D \to \mathbb{Z}^I/D$ making $(F_2^I/D, (\mathbb{Z}^I/D, <), {}^*\lambda)$ into a normed group whenever I is an index set and D is an ultrafilter on I. The restriction of $^*\lambda$ to the subgroup G makes G into a normed group. Thus, every model of Φ is also a model of Θ. In symbols -

THEOREM 2. $\mathbb{M}(\Phi) \subseteq \mathbb{M}(\Theta)$.

Brignole and Ribeiro [4] have given a proof of a theorem of Gurevic and Kokorin asserting that any two ordered Abelian groups have the same universal theory. Since any ordered Abelian group $\Lambda = (\mathbf{A}, <)$ contains an ordered subgroup isomorphic to $(\mathbb{Z}, <)$, it follows that every ordered Abelian group Λ is embeddable in some ultrapower

$$(\mathbb{Z}, <)^I/D = (\mathbb{Z}^I/D, <).$$

Thus, every normed group admits a norm with values in an ordered Abelian group of the form $(\mathbb{Z}^I/D, <)$.

Let L be a first-order language with equality. A sentence of L of the form $\forall \overline{x} \exists \overline{y} \phi(\overline{x}, \overline{y})$ where $\overline{x}$ and $\overline{y}$ are disjoint tuples of variables, $\phi(\overline{x}, \overline{y})$ contains no quantifiers and $\phi(\overline{x}, \overline{y})$ contains free at most the variables in $\overline{x}$ and $\overline{y}$ is a *universal-existential* sentence of L. Any sentence of L logically equivalent to a universal-existential sentence of L is a Π_2-*sentence* of L. Since vacuous quantifications are permitted, every universal sentence of L and every existential sentence of L is also a Π_2-sentence of L.

THEOREM 3. Θ *may be taken to be a set of* Π_2-*sentences of* L_o.

PROOF. In view of Theorem 2, p. 279 of Grätzer [10], it suffices to show that the union $G = \bigcup_{n < \omega} G_n$ of a chain $(G_n)_{n < \omega}$ of non-Abelian, normable subgroups $G_0 \subseteq G_1 \subseteq \ldots \subseteq G_n \subseteq \ldots$ is normable. To that end, suppose that $(G_n, \Lambda_n, \lambda_n)$ is a normed group. Let D be a non-principal ultrafilter on ω and let $\Lambda = (\prod_{n < \omega} \Lambda_n)/D$ be the ultraproduct of the family $(\Lambda_n)_{n < \omega}$ of ordered Abelian groups with respect to the ultrafilter D. Let $\Lambda = (A, <)$. For each $g \in G$, let $\deg(g) = \min\{n \in \omega \mid g \in G_n\}$. Finally, let $\lambda : G \to A$ be given by $g \xmapsto{\lambda} L_g/D$ where

$$\begin{cases} L_g(n) = 0, & \text{if } n < \deg(g), \\ L_g(n) = \lambda_g(n), & \text{otherwise} . \end{cases}$$

Then it is straightforward to verify that (G, Λ, λ) is a normed group. ∎

QUESTION. *Is* $\mathbb{M}(\Phi) = \mathbb{M}(\Theta)$? *Equivalently: Does every non-Abelian, normable group have the same universal theory as the non-Abelian free groups?*

By Alperin and Bass [1, (5.3),(5.4),(6.4)], we may also pose the

QUESTION. *Let G be a non-Abelian group. Is it the case that G is a model of* Φ *if and only if there is an ordered Abelian group* Λ *and a* Λ-*tree T such that G acts freely on* Λ *without inversions, i.e., G is tree-free in the sense of Bass (see [2])?*

ADDENDUM

Since the original preparation of the manuscript, the authors have learned of [14]. In that work, Remeslennikov also shows that every model of Φ is normable. Moreover, a negative answer to our question is given independently in [9] and in [14].

The authors wish to thank the referee for helpful comments.

REFERENCES

[1] Alperin, R. and Bass, H., "Length functions of group actions on Λ-trees," *Annals of Mathematical Studies, Combinatorial Group Theory and Topology,* Gersten, S. M. and Stallings, J. R., Editors, Princeton University Press, Princeton, NJ, 1987, 265-378.

[2] Bass, H., "Group actions on non-Archimedean trees," *Mathematical Sciences Research Institute Publications, Arboreal Group Theory,* Alperin, R. C., Editor, Springer-Verlag, New York, NY, 1991, 69-131.

[3] Bell, J. L. and Slomson, A. B., *Models and Ultraproducts,* (Second revised printing), North-Holland, 1971, Amsterdam.

[4] Brignole, D. and Ribeiro, H., "On the universal equivalence for ordered abelian groups," *Algebra i Logika* Sem. 4 (1965), no. 2, 51-55.

[5] Chiswell, I. M., "Abstract length functions in groups," *Math. Proc. Camb. Phil. Soc.* 80 (1976), 451-463.

[6] Gaglione, A. M. and Spellman,D., "More model theory of free groups," *Houston J. Math.,* to appear

[7] Gaglione, A. M. and Spellman, D., "Some model theory of free groups and free algebras," *Houston J. Math.* 19(1993),327-356.

[8] Gaglione, A. M. and Spellman, D., "Even more model theory of free groups," *Proc. of Infinite Groups and Group Rings, Tuscaloosa, March, 1992,* Corson, J., Dixon, M., Evans, M., and Röhl, F., Editors, World Scientific 1993, New Jersey, 37-40

[9] Gaglione, A.M. and Spellman, D., "Every 'universally free' group is tree-free," *Proc. Ohio State-Denison Conf. for H. Zassenhaus,* Sehgal, S. and Solman, R., Editors, World Scientific, 1993, New Jersey, 149-154

[10] Grätzer, G., *Universal Algebra,* Van Nostrand, 1968, USA.

[11] Harrison, N., "Real length functions in groups," *Trans. Amer. Math. Soc.* 174 (1972), 77- 106.

[12] Lyndon, R. C., "Length functions in groups," *Math. Scand.* 12 (1963), 209-234.

[13] Morgan, J. W., "Λ-trees and their applications," *Bull. Amer. Math. Soc.* 26 (1992), 87-112.

[14] Remeslennikov, V.N., "$\exists$-Free groups and groups with a length function," Preprint.

[15] Shelah, S., "Every two elementary equivalent models have isomorphic ultrapowers," *Israel J. Math.* 10 (1971) 224-233.

Department of Mathematics, U.S. Naval Academy, Annapolis, $MD, 21402-5002$
E-mail address: amg@sma.usna.navy.mil
Philadelphia, PA 19124-3036

Contemporary Mathematics
Volume **169**, 1994

SCHOTTKY GROUPS
AND THE BOUNDARY OF
TEICHMÜLLER SPACE: GENUS 2

DANIEL M. GALLO

ABSTRACT. We show that, in genus 2, every projective structure which lies on the boundary of Teichmüller space is a limit of structures which correspond to Schottky groups.

1. Introduction

Let Γ be a Fuchsian group acting on the upper half plane U and covering a compact Riemann surface U/Γ. It is well known that the space $B_2(\Gamma)$ of of holomorphic, quadratic differentials for Γ can be identified with projective structures on U/Γ. Moreover, there is a canonical embedding of the Teichmüller space $T(\Gamma)$ as a bounded domain in $B_2(\Gamma)$. In this paper we will show that, for U/Γ a compact surface of genus 2, the maximal cusps on the boundary $\partial T(\Gamma)$ can be approximated by differentials corresponding to Schottky structures. It then follows from a recent result of McMullen [11] that all points of $\partial T(\Gamma)$ can be approximated by Schottky structures.

Starting with a maximal cusp $\psi \in \partial T(\Gamma)$, we will construct a convergent sequence $\varphi_n \in B_2(\Gamma)$ of Schottky structures on U/Γ, $\varphi_n \to \varphi$; the sequence φ_n is determined by the topological dissection of U/Γ associated to ψ. We then show that φ corresponds to a torsion-free Kleinian group which uniformizes a compact surface of genus 2 and contains three distinct, maximal parabolic subgroups; this implies that φ corresponds to a maximal cusp. Since the maximal cusp associated to a fixed dissection of U/Γ is unique, we then obtain that $\varphi = \psi$.

In Section 2 we establish preliminaries and state the main result. Sections 3 and 4 are devoted, respectively, to treating the two topological classes of maximal dissections encountered in genus 2. At the end of Section 4 we give a formal proof of the main result.

1991 *Mathematics Subject Classification.* Primary 32G15; Secondary 30F10.
This paper is in final form and no version of it will be submitted for publication elsewhere.

2. Preliminaries

2.1. Let Γ be a torsion free Fuchsian group acting on the upper half plane U, with U/Γ a compact Riemann surface. The set $B_2(\Gamma)$ of holomorphic, quadratic differentials defined on U for Γ is a vector space of complex dimension $3g-3$, where g is the genus of U/Γ. $B_2(\Gamma)$ is given the norm $\|\varphi\| = sup_{z \in U}\{|\varphi(z)|(2Im\ z)^2\}$, for $\varphi \in B_2(\Gamma)$.

Starting with $\varphi \in B_2(\Gamma)$, one associates a pair $(f_\varphi, \chi_\varphi)$, called a *projective structure* on U/Γ. Here, $f_\varphi : U \to \hat{\mathbf{C}}$ is a meromorphic, local homeomorphism normalized by the conditions $f_\varphi(i) = 0$, $f'_\varphi(i) = 1$, $f''_\varphi(i) = 0$, and $\chi_\varphi : \Gamma \to PSL = PSL(2, \mathbf{C})$ is a homomorphism which satisfies $f_\varphi \circ \gamma(z) = \chi_\varphi(\gamma) \circ f_\varphi(z)$, for all $z \in U$, $\gamma \in \Gamma$. Moreover, $\varphi = S(f_\varphi)$, where $S(f)$ is the *Schwarzian derivative* given by

$$S(f) = (f''/f')' - \frac{1}{2}(f''/f')^2.$$

Conversely, if (f, χ) is a pair satisfying the conditions above, $S(f) \in B_2(\Gamma)$; this establishes a bijective correspondence between normalized projective structures and $B_2(\Gamma)$ (see [8] or [10] for details). In this paper, we will work exclusively with the subset

$$C_2(\Gamma) = \{\varphi \in B_2(\Gamma) : f_\varphi \text{ is a covering map}\}.$$

It is well known that, for $\varphi \in C_2(\Gamma)$, $\chi_\varphi(\Gamma)$ is a Kleinian group with $f_\varphi(U)$ an invariant component (see [7]).

We will need the following fundamental results:

THEOREM 1(KRA-MASKIT [10]). $C_2(\Gamma)$ *is a closed and bounded subset of* $B_2(\Gamma)$.

THEOREM 2 (JORGENSEN-KLEIN [7]). *Let* $\varphi_n \in C_2(\Gamma)$ *with* $\varphi_n \to \varphi$, *and define* $\theta_n : \chi_\varphi(\Gamma) \to \chi_{\varphi_n}(\Gamma)$ *by* $\theta_n(\chi_\varphi(\gamma)) = \chi_{\varphi_n}(\gamma)$, *for* $\gamma \in \Gamma$. *Then* θ_n *is a homomorphism for almost all* n. *In particular,* $ker\ \chi_\varphi \subset ker\ \chi_{\varphi_n}$ *for almost all* n.

Remark. Theorem 2 is valid in a more general context, see [6] or [9] for details.

2.2. The *Teichmüller space* $T(\Gamma) \subset C_2(\Gamma)$ consists of those $\varphi \in C_2(\Gamma)$ for which χ_φ is an isomorphism onto a quasifuchsian group. In this case, f_φ is a homeomorphism onto one of the invariant components of $\chi_\varphi(\Gamma)$. It is a basic result of the subject (Nehari [13]) that $T(\Gamma)$ is a bounded subset of $B_2(\Gamma)$. In a seminal paper, Bers [2] showed that $T(\Gamma)$ is an open subset of $B_2(\Gamma)$. Let $\partial T(\Gamma)$ be the boundary of $T(\Gamma)$ relative to $B_2(\Gamma)$. Necessarily, $\partial T(\Gamma)$ consists of differentials $\varphi \in C_2(\Gamma)$ for which f_φ is a homeomorphism and χ_φ is an isomorphism, see Bers [4].

A differential $\varphi \in \partial T(\Gamma)$ is a *cusp* if there exists a hyperbolic transformation $\gamma \in \Gamma$ for which $\chi_\varphi(\gamma)$ is parabolic; the element $\chi_\varphi(\gamma)$ is called an *accidental* parabolic. A cusp φ is called *maximal* if there exists a maximal collection α_i, $i = 1, ..., 3g - 3$, of non-homotopic, disjoint, simple closed curves in U/Γ, called a *maximal dissection* of U/Γ, each of which is represented by a hyperbolic element γ_i with $\chi_\varphi(\gamma_i)$ parabolic. An element $\gamma \in \Gamma$ *represents* a curve α if the axis of γ covers a curve which is homotopic to α.

A finitely generated Kleinian group is *geometrically finite* if it has a finite sided fundamental polyhedron for its action on $\mathbf{H}^3$.

THEOREM 3 (ABIKOFF [1]). *Let $\varphi \in B_2(\Gamma)$ be such that χ_φ is an isomorphism with f_φ a homeomorphism; further, suppose $\chi_\varphi(\Gamma)$ is geometrically finite and contains accidental parabolics. Then $\varphi \in \partial T(\Gamma)$.*

THEOREM 4 (MCMULLEN [11]). *Maximal cusps are dense in $\partial T(\Gamma)$.*

We note that maximal cusps are uniquely determined by their corresponding maximal dissections (see Maskit [12])..

2.3. Let $C_1, C'_1, ..., C_n, C'_n$, $n > 1$, be a collection of mutually disjoint Jordan curves contained in the Riemann sphere $\hat{\mathbf{C}}$ which bound a connected region F; let $\gamma_i \in PSL$ be a loxodromic transformation which maps the interior of C_i (that region not contained in F) onto the exterior of C'_i, $i = 1, ..., n$. Then $S = \langle \gamma_1, ..., \gamma_n \rangle$ is a purely loxodromic, free group which is Kleinian. The region of discontinuity of S is $\Omega(S) = \cup_{\gamma \in S} \gamma(F) \subset \hat{\mathbf{C}}$, and S is called a *Schottky* group.

For α, β, two elements in a group, we will write $\beta\alpha$ to mean α followed by β. The following is a classical result (see [3]):

Let $A_1, B_1, ..., A_g, B_g$ be canonical generators for Γ; that is, $\prod_{i=1}^g [A_i, B_i] = id$, where $[A_i, B_i] = B_i^{-1} A_i^{-1} B_i A_i$. Let $N = \langle\langle A_1, ..., A_g \rangle\rangle \subset \Gamma$ be the normal subgroup of Γ generated by the elements A_i, $i = 1, ..., g$. Then Γ/N is canonically isomorphic to a Schottky group S with the following properties:

i) $\Omega(S)/S = U/\Gamma$, and

ii) there exists a normalized meromorphic covering map $f : U \to \hat{\mathbf{C}}$ such that $f \circ \gamma = \chi(\gamma) \circ f$, for all $\gamma \in \Gamma$, where $\chi : \Gamma \to \Gamma/N = S$ is the natural homomorphism, and

iii) $\chi(B_i)$, $i = 1, ..., g$, are free generators for S.

We will prove:

THEOREM 5. *Let $\varphi \in \partial T(\Gamma)$, where the genus of U/Γ is 2. Then there exists a sequence $\varphi_n \in C_2(\Gamma)$, with $\chi_{\varphi_n}(\Gamma)$ a Schottky group and $\varphi_n \to \varphi$.*

In view of Theorem 4, the proof of Theorem 5 will be complete once we show that every maximal cusp can be approximated by Schottky structures. (We include a formal proof of Theorem 5 at the end of Section 4.) There are two types of maximal dissections in genus 2: those that contain a dividing curve, and those that don't. We will treat these two cases separately; in Section 3, we will show that cusps defined by maximal dissections without a dividing curve can be approximated by Schottky structures (Proposition 3.10). In Section 4, we solve the corresponding problem for maximal dissections with a dividing curve (Proposition 4.16).

3. Non-dividing maximal cusps

3.1. From now on, Γ will denote a Fuchsian group acting on the upper half plane U and covering a compact surface of genus 2.

Starting with a non-dividing maximal cusp, we will apply iterated Dehn twists to the surface along the defining curves of the corresponding dissection. This process will allow us, using the induced map on Γ, to extract a convergent sequence of Schottky structures on U/Γ. We use Theorem 2 to discern several algebraic properties of the limiting structure (Lemmas 3.1-4). We then note a property of certain convergent sequences in $SL(2, \mathbf{C})$ (Lemma 3.6) that allows us to prove that the images of the representatives of the defining curves become parabolic in the limit group (Lemma 3.7). It then follows from structural properties of Kleinian groups that the limiting structure corresponds to the original cusp (Lemma 3.9).

Let $\psi_1 \in \partial T(\Gamma)$ be a maximal cusp, and assume the corresponding maximal dissection contains no dividing curve. Let b_1, b_2, b_3 be the curves of the dissection, Figure 1.

It is well known that we may choose B_1, $B_2 \in \Gamma$ hyperbolic elements representing b_1, b_2, respectively, and A_1, $A_2 \in \Gamma$ so that $\{A_1, B_1, A_2, B_2\}$ is a canonical set of generators for Γ, with $W = C_2 B_1^{-1}$ representing b_3. Here, $C_i = A_i^{-1} B_i A_i$, $i = 1, 2$. See Figure 1, where we have identified A_i, B_i, $i = 1, 2$, with elements of $\pi_1(U/\Gamma, z_0)$, for $z_0 \in U/\Gamma$.

For each integer $n \geq 0$, we consider the canonical set of generators for Γ given by $\{W^{-n} B_1^n A_1,\ W^{-n} B_1 W^n,\ A_2 C_2^n W^n,\ B_2\}$.

It is straightforward to verify that $A_{(1,n)} = W^{-n} B_1^n A_1$, $B_{(1,n)} = W^{-n} B_1 W^n$, $A_{(2,n)} = A_2 C_2^n W^n$, and $B_{(2,n)} = B_2$ satisfy the commutator relation. To see that they generate Γ, let $C_{(2,n)} = A_{(2,n)}^{-1} B_{(2,n)} A_{(2,n)}$ and note that $W = C_{(2,n)} B_{(1,n)}^{-1}$, with

$$A_1 = W^n B_{(1,n)}^{-n} A_{(1,n)},$$

$$B_1 = W^n B_{(1,n)} W^{-n},$$

$$A_2 = A_{(2,n)} C_{(2,n)}^{-n} W^{-n},$$

$$B_2 = B_{(2,n)}.$$

Remark. These generators are obtained from the canonical set of generators $\{A_1, B_1, A_2, B_2\}$ by applying Dehn twists of order n along the curves b_1, b_2, b_3 (in order) and taking the induced map on Γ (see Figure 1).

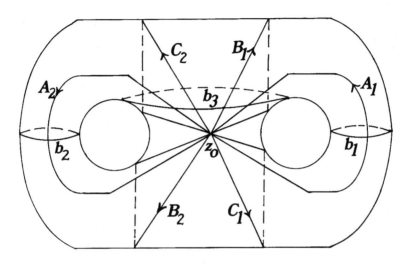

Figure 1

Let $N_n = \; << W^{-n}B_1^n A_1, \; A_2 C_2^n W^n >>$. By our previous observations, $\chi_n(\Gamma) = \Gamma/N_n$ is a Schottky group, with the corresponding set of free generators $\{\chi_n(W^{-n}B_1 W^n), \; \chi_n(B_2)\}$. Moreover, there exists a normalized, meromorphic covering map $h_n : U \to \hat{\mathbf{C}}$ so that (h_n, χ_n) is a projective structure on U/Γ. Let $S(h_n) = \varphi_n \in C_2(\Gamma)$.

Applying Theorem 1, we choose a convergent subsequence of φ_n, which we will also denote φ_n. Let $\lim \varphi_n = \varphi$. Note that $h_n \to f_\varphi$ uniformly on compact subsets of U, and $\chi_n \to \chi_\varphi$ (see [7]). Since $\varphi \in C_2(\Gamma)$, $\chi_\varphi(\Gamma)$ is a Kleinian group with $f_\varphi(U)$ an invariant component. The structures (h_n, χ_n) and $(f_\varphi, \chi_\varphi)$ will remain fixed for the remainder of this section. We will show, Lemma 3.9, that $\varphi = \psi_1$.

LEMMA 3.1. $\chi_\varphi(\Gamma)$ *has no elliptic elements.*

PROOF. Let $\gamma \in \Gamma$ and suppose $(\chi_\varphi(\gamma))^m = id.$ for some integer $m > 0$. By Theorem 2, $(\chi_n(\gamma))^m = id.$ for almost all n. But, since $\chi_n(\Gamma)$ is a Schottky group, which contains no elliptic elements, $\chi_n(\gamma) = id$, for almost all n. It follows that $\chi_\varphi(\gamma) = id$, and $\chi_\varphi(\Gamma)$ has no elliptic elements.

LEMMA 3.2. *The elements* $\chi_\varphi(B_1)$, $\chi_\varphi(B_2)$, *and* $\chi_\varphi(W)$ *are nontrivial.*

PROOF. We observe that, for each n, $\chi_n(B_1)$ is a conjugate of a free generator for $\chi_n(\Gamma)$; hence, $\chi_n(B_1) \neq id.$ It follows from Theorem 2 that $\chi_\varphi(B_1) \neq id.$ Similarly, one shows $\chi_\varphi(B_2) \neq id.$

Denote by G_n the free abelian group on the generators $[\chi_n(B_1)]$, $[\chi_n(B_2)]$ obtained by taking $\chi_n(\Gamma)$ modulo its commutator subgroup. Now,

$$\chi_n(W) = \chi_n(A_2^{-1})\chi_n(B_2)\chi_n(A_2)\chi_n(B_1^{-1}).$$

Hence, in G_n, $[\chi_n(W)] = [\chi_n(B_2)][\chi_n(B_1^{-1})] \neq id$. We conclude that $\chi_n(W) \neq id$, and it follows from Theorem 2 that $\chi_\varphi(W) \neq id$.

LEMMA 3.3. $\chi_\varphi(W)$ does not commute with any conjugate of $\chi_\varphi(B_1)$ or of $\chi_\varphi(B_2)$.

PROOF. Suppose that there exists $\gamma \in \Gamma$ such that $\chi_\varphi(W)$ commutes with $\chi_\varphi(\gamma)\chi_\varphi(B_1)\chi_\varphi(\gamma^{-1})$. Let $V = \gamma B_1 \gamma^{-1}$. It follows from Theorem 2 that $\chi_n(W)$ commutes with $\chi_n(V)$ for almost all n. Since neither $\chi_n(V)$ nor $\chi_n(W)$ is trivial, and $\chi_n(\Gamma)$ is a purely loxodromic group, there exist integers $p, q \neq 0$ such that $(\chi_n(W))^p(\chi_n(V))^q = id$. Passing to G_n, we have $[\chi_n(W)]^p[\chi_n(V)]^q = id$. Hence, $([\chi_n(B_2)][\chi_n(B_1)])^p[\chi_n(B_1)]^q = id$. This is a contradiction, since $[\chi_n(B_1)], [\chi_n(B_2)]$ are free generators for G_n. A similar argument shows that $\chi_\varphi(W)$ does not commute with any conjugate of $\chi_\varphi(B_2)$.

The proof of the next lemma is similar to the proof of Lemma 3.3 and will be omitted.

LEMMA 3.4. $\chi_\varphi(B_2)$ does not commute with any conjugate of $\chi_\varphi(B_1)$.

3.2. We will denote by $SL = SL(2, \mathbf{C})$ the group of 2×2 complex matrices with determinant 1. Let $p : SL \to PSL$ be the canonical twofold cover.

It is a basic result of the subject (see Gunning [5]) that, for (f, χ) a projective structure on U/Γ, the homomorphism χ *lifts* to SL in the following sense: there exists a homomorphism $\chi' : \Gamma \to SL$ which completes the commutative diagram

$$
\begin{array}{ccc}
\Gamma & \xrightarrow{\;\chi'\;} & SL \\
{\scriptstyle Id.}\downarrow & & \downarrow{\scriptstyle p} \\
\Gamma & \xrightarrow{\;\chi\;} & PSL,
\end{array}
$$

where $Id. : \Gamma \to \Gamma$ is the identity map.

We will need the following:

LEMMA 3.5. *There exist lifts, $\chi_n' : \Gamma \to SL$ and $\chi_\varphi' : \Gamma \to SL$, of χ_n and χ_φ, respectively, such that $\chi_n' \to \chi_\varphi'$.*

PROOF. Choose $\chi_\varphi' : \Gamma \to SL$ a lift of χ_φ, and let $D_j \in \Gamma$, $j = 1, ..., 4$, be canonical generators for Γ satisfying the single relation $D_4^{-1}D_3^{-1}D_4D_3D_2^{-1}D_1^{-1}D_2D_1 = id$. Let V_j and E be mutually disjoint neighborhoods of $\chi_\varphi'(D_j)$ and id, with the properties $V_j \cap -V_j = \emptyset$, $E \cap -E = \emptyset$. Now define $\prod : SL^4 \to SL$ by

$$\prod(a_1, ..., a_4) = a_4^{-1}a_3^{-1}a_4a_3a_2^{-1}a_1^{-1}a_2a_1,$$

and let
$$\hat{W} = (V_1 \times \cdots \times V_4) \bigcap {\prod}^{-1}(E).$$

Note that $\hat{W} \subset SL^4$ is an open set with $(\chi'_\varphi(D_1), ..., \chi'_\varphi(D_4)) \in \hat{W}$. Let $\pi_j : SL^4 \to SL$, be the projection on the j-th factor. Then $\chi'_\varphi(D_j) \in \pi_j(\hat{W}) \subset V_j$. Choose N an integer so that $n \geq N$ implies $\chi_n(D_j) \in p(\pi_j(\hat{W}))$ for all j. Then, for $n \geq N$, define $\chi'_n(D_j)$ to be the unique element of $\pi_j(\hat{W})$ for which $p(\chi'_n(D_j)) = \chi_n(D_j)$. Thus, $\prod(\chi'_n(D_1), ..., \chi'_n(D_4)) = \pm id$. Now, $(\chi'_n(D_1), ..., \chi'_n(D_4)) \in \hat{W}$, and $\prod(\hat{W}) \subset E$. Since $E \cap -E = \emptyset$, we must have $\prod(\chi'_n(D_1), ..., \chi'_n(D_4)) = id$. It follows that we may extend χ'_n uniquely to a homomorphism $\chi'_n : \Gamma \to SL$. Clearly, χ'_n is a lift of χ_n. For $n < N$, let χ'_n be an arbitrary lift of χ_n. Then, since $\chi_n \to \chi_\varphi$, we have $\chi'_n \to \chi'_\varphi$. This concludes the proof of the lemma.

We define the following subsets of $\mathbf{C}$:
$$U_1 = \{z \in \mathbf{C} : x > 0, \ 0 \leq y < \pi\},$$
$$U_2 = \{z \in \mathbf{C} : x \geq 0, \ 0 \leq y < 2\pi\},$$
for $z = x + iy$.

As is well known, given $X, Y \in SL$ loxodromic elements sharing no fixed points, there is a transformation $T \in SL$ (unique up to sign) such that
$$X' = T^{-1}XT = \begin{pmatrix} \cosh \alpha & \sinh \alpha \\ \sinh \alpha & \cosh \alpha \end{pmatrix},$$
and
$$Y' = T^{-1}YT = \begin{pmatrix} \cosh \gamma & e^\beta \sinh \gamma \\ e^{-\beta} \sinh \gamma & \cosh \gamma \end{pmatrix}.$$

Here, $\alpha, \gamma \in U_1$, and $\beta \in U_2$ are uniquely determined by X, Y. The fixed points of Y' (resp. X') correspond to $\pm e^\beta$ (resp. ± 1). Since X, Y share no fixed points, $Im \ \beta \neq 0, \pi$ whenever $Re \ \beta = 0$; in particular, $\cosh \beta \neq -1$.

Now, if $X_n, Y_n \in SL$ are sequences of loxodromic transformations with $X_n \to X, Y_n \to Y$, let $T_n \in SL$ be a transformation for which
$$X'_n = T_n^{-1}X_nT_n = \begin{pmatrix} \cosh \alpha_n & \sinh \alpha_n \\ \sinh \alpha_n & \cosh \alpha_n \end{pmatrix},$$
and
$$Y'_n = T_n^{-1}Y_nT_n = \begin{pmatrix} \cosh \gamma_n & e^{\beta_n} \sinh \gamma_n \\ e^{-\beta_n} \sinh \gamma_n & \cosh \gamma_n \end{pmatrix},$$
$\alpha_n, \gamma_n \in U_1, \beta_n \in U_2$. Since the entries of T_n may be chosen continuously in a neighborhood of $(X, Y) \in SL \times SL$, one may also choose the signs of T_n so that $T_n \to T$. Hence, $X'_n \to X'$, and $Y'_n \to Y'$. One then verifies easily that $\alpha_n \to \alpha$, $\gamma_n \to \gamma$, and $\beta_n \to \beta$.

If X is a loxodromic transformation sharing no fixed points with the parabolic transformation Y, then there is a transformation $T \in SL$ (unique up to sign) such that

$$X' = T^{-1}XT = \begin{pmatrix} \cosh \alpha & \sinh \alpha \\ \sinh \alpha & \cosh \alpha \end{pmatrix},$$

and

$$Y' = T^{-1}YT = \begin{pmatrix} \epsilon & \gamma \\ 0 & \epsilon \end{pmatrix}.$$

Here, $\alpha \in U_1$, $\gamma \in \mathbf{C} - \{0\}$, and $\epsilon = \pm 1$ are uniquely determined by X, Y. Now if X_n, $Y_n \in SL$ are sequences of loxodromic transformations with $X_n \to X$, $Y_n \to Y$, let $T_n \in SL$ be a transformation for which

$$X_n' = T_n^{-1}X_nT_n = \begin{pmatrix} \cosh \alpha_n & \sinh \alpha_n \\ \sinh \alpha_n & \cosh \alpha_n \end{pmatrix},$$

and

$$Y_n' = T_n^{-1}Y_nT_n = \begin{pmatrix} \cosh \gamma_n & e^{\beta_n} \sinh \gamma_n \\ e^{-\beta_n} \sinh \gamma_n & \cosh \gamma_n \end{pmatrix},$$

α_n, $\gamma_n \in U_1$, $\beta_n \in U_2$.

Again, one may choose the transformations T_n so that $T_n \to T$. It follows that $X_n' \to X'$, $Y_n' \to Y'$. One verifies easily that $\alpha_n \to \alpha$, $\cosh \gamma_n \to \epsilon$, and $e^{\beta_n} \sinh \gamma_n \to \gamma$, with $\mathrm{Re}\, \beta_n \to \infty$.

We will use these normalizations repeatedly throughout the text.

LEMMA 3.6. *Let $X_n, Y_n \in SL$ be loxodromic transformations and suppose $X_n \to X, Y_n \to Y$, where X, Y are neither elliptic nor trivial. Furthermore, suppose X, Y share no fixed points and $X_n^n Y_n^n$ converges. Then X, Y are parabolic.*

PROOF. We will assume at least one of X, Y is not parabolic in order to obtain a contradiction.

Case 1. X is loxodromic and Y parabolic.

Choose $T \in SL$ such that

$$X' = T^{-1}XT = \begin{pmatrix} \cosh \alpha & \sinh \alpha \\ \sinh \alpha & \cosh \alpha \end{pmatrix},$$

and

$$Y' = T^{-1}YT = \begin{pmatrix} \epsilon & \gamma \\ 0 & \epsilon \end{pmatrix}.$$

Here, $\alpha \in U_1$, $\gamma \in \mathbf{C} - \{0\}$, and $\epsilon = \pm 1$. Then choose a convergent sequence $T_n \in SL$ with $T_n \to T$ so that

(3.1) $$X_n' = T_n^{-1}X_nT_n = \begin{pmatrix} \cosh \alpha_n & \sinh \alpha_n \\ \sinh \alpha_n & \cosh \alpha_n \end{pmatrix},$$

and

$$(3.2) \qquad Y'_n = T_n^{-1} Y_n T_n = \begin{pmatrix} \cosh \gamma_n & e^{\beta_n} \sinh \gamma_n \\ e^{-\beta_n} \sinh \gamma_n & \cosh \gamma_n \end{pmatrix}.$$

Here, $X'_n \to X'$, $Y'_n \to Y'$ so that $\alpha_n \to \alpha$, $\cosh \gamma_n \to \epsilon$, $Re\ \beta_n \to \infty$, and $e^{\beta_n} \sinh \gamma_n \to \gamma$.

Let

$$(X'_n)^n (Y'_n)^n = \begin{pmatrix} a_n & b_n \\ c_n & d_n \end{pmatrix}.$$

Since $(X'_n)^n$ (resp.$(Y'_n)^n$) may be obtained from X'_n (resp. Y'_n) by replacing α_n (resp. γ_n) by $n\alpha_n$ (resp. $n\gamma_n$) in (3.1) (resp. (3.2)), one has

$$a_n = \cosh n\alpha_n \cosh n\gamma_n + e^{-\beta_n} \sinh n\alpha_n \sinh n\gamma_n,$$

and

$$d_n = \cosh n\alpha_n \cosh n\gamma_n + e^{\beta_n} \sinh n\alpha_n \sinh n\gamma_n.$$

There are two possibilities:

i) $\cosh n\alpha_n \cosh n\gamma_n$ (or some subsequence) converges. In this case, since $Re\ n\alpha_n \to \infty$, $\cosh n\alpha_n \to \infty$, and we must have $\cosh n\gamma_n \to 0$. We can now deduce that $\sinh n\gamma_n \to \pm i$. Thus, since $e^{\beta_n} \to \infty$ and $\sinh n\alpha_n \to \infty$,

$$\lim d_n = \lim \cosh n\alpha_n \cosh n\gamma_n + \lim e^{\beta_n} \sinh n\alpha_n \sinh n\gamma_n = \infty.$$

ii) $\cosh n\alpha_n \cosh n\gamma_n \to \infty$. We write

$$a_n = \cosh n\alpha_n \cosh n\gamma_n (1 + e^{-\beta_n} \tanh n\alpha_n \tanh n\gamma_n),$$

and

$$d_n = \cosh n\alpha_n \cosh n\gamma_n (1 + e^{\beta_n} \tanh n\alpha_n \tanh n\gamma_n).$$

Note that since $Re\ n\alpha_n \to \infty$, $\tanh n\alpha_n \to 1$. Hence, if $\tanh n\gamma_n$ (or some subsequence, also denoted by $\tanh n\gamma_n$) converges, $a_n \to \infty$; if $\tanh n\gamma_n \to \infty$, $d_n \to \infty$.

For both possibilities, we have shown that $(X'_n)^n (Y'_n)^n$ cannot converge; hence, $X_n^n Y_n^n$ cannot converge either, contradicting our hypothesis.

Case 2. X is parabolic and Y loxodromic.

Interchanging the roles of X and Y in the normalizations of the previous case, one reaches a contradiction after a parallel calculation. (In obtaining the contradiction, the roles of a_n and d_n are interchanged.) We leave the verification to the reader.

Case 3. Both X and Y are loxodromic.

Choose $T \in SL$ such that

$$X' = T^{-1} X T = \begin{pmatrix} \cosh \alpha & \sinh \alpha \\ \sinh \alpha & \cosh \alpha \end{pmatrix},$$

and

$$Y' = T^{-1}YT = \begin{pmatrix} \cosh\gamma & e^\beta \sinh\gamma \\ e^{-\beta} \sinh\gamma & \cosh\gamma \end{pmatrix}.$$

Here, α, $\gamma \in U_1$, and $\beta \in U_2$, with $\cosh\beta \neq -1$.

Then chose $T_n \in SL$ with $T_n \to T$ so that

$$X'_n = T_n^{-1}X_nT_n = \begin{pmatrix} \cosh\alpha_n & \sinh\alpha_n \\ \sinh\alpha_n & \cosh\alpha_n \end{pmatrix},$$

and

$$Y'_n = T_n^{-1}Y_nT_n = \begin{pmatrix} \cosh\gamma_n & e^{\beta_n} \sinh\gamma_n \\ e^{-\beta_n} \sinh\gamma_n & \cosh\gamma_n \end{pmatrix}.$$

Here, $\alpha_n \to \alpha$, $\gamma_n \to \gamma$, and $\beta_n \to \beta$.

Thus,

$$tr(X'_n)^n(Y'_n)^n = 2(\cosh n\alpha_n \cosh n\gamma_n + \cosh\beta_n \sinh n\alpha_n \sinh n\gamma_n)$$
$$= 2\cosh n\alpha_n \cosh n\gamma_n(1 + \cosh\beta_n \tanh n\alpha_n \tanh n\gamma_n).$$

Since both $Re\, n\alpha_n \to \infty$ and $Re\, n\gamma_n \to \infty$, $\tanh n\alpha_n \to 1$ and $\tanh n\gamma_n \to 1$. Hence,

$$\lim tr(X'_n)^n(Y'_n)^n = \lim \cosh n\alpha_n \cosh n\gamma_n(1 + \cosh\beta).$$

But $\cosh\beta \neq -1$, and $\lim tr(X'_n)^n(Y_n')^n = \infty$. It follows that $(X'_n)^n(Y'_n)^n$ cannot converge; hence, $X_n^nY_n^n$ cannot converge either, contradicting our hypothesis.

In all three cases, we have reached a contradiction. Hence, both X and Y are parabolic.

LEMMA 3.7. *The elements $\chi_\varphi(W)$, $\chi_\varphi(B_1)$, $\chi_\varphi(B_2)$ are parabolic.*

PROOF. For all n, one has that $W^{-n}B_1^nA_1 \in N_n$. Consequently, one also has $\chi_n(W^{-n}B_1)\chi_n(A_1) = id$. Hence, $\chi'_n(W^{-n}B_1)\chi'_n(A_1) = \pm id$, and $\chi'_n(W^{-n}B_1) = \pm\chi'_n(A_1^{-1})$. It follows, since $\chi'_n(A_1^{-1}) \to \chi'_\varphi(A_1^{-1})$, that we may choose a subsequence of $\pm\chi'_n(A_1^{-1})$ (also denoted by $\pm\chi'_n(A_1^{-1})$) which converges. Hence, $\chi'_n(W^{-n})\chi'_n(B_1^n)$ converges. We will apply Lemma 3.6 to $\chi'_n(W^{-n})\chi'_n(B_1^n)$.

Lemmas 3.1 and 3.2 imply that neither $\chi'_\varphi(W^{-1})$ nor $\chi'_\varphi(B_1)$ is trivial or elliptic; Lemma 3.3 implies that they share no fixed points. Thus, since

$$\chi'_n(W^{-1}) \to \chi'_\varphi(W^{-1})$$

and

$$\chi'_n(B_1) \to \chi'_\varphi(B_1),$$

the elements $\chi'_n(W^{-1})$, $\chi'_n(B_1)$ satisfy the hypothesis of Lemma 3.6, and we deduce that $\chi'_\varphi(W^{-1})$ and $\chi'_\varphi(B_1)$ are parabolic. Hence, $\chi_\varphi(W)$ and $\chi_\varphi(B_1)$ are parabolic, also.

Since $A_2C_2^nW^n \in N_n$ for all n, we deduce by similar reasoning that $\chi_\varphi(C_2)$ is parabolic. It follows that its conjugate $\chi_\varphi(B_2)$ is also parabolic.

COROLLARY 3.8. $\chi_\varphi(W)$, $\chi_\varphi(B_1)$, and $\chi_\varphi(B_2)$ belong to distinct, non-conjugate, maximal parabolic subgroups of $\chi_\varphi(\Gamma)$.

PROOF. Since the only parabolic subgroups of a Kleinian group are either cyclic or rank 2 abelian, Lemmas 3.3, 3.4, 3.6, and 3.7 imply the desired result.

LEMMA 3.9. $\varphi = \psi_1$

PROOF. We consider the standard commutative diagram for the projective structure $(f_\varphi, \chi_\varphi)$:

$$
\begin{array}{ccc}
U & \xrightarrow{\ f_\varphi\ } & f_\varphi(U) \\
p_1 \downarrow & & \downarrow p_2 \\
U/\Gamma & \xrightarrow{\ g_\varphi\ } & f_\varphi(U)/\chi_\varphi(\Gamma),
\end{array}
$$

where p_1, p_2 are the natural projections and g_φ is the map induced by f_φ. Since $\chi_\varphi(\Gamma)$ has no elliptic elements, p_2 is a covering map. Hence, g_φ is a covering map, and $f_\varphi(U)/\chi_\varphi(\Gamma)$ is covered by U/Γ, a compact surface of genus 2. It follows that $f_\varphi(U)/\chi_\varphi(\Gamma)$ also has genus 2.

From the algebraic classification of torsion-free Kleinian groups with an invariant component, see Maskit [12], we then have the following four possibilities:

i) $\chi_\varphi(\Gamma) = < \gamma_1 > \times < \gamma_2 >$, where γ_1, γ_2 are loxodromic transformations, and $\chi_\varphi(\Gamma)$ is a Schottky group, or

ii) $\chi_\varphi(\Gamma) = \Gamma_1 \times < \gamma >$, where γ is a loxodromic transformation, and Γ_1 is a rank 2 parabolic group, or

iii) $\chi_\varphi(\Gamma) = \Gamma_1 \times \Gamma_2$, where Γ_1, Γ_2 are rank 2 parabolic groups, or

iv) $\chi_\varphi(\Gamma)$ is isomorphic to the covering group of a compact surface of genus 2.

Each of the first three possibilities is eliminated since it does not contain three non-conjugate, maximal parabolic subgroups. It follows that $\chi_\varphi(\Gamma)$ must satisfy iv).

Thus, $\chi_\varphi : \Gamma \to PSL$ is an isomorphism. It then follows, since f_φ is a covering, that f_φ is a homeomorphism, see Kra [9]. Moreover, $\chi_\varphi(\Gamma)$ contains parabolic elements, and these must be accidental. Since χ_φ is an isomorphism, $\chi_\varphi(B_1)$, $\chi_\varphi(B_2)$, and $\chi_\varphi(W)$ are generators for the respective maximal parabolic groups containing them. Thus, b_1, b_2, b_3 are the curves of a maximal dissection corresponding to φ. It follows easily from the results of [12] that $\chi_\varphi(\Gamma)$ is geometrically finite. From Theorem 3, we deduce that $\varphi = S(f_\varphi)$ is a cusp. But maximal cusps are determined by their corresponding dissections, and $\varphi = \psi_1$.

We gather the results of this section in the

PROPOSITION 3.10. Let $\psi_1 \in \partial T(\Gamma)$ be a maximal cusp and suppose the associated dissection of U/Γ contains no dividing curve. Then there exists a sequence $\varphi_n \in C_2(\Gamma)$, with $\chi_{\varphi_n}(\Gamma)$ a Schottky group and $\varphi_n \to \psi_1$.

4. Dividing maximal cusps

4.1. The overall method we will apply is similar to the previous case. Starting with a dividing maximal cusp we apply Dehn twists to the surface in order to obtain a convergent sequence of Schottky structures. Theorem 2 is then used to obtain algebraic properties of the limiting structure (Lemmas 4.1-4). We then note certain properties of convergent sequences in SL (Lemmas 4.5-7 and 4.10-11) that allow us to prove the existence of accidental parabolics in the limit group (Lemmas 4.9 and 4.12-13). As in the previous section, it follows from structural properties of Kleinian groups that the limiting structure corresponds to the original cusp (Lemma 4.15).

Let $\psi_2 \in \partial T(\Gamma)$ be a maximal cusp, and assume the corresponding maximal dissection contains a dividing curve, say b_3. Let b_1, b_2 be the remaining curves of the dissection, Figure 2.

It is well known that we may choose B_1, $B_2 \in \Gamma$ hyperbolic elements representing b_1, b_2, respectively, and A_1, $A_2 \in \Gamma$ so that $\{A_1, B_1, A_2, B_2\}$ is a canonical set of generators for Γ, where $W = B_1^{-1}A_1^{-1}B_1A_1$ represents b_3. Define $C_i = A_i^{-1}B_iA_i$, $i = 1, 2$ and $V_n = C_2W^{-n}B_1^{-1}W^n$. See Figure 2, where we have identified A_i, B_i, $i = 1, 2$, with elements of $\pi_1(U/\Gamma, z_0)$, for $z_0 \in U/\Gamma$.

For each integer $n \geq 0$, we consider the canonical set of generators
$$\{C_2W^{-n}B_1^nA_1W^n, V_nW^{-n}B_1W^nV_n^{-1}, A_2C_2^nW^{-n}B_1W^nC_2^{-1}, B_2\}.$$

It is straightforward to verify that $A_{(1,n)} = C_2W^{-n}B_1^nA_1W^n$, $B_{(1,n)} = V_nW^{-n}B_1W^nV_n^{-1}$, $A_{(2,n)} = A_2C_2^nW^{-n}B_1^{-1}W^nC_2^{-1}$, $B_{(2,n)} = B_2$ satisfy the commutator relation. To see that they generate Γ, let

$$C_{(i,n)} = A_{(i,n)}^{-1}B_{(i,n)}A_{(i,n)}, \ i = 1, 2,$$

and $D_n = C_{(2,n)}B_{(1,n)}^{-1}$. Note that $W = D_n^{-1}B_{(1,n)}^{-1}D_nC_{(1,n)}$, and

$$A_1 = W^nD_n^{-1}B_{(1,n)}^{-(n+1)}A_{(1,n)}W^{-n},$$

$$B_1 = W^nD_n^{-1}B_{(1,n)}D_nW^{-n},$$

$$A_2 = A_{(2,n)}C_{(2,n)}^{-n}D_n,$$

$$B_2 = B_{(2,n)}.$$

Remark. These generators are obtained from the canonical set of generators $\{A_1, B_1, A_2, B_2\}$ by applying a Dehn twist along a curve homotopic to $C_2 B_1^{-1}$ and then Dehn twists of order $n+1$, n, and n, respectively, about the curves b_1, b_2, b_3 (in order) and taking the induced map on Γ (see Figure 2).

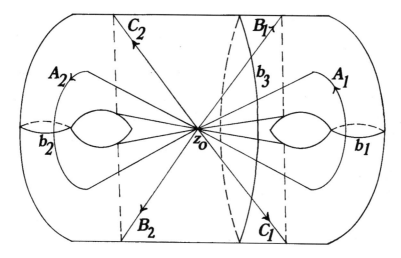

Figure 2

Let $N_n = \ll C_2 W^{-n} B_1^n A_1 W^n$, $A_2 C_2^n W^{-n} B_1 W^n C_2^{-1} \gg$. By our previous observations, $\chi_n(\Gamma) = \Gamma/N_n$ is a Schottky group, and

$$\{\chi_n(V_n^{-1} W^{-n} B_1 W^n V_n), \ \chi_n(B_2)\}$$

is a set of free generators for $\chi_n(\Gamma)$. Moreover, there exists a normalized meromorphic covering map $h_n : U \to \hat{\mathbf{C}}$ so that (h_n, χ_n) is a projective structure on U/Γ. Let $S(h_n) = \varphi_n \in C_2(\Gamma)$.

Applying Theorem 1, we choose a convergent subsequence of φ_n which we will also denote φ_n. Let $\lim \varphi_n = \varphi$. Again, note that $h_n \to f_\varphi$ uniformly on compact subsets of U, and $\chi_n \to \chi_\varphi$. Since $\varphi \in C_2(\Gamma)$, $\chi_\varphi(\Gamma)$ is a Kleinian group with $f_\varphi(U)$ an invariant component. The structures (h_n, χ_n) and $(f_\varphi, \chi_\varphi)$ will remain fixed from now on. We will show, Lemma 4.15, that $\varphi = \psi_2$.

4.2. The proof of the following lemma is the same as the proof of Lemma 3.1 and will be omitted.

LEMMA 4.1. $\chi_\varphi(\Gamma)$ *has no elliptic elements.*

LEMMA 4.2. *The elements* $\chi_\varphi(B_1)$, $\chi_\varphi(B_2)$, *and* $\chi_\varphi(W)$ *are nontrivial.*

PROOF. We observe that, for each n, $\chi_n(B_1)$ is a conjugate of a free generator for $\chi_n(\Gamma)$; hence, the same argument of Lemma 3.2 may be used to show that $\chi_\varphi(B_1) \neq id$, $\chi_\varphi(B_2) \neq id$.

Again, denote by G_n the free abelian group on the generators $[\chi_n(B_1)]$, $[\chi_n(B_2)]$ obtained by taking $\chi_n(\Gamma)$ modulo its commutator subgroup.

Now, $\chi_n(W) = id.$ only if $\chi_n(A_1)$ and $\chi_n(B_1)$ commute. Since $\chi_n(\Gamma)$ is a Schottky group, there exist integers $p, q \neq 0$ such that $(\chi_n(A_1))^p(\chi_n(B_1))^q = id.$ But, from the definition of N_n,

$$\chi_n(A_1) = \chi_n(B_1^n)\chi_n(W^n)\chi_n(C_2^{-1})\chi_n(W^{-n}).$$

Hence, in G_n, $([\chi_n(B_1^n)][\chi_n(B_2^{-1})])^p([\chi_n(B_1)])^q = id.$, which is a contradiction. We conclude that $\chi_n(W) \neq id.$, and it follows from Theorem 2 that $\chi_\varphi(W) \neq id.$

The following lemmas are proved by arguments similar to those of the proofs of Lemmas 3.1, 3.2 and 3.3, and we omit their proofs.

LEMMA 4.3. $\chi_\varphi(W)$ *does not commute with any conjugate of* $\chi_\varphi(B_1)$ *or of* $\chi_\varphi(B_2)$.

LEMMA 4.4. $\chi_\varphi(B_1)$ *does not commute with any conjugate of* $\chi_\varphi(B_2)$.

As in Section 3 (Lemma 3.5), we choose lifts, $\chi_n' : \Gamma \to SL$ and $\chi_\varphi' : \Gamma \to SL$, of χ_n and χ_φ, respectively, such that $\chi_n' \to \chi_\varphi'$.

LEMMA 4.5. *Let* X_n, $Y_n \in SL$ *with* $X_n \to X$, $Y_n \to Y$, *where* X *is loxodromic. Let* $Z_n = X_n^n Y_n X_n^{-n}$ *and suppose* $Z_n \to Z$. *Then* Z *shares a fixed point with* X.

PROOF. Choose $T \in SL$ such that

$$X' = T^{-1}XT = \begin{pmatrix} \lambda & 0 \\ 0 & \lambda^{-1} \end{pmatrix},$$

and $|\lambda| > 1$. Then choose a convergent sequence $T_n \in SL$ with $T_n \to T$ so that

$$X_n' = T_n^{-1}X_n T_n = \begin{pmatrix} \lambda_n & 0 \\ 0 & \lambda_n^{-1} \end{pmatrix},$$

and $\lambda_n \to \lambda$.

Let

(4.1)
$$Y_n' = T_n^{-1}Y_n T_n = \begin{pmatrix} a_n & b_n \\ c_n & d_n \end{pmatrix}$$

with $a_n d_n - b_n c_n = 1$. Then

(4.2)
$$Z_n' = T_n^{-1}Z_n T_n = \begin{pmatrix} a_n & \lambda_n^{2n}b_n \\ \lambda_n^{-2n}c_n & d_n \end{pmatrix}$$

Since $Y_n' \to Y' = T^{-1}YT$, (4.1) converges, and each of the sequences a_n, b_n, c_n, d_n must converge. Let $a_n \to a$, $d_n \to d$. Now (4.2) also converges, and since $\lambda_n^{-2n} \to 0$, we have

$$Z_n' \to Z' = \begin{pmatrix} a & \gamma \\ 0 & d \end{pmatrix}.$$

Where, necessarily, $\lambda_n^{2n}b_n \to \gamma \in \mathbf{C}$. Note that $ad = 1$. Thus, ∞ is a fixed point of both Z' and X'. The result follows.

LEMMA 4.6. *Let* $X_n, Y_n \in SL$ *be loxodromic transformations where* $X_n \to X$ *with* X *loxodromic. Let* $Z_n = X_n^n Y_n X_n^{-n}$ *and suppose* $Z_n \to Z$. *Furthermore, suppose there exists a sequence* y_n *of fixed points of* Y_n *with* $y_n \to y$ *where* y *is not a fixed point of* X. *Then* Z *shares a fixed point with* X.

PROOF. Choose $T \in SL$ such that

$$X' = T^{-1}XT = \begin{pmatrix} \lambda & 0 \\ 0 & \lambda^{-1} \end{pmatrix},$$

and $|\lambda| > 1$. Then choose a convergent sequence $T_n \in SL$ with $T_n \to T$ so that

$$X_n' = T_n^{-1}X_nT_n = \begin{pmatrix} \lambda_n & 0 \\ 0 & \lambda_n^{-1} \end{pmatrix},$$

and $\lambda_n \to \lambda$.

Let $Y' = T^{-1}YT$ and $Y_n' = T_n^{-1}Y_nT_n$, with $T_n^{-1}(y_n) = y_n'$ so that $y_n' \to y' = T^{-1}(y) \neq \infty$

Then, $a_n' = (X_n')^{-n}(y_n')$ is given by the formula $a_n' = \lambda_n^{-2n}y_n'$, and, since $\lim y_n' \neq \infty$, $\lim a_n' = 0$.

Thus, $X_n^{-n}(y_n) \to x = T(0)$, a fixed point of X. But $X_n^{-n}(y_n)$ is a fixed point of Z_n, and one sees easily that any convergent sequence of fixed points of Z_n must converge to a fixed point of Z. Hence, x is a fixed point of Z, and X, Z share a fixed point.

LEMMA 4.7. *Let* $X_n, Y_n \in SL$ *be loxodromic transformations, with* $X_n \to X$, $Y_n \to Y$, *where* X, Y *are neither elliptic nor trivial and share no fixed points. Suppose* $X_n^n Y_n$ *has no convergent subsequence. Then there exists a subsequence* $X_m^m Y_m$ *(of* $X_n^n Y_n$*) and fixed points* x_m *of* $X_m^m Y_m$ *with* x_m *converging to a fixed point of* X.

PROOF. We analyze the two possibilities for X.

Case 1. X is loxodromic.

Choose $T \in SL$ such that

$$X' = T^{-1}XT = \begin{pmatrix} \lambda & 0 \\ 0 & \lambda^{-1} \end{pmatrix},$$

where $|\lambda| > 1$, and let

$$Y' = T^{-1}YT = \begin{pmatrix} a & b \\ c & d \end{pmatrix},$$

where $ad - bc = 1$. Since X, Y share no fixed points, b, $c \neq 0$.

Then choose a convergent sequence $T_n \in SL$ with $T_n \to T$ so that

$$X_n' = T_n^{-1}X_nT_n = \begin{pmatrix} \lambda_n & 0 \\ 0 & \lambda_n^{-1} \end{pmatrix},$$

and $\lambda_n \to \lambda$.

Let

$$Y_n' = T_n^{-1}Y_nT_n = \begin{pmatrix} a_n & b_n \\ c_n & d_n \end{pmatrix},$$

with $a_n d_n - b_n c_n = 1$.

Now,

$$(4.3) \qquad (X_n')^n Y_n' = \begin{pmatrix} \lambda_n^n a_n & \lambda_n^n b_n \\ \lambda_n^{-n} c_n & \lambda_n^{-n} d_n \end{pmatrix}.$$

Consider the two sequences of fixed points of (4.3) which are given by

$$(4.4) \qquad s_n = \frac{(\lambda_n^n a_n - \lambda_n^{-n} d_n) + ((\lambda_n^n a_n + \lambda_n^{-n} d_n)^2 - 4)^{\frac{1}{2}}}{\lambda_n^{-n} 2 c_n},$$

and

$$(4.5) \qquad t_n = \frac{(\lambda_n^n a_n - \lambda_n^{-n} d_n) - ((\lambda_n^n a_n + \lambda_n^{-n} d_n)^2 - 4)^{\frac{1}{2}}}{\lambda_n^{-n} 2 c_n}.$$

(here we have chosen a fixed branch of the square root).

Let

$$(4.6) \qquad A_n' = \lambda_n^n a_n + ((\lambda_n^n a_n)^2 - 4)^{\frac{1}{2}},$$

$$(4.7) \qquad B_n' = \lambda_n^n a_n - ((\lambda_n^n a_n)^2 - 4)^{\frac{1}{2}}.$$

Since $Y_n' \to Y'$, we have that $d_n \to d$ and $\lambda_n^{-n} d_n \to 0$, so that

$$(4.8) \qquad \lim s_n = \lim \frac{A_n'}{2\lambda_n^{-n} c_n},$$

$$(4.9) \qquad \lim t_n = \lim \frac{B_n'}{2\lambda_n^{-n} c_n},$$

whenever the numerator of (4.8) ,(4.9) does not tend to zero.

There are two possibilities:

i) $\lambda_n^n a_n$ has a convergent subsequence (also denoted by $\lambda_n^n a_n$). Then, clearly, (4.6) and (4.7) imply that $\lim A_n' \neq 0$, $\lim B_n' \neq 0$. Since $\lim \lambda_n^{-n} c_n = 0$, it follows from (4.8) and (4.9) that $\lim s_n = \lim t_n = \infty$.

ii) $\lim \lambda_n^n a_n = \infty$. We deduce from (4.6) , (4.7) that at least one of the following holds: $\lim A_n' = \infty$, or $\lim B_n' = \infty$. Thus, at least one of the limits (4.8) or (4.9) is equal to ∞.

We have shown (for both possibilities) that at least one of the sequences of (4.4) or (4.5) has a subsequence converging to ∞, a fixed point of X'. Denote this subsequence by x_m'; then $T_m^{-1}(x_m') = x_m$ is a sequence of fixed points of $X_m^m Y_m$ converging to $T^{-1}(\infty)$, a fixed point of X.

Case 2. X is parabolic.

Choose $T \in SL$ such that

$$X' = T^{-1}XT = \begin{pmatrix} \epsilon & \gamma \\ 0 & \epsilon \end{pmatrix},$$

and let

$$Y' = T^{-1}YT = \begin{pmatrix} a & b \\ c & d \end{pmatrix},$$

where $\gamma \in \mathbf{C} - \{0\}$, $\epsilon = \pm 1$, and $c \neq 0$, with $ad - bc = 1$.

Then, choose a convergent sequence $T_n \in SL$, with $T_n \to T$, so that

$$X'_n = T_n^{-1}X_nT_n = \begin{pmatrix} \cosh \gamma_n & e^{\beta_n} \sinh \gamma_n \\ e^{-\beta_n} \sinh \gamma_n & \cosh \gamma_n \end{pmatrix}.$$

Here $Re\ \beta_n \to \infty$, and $e^{\beta_n} \sinh \gamma_n \to \gamma$.

Let

$$Y'_n = T_n^{-1}Y_nT_n = \begin{pmatrix} a_n & b_n \\ c_n & d_n \end{pmatrix}$$

with $a_nd_n - b_nc_n = 1$. Note that $Y'_n \to Y'$.

Now,

(4.10)
$$(X'_n)^nY'_n = \begin{pmatrix} a'_n & b'_n \\ c'_n & d'_n \end{pmatrix},$$

where

$$a'_n = a_n \cosh n\gamma_n + c_n e^{\beta_n} \sinh n\gamma_n,$$

$$b'_n = b_n \cosh n\gamma_n + d_n e^{\beta_n} \sinh n\gamma_n,$$

$$c'_n = a_n e^{-\beta_n} \sinh n\gamma_n + c_n \cosh n\gamma_n,$$

$$d'_n = b_n e^{-\beta_n} \sinh n\gamma_n + d_n \cosh n\gamma_n.$$

Observe that, if $e^{\beta_n} \sinh n\gamma_n$ has a convergent subsequence, (also denoted by $e^{\beta_n} \sinh n\gamma_n$) then, necessarily, $\sinh n\gamma_n \to 0$ so that $\cosh n\gamma_n$ converges. Thus, (4.10) converges, contrary to our hypothesis. It follows that

(4.11)
$$e^{\beta_n} \sinh n\gamma_n \to \infty.$$

Let

$$B'_n = \frac{a'_n - d'_n}{2c'_n},$$

and consider the two sequences of fixed points of (4.10) which are given by

(4.12)
$$s_n = B'_n + (\frac{(a'_n + d'_n)^2 - 4}{4c'^2_n})^{\frac{1}{2}},$$

(4.13)
$$t_n = B'_n - (\frac{(a'_n + d'_n)^2 - 4}{4c'^2_n})^{\frac{1}{2}}.$$

Now,

$$(4.14) \qquad B'_n = \frac{(a_n - d_n) + \tanh n\gamma_n(c_n e^{\beta_n} - b_n e^{-\beta_n})}{2c_n}.$$

Recall that $a_n \to a$, $b_n \to b$, $c_n \to c \neq 0$, $d_n \to d$, and $e^{\beta_n} \to \infty$.

There are now three possibilities:

i) $\tanh n\gamma_n$ has a convergent subsequence with $\lim \tanh n\gamma_n \neq 0$. Then, since $(c_n e^{\beta_n} - b_n e^{-\beta_n}) \to \infty$, $\tanh n\gamma_n(c_n e^{\beta_n} - b_n e^{-\beta_n}) \to \infty$, and it follows from (4.14) that $\lim B'_n = \infty$.

ii) $\tanh n\gamma_n$ has a convergent subsequence with $\lim \tanh n\gamma_n = 0$. Then since $\sinh n\gamma_n \to 0$, $\cosh n\gamma_n$ converges; it follows from (4.11) that $e^{\beta_n} \tanh n\gamma_n \to \infty$. Thus,

$$\lim(a_n - d_n) + \lim c_n e^{\beta_n} \tanh n\gamma_n - \lim b_n e^{-\beta_n} \tanh n\gamma_n = \infty,$$

and it follows from (4.14) that $\lim B'_n = \infty$.

iii) $\lim \tanh n\gamma_n = \infty$. Then, clearly, since $(c_n e^{\beta_n} - b_n e^{-\beta_n}) \to \infty$, the equation (4.14) implies that $\lim B'_n = \infty$.

We have shown (by considering all three possibilities) that there is always a subsequence with $\lim B'_n = \infty$; thus, for that subsequence, one of $\lim s_n$ or $\lim t_n$ (given by (4.12) or (4.13), respectively) will be equal to ∞, a fixed point of X'. As in the previous case, we can now obtain the desired subsequence x_m. This concludes the proof of the lemma.

COROLLARY 4.8. *Let X_n, R_n, $Y_n \in SL$ be loxodromic transformations with $X_n \to X$, $R_n \to R$, $Y_n \to Y$. Suppose X is neither trivial nor elliptic, and Y is loxodromic, sharing no fixed points with X. Furthermore, suppose X shares no fixed points with R. Let $Z_n = Y_n^{-n} X_n^n R_n Y_n^n \to Z$. Then Y shares a fixed point with Z.*

PROOF. We consider two cases.

Case 1. $X_n^n R_n$ has a convergent subsequence (also denoted by $X_n^n R_n$). Since $X_n^n R_n$, Y_n satisfy the hypothesis of Lemma 4.5, we conclude that Z shares a fixed point with Y.

Case 2. $X_n^n R_n$ has no convergent subsequence. Since X_n, R_n satisfy the hypothesis of Lemma 4.7, we choose a subsequence of $X_n^n R_n$ (also denoted by $X_n^n R_n$) and fixed points x_n of $X_n^n R_n$ converging to x, a fixed point of X. Since X, Y share no fixed points, x is not a fixed point of Y. Thus $X_n^n R_n$ and Y_n satisfy the hypothesis of Lemma 4.6, and we conclude that Z shares a fixed point with Y.

LEMMA 4.9. $\chi_\varphi(W)$ is parabolic.

PROOF. It follows from Lemma 4.2 that $\chi_\varphi(W) \neq id$. Hence, $\chi_\varphi(B_1)$ and $\chi_\varphi(A_1)$ do not commute and, consequently, they share no fixed points. Thus, $\chi'_\varphi(B_1)$, $\chi'_\varphi(A_1)$ share no fixed points either.

From the definition of N_n, it follows that

$$\chi_n(C_2)\chi_n(W^{-n})\chi_n(B_1^n)\chi_n(A_1)\chi_n(W^n) = id.$$

Thus,

$$\chi'_n(C_2)\chi'_n(W^{-n})\chi'_n(B_1^n)\chi_n(A_1)\chi'_n(W^n) = \pm id,$$

and $\chi'_n(W^{-n})\chi'_n(B_1^n)\chi'_n(A_1)\chi'_n(W^n) = \pm\chi'_n(C_2^{-1})$.

Since $\chi'_n(C_2^{-1}) \to \chi'_\varphi(C_2^{-1})$, we choose a convergent subsequence of $\pm\chi'_n(C_2^{-1})$ (also denoted by $\pm\chi'_n(C_2^{-1})$). We thus have that

$$\chi'_n(W^{-n})\chi'_n(B_1^n)\chi'_n(A_1)\chi'_n(W^n)$$

converges. Suppose $\chi'_\varphi(W)$ is loxodromic. Then, the transformations $\chi'_n(B_1)$, $\chi'_n(A_1)$, $\chi'_n(W)$, and $\chi'_n(W^{-n})\chi'_n(B_1^n)\chi'_n(A_1)\chi'_n(W^n)$ satisfy the hypothesis of Lemma 4.8. We conclude that $\chi'_\varphi(W)$ shares a fixed point with $\chi'_\varphi(C_2^{-1})$ and, consequently, $\chi_\varphi(W)$, $\chi_\varphi(C_2)$ commute. This contradicts Lemma 4.3. Hence, $\chi_\varphi(W)$ is parabolic.

LEMMA 4.10. Let $X_n, Y_n \in SL$ be loxodromic with $X_n \to X$, $Y_n \to Y$, where X is parabolic and shares no fixed point with Y. Suppose X_n^n has no convergent subsequence. Then the attracting (resp. repelling) fixed points of $Z_n = X_n^{-n}Y_nX_n^n$ converge to the fixed point of X.

PROOF. Choose $T \in SL$ such that

$$X' = T^{-1}XT = \begin{pmatrix} \epsilon & \gamma \\ 0 & \epsilon \end{pmatrix},$$

where $\gamma \in \mathbf{C} - \{0\}$, and $\epsilon = \pm 1$. Then choose a convergent sequence $T_n \in SL$ with $T_n \to T$ so that

$$X'_n = T_n^{-1}X_nT_n = \begin{pmatrix} \cosh\gamma_n & e^{\beta_n}\sinh\gamma_n \\ e^{-\beta_n}\sinh\gamma_n & \cosh\gamma_n \end{pmatrix}.$$

Here $Re\ \beta_n \to \infty$, $\cosh\gamma_n \to \epsilon$ and $e^{\beta_n}\sinh\gamma_n \to \gamma$. Then

$$(X'_n)^n = \begin{pmatrix} \cosh n\gamma_n & e^{\beta_n}\sinh n\gamma_n \\ e^{-\beta_n}\sinh n\gamma_n & \cosh n\gamma_n \end{pmatrix}.$$

Let a_n be the sequence of attracting (resp. repelling) fixed points of $Y'_n = T_n^{-1}Y_nT_n$. One shows easily, since $Y'_n \to Y' = T^{-1}YT$, that a_n converges to a fixed point of Y'. Let $a_n \to a$, and note that $a \neq \infty$ since Y' shares no fixed point with X'.

Then the fixed points $a'_n = (X'_n)^{-n}(a_n)$ of $Z'_n = T_n^{-1}Z_nT_n$ are given by

(4.15)
$$a'_n = \frac{(\cosh n\gamma_n)a_n - e^{\beta_n}\sinh n\gamma_n}{-(e^{-\beta_n}\sinh n\gamma_n)a_n + \cosh n\gamma_n},$$

(4.16)
$$a'_n = \frac{a_n - e^{\beta_n} \tanh n\gamma_n}{-(e^{-\beta_n} \tanh n\gamma_n)a_n + 1}.$$

There are two possibilities:

i) $\cosh n\gamma_n$ has a convergent subsequence. Then $e^{\beta_n} \sinh n\gamma_n$ must diverge (otherwise $(X'_n)^n$ converges), and $\lim e^{\beta_n} \tanh n\gamma_n = \infty$. We obtain from (4.15), since $\lim e^{-\beta_n} \sinh n\gamma_n = 0$, that

$$\lim a'_n = \lim \frac{(\cosh n\gamma_n)a_n - e^{\beta_n} \sinh n\gamma_n}{\cosh n\gamma_n}$$

$$= \lim(a_n - e^{\beta_n} \tanh n\gamma_n) = \infty.$$

ii) $\lim \cosh n\gamma_n = \infty$. We now have $Re \, n\gamma_n \to \infty$, and it follows that

$$\lim \tanh n\gamma_n = 1.$$

We obtain from (4.16)

$$\lim a'_n = \lim(a_n - e^{\beta_n} \tanh n\gamma_n) = \infty.$$

Thus, the attracting (resp. repelling) fixed points of Z'_n converge to ∞, the fixed point of X'. The result follows.

LEMMA 4.11. *Let* $X_n, Y_n, R_n \in SL$ *be loxodromic with* $X_n \to X$, $Y_n \to Y$, *and* $R_n \to R$, *where* R *is neither trivial nor elliptic. Let* X *be parabolic and suppose neither* R *nor* Y *shares a fixed point with* X. *Furthermore, suppose* $Z_n = Y_n^n X_n^{-n} R_n X_n^n$ *converges. Then* Y *is parabolic.*

PROOF. Suppose it is loxodromic. We will consider two cases:
Case 1. X_n^n has a convergent subsequence (also denoted by X_n^n).
Choose $T \in SL$ such that

$$Y' = T^{-1}YT = \begin{pmatrix} \lambda & 0 \\ 0 & \lambda^{-1} \end{pmatrix},$$

and $|\lambda| > 1$. Then choose a convergent sequence $T_n \in SL$ with $T_n \to T$ so that

$$Y'_n = T_n^{-1}Y_nT_n = \begin{pmatrix} \lambda_n & 0 \\ 0 & \lambda_n^{-1} \end{pmatrix},$$

and $\lambda_n \to \lambda$.
Let

(4.17)
$$T_n^{-1}X_n^{-n}R_nX_n^nT_n = \begin{pmatrix} a_n & b_n \\ c_n & d_n \end{pmatrix}$$

with $a_n d_n - b_n c_n = 1$. From the hypothesis of the lemma, R_n converges; hence (4.17) converges. Clearly,

$$T_n^{-1} Z_n T_n = \begin{pmatrix} \lambda_n^n a_n & \lambda_n^n b_n \\ \lambda_n^{-n} c_n & \lambda_n^{-n} d_n \end{pmatrix}$$

cannot converge. This contradicts the assumption that Z_n converges.

Case 2. X_n^n has no convergent subsequence.

Since X and Y share no fixed points, we choose $T \in SL$ such that

$$Y' = T^{-1} Y T = \begin{pmatrix} \cosh \alpha & \sinh \alpha \\ \sinh \alpha & \cosh \alpha \end{pmatrix},$$

and

$$X' = T^{-1} X T = \begin{pmatrix} \epsilon & \gamma \\ 0 & \epsilon \end{pmatrix},$$

where $\alpha \in U_1$, $\gamma \neq 0$, and $\epsilon = \pm 1$

It follows from Lemma 4.8 that the attracting (resp. repelling) fixed points of the hyperbolic transformation $U_n = X_n^{-n} R_n X_n^n$ converge to the fixed point of the parabolic transformation X. Hence, we choose a convergent sequence $T_n \in SL$ with $T_n \to T$ so that

$$Y_n' = T_n^{-1} Y_n T_n = \begin{pmatrix} \cosh \alpha_n & \sinh \alpha_n \\ \sinh \alpha_n & \cosh \alpha_n \end{pmatrix},$$

and

$$U_n' = T_n^{-1} U_n T_n = \begin{pmatrix} \cosh \gamma_n & e^{\beta_n} \sinh \gamma_n \\ e^{-\beta_n} \sinh \gamma_n & \cosh \gamma_n \end{pmatrix}.$$

Here, $\alpha_n \to \alpha$, and $Re \, \beta_n \to \infty$. Note that e^{β_n} (resp. $-e^{\beta_n}$) corresponds to the attracting (resp. repelling) fixed point of U_n'.

Then,

$$(Y_n')^n U_n' = \begin{pmatrix} a_n & b_n \\ c_n & d_n \end{pmatrix},$$

where

(4.18) $a_n = \cosh n\alpha_n \cosh \gamma_n (1 + e^{-\beta_n} \tanh n\alpha_n \tanh \gamma_n).$

Since $Re \, n\alpha_n \to \infty$, we have $\lim \tanh n\alpha_n = 1$. Moreover, since $2 \cosh \gamma_n = tr \, U_n' = tr \, R_n \to tr \, R \neq 0$ (recall that R is not elliptic), we have $\lim \cosh \gamma_n \neq 0$ and $\lim \tanh \gamma_n \neq \infty$. Thus,

$$\lim e^{-\beta_n} \tanh n\alpha_n \tanh \gamma_n = 0.$$

We obtain from (4.18) $\lim a_n = \lim \cosh n\alpha_n \cosh \gamma_n = \infty$.

It follows that $(Y_n')^n U_n'$ cannot converge. Thus, $Z_n = Y_n^n U_n$ cannot converge either, contradicting our hypothesis.

In both cases we have reached a contradiction. Thus, Y must be parabolic.

LEMMA 4.12. $\chi_\varphi(B_2)$ is parabolic.

PROOF. From the definition of N_n, we have

$$A_2 C_2^n W^{-n} B_1 W^n C_2^{-1} \in N_n.$$

Hence, $C_2^{-1} A_2 C_2^n W^{-n} B_1 W^n \in N_n$, and

$$\chi_n(C_2^{-1})\chi_n(A_2)\chi_n(C_2^{n+1})\chi_n(W^{-n})\chi_n(B_1^{-1})\chi_n(W^n) = id.$$

Thus, $\chi'_n(C_2^{-1})\chi'_n(A_2)\chi'_n(C_2^n)\chi'_n(W^{-n})\chi'_n(B_1)\chi'_n(W^n) = \pm id$, and

$$\chi'_n(C_2^n)\chi'_n(W^{-n})\chi'_n(B_1^{-1})\chi'_n(W^n) = \pm \chi'_n(C_2)\chi'_n(A_2^{-1}).$$

Since $\chi'_n(C_2)\chi'_n(A_2^{-1}) \to \chi'_\varphi(C_2)\chi'_\varphi(A_2^{-1})$, we choose a convergent subsequence of $\pm\chi'_n(C_2)\chi'_n(A_2^{-1})$ (also denoted by $\pm\chi'_n(C_2)\chi'_n(A_2^{-1})$). It follows that

$$\chi'_n(C_2^n)\chi'_n(W^{-n})\chi'_n(B_1)\chi'_n(W^n)$$

converges. From Lemma 4.7, we have that $\chi'_\varphi(W)$ is parabolic. From Lemma 4.3, we deduce that $\chi'_\varphi(W)$ shares no fixed points with either $\chi'_\varphi(C_2)$ or $\chi'_\varphi(B_1)$. Hence, the hypothesis of Lemma 4.11 is satisfied, and we conclude that $\chi_\varphi(C_2)$ is parabolic. It follows that its conjugate $\chi_\varphi(B_2)$ is also parabolic.

LEMMA 4.13. $\chi_\varphi(B_1)$ is parabolic.

PROOF. From the definition of N_n, we have $C_2 W^{-n} B_1^n A_1 W^n \in N_n$. Thus, $A_1^{-1} B_1^{-n} W^n C_2^{-1} W^{-n} \in N_n$. Reasoning as in the previous lemma, we obtain a subsequence for which

$$(4.19) \qquad \chi'_n(B_1^{-n})\chi'_n(W^n)\chi'_n(C_2^{-1})\chi'_n(W^{-n})$$

converges. Now, $\chi'_\varphi(W)$ is parabolic and shares no fixed point with either $\chi'_\varphi(C_2)$ or $\chi'_\varphi(B_1)$ (Lemma 4.3). We deduce from the convergence of (4.19) and Lemma 4.11 that $\chi_\varphi(B_1)$ is parabolic.

COROLLARY 4.14. $\chi_\varphi(W)$, $\chi_\varphi(B_1)$, and $\chi_\varphi(B_2)$ belong to distinct, non-conjugate, maximal parabolic subgroups of $\chi_\varphi(\Gamma)$.

PROOF. As in the proof of Corollary 3.8, we note that the only parabolic subgroups of a Kleinian group are either cyclic or rank 2 abelian; hence, since the three elements in question are parabolic, Lemmas 4.3, and 4.4 imply the desired result.

LEMMA 4.15. $\varphi = \psi_2$

PROOF. The proof is precisely the same as the proof of Proposition 3.9. We first deduce that $\chi_\varphi : \Gamma \to PSL$ is an isomorphism. Hence, $\varphi = S(f_\varphi)$ is a cusp and, since it is maximal, sharing the same associated dissection as ψ_2, $\varphi = \psi_2$.

We gather the results of this section in the

PROPOSITION 4.16. *Let $\psi_2 \in \partial T(\Gamma)$ be a maximal cusp and suppose the associated dissection of U/Γ contains a dividing curve. Then there exists a sequence $\varphi_n \in C_2(\Gamma)$ with $\chi_{\varphi_n}(\Gamma)$ a Schottky group and $\varphi_n \to \psi_2$.*

PROOF OF THEOREM 5. Let $\psi \in \partial T(\Gamma)$ be a maximal cusp. Then, according as the associated dissection does or does not contain a dividing curve, we apply either Proposition 3.10 or 4.16 to obtain a sequence $\psi_n \to \psi$, where $\chi_{\psi_n}(\Gamma)$ is a Schottky group. The result now follows trivially from Theorem 4.

REFERENCES

1. W. Abikoff, *On boundaries of Teichmüller spaces and on Kleinian groups: III*, Acta Math. **134** (1975), 211-237.
2. L. Bers, *A non-standard integral equation with applications to quasiconformal mappings*, Acta Math. **116** (1960), 113-134.
3. L. Bers, *Automorphic forms for Schottky groups*, Adv. in Math. **16** (1975), 332-361.
4. L. Bers, *On boundaries of Teichmüller space and on Kleinian groups, I*, Ann. of Math. **91** (1977), 570-600.
5. R. C. Gunning, Lectures on Vector Bundles over Riemann Surfaces, Princeton University Press, (Mathematical Notes 6), 1967.
6. T. Jorgensen, *On discrete groups of Mobius transformations*, Am. J. Math. **98** (1976), 739-749.
7. T. Jorgensen and P. Klein, *Algebraic convergence of finitely generated Kleinian groups*, Quart. J. Math. Ser. **33** (1982), 325-332.
8. I. Kra, *Deformations of Fuchsian groups II*, Duke Math. J. **38** (1971), 499-508.
9. I. Kra, *Families of univalent functions and Kleinian groups*, Isr. J. Math. **60** (1987), 89-127.
10. I. Kra and B. Maskit, *Remarks on projective structures*, Proceedings of the 1978 Stony Brook Conference, Ann. of Math. Studies, Princeton University Press Vol. 97, Princeton, New Jersey, 1981, pp. 343-359.
11. C. McMullen, *Cusps are dense*, Ann. Math. **133** (1991), 217-247.
12. B. Maskit, *Decomposition of certain Kleinian groups*, Acta Math. **130** (1973), 243-263.
13. Z. Nehari, *Schwarzian derivatives and schlicht functions*, Bull. Amer. Math. Soc. **55** (1949), 545-551.

ST. JOHN'S UNIVERSITY, STATEN ISLAND NY, 10301

E-mail address: gallod at sjuvm.bitnet

Contemporary Mathematics
Volume **169**, 1994

Lacunary Series as Quadratic Differentials in Conformal Dynamics

Frederick P. Gardiner and Dennis Sullivan[1]

Introduction

Given a polynomial-like mapping P acting on a plane domain, we denote the forward iterates of P under composition by P^n. These iterates determine a dynamical system acting on a neighborhood of the Julia set. If we assume the Julia set is connected, we may uniformize the complement of the Julia set by a Riemann mapping r, mapping the complement to the exterior of the unit disk. The forward iterates P^n are conjugated to iterates $r \circ P^n \circ r^{-1}$, which form a dynamical system acting on the exterior of the unit disk and which extend to an action on the unit circle. This conjugated dynamical system is expanding along the unit circle, which is a repeller.

The chief result of this paper is that such dynamical systems, for a given degree, form part of a Teichmüller space. It is an infinite dimensional complex manifold modeled on a Banach space and it has a Teichmüller's metric which is complete. The Banach space is a space of quadratic differentials and, for a special case, this space is identifiable with certain lacunary series. There is an associated infinite dimensional Banach space for every polynomial-like mapping and these Banach spaces generalize lacunary series. The homogeneity of smoothness of lacunary series and of more general vector fields for the dynamical system can be seen as a form of Mostow rigidity.

In order to create this complete Teichmüller space, we pick out a natural class which is larger than the class of polynomial-like mappings. It is the class UAC of *uniformly asymptotically conformal dynamical systems* with a fixed topological type, acting on an open neighborhood of the repeller and factored by asymptotically conformal equivalence.

These concepts are defined in section 1. For now we say only that there is a type of uniform almost isometric behaviour, with respect to the Poincaré metric, satisfied by the forward powers of a polynomial-like mapping P. This almost isometric behaviour is shared by elements of UAC. That polynomial-like mappings have almost isometric behaviour is the content of the *hyperbolic*

---

[1]Partially supported by a grant from the National Science Foundation
1991 *Mathematics Subject Classification*. Primary 32G15; Secondary 30C60

distortion lemma, (Lemma 3). The lemma shows that single-valued branches of an P^n are nearly isometric and, in this sense, the dynamical system for P^n resembles a Fuchsian group. Of course, the transformations of a Fuchsian group are exact isometries.

Teichmüller's metric on the factor space T of UAC becomes the dynamical boundary dilatation metric, a metric which is defined in [5]. This metric identifies two elements of UAC if they have the same behaviour at fine scales; in particular, if two elements are identified, the limits of their scaling ratios will be identical. Elements of T have well-defined eigenvalues at repelling periodic fixed points. The fibers of the tangent space to T are spaces of holomorphic quadratic differentials for analytic realizations of the quasiconformal dynamical system. The fibers over the special polynomial-like maps of the form $P(z) = z^d$ are given by certain lacunary series. Our proof of the existence of fibers over the other polynomial-like mappings depends on a discussion of Bers' $\mathcal{L}$-operator [1] and another distortion lemma, called the *Schwarzian distortion lemma*, (Lemma 7). Its proof is a familiar chain rule and geometric series argument.

There is a Riemann surface lamination associated to the expanding mapping P. It is obtained from the germ of the action of the iterates of P on small open neighborhoods of the repeller, which is the intersection of the descending sequence of sets, $P^{-n}(V)$. Elements of the lamination are tails of strings of points $(..., z_3, z_2, z_1, z_0)$ which have the property that z_0 is in $V \setminus \{\text{the repeller}\}$ and $P(z_{n+1}) = z_n$ for each n. This space can be viewed locally as the direct product of a Cantor set with complex disks. The global form of the space is determined by the dynamics of P and we call it the Riemann surface lamination L for the mapping P. The theory of the Teichmüller space for a general Riemann surface lamination all of whose leaves are hyperbolic is developed in [10]. That theory encompasses the Teichmüller theory we consider here for UAC systems and some of the lemmas used in this article could be deduced from the more general theory. In the appendix, we give some further explanation for the correspondences between the UAC theory, the Riemann surface lamination theory of [10] and the UAA theory of [7].

We are indebted to Adam Epstein, Nikola Lakic and Mitsu Shishikura for several helpful conversations.

Section 1. Uniformly asymptotically conformal dynamical systems

Let Δ^* be the set of points z in complex plane for which $|z| > 1$ and let U be an arbitrary annular neighborhood in Δ^* of the unit circle. By the annular neighborhood U we mean a doubly connected open subset of Δ^* for which there are numbers r_1 and r_2 greater than 1 such that $\{z : 1 < |z| < r_1\} \subseteq U \subseteq \{z : 1 < |z| < r_2\}$. As a matter of technical convenience, we assume that P is defined on all of Δ^*, but the ingredients of P important to us depend only on its values on small annular neighborhoods of the unit circle.

We make the following assumptions on P. It is a quasiregular mapping defined on Δ^*, it is expanding of degree d in an annular neighborhood of the unit circle and it is *uniformly asymptotically conformal.*

By quasiregular we mean that P factors into a quasiconformal homeomorphism from Δ^* onto Δ^* followed by a holomorphic mapping.

By expanding of degree d in an annular neighborhood of the unit circle, we mean that there are annular neighborhoods U_1 and U_0 with $U_1 \subset U_0$ such that $U_0 - U_1$ is a topological annulus with positive modulus and such that P restricted to U_1 is an unbranched degree d covering of U_0. We also assume that the sets $U_n = P^{-n}(U_0)$ form a descending chain of annuli $U_0 \supset U_1 \cdots \supset U_n \cdots$ whose intersection is empty and that each ring $\omega_n = U_n - U_{n+1}$ is an annulus of positive modulus surrounding the unit circle. From the geometric neighborhood lemma (Lemma 5), the rings ω_n are contained in $\{z : 1 < |z| < 1+\varepsilon\}$ where ε converges to zero exponentially as n converges to ∞. The family of rings ω_n related to P in this way is called a sequence of ring domains for P.

By *uniformly asymptotically conformal*, we mean that the branches of P^{-n} are nearly conformal on sufficiently small annular neighborhoods for all nonnegative integers n. More precisely, for every $\varepsilon > 0$, there is an annular neighborhood U, such that for every integer $n \geq 0$ and every z in $P^{-n}(U)$, the dilatation of P^n at z is less than $1 + \varepsilon$.

Definition. *Let UAC be the set of all mappings P which are quasiregular and defined on Δ^*, which are expanding and of degree d in an annular neighborhood of the unit circle and which are uniformly asymptotically conformal.*

The most obvious examples of degree 2 mappings P in UAC are $P(z) = z^2$ or $P(z) = z \left(\frac{z-a}{1-\bar{a}z} \right)$ with $|a| < 1$. Other examples can be obtained by conjugating the action of a polynomial-like mapping on the exterior of the filled-in Julia set to an action on the exterior of the unit circle. We will eventually see that there are elements of UAC which are not in the same conjugacy class as a polynomial-like map.

Let P and $\hat{P}$ be elements of UAC. Then there are annular neighborhoods U_1 and U_0 for P and $\hat{U}_1$ and $\hat{U}_0$ for $\hat{P}$ on which P and $\hat{P}$ are unbranched degree d coverings. By trimming the boundaries of $U_1, U_0, \hat{U}_1$ and $\hat{U}_0$, we can assume that the outer boundaries of these annular neighborhoods are quasicircles. Then we can construct a quasiconformal homeomorphism f from ω_0 to $\hat{\omega}_0$ such that $f \circ P(z) = \hat{P} \circ f(z)$ for values of z on the inner boundary of ω_0. By pulling back the quasiconformal mapping f defined in ω_0, we extend f so that it maps $\omega_1 \cup \omega_0$ onto $\hat{\omega}_1 \cup \hat{\omega}_0$ and satisfies $f \circ P(z) = \hat{P} \circ f(z)$. The maximal dilatation of f on $\omega_1 \cup \omega_0$ is bounded by the product of the dilatation of f restricted to ω_0 and of the dilatations of P and $\hat{P}$. In a similar manner, f can be extended to the union of all the ω_k's satisfying the relation $f \circ P(z) = \hat{P} \circ f(z)$. We call a homeomorphism f constructed in this manner a pullback homeomorphism. The hypotheses that the backwards branches of P^n and $\hat{P}^n$ uniformly asymptotically conformal implies, in particular, that they are uniformly quasiconformal and, thus, the pullback homeomorphism f is quasiconformal on U_0.

We see that any two elements in UAC are conjugate in some annular neighborhood by a quasiconformal homeomorphism f. The conjugating mapping f is called asymptotically conformal if, for every $\varepsilon > 0$, there exists a number $r > 1$, such that the dilatation of f in the annulus $1 < |z| < r$ is less that $1 + \varepsilon$. By demanding that f be asymptotically conformal, we obtain an equivalence relation on UAC.

Definition. *Two elements P and $\hat{P}$ in UAC are equivalent if there is a quasiconformal homeomorphism f asymptotically conformal at the boundary of Δ^* and defined on some annular neighborhood U such that $f \circ P(z) = \hat{P} \circ f(z)$ for all z with $|z| = 1$. The Teichmüller space T of uniformly asymptotically conformal expanding degree d mappings is the space UAC factored by this equivalence relation.*

Remarks. 1. Every element P of UAC can be symmetrized by Schwarz reflection. It becomes a quasiregular mapping P of $\overline{\mathbb{C}}$ onto $\overline{\mathbb{C}}$ which fixes the unit circle and is invariant under conjugation by the reflection $j(z) = 1/\bar{z}$. Therefore, it makes sense to speak of the values of $P(z)$ and $f(P(z))$ for values of z on the unit circle.

2. Even though mappings P in UAC are not differentiable, if q is a periodic point of P of period m on the unit circle, then the notion of the eigenvalue of P at q is well-defined on the equivalence class of P in T. Since P^m fixes q, it must map a sufficiently small disk neighborhood of q over itself leaving an intervening annulus. The hypothesis that P^m is asymptotically conformal implies that the modulus of this annulus, viewed as a function of the selected disk neighborhood, has a limit as the disk neighborhood shrinks towards the point q. If we define the eigenvalue of P at the periodic point q to be the exponential function applied to 2π times this limiting modulus, then it coincides with the usual notion of eigenvalue, namely, $\frac{dP^m}{dx}(q)$, when P is holomorphic at q.

Choose an element P in UAC and annular neighborhoods U_1 and U_0 such that P is a degree d, unbranched cover of U_1 over U_0. Define $QS(P)$ to be the set of quasiconformal homeomorphisms h mapping U_0 onto annular neighborhoods of the boundary of the unit disk such that the dynamical system $h \circ P^n \circ h^{-1}$ is uniformly asymptotically conformal. The Beltrami coefficient of a mapping h, which we denote by the symbol Beltr(h), is defined to be $\frac{\partial h}{\partial \bar{z}} / \frac{\partial h}{\partial z}$.

For h in $QS(P)$, we consider three different Beltrami coefficients:
i) the Beltrami coefficient μ of h,
ii) the Beltrami coefficient σ_n of $h \circ P^n$,
iii) the Beltrami coefficient $\mu_n = P^{n^*}\mu(z)$ defined by

$$\mu_n(z) = \mu(P^n(z))\bar{q}/q \quad \text{where} \quad q = \frac{\partial}{\partial z}P^n(z).$$

Lemma 1. (*P*-**equivariant deformations**) *Assume h is a quasiconformal homeomorphism of Δ^*. The following conditions on h are equivalent:*

i) *h is an element of $QS(P)$,*

ii) *for every $\varepsilon > 0$, there exists an annular neighborhood U such that for all positive integers n and for all z in $P^{-n}(U)$,*

$$|\sigma_n(z) - \mu(z)| < \varepsilon.$$

iii) *for every $\varepsilon > 0$, there exists an annular neighborhood U such that for all positive integers n and for all z in $P^{-n}(U)$,*

$$|\mu_n(z) - \mu(z)| < \varepsilon.$$

Proof. The Beltrami coefficient of $(h \circ P^n) \circ h^{-1}$ is $\sigma_n - \mu$ divided by a term which is bounded away from zero. Therefore, the assertion that the dilatation of $h \circ P^n \circ h^{-1}$ is uniformly near to 1 for z in $P^{-n}(U)$ is equivalent to ii). Let ν_n be the Beltrami coefficient of P^n. Observe that μ_n, ν_n and σ_n are related by the equation

$$\sigma_n = \frac{\nu_n + \mu_n}{1 + \overline{\nu_n}\mu_n}$$

and, therefore,

$$\sigma_n - \mu_n = \frac{\nu_n - \mu_n^2\overline{\nu_n}}{1 + \overline{\nu_n}\mu_n}.$$

Since we can make $|\nu_n(z)|$ as small as we like by suitably choosing U and letting z be in $P^{-n}(U)$ and since $|\mu_n(z)|$ is uniformly bounded for these z, the equivalence of i) and iii) is a consequence of this last formula.   □

We denote by $M(P)$ the space of Beltrami coefficients of mappings h in $QS(P)$. By the preceding lemma any element of $M(P)$ is a measurable complex valued function defined on U_0 satisfying the following properties:

a) $\operatorname{esssup}_{z \text{ in } U_0}|\mu(z)| < 1$

b) for every $\varepsilon > 0$, there exists an annular neighborhood U of the boundary of the unit disk, such that for all positive integers n and all z in $P^{-n}(U)$,

$$|P^{n^*}\mu(z) - \mu(z)| < \varepsilon.$$

Conversely, if μ satisfies these two properties, then by solving the Beltrami equation for a mapping with Beltrami coefficient μ, we obtain a mapping h such that $h \circ P \circ h^{-1}$ is uniformly asymptotically conformal.

Let S be the group of quasiconformal homeomorphisms of Δ^* onto Δ^* which are asymptotically conformal. This group determines an equivalence relation on $QS(P)$ by declaring two elements h_1 and h_2 of $QS(P)$ to be equivalent if there is an element s of S such that $s \circ h_1(z) = h_2(z)$ for $|z| = 1$. We let $QS(P)/S$ be the space of equivalence classes for this equivalence relation.

There is a mapping from $QS(P)$ to UAC; for any h in $QS(P)$, the mapping is $h \mapsto \hat{P}$ where $\hat{P} = h \circ P \circ h^{-1}$. By the pullback construction, we see that this mapping is surjective and, since T is a quotient space of UAC, we obtain a mapping from $QS(P)$ onto T. Take two elements of h_0 and h_1 of $QS(P)$ which conjugate P into P_0 and P_1, respectively. Assume that h_0 and h_1 are equivalent modulo S. Then there is an asymptotically conformal mapping s in S such that $s \circ h_0(z) = h_1(z)$ for $|z| = 1$. Then $s \circ P_0 \circ s^{-1}(z) = P_1(z)$ for $|z| = 1$ and, consequently, P_0 is equivalent to P_1 in UAC.

Conversely, if P_0 and P_1 are equivalent in UAC, there is an asymptotically conformal mapping s such that $s \circ P_0 \circ s^{-1}(z) = P_1(z)$ for all z on the boundary of the unit circle. If h_0 and h_1 conjugate P to P_0 and P_1, respectively, then we find that $s \circ h_0 \circ P^n \circ (s \circ h_0)^{-1} = h_1 \circ P^n \circ h_1^{-1}$ for all positive integers n. Since repelling periodic points of P are dense in the unit circle, we conclude that $s \circ h_0(z) = h_1(z)$ for z on the boundary of the unit circle and, hence, h_0 and h_1 are in the same equivalence class for $QS(P)/S$. We have proved the following theorem.

Theorem 1. *The natural mapping from $QS(P)$ onto T induces an isomorphism from $QS(P)/S$ to T.*

Theorem 2. *The boundary dilatation metric on $QS(P)/S$ makes T into a complete metric space and the metric is independent of the base point P.*

Proof. In [6] it is shown that the boundary dilatation metric on $QS\bmod S$ is complete. Here QS is the space of all quasisymmetric homeomorphisms of a circle and S is the closed subgroup of symmetric homeomorphisms. The issue here is to show that $QS(P)$ is a closed subset of QS. Assume that h is a quasiconformal mapping and that h_j is a sequence of quasiconformal mappings in $QS(P)$ such that $\|Beltr(h_j) - Beltr(h)\|_\infty < \varepsilon$ for sufficiently large j. It follows that for z in $P^{-n}(U_0)$,

$$|Beltr(h_j \circ P^n)(z) - Beltr(h \circ P^n)(z)| < \varepsilon',$$

where $\varepsilon' = \varepsilon/(1 - k^2)$ and k is a uniform bound on the absolute values of the Beltrami coefficients of P^n. On the other hand, since each h_j is in $QS(P)$, we know that for every $\varepsilon > 0$, there exists j_0, such that for all $j \geq j_0$ and for all z in $P^{-n}(U_j)$,

$$|Beltr(h_j \circ P^n)(z) - Beltr(h_j)(z)| < \varepsilon.$$

Therefore, for all n and z in $P^{-n}(U_{j_0})$,

$$|Beltr(h \circ P^n)(z) - Beltr(h)| < 2\varepsilon + \varepsilon'.$$

The fact that the metric on $QS(P)/S$ is independent of base point follows from the fact that composition on the right induces an isometry for the respective Teichmüller metrics.   □

Section 2. Polynomial-like mappings

An element of UAC which is conformal in some annular neighborhood U of the boundary of Δ^* is called polynomial-like. First we prove that polynomial-like mappings are dense in T.

Given an element P in UAC and a positive number ε, select annular neighborhoods U_1 and U_0 such that P restricted to U_1 is a degree d cover over U_0 and the dilatation of P on U_1 is less than $1 + \varepsilon$. As before, we let $U_n = P^{-n}(U_0)$. Let μ be the Beltrami coefficient of P in U_n and extend μ to be identically zero outside of U_n. Form the quasiconformal homeomorphism g of Δ^* which fixes ∞ and which has Beltrami coefficient μ. The mapping g is conformal in $|z| > 1 + \delta$ provided that the circle of radius $1 + \delta$ contains U_n. Moreover, g is $1 + \varepsilon$-quasiconformal in $|z| > 1$. Since δ approaches zero as n approaches ∞, by the Hölder continuity of g, we may choose n large enough so that $g(U_1)$ is contained in U_0 and so that the complement of $g(U_1)$ in U_0 is an annulus of positive modulus.

Let $\hat{P} = P \circ g^{-1}$ and $\hat{U}_1 = g(U_1)$. Then $\hat{P}(\hat{U}_1) = P(U_1) = U_0$ and $\hat{P}$ is a degree d unbranched covering of $\hat{U}_1$ over U_0 and the closure of $\hat{U}_1$ is compact in U_0. Moreover, $\hat{P}$ is conformal in $\hat{U}_1$. Therefore, $\hat{P}$ is a polynomial-like element of UAC. Now, we must repeat the pullback argument used to prove Theorem 1 to show that there is a quasiconformal map conjugating P to $\hat{P}$ which, on a sufficiently small annular neighborhood, has dilatation less than $1 + \varepsilon'$. We obtain the following result.

Lemma 2. (Density of polynomial-like mappings) *The set of polynomial-like mappings is dense in the Teichmüller space of uniformly asymptotically expanding degree d mappings with Teichmüller's boundary dilatation metric.*

The Poincaré metric for Δ^* is $\lambda(z)|dz| = \left(|dz|/|z|log|z| \right)$. We are concerned only with how this metric measures the sizes of objects which are very near to the boundary of the unit circle. If $\delta(z)$ is the distance from z to the boundary, the formula for $\lambda(z)$ is asymptotically equal to $(1/\delta(z))$ for $|z|$ near to 1. The Poincaré metric for $\Delta^* \cup \{\infty\}$ is $\left[2|dz|/(|z|^2 - 1) \right]$ and this has the same asymptotic values for $|z|$ near to 1. For the purposes of the next lemma, we could use either one of these two metrics. The lemma says that if P is polynomial-like, the branches of P^n, restricted to neighborhoods sufficiently near to the boundary, are approximate isometries in the Poincaré metric. It is a kind of distortion lemma, because it gives a bound on distortion for branches of P^n, which is independent of n.

Lemma 3. (Hyperbolic distortion) *Assume P is a polynomial-like mapping acting on Δ^*. Then for every ε, there exists an annular neighborhood U, such that for every positive integer n and every z in $P^{-n}(U)$,*

$$\frac{1}{1+\varepsilon} \leq \frac{\lambda(P^n(z))|P^{n'}(z)|}{\lambda(z)} \leq 1 + \varepsilon.$$

Proof. (We are grateful for a helpful discussion with Mitsu Shishikura concerning this proof.) One views $\Delta^* - \{\infty\}$ as a half-cylinder, the upper half plane factored by the cyclic group generated by $z \mapsto z + 1$. Let $\lambda(z)$ by the Poincaré metric for this half plane. One lifts the mapping P restricted to U_1 to a mapping $\hat{P}$ defined on a periodic strip S_1 in the upper half plane bounded by the real axis and a periodic curve which is the lift of the outer boundary of U_1. This lifted mapping maps S_1 over itself and onto a domain S_0 which is the lift of the domain U_0. $\hat{P}$ restricted to S_1 is one-to-one and $\hat{P}^{-1}$ is a mapping from the strip S_0 into itself. The natural extension of $\hat{P}^{-1}$ acts as an isometry of a hyperbolic plane. Let ρ be the hyperbolic metric for this plane. This hyperbolic plane contains the strip S_0. Let λ_0 be the hyperbolic metric for S_0. The inequalities $\lambda < \lambda_0$ and $\rho < \lambda_0$ follow from Schwarz's inequality. On the other hand by shrinking to strips smaller than S_0 whose boundaries are straight lines, we can use the exact formulas for the Poincaré metrics to deduce the following fact; given any $\varepsilon > 0$, there exists $\delta > 0$, such that for any $z = x + iy$ with $0 < y < \delta$, $(1+\varepsilon)^{-1} \leq \rho(z)|dz|/\lambda(z)|dz| \leq 1+\varepsilon$. Because the natural extension of $\hat{P}^{-1}$ is a non-Euclidean isometry for the ρ-metric, we obtain the uniformity condition of the lemma. $\qquad\square$

Section 3. Quadratic differentials for polynomial-like mappings

For the purposes of clarity of exposition, in this section we assume the the mapping has degree 2; the case for general degree follows by trivial modifications, usually no more complicated than the replacement of the symbol 2 by d.

The first requirement is to construct a set of domains for the dynamical system generated by the branches of the mappings P^n which are analogous to a fundamental domain for a Fuchsian group. By assumption, P is a cover of a domain U_1 over U_0, U_1 is properly contained in U_0 and the domain $\omega_0 = U_0 - U_1$ is an annulus of positive modulus. We arbitrarily choose a cross-cut of ω_0, that is, a simple arc which joins a point Q_1 on the inner boundary component of ω_0 to a point Q_0 on the outer boundary component of ω_0 with the property that $P(Q_1) = Q_0$. We let β_0 equal ω_0 with this arc deleted and call the simply connected set β_0 the Carleson box at level zero. Since we assume P is a regular covering of degree 2, this arc pulls back to two disjoint arcs joining the boundary contours of ω_1; these two arcs divide ω_1 into two boxes at level one. Continuing in this manner, $\omega_n = U_n - U_{n-1}$, where $U_n = P^{-n}(U_0)$, is divided into 2^n boxes at level n. Branches of the mapping P^n , for different values of n, give holomorphic homeomorphisms between boxes in the ring ω_{n+k} and boxes in the ring ω_k. As a consequence of the hyperbolic distortion lemma, if we take any two boxes, one from ω_m and one from ω_n for sufficiently large m and n, the homeomorphism of these two boxes will be almost an isometry. Even if m and n are not large, each box is quasi-isometric to a square with unit area and unit side length measured in the Poincaré metric. A bound on the constant of quasi-isometry depends only on the geometry of P. In contrast, the side length of any box in a ring w_n measured in the Euclidean metric and its distance from the boundary of the unit circle are bounded below and above by positive constants times ε_0^n and ε_1^n, where $0 < \varepsilon_0 < \varepsilon_1 < 1$. This fact follows from Lemma 5, the geometric neighborhoods lemma.

The Carleson boxes, so constructed, are quite arbitrary; nonetheless, they lead to definitions and properties which are invariant under the choices made in their construction.

Consider the Banach space B of bounded holomorphic functions φ defined in Δ^* for which

$$\|\varphi\|_B = \sup_{z \text{ in } \Delta^*} |\lambda^{-2}(z)\varphi(z)| < \infty.$$

It is a matter of technical convenience that φ is defined in the whole domain Δ^*; in the end, we only care about its values near the boundary of the unit circle.

B contains a closed subspace B_0 consisting of those φ such that $|\lambda^{-2}(z)\varphi(z)|$ vanishes at the boundary of the unit circle. More precisely, φ is in B_0 if for every $\varepsilon > 0$, there exists a number $r > 1$, such that

$$\sup_{1<|z|<r} |\lambda^{-2}(z)\varphi(z)| < \varepsilon.$$

For a quadratic-like mapping P, there is an annular neighborhood U_1 inside which it is possible to take the derivative, $\frac{dP}{dz}$, and the derivatives $\frac{dP^n}{dz}$ are defined inside the sets $P^{-n}(U_0)$. Define $P^{n^*}(\varphi)(z)$ to be $\varphi(P^n(z))\left(\frac{dP^n}{dz}(z)\right)^2$ for z in the set $P^{-n}(U_0)$. We can now define quadratic differentials for P.

Definition. *The space $B(P)$ of bounded holomorphic quadratic differentials for P consists of all functions φ holomorphic in the exterior of the unit circle and such that*

i) *φ is contained in B and*

ii) *for every ε, there exists an annular neighborhood $U \subset U_0$, such that for all n and for all z in $P^{-n}(U)$,*

$$|P^{n^*}(\varphi)(z) - \varphi(z)| \leq \varepsilon\lambda^2(z).$$

The key element of this definition is that the inequality is uniform in n; it says that if we look in a deep enough annular neighborhood U_k, φ on the ring w_{n+k} is nearly the same as the pullback by P^n of φ on w_k and the amount by which φ and the pullback of φ differ is independent of n.

Proposition. B_0 *is a closed subspace of* $B(P)$ *and* $B(P)$ *is a closed subspace of* B. *Moreover, the natural Banach norm for the quotient space* $B(P)/B_0$ *is*

$$||\varphi||_{B(P)} = \lim_{U} \sup_{z \text{ in } U} |\lambda^{-2}(z)\varphi(z)|,$$

where the limit is taken as the annular neighborhood U *shrinks to the boundary of the unit circle.*

Proof. First we show that B_0 is contained in $B(P)$. If φ is in B_0, then for any $\varepsilon > 0$, there exists $r > 1$ so that

$$\sup_{1 < |z| < r} |\lambda^{-2}(z)\varphi(z)| < \varepsilon.$$

Thus, for z is in $P^{-n}(U)$, $|\lambda^{-2}(P^n(z))\varphi(P^n(z))| < \varepsilon$. But by the hyperbolic distortion lemma, we can shrink the annular neighborhood U sufficiently so that $\lambda(P^n(z))|P^{n'}(z)| < (1 + \varepsilon)\lambda(z)$. We obtain $|\lambda^{-2}(z)(P^{n^*}\varphi)(z)| < \varepsilon(1 + \varepsilon)$ and, finally, $|P^{n^*}(\varphi)(z) - \varphi(z)| \leq \{\varepsilon(1 + \varepsilon) + \varepsilon\}\lambda^2(z)$, which implies that $\varphi \in B(P)$.

We omit the proof that $B(P)$ is a closed subspace of B since it is so similar to the proof of Theorem 2; the one new ingredient is the hyperbolic distortion lemma (Lemma 3). □

Section 4. An L_1-norm on for $B(P)$ when $P(z) = z^2$

We now construct a second norm for the Banach space of holomorphic quadratic differentials which are automorphic for the dynamical system determined by P. Our definition will depend on the choice of Carleson boxes, but since the norm will turn out to be equivalent to the natural norm for $B(P)/B_0$, this dependence is only apparent.

We assume that $P(z) = z^2$ and that a system of Carleson boxes for P has been constructed. Consider the space $A(P)$ of functions φ holomorphic in Δ^* satisfying

i) $\sup_\beta \int \int_\beta |\varphi| dx dy < \infty$, where the supremum is over all Carleson boxes β, and

ii) for every $\varepsilon > 0$, there exists an annular neighborhood U, such that for all positive integers n and for all z in $P^{-n}(U)$,

$$\sup_\beta \int \int_\beta |P^{n*}(\varphi)(z) - \varphi(z)| dx dy < \varepsilon.$$

Let $area(\beta)$ denote the Poincaré area of β. We define the $A(P)$ norm of φ to be

$$||\varphi||_{A(P)} = \lim_{U} \sup_{\beta \text{ in } U} \frac{1}{area(\beta)} \int \int_\beta |\varphi(z)| dx dy,$$

where the supremum is over all Carleson boxes in U and the limit is taken as the annular neighborhood U shrinks towards the boundary. Since the Poincaré areas of all of the Carleson boxes are comparable, the area factor in this definition is superfluous. However, the area factor is necessary if we expect to show that the norm, up to a constant factor, is independent of the choice of Carleson boxes. From the inequality

$$\frac{1}{\text{area}(\beta)} \int \int_\beta |\varphi(z)| dx dy \leq \sup_{z \text{ in } \beta} |\varphi(z)\lambda^{-2}(z)| \left(\frac{1}{\text{area}(\beta)} \int \int_\beta \lambda^2(z) dx dy\right),$$

it follows that $||\varphi||_{A(P)} \leq ||\varphi||_{B(P)}$ since the second factor on the right hand side of this inequality is equal to 1. The same type of inequality shows that if φ satisfies condition ii) for $B(P)$ then it satisfies condition ii) for $A(P)$. It follows that restriction of a function φ defined on Δ^* to a function defined on U_0 defines a mapping from $B(P)$ into $A(P)$ which is continuous and that the $A(P)$-norm of any element of B_0 is equal to zero. Provided we show that the automorphy condition ii) for $A(P)$ implies condition ii) for $B(P)$, then the restriction mapping induces a surjection from $B(P)/B_0$ to $A(P)/B_0$ because any function holomorphic in U_0 differs by an element of B_0 from a function which is holomorphic in Δ^* and vanishes of order $|z|^{-4}$ as z approaches ∞.

Any Carleson box is surrounded by eight adjacent Carleson boxes. If these boxes are at level n, the Euclidean diameter of each of these eight neighbors is on the order of a constant times $1/2^n$, which is the same as the order of the value of the Poincaré metric in these boxes. From the areal mean value theorem for the subharmonic function $|\varphi(z)|$, we obtain a constant C such that

$$\sup_{z \text{ in } \beta} |\varphi(z)\lambda^{-2}(z)| \leq C \int \int_\beta |\varphi(z)| dx dy$$

and it follows that $||\varphi||_{B(P)} \leq C||\varphi||_{A(P)}$. The same argument shows that the automorphy condition ii) for $A(P)$ implies the automorphy condition for $B(P)$.

Theorem 3. *In the case that $P(z) = z^2$ the mapping from $B(P)/B_0$ to $A(P)/B_0$ induced by the restriction of an element of $B(P)$ to $A(P)$ is an isomorphism of Banach spaces.*

Section 5. Lacunary series as quadratic differentials

Theorem 3 gives us equivalent norms for the Banach space $B(P)/B_0$. At this point, we have no examples of nontrivial elements in this Banach space. In this section, we show that $B(P)/B_0$ is infinite dimensional when $P(z) = z^2$.

From $P(z) = z^2$, it follows that $P^k(z) = z^{2^k}$ and $\frac{d}{dz}P^k(z) = 2^k z^{2^k-1}$. For any holomorphic function Φ, considered as a quadratic differential, define $P^{k^*}\Phi(z)$ to be equal to $\Phi(P^k(z))\left(\frac{d}{dz}P^k(z)\right)^2$. If we let $\varphi(z)$ be a function holomorphic in Δ^*, by looking at a power series for φ, we find that it is impossible to solve the equation $P^{k^*}\varphi(z) = \varphi(z)$. On the other hand, if we only require equality modulo B_0, this equation has many solutions. They take the form of any linear combination of

$$\varphi_j(z) = \sum_{k=0}^{\infty} \frac{2^{2k}}{z^{2+j2^k}}, \text{ where } j \text{ is an odd integer} \geq 1.$$

The exponents $2 + j2^k$, for $k \geq 0$, coincide with the orbit of $j + 2$ under the mapping $\alpha \mapsto 2\alpha - 2$ acting on the positive integers. Therefore, if we take two unequal odd values of $j \geq 1$, say j_1 and j_2, the exponents of z appearing in the summation for φ_{j_1} will all be distinct from the exponents of z appearing in the summation for φ_{j_2}.

Notice that $\varphi_j(z)$ also can be written in the form

$$\varphi_j(z) = \sum_{k=0}^{\infty} P^{k^*} \Phi_j(z) \text{ where } \Phi_j(z) = \frac{1}{z^{j+2}}$$

and so φ_j is a theta-series of the holomorphic function Φ_j whose absolute value is integrable over Δ^*. Thus

$$\varphi_j(z) - P^{n^*}\varphi_j(z) = \sum_{k=0}^{n-1} P^{k^*}\Phi(z)$$

and, therefore, if β is a Carleson box for P in $P^{-n}(U)$,

$$\int\int_{\beta} |P^{n^*}\varphi_j(z) - \varphi_j(z)| \, dx dy \leq \int\int_{\beta \cup P(\beta)\cdots\cup P^{n-1}(\beta)} |\Phi_j(z)| \, dx dy.$$

Since Φ_j is integrable, this inequality shows that φ_j satisfies the automorphy condition for $B(P)$.

The function φ_j is an element of $B(\Delta^*)$ because its $A(P)$ norm is bounded by a constant times $\int_{\Delta^*} \int |\Phi_j| dx dy$.

A second way to see that φ_j is bounded in $B(\Delta^*)$ is to use the Banach space isomorphism between $B(\Delta^*)$ and the Zygmund bounded vector fields on the unit circle, (see [6]). The isomorphism involves two steps. The first is to integrate φ_j in Δ^* three times to a vector field $V_j(z)\frac{\partial}{\partial z}$. The second step is take the real part of this vector field and restrict its values to the unit circle. For the substitution $z = e^{i\theta}$, restricting z to the unit circle is the same as restricting θ to the real axis. We find that

$$\text{real part of } (V_j(e^{i\theta})/ie^{i\theta}d\theta) = \sum_{k=0}^{\infty} \frac{2^{2k} \sin(j2^k\theta)}{(j2^k + 1)(j2^k)(j2^k - 1)}.$$

This function is an element of the Zygmund class Λ^* and, on making the substitution $m = j2^k$ and using the identity

$$\frac{m^2}{(m+1)(m)(m-1)} = \frac{1}{m}\{1 + \frac{1}{(m^2-1)}\},$$

we obtain

$$\text{real part of } (V_j(e^{i\theta})/ie^{i\theta}d\theta) = \frac{1}{j^2} \sum_{k=0}^{\infty} \frac{1}{j2^k}\{1 + \frac{1}{(j^2 2^{2k}-1)}\} \sin(j2^k\theta).$$

The second term inside the curly bracket approaches zero as k approaches ∞ and, therefore, by the Zygmund-Jackson theorem [13], the contribution of this second term to the summation corresponds to an element of λ^*. Under the isomorphism between Λ^* and $B(\Delta^*)$, this means it corresponds to a cusp form which vanishes at the boundary, namely, to an element of B_0. Since we only care about the class of this function modulo λ^*, which is isomorphic to B_0, we may neglect the second term inside the curly bracket, and the summation takes exactly the form of the classical Weierstrass example, [12, page 48-50]. Our computation shows that the Zygmund norm of V_j is asymptotic to j^{-2}.

Theorem 4. *For the case $P(z) = z^2$, the functions φ_j for j equal to an odd integer bigger than or equal to 1 form a linearly independent set in the Banach space $B(P)/B_0$.*

Proof. We have already shown that the φ_j determine elements of $B(P)/B_0$. We now show that the φ_j's are linearly independent in the quotient Banach space. That is, we show, if a linear combination of the φ_j's is in B_0, then each constant in the linear combination is zero. We actually show much more, namely, that there is a convergent notion of inner product defined for pairs of elements of $B(P)/B_0$ and that, with respect to this inner product, the functions φ_j form an orthogonal family of elements not contained in B_0. This notion of inner product is defined when P is any polynomial-like mapping.

Lemma 4. (**Convergence of inner product**) *Let μ be a P-automorphic Beltrami differential and φ be in $A(P)$. Let w_n be a sequence of ring domains determined by P. Then $\lim_{n\to\infty} (\text{area}(w_n))^{-2} \int\int_{w_n} \mu\varphi\, dxdy$ converges.*

Proof. Because μ and φ are automorphic, we know that for every $\varepsilon > 0$, there exists an annular neighborhood U, such that for all integers $k \geq 0$ and for all z in $P^{-k}(U)$,

$$|P^{k*}\mu(z) - \mu(z)| < \varepsilon \text{ and } |P^{k*}\varphi(z) - \varphi(z)| < \varepsilon\lambda^2(z).$$

To show the sequence of the lemma is a Cauchy sequence, we must show that for every $\varepsilon > 0$, there exists an n, such that for all $k > 0$,

$$|\int\int_{w_{n+k}} \mu\varphi - 2^k\int\int_{w_n} \mu\varphi| = |\int\int_{w_{n+k}} \mu\varphi - \int\int_{w_{n+k}} P^{k*}(\mu\varphi)| < (const)\varepsilon 2^{n+k}.$$

Inside the absolute value we subtract and add the term $\int\int P^{k*}(\mu)\varphi$, with integration over the domain w_{n+k}, and then apply the triangle inequality. The resulting two terms are seen to be less than or equal to

$$\varepsilon \int \int_{\omega_{n+k}} |\varphi| + \left(\sup_{z \ in \ \omega_{n+k}} |P^{k^*} \mu(z)| \right) \varepsilon \int \int_{\omega_{n+k}} \lambda^2.$$

The first term is less than a constant times $\varepsilon \ 2^{n+k}$ because φ is in $A(P)$. The second term is bounded similarly because μ is bounded and because of the hyperbolic distortion lemma. □

From the hyperbolic distortion lemma, one can show that for ψ in $A(P)$, $\lambda^{-2}\overline{\psi}$ is P-automorphic Beltrami differential, (see Lemma 3 of section 2). From the above lemma we conclude that the limit in the following inner product converges:

$$< \varphi, \psi >= \lim_{n \to \infty} \frac{1}{\text{area}(\omega_n)} \int \int_{\omega_n} \varphi(z) \lambda^{-2}(z) \overline{\psi(z)} \ dx dy$$

It is easy to see that $< \varphi, \varphi >^{1/2} \le ||\varphi||_{B(P)/B_0}$. The ring domains for $P(z) = z^2$ can be taken to be bounded by concentric circles. If we take two P-automorphic forms φ_{j_1} and φ_{j_2} with j_1 not equal to j_2 we obtain $< \varphi_{j_1}, \varphi_{j_2} >= 0$ since none of the exponents of z occurring in the series for φ_{j_1} coincides with any exponent of z occurring in the series for φ_{j_2}.

We now show that $< \varphi_j, \varphi_j >$ is asymptotic to a positive constant times j^{-4} and this will complete the proof of the theorem. We take ω_n to be the region between $(1 + \varepsilon)$ and $(1 + \varepsilon)^2$. Then the Poincaré area of ω_n is asymptotic to $2\pi/\varepsilon$ and

$$< \varphi_j, \varphi_j >= \lim_{\varepsilon \to 0} \varepsilon \int_{1+\varepsilon}^{(1+\varepsilon)^2} \sum_{k=0}^{\infty} \frac{2^{4k} (r \log r)^2}{r^{4+j2^{k+1}}} \ r dr = \lim_{\varepsilon \to 0} 4 \ \varepsilon^4 \sum_{k=0}^{\infty} \frac{2^{4k}}{(1 + \varepsilon)^{j2^k}}.$$

Since the omission of any finite number of terms in this summation does not affect the limit as ε approaches zero, we see that the inner product norm of φ_j depends only on the tail of the infinite series of its definition. To estimate this limit, we replace the summation by an integral and integrate by parts three times. We obtain $< \varphi_j, \varphi_j >= (\text{constant}) \ j^{-4}$ and so the Zygmund norm of V_j and the inner product norm of φ_j are both asymptotic to j^{-2}. □

Section 6. The Bers' $\mathcal{L}$-operator

The Bers' $\mathcal{L}$-operator [1,4] is a device which converts Beltrami differentials into holomorphic quadratic differentials. The $\mathcal{L}$-operator arises from analyzing the derivative of composition of mappings representing points in Teichmüller space. More precisely, for the quasisymmetric mappings f and h, if we hold h fixed and let f vary, the mapping $[f] \mapsto [f \circ h]$ is differentiable. Calculating the expression for its derivative, followed through the conformal welding process, gives rise to the $\mathcal{L}$-operator.

In the next section, we use the $\mathcal{L}$-operator and the existence of quadratic differentials for the special polynomial mapping $P(z) = z^d$ which was shown in the previous section to imply the existence of quadratic differentials for arbitrary polynomial-like mappings of degree d. In this section we show that Bers' $\mathcal{L}$-operator preserves P-automorphic forms.

It is convenient to transport our spaces of differentials to spaces of differential forms on the logarithmic covering space. The lifting of a degree d mapping to this covering is an injective mapping. This convenient fact was already used in section 2 to prove the hyperbolic distortion lemma. As before, for simplicity of notation, we assume that $d = 2$, but all the theorems apply for arbitrary positive integers $d \geq 2$.

In our set-up, we assume we have a quasiconformal conjugation of the lifting of a quadratic-like mapping. The picture is of a quasicircle C, which is periodic in the sense that $C + 2\pi i = C$. (Think of C as the imaginary axis or a periodic quasiconformal distortion of the imaginary axis.) The complement of C consists of two simply connected domains, L and R, which we refer to as the left side and the right side. There is a one-to-one conformal mapping α, the lift of the quadratic-like mapping P, which is defined in a neighborhood of C, periodic in the sense that $\alpha(z + 2\pi i) = \alpha(z) + 4\pi i$, leaves C invariant, and is expanding in the following sense. There are periodic open sets U_0 and U_1 contained in L and V_0 and V_1 contained in R such that $U_1 \cup C \cup V_1$ is an open neighborhood of C in the complex plane. The mapping α fixes the set C and maps U_1 onto U_0 and V_1 onto V_0. The topological periodic strip domains $U_0 - U_1$ and $V_0 - V_1$ factored by the translation $z \mapsto z + 2\pi i$ are conformal cylinders with positive moduli. By following the backwards images of U_0 under iterates of α, we obtain a sequence of strips $U_0 \supset U_1 \supset \cdots \supset U_n \supset \cdots$ and the modulus of each cylinder $(U_{n-1} - U_n)/(z \mapsto z + 2\pi i)$ has twice the modulus of the preceding cylinder $(U_n - U_{n+1})/(z \mapsto z + 2\pi i)$. We have a similar picture on the right hand side R where there is a sequence of strip domains $V_0 \supset V_1 \supset \cdots \supset V_n \cdots$, which are the backwards images of V_0 under the iterates of α. It is convenient to let $W_n = U_n \cup C \cup V_n$, so α^n maps W_n injectively and holomorphically onto W_0. The log of the absolute value of the derivative of α is bounded above and below by positive constants throughout the closure of W_1.

We need a lemma concerning strip neighborhood systems. For any z, let $\delta(z)$ be the minimum Euclidean distance from z to C. Consider the sequence of neighborhoods of C given by $U_n(\varepsilon) = \{z : z \in R \text{ and } \delta(z) < \varepsilon^n\}$ We call such a sequence $U_n(\varepsilon)$ a geometric system of neighborhoods.

Lemma 5. (**Geometric neighborhoods**) *Given any strip neighborhood U contained in R, there exists geometric systems of neighborhoods $U_n(\varepsilon_0)$ and $U_n(\varepsilon_1)$ with $0 < \varepsilon_0 < \varepsilon_1 < 1$ and an integer k such that $U_{n+k}(\varepsilon_0) \subseteq \alpha^{-n}(U) \subseteq U_{n-k}(\varepsilon_1)$ for all $n \geq k$.*

Proof. The lemma is obviously true in the case that $P_0(z) = z^2$ with $\varepsilon_0 = \varepsilon_1 = 1/2$. Any P is related to P_0 by the equation $h \circ P_0 \circ h^{-1} = P$, where h is quasiconformal and h preserves the unit circle. The lemma follows from the observation that h is quasiconformal and, hence, both h and h^{-1} are Hölder continuous. $\square$

In the right hand side R, we consider the Banach space $L_\infty(\alpha)$ of Beltrami differentials for α. It consists of L_∞-complex-valued functions μ defined on R satisfying
i) $\mu(z + 2\pi i) = \mu(z)$ and
ii) for every $\varepsilon > 0$, there exists a periodic strip neighborhood V of C in R, such that for all $n \geq 0$ and for all z in $\alpha^{-n}(V)$,

$$|\mu(\alpha^n(z))\frac{\overline{\alpha^{n'}(z)}}{\alpha^{n'}(z)} - \mu(z)| < \varepsilon.$$

On the left side L, we consider the Banach space $B(\alpha)$ of holomorphic functions φ which are bounded cusp forms in the sense that, if λ is the Poincaré metric for L, then $\sup_{z\ in\ L} |\varphi(z)\lambda^{-2}(z)| < \infty$,
i) $\varphi(z + 2\pi i) = \varphi(z)$ and
ii) for every $\varepsilon > 0$, there exists a periodic strip neighborhood U of C in L, such that for all $n \geq 0$ and for all z in $\alpha^{-n}(U)$,

$$|\varphi(\alpha^n(z))\alpha^{n'}(z)^2 - \varphi(z)| < \varepsilon\lambda^2(z).$$

The next lemma enables us to view condition ii) in an apparently weaker form.

Lemma 6. (**Bootstrapping**) *Suppose ψ is a function holomorphic in L such that $\psi(z + 2\pi i) = \psi(z)$ and the $\sup |\lambda^{-2}(z)\psi(z)|$ over z in L is bounded. Suppose further that there exists a positive integer j, such that for every $\varepsilon > 0$, there exists a strip neighborhood U, such that for all positive integers n and for all z in $\alpha^{-nj}(U)$, $|\psi(\alpha^n(z))\alpha^{n'}(z)^{-2} - \psi(z)| < \varepsilon\lambda^2(z)$. Then ψ is in $B(\alpha)$.*

Proof. We apply the hypothesis j times. For z in $\alpha^{-nj}(U)$,

$$|\alpha^{jn^*}\psi - \alpha^{j(n-1)^*}\psi| \leq \varepsilon\alpha^{n(j-1)^*}\lambda^2,$$

$$|\alpha^{j(n-1)^*}\psi - \alpha^{j(n-2)^*}\psi| \leq \varepsilon\,\alpha^{n(j-2)^*}\lambda^2, \ldots$$

$$|\alpha^{j^*}\psi - \psi| \leq \varepsilon\,\lambda^2.$$

Summing these inequalities and using the hyperbolic distortion lemma, we find that for every z in $\alpha^{-jn}(U)$, $|\alpha^{jn^*}\psi(z) - \psi(z)| \leq (\text{const})j\varepsilon\lambda^2(z)$. A simple modification of this argument applies to arbitrary integers of the form $jn + m$, where m is between 0 and $j - 1$. $\qquad\qquad\qquad\qquad\qquad\qquad\qquad\qquad\square$

Lemma 7. (**Schwarzian distortion**) *There is a constant ε_0 with $0 < \varepsilon_0 \leq 1/2$ depending on α, such that for every ε with $0 < \varepsilon < \varepsilon_0$ and every z and ζ in W_n with $|z - \zeta| < \varepsilon\varepsilon_0^n$,*

$$A) \quad \left| \frac{\alpha^{n\prime}(z)\alpha^{n\prime}(\zeta)}{(\alpha^n(z) - \alpha^n(\zeta))^2} - \frac{1}{(z - \zeta)^2} \right| < \frac{\varepsilon}{|z - \zeta|^2}.$$

Moreover,

$$B) \quad \left| \frac{\alpha^{n\prime}(z)^2\alpha^{n\prime}(\zeta)^2}{(\alpha^n(z) - \alpha^n(\zeta))^4} - \frac{1}{(z - \zeta)^4} \right| < \frac{\varepsilon}{|z - \zeta|^4}.$$

Proof. The Schwarzian derivative $\mathcal{S}f$ of a holomorphic function f is equal to

$$\mathcal{S}f(z) = 6 \lim_{\zeta \to z} \frac{\partial^2}{\partial z \partial \zeta} \log \frac{f(z) - f(\zeta)}{z - \zeta} = 6 \lim_{\zeta \to z} \left(\frac{f'(z)f'(\zeta)}{(f(z) - f(\zeta))^2} - \frac{1}{(z - \zeta)^2} \right).$$

We define the bi-Schwarzian derivative S to be the same expression without the constant 6 and without taking the limit as ζ approaches z:

$$Sf(z, \zeta) = \frac{\partial^2}{\partial z \partial \zeta} \log \frac{f(z) - f(\zeta)}{z - \zeta}.$$

The composition law for the bi-Schwarzian is

$$S(f \circ g)(z, \zeta) = (Sf)(g(z), g(\zeta))g'(z)g'(\zeta) + Sg(z, \zeta).$$

Since α is assumed to be univalent in an open neighborhood of W_1, $S\alpha(z)$ is bounded on W_1 and the bi-Schwarzian $S\alpha(z, \zeta)$ is bounded on $W_1 \times W_1$. On taking the bi-Schwarzian of α^n, we obtain

$$(S\alpha)(\alpha^{n-1}(z), \alpha^{n-1}(\zeta))\alpha^{n-1\prime}(z)\alpha^{n-1\prime}(\zeta)$$

$$+(S\alpha)(\alpha^{n-2}(z), \alpha^{n-2}(\zeta))\alpha^{n-2\prime}(z)\alpha^{n-2\prime}(\zeta) + \cdots + S\alpha(z, \zeta)$$

and, on letting K_1 be a bound for $S\alpha$ on $W_1 \times W_1$ and k_1 a bound for α' on W_1, we see that if z and ζ are in W_n this expression is bounded by

$$K_1\{1 + k_1^2 + \cdots + (k_1^2)^{n-1}\} \leq K_1 \frac{k_1^{2n} - 1}{k_1^2 - 1} \leq (\text{const})k_1^{2n}.$$

Thus, if $|z - \zeta| < \varepsilon\varepsilon_0^n$, then $|z - \zeta|^2$ multiplied by the left hand side of A) is less than $(\text{const})\varepsilon^2\varepsilon_0^{2n}k_1^{2n}$. Part A) follows by choosing ε_0 small enough so that $\varepsilon_0 k_0 \leq 1$ and $(\text{const})\varepsilon_0 \leq 1$.

To prove part B), we apply the factorization $C^2 - D^2 = (C - D)(C + D)$ and the result of part A). On multiplying both sides of the inequality in B) by $|z - \zeta|^4$, the result follows from part A and the bound

$$\left| \frac{\alpha^{n'}(z)\alpha^{n'}(\zeta)(z - \zeta)^2}{(\alpha^n(z) - \alpha^n(\zeta))^2} \right| \le k_2^{2n}$$

where k_2 is chosen so that $|\alpha'(p)/\alpha'(q)| < k_2$ for every p and q in W_1. Of course, the maximum value of $|\alpha''/\alpha'|$ times the Euclidean diameter of $W_1 \cap$ (the horizontal strip between $y = 0$ and $y = 2\pi$) is a bound for $\log k_2$. To guarantee the inequality in part B, we choose ε_0 so that $\varepsilon_0 k_1 k_2 \le 1$ and (const)$\varepsilon_0 \le 1$. This completes the proof of the Schwarzian distortion lemma. □

The $\mathcal{L}$-operator is defined by the formula

$$\mathcal{L}\mu(z)dz^2 = \psi(z)dz^2 = \int\int_{V_0} \frac{\mu(\zeta)d\zeta d\bar{\zeta}}{(\zeta - z)^4}dz^2.$$

In this definition we could take the domain of integration to be the lift of any annular neighborhood of the circumference of the unit circle, since, in the end, we only care about the equivalence class of ψ modulo B_0.

Theorem 5 *The $\mathcal{L}$-operator is a bounded linear mapping from the Banach space $L_\infty(\alpha)$ to the Banach space $B(\alpha)/B_0$.*

Proof. In the classical case, the domain of integration is the whole right hand side, R, and α is a Möbius transformation preserving R. The conditions ii) in the definitions of $L_\infty(\alpha)$ and $B(\alpha)$ are replaced by exact automorphy, (the inequalities are true with $\varepsilon = 0$). Then this Theorem follows by changing variables and using the identity $(\alpha(z) - \alpha(\zeta))^2 = \alpha'(z)\alpha'(\zeta)(z - \zeta)^2$, which is true if α is a Möbius transformation. In the situation at hand we know only that the holomorphic mapping α arises from an expanding polynomial-like mapping. Our substitute for the exact identity valid for Möbius transformations is the Schwarzian distortion lemma.

Let $K(z, \zeta) = (z - \zeta)^{-4}, \alpha^* K(z, \zeta) = \alpha'(z)^2 \alpha'(\zeta)^2 \left[\alpha(z) - \alpha(\zeta) \right]^4$ and $\alpha^*\mu(z) = \mu(\alpha(z))\frac{\overline{\alpha'(z)}}{\alpha'(z)}$. To prove the theorem we must estimate

$$\mathcal{L}\varphi(\alpha^n(z))\alpha^{n'}(z)^2 - \mathcal{L}\varphi(z).$$

We write this difference as a sum of three terms:

$$I(z) = \int\int_{V_n} (\alpha^{n^*}\mu)(\zeta)(\alpha^{n^*}K(z, \zeta) - K(z, \zeta))d\xi d\eta,$$

$$II(z) = \int\int_{V_n} \left[(\alpha^{n^*}\mu)(\zeta) - \mu(\zeta) \right] K(z, \zeta)d\xi d\eta \text{ and}$$

$$III(z) = \int\int_{V_0 - V_n} \mu(\zeta)K(z, \zeta)d\xi d\eta.$$

Each of these integrals is a periodic function of z and holomorphic for values of z in L. To prove ii) in the definition of $B(\alpha)$ we repeatedly use the fact that

$$\int\int_{|\zeta - z| > \gamma} |\zeta - z|^{-4} d\xi d\eta = \pi\gamma^{-2}.$$

From the bootstrapping lemma, it suffices to find j such that for all $\varepsilon > 0$, there exists a k, such that for all $n \geq 0$ and for all z in $\alpha^{-jn}(U_k) = U_{jn+k}$, $|\alpha^{n^*}\psi(z) - \psi(z)| < \varepsilon\lambda(z)^2$. From the geometric neighborhood lemma, select β_1 and k so that $U_m \subseteq U_{m-k}(\beta_1)$. If z is in U_{jn+k}, then $\delta(z) < \beta_1^{jn}$. Select j so that $\beta_1^j < \varepsilon_0$, where ε_0 is the value in the Schwarzian distortion lemma. To estimate the integral $I(z)$, take a disk of radius ε_0^n centered at z and write the integral as a sum of two integrals, the first over the $V_n \cap$(the disk) and the second over $V_n \cap$(the complement of the disk). By the Schwarzian distortion lemma, the first integral is bounded by

$$\varepsilon||\mu||_\infty 2\pi \int_{\delta(z)}^\infty \frac{dr}{r^3} \leq \varepsilon\pi||\mu||_\infty \delta(z)^{-2}.$$

The second integral is bounded by $\pi||\mu||_\infty(\varepsilon_0^{-2n} + \varepsilon_0^{-2n}k_2^{2n})$, where k_2 is the number in the proof of part B) of Lemma 7. Now, put the additional condition on j that $\beta_1^j < \varepsilon_0/k_2$ and we obtain the desired estimate.

To estimate $II(z)$, the assumption ii) in the definition of μ implies, for every $\varepsilon > 0$, there exists an n_0, such that for all $n \geq 0$ and for all z with $\delta(z) < \varepsilon_0^{n+n_0}$, $|P^{n^*}\mu(z) - \mu(z)| < \varepsilon$. Consider a disk centered at a point z in L and of radius $\varepsilon_0^{n+n_0}$. The integral $II(z)$ is bounded by the sum of an integral over $V_n \cup$(this disk) plus an integral over $V_n \cup$(the complement of this disk). The first integral is bounded by a constant times

$$\varepsilon \int_{\delta(z)}^{\varepsilon_0^{n+n_0}} \frac{dr}{r^3} \leq \varepsilon \text{ (const) } \delta(z)^{-2}.$$

The second integral is bounded by $2||\mu||_\infty(1/\varepsilon_0^{2n+2n_0})$. Pick β_1 so that

$$\alpha^{-n}(U_0) \subseteq U_{n-k}(\beta_1).$$

If j is so large that $\beta_1^j < \varepsilon_0$ and if $\delta(z) < \beta_1^{nj}$, then this term grows much more slowly than $\delta(z)^{-2}$ and we again obtain the desired estimate.

To estimate $III(z)$, select systems of geometric neighborhoods $V_n(\varepsilon_0)$ and $U_n(\beta_1)$ and a positive integer k, such that $V_{n+k}(\varepsilon_0) \subseteq V_n$ and $U_n \subseteq U_{n-k}(\beta_1)$. Then $III(z)$ is bounded by $\pi||\mu||_\infty/(\text{dist}(z, \text{right hand bdry of } V_n))^2$ which is less than $\pi||\mu||_\infty/(\delta(z) + \varepsilon_0^{n+k})^2 = \pi||\mu||_\infty/\delta(z)^2(1 + \varepsilon_0^{n+k}/\delta(z))^2$. If we pick z in U_{jn+k}, then z is in $U_{jn}(\beta_1)$ then $\varepsilon_0^{n+k}/\delta(z) > \varepsilon_0^{n+k}/\beta_1^{jn}$ and, by choosing j large enough, this fraction is as large as we like. We obtain the desired inequality and this concludes the proof of the theorem. $\square$

Theorem 6. *The Bers' $\mathcal{L}$-operator induces an isomorphism from $B(\alpha, R)/B_0$ onto $B(\alpha, L)/B_0$, where B_0 is the Banach space of periodic bounded cusp forms which vanish at C.*

Proof of theorem. We define a modified $\mathcal{L}$-operator, $\hat{\mathcal{L}}$ by

$$\hat{\mathcal{L}}\psi(z) = \int\int_{V_0} \frac{\lambda^{-2}(\zeta)\overline{\psi(\zeta)}d\xi d\eta}{(\zeta - z)^4}.$$

We have already observed that the hyperbolic distortion lemma implies that if ψ is an element of $B(\alpha, R)$ then $\lambda^{-2}\overline{\psi}$ is an element of $L_\infty(\alpha, R)$. The previous theorem then implies that $\hat{\mathcal{L}}\psi$ is an element of $B(\alpha, L)$. It is obvious that $\hat{\mathcal{L}}$ preserves the periodic bounded cusp forms which vanish at C. It is shown by Bers in [1] that $\mathcal{L}$ from $L_\infty(R)$ to $B(\alpha, L)$ and that $\hat{\mathcal{L}}$ from $B(R)$ to $B(L)$ is an injection. Thus $\hat{\mathcal{L}}$ restricted to $B(\alpha, R)$ is injective. Bers' method of proof relies on a reproducing formula and an antiquasiconformal involution χ which fixes the curve C pointwise, which is periodic and which commutes with α. Thus, the same proof, using his reproducing formula, shows that $\hat{\mathcal{L}}$ gives a bijection from $B(\alpha, R)/B_0$ to $B(\alpha, L)/B_0$.   □

Section 7. Quadratic differentials for arbitrary P

The goal of this section is to show that, for all quadratic-like mappings P, the Banach spaces $B(P)/B_0$ are isomorphic. We create a natural isomorphism between any two of these spaces by using the $\mathcal{L}$-operator and imitating the construction of quasifuchsian groups by conformal welding.

Consider the universal covering of the punctured disk $\Delta^* = \mathbb{C} - \overline{\Delta}$. This covering can be realized by the right half plane, $Y = \{\zeta = \xi + i\eta : \xi > 0\}$, with covering mapping $\pi(\zeta) = \exp(\zeta) = z$. The mapping $P_0(z) = z^2$ lifts to $\alpha_0(\zeta) = 2\zeta$. The quadratic differentials φ in $B(P_0)$ lift to differential forms $\psi(\zeta) = \varphi(\pi(\zeta))\pi'(\zeta)^2$. The differential forms $\psi(\zeta)$ corresponding to elements of $B(P_0)$ by this formula form the Banach space $B(\alpha_0, Y)$. We use the notation B_0 to denote the cusp forms which vanish at the boundary. In the context of $B(P)$, the space B_0 consists of holomorphic functions such that $|\lambda^{-2}(z)\varphi(z)|$ approaches zero as z approaches $|z| = 1$. In the context of $B(\alpha_0, Y)$, B_0 consists of periodic functions such that $|\psi(\zeta)\xi^{-2}|$ approaches zero as ξ approaches zero. With this understanding, it is obvious that lifting induces an isomorphism from $B(P_0)/B_0$ onto $B(\alpha_0, Y)/B_0$.

Let $P(z)$ be any quadratic-like mapping giving a double covering of $\mathcal{A}_1$ over $\mathcal{A}_0$ where $\mathcal{A}_1$ and $\mathcal{A}_0$ are annular neighborhoods of $|z| = 1$ in Δ^* and $\mathcal{A}_1 \subset \mathcal{A}_0$. Let $\mathcal{B}_0$ and $\mathcal{B}_1$ be equal to $\pi^{-1}(\mathcal{A}_0)$ and $\pi^{-1}(\mathcal{A}_1)$, respectively. The lift of P is a one-to-one mapping α from $\mathcal{B}_1$ to $\mathcal{B}_0$. There is a quasiconformal mapping which conjugates P_0 into P in an annular neighborhood of the boundary of Δ^*. This mapping lifts to a quasiconformal mapping f from a strip neighborhood of the y-axis in the right half plane to another strip neighborhood of the y-axis in the right half plane which conjugates α_0 to α_1.

Let ν be equal to $f_{\bar{z}}/f_z$ in $\mathcal{B}_0$, which is a periodic strip domain bordering the y-axis and contained in the right half plane, Y. Let ν be identically zero elsewhere. Let g be a quasiconformal homeomorphism of the whole plane with Beltrami coefficient ν and which is normalized to fix 0, i and ∞. If r is the Riemann mapping from $g(Y)$ onto Y, suitably normalized at 0, i, and ∞, then $r \circ g = f$ in Y and g conjugates α_0 to an expanding mapping α mapping a strip neighborhood $W_1 = U_1 \cup C \cup V_1$ over a strip neighborhood $W_0 = U_0 \cup C \cup V_0$, where C is the periodic quasicircle equal to g(imaginary axis). Moreover, r conjugates α to α_1.

We have an isomorphism from $B(\alpha_0, Y)/B_0$ onto $B(\alpha_1, Y)/B_0$ induced by g and r in the following way. The conformal mapping r induces by pullback an isomorphism from $B(\alpha_1, Y)/B_0$ onto $B(\alpha, R)/B_0$. The operator $\hat{\mathcal{L}}$ is an isomorphism from $B(\alpha, R)/B_0$ onto $B(\alpha, L)/B_0$. The mapping g, which is conformal in the left half plane, induces by pullback an isomorphism from $B(\alpha, L)/B_0$ onto $B(\alpha_0, \mathbb{C} - Y)/B_0$ and reflection gives an isomorphism from $B(\alpha_0, \mathbb{C} - Y)/B_0$ onto $B(\alpha_0, Y)/B_0$. We deduce the following theorem.

Theorem 7. *The Banach space $B(P)/B_0$ is isomorphic to $B(P_0)/B_0$ for all polynomial-like mappings P in the Teichmüller space $QS(P_0)$ mod S and, thus, for any for any polynomial-like P, the space $B(P)/B_0$ is infinite dimensional.*

Section 8. The complex manifold structure on T

We have created all the background necessary to copy the Ahlfors-Bers technique for providing complex coordinate charts for the Teichmüller space $T = QS(P)/S$.

Theorem 8. *Assume P is a locally polynomial-like element of UAC and that h represents an element in $QS(P)/S$ and that the Teichmüller distance from h to the identity is less than $\log 2$. Then $h = h^{\mu}$ for some unique Beltrami differential of the form $\lambda^{-2}\bar{\varphi}$, where φ is an element of $B(P)/B_0$. These correspondences give local charts centered at the quadratic-like elements of UAC and make T into a complex manifold modeled on the Banach space $B(P)/B_0$.*

Proof. In [6] it has already been shown that the Ahlfors-Weill section makes QS mod S into a manifold modelled on the Banach space B/B_0. The proof of this theorem involves repeating the same argument applied to the logarithmic coverings of the mapping P. The Ahlfors-Weill charts constructed in this manner are centered at every polynomial-like element of the Teichmüller space. But by lemma 2, such points are dense in the Teichmüller space and we have a uniform lower bound on the Teichmüller radius of the open sets on which these charts are defined. $\qquad\square$

Recently, Jeremy Kahn has shown that any of the coordinate chart mappings described in this theorem are globally one-to-one. They therefore embed T as a domain in $B(P)/B_0$. The argument is elementary and applies to the $QS mod S$ Teichmüller space described in [6].

Appendix. Correspondences between the UAC theory, the Riemann surface lamination theory and the $U\tilde{A}A$ theory

There is a dictionary between the uniformly asymptotically conformal (uac) theory described here, the Riemann surface lamination theory described in [10], and the one real variable uniformly asymptotically affine (uaa) theory of [7].

1) Here we begin with the germ of an expanding conformal mapping F of degree $d \geq 2$ in the exterior of $|Z| = 1$. The inverse limit of $\ldots \to F^{-2}U \to F^{-1}U \to U$ is a solenoidal surface $\tilde{L}$ with an injective onto mapping $\tilde{F} = \underleftarrow{\lim} F$ defined on a subset of $\tilde{L}$ to $\tilde{L}$. The space of orbits of $\tilde{F}^{-1}$ acting on $\tilde{L}$ is a compact solenoidal surface which is studied in [10].

The ideal boundary of $\tilde{L}$ is a one dimensional solenoid which is the inverse limit of $f = F$ restricted to $|Z| = 1$. The solenoid $\tilde{S}$ has a self-mapping $\tilde{f} = \underleftarrow{\lim} f$. These are the topological objects studied in [7].

2) A quasiconformal structure preserved by F in the UAC setting here determines a quasisymmetric structure on $\tilde{S}$ preserved by $\tilde{f}$ and a quasiconformal structure on $L = \tilde{L}/orbits\ of\ \tilde{F}$. The charts of the latter quasiconformal structure preserve the germs of fibres of the projections $\tilde{L} \to F^{-n}U$ which is called a topological or quasiconformal TLC structure for $\tilde{L}$ and L. TLC is short for transversally locally constant, because this structure gives meaning to the idea of objects on L which are locally constant in the transverse direction.

3) A UAC conformal structure for $F : F^{-1}U \to U$ (and all those at zero distance, in the sense of this paper, from the given one) determines a precise conformal structure for L which is a) bounded measurable on the leaves of L relative to the quasiconformal structure of 2), and b) transversally continuous in the essential sup norm topology on the bounded measurable conformal structures. The latter topology makes sense in the presence of TLC structure on L mentioned in 2) arising from the inverse limit projections. This TLC discussion and an analagous smooth discussion appear in [10]. The smooth theory is used there to show that any two TLC structures on L, topological or quasiconformal, are homeomorphic or quasiconformally homeomorphic.

The UAC structure for $F : F^{-1}U \to U$ also determines a uaa structure uniformly asymptotically affine for the boundary map $f : \partial_+ U \to \partial_+ U$ where $\partial_+ U$ means the circle $\{|Z| = 1\}$. The concept of uaa structure for f is discussed in [7] and abstracts the basic distortion property familiar for $C^{1+\alpha}$ – smooth expanding mappings of the circle. It says that relative to a symmetric structure the branches of f^{-n} are uniformly close to preserving midpoints at fine scales (cf UAC here).

4) The hyperbolic distortion lemma, which is Lemma 3 in this paper, gives a continuous family of hyperbolic structures on the leaves of the solenoidal Riemann surfaces $\tilde{L}$ and L. Such hyperbolic structures exist in general thanks to A. Verjovsky [11] and A. Candel [2]. Passing to the boundary of $\tilde{L}$ which is the solenoid $\tilde{S}$ one obtains an $\tilde{f}$ invariant family of affine structures on the leaves of $\tilde{S}$ discussed in [7]. These affine structures are transversally continuous and are labeled by scaling functions in [7]. Transverse Hölder continuity for the

affine structures or scaling functions on the circle corresponds to $C^{1+\alpha}$, $0 < \alpha$, expanding systems. It is known $C^{1+\alpha}$ systems are classified by the Holderian Gibbs interaction potentials (or cocycles), see [8]. It is also known that the eigenvalues at periodic points comprise a complete system of invariants for the $C^{1+\alpha}$ systems, [9]. Recall then eigenvalues were mentioned above as the moduli of the annular leaves of L. Thus conformal structures on the solenoidal surface L or scaling functions on the solenoid $\tilde{S}$ generalize the Gibbs theory and the corresponding eigenvalue invariants. The new theory, which is called the UAC theory here, the lamination theory in [10] or the solenoidal theory in [7] is a natural closure or completion of the Gibbs theory.

5) Following the laws of the QS/S proof in [6] we have here shown the set of Teichmüller eigenvalence classes of conformal structures (either UAC for F or on L after we verify the identification alluded to 2)) form a Banach complex manifold and the Teichmüller metric is infinitessimally given by a natural Finsler metric. In [9], a different scheme was employed for proving this result. The latter proof scheme was possible because of the existence of the laminations. It is based on the idea that the Teichmüller equivalence classes of conformal structures are the leaves of a locally trivial holomorphic foliation. In the QS/S context the foliation is topological with holomorphic leaves and holomorphic quotient but there is no holomorphic cross section showing holomorphic local triviality of the foliation by Teichmüller equivalence classes. This distinction is not well known and shortens the basic argument for the classical Teichmüller spaces.

6) The UAC presentation of the Teichmüller theory of conformal structures on the solenoidal surface L fits well with the technology requirements of the renormalization theory of [3], [9] and [10]. Namely, the topological picture for the Julia sets explaining the little copies of the Mandelbrot set inside the big one leads to uncountably many holomorphic mappings between the Teichmüller space of L. These are all naturally defined at the level of measurable conformal structure on L transversally continuous in the ess sup norm topology. At this level the mapping is a restriction to the complement of a fractal set, followed by an extension by standard structure on the fractal set.

To prove universality properties generalizing Feigenbaum's discovery one uses apriori complex bounds for the renormalization of TLC points of the Teichmüller space of L and the almost geodesic property for very coherent Beltrami paths, cf [3,5,9 and 10].

7) Here is a list of how the lemmas of this paper correspond to statements about Riemann surface laminations. Lemma 1 says that the Beltrami coefficients of allowable deformations vary continuously in the essential supremum topology across the leaves of the lamination. Lemma 2 establishes that the leaves with TLC structure are dense in the Teichmüller space. Lemma 3 asserts that for laminations with TLC structure, the natural hyperbolic structure on the leaves of the lamination varies continuously in the Cantor direction. Lemma 4 gives a way to recognize linearly independent tangent vectors in the Teichmüller space. Lemmas 5, 6 and 7 show that when L has TLC structure it is simultaneously uniformized with any other given lamination in the same Teichmüller space. The two laminations are glued together along a solenoidal locus.

BIBLIOGRAPHY

1. Bers, L., "A non-standard integral equation with applications to quasi-conformal mapping," *Acta Math.* **116** (1966) 113-134.

2. Candel, A., "The uniformization theorem for surface laminations," Ann. Sci. Ecole Normale Superiere, to appear, 1993.

3. De Melo, W. and Van Strein, S., *One-dimensional Dynamics*, Springer-Verlag, 1993.

4. Gardiner, F. P., "An analysis of the group operation in universal Teichmüller space," *Trans. Am. Math. Soc.* **132** (1968) 471-486.

5. ___,"On Teichmüller contraction," Proc. Amer. Math. Soc., **118**, no.3 (1993) 865-875.

6. Gardiner, F. P. and Sullivan, D. P., "Symmetric and quasisymmetric structures on a closed curve," accepted by *Am. J. of Math.*, vol. **114** (1992) 683-736.

7. Pinto, A. and Sullivan, D., "The circle and the solenoid," to appear.

8. Sullivan, D. P., "Quasiconformal homeomorphisms in dynamics, topology and geometry," Proc. of the International Congress of Mathematicians, Berkeley, CA, **2** (1986) 1216-1228.

9. ___, "Bounds, quadratic differentials, and renormalization conjectures," to appear in *A.M.S. Centennial Publications* **2** (1992).

10. ___, "Linking the universalities of Milnor-Thurston, Feigenbaum and Ahlfors-Bers," *Topological Methods in Modern Mathematics*, Publish or Perish, 1992.

11. Verjovsky, A., "A uniformization theorem for holomorphic foliations," *Contemporary Mathematics*, **58** (1987).

12. Zygmund, A., *Trigonometric Series*, 3rd edition, Cambridge University Press, 1987.

13. ___, "Smooth functions," *Duke Math. J.*, **12** (1945) 47-76.

Brooklyn College, CUNY, Brooklyn, NY 11210
(FPGBC@CUNYVM.BITNET)

and

Graduate Center of CUNY, 33 West 42nd Street, NY, NY, 10036
(DPSGC@CUNYVM.BITNET)

Contemporary Mathematics
Volume **169**, 1994

The Geometry of Cycles in the Cayley Diagram of a Group

ROBERT H. GILMAN

Dedicated to the memory of Wilhelm Magnus

ABSTRACT. A study of triangulations of cycles in the Cayley diagrams
of finitely generated groups leads to a new geometric characterization of
hyperbolic groups.

1. Introduction

During the past several years combinatorial group theory has received an
infusion of ideas both from topology and from the theory of formal languages.
The resulting interplay between groups, the geometry of their Cayley diagrams,
and associated formal languages such as the language of all words defining the
identity has led to several developments including the introduction of automatic
groups [**Eps**], hyperbolic groups [**Gro**], and geometric and language-theoretic
characterizations of finitely generated virtually free groups [**MS1**] [**MS2**]. A
group is virtually free if it has a free subgroup of finite index and in particular if it
is finite. In [**MS1**] (together with [**Dun**]) virtually free groups are characterized
as those groups for which a finite set of diagonals suffices to triangulate all cycles
in the Cayley diagram. Our goal here is to investigate triangulations of cycles
for arbitrary finitely generated groups.

From now on all groups under discussion are understood to be finitely generated and all sets of generators finite. G will be a group, S a set of generators
closed under inverse, and Γ the corresponding Cayley diagram. A path γ of
length $n \geq 1$ in Γ is a sequence of group elements $g_0, \ldots, g_n$ with an edge of Γ

1991 *Mathematics Subject Classification*. Primary 20F32; Secondary 20F05, 20F06.
Key words and phrases. Group, triangulation, hyperbolic group, virtually free group.
I thank the Institute for Advanced Study for its hospitality while part of this work was
being done.
This paper is in final form and no version of it will be submitted for publication elsewhere.

from each g_{i-1} to g_i. The label of γ is the product in order of the labels of its edges. If $g_n = g_0$, then γ is a cycle. A word in S represents the identity in G if and only if it is the label of a cycle. Finally $\lceil x \rceil$ stands for the least integer not less than x, and $\log x$ is to the base 2.

DEFINITION 1. *A diagonal triangulation of a circle in the plane is obtained by distinguishing one or more points on the circle and joining them by chords in such a way that*

(1) *No two chords meet in the interior of the circle;*
(2) *The interior of the circle is divided into triangles;*
(3) *Each arc of the circle between two neighboring distinguished points is one side of a triangle.*

Circles with one, two, or three distinguished points are considered to be triangulated without adding any chords.

DEFINITION 2. *A diagonal triangulation of a cycle $\gamma = g_0 \ldots g_n$ in Γ is a diagonal triangulation of a circle with points $p_1, \ldots, p_n$ distinguished and a corresponding labeling of these points, $g_1, \ldots, g_n$, counterclockwise around the circle. For any chord C with endpoints p_i, p_j, a word of minimum length representing $g_i^{-1} g_j$ is called a label of C in the direction from p_i to p_j. The label of the arc of the circle from p_i to p_j is defined to be the label of the corresponding subpath of γ. A triangulation of word w in S representing the identity in G is a triangulation of any cycle with label w.*

From now on we will say simply triangulation instead of diagonal triangulation. A triangulation of γ makes the triangulated circle into a directed labeled graph in which arcs of the circle between adjacent distinguished points are directed counterclockwise, and each chord is construed as two associated edges, one in each direction. The label of any path in this directed graph from a point with label g to one with label g' is a word representing $g^{-1} g'$.

Recall that every choice of generators S determines a metric on G with distance $d(g, g')$ equal to the length of the shortest word in S which represents $g^{-1} g'$. This metric is extended to Γ by making each edge isometric to the unit interval.

DEFINITION 3. *The length of a chord in a triangulation of a cycle is the distance in G between the labels of its endpoints. A k-triangulation is one in which all chords have length at most k, and $\Delta(n)$ is the minimum value of k such that all cycles of length at most n can be k-triangulated. In particular $\Delta(n) = 0$ for $1 \leq n \leq 3$. To display the dependence of $\Delta(n)$ on the generators S write $\Delta_S(n)$.*

Now let us consider triangulations of cycles in Cayley diagrams of arbitrary finitely generated groups.

THEOREM A. *For any group G and set of generators of G, $\Delta(n) \leq \lceil n/3 \rceil$. If, for all sufficiently large n, $\Delta(n) < \lceil n/3 \rceil$, then G is finitely presented and satisfies an exponential isoperimetric inequality.*

The meaning of the last assertion of Theorem A is that there is a constant c such that every word w (in the generators of G) of length n which defines the identity in G is freely equivalent to the product of at most c^n conjugates of the defining relators and their inverses. If this condition holds for one presentation of G, then it holds for all although the value of c depends on the presentation. See [**Ger**] for details.

THEOREM B. *For any group G the following conditions are equivalent.*

(1) *For some set of generators and constant K, $\Delta(n) \leq n/6 + K$;*
(2) *G is a hyperbolic group;*
(3) *For any set of generators there are constants Q and R such that $\Delta(n) \leq Q \log(n) + R$.*

The proof of this theorem relies on the characterization of hyperbolic groups by subquadratic isoperimetric inequalities [**Ol**], [**Pa**]. By hyperbolic groups we mean the word hyperbolic groups of Gromov [**Gro**]. G is hyperbolic if there is a constant δ such that every geodesic triangle in Γ (i.e., triangle whose sides are geodesic segments) has the property that each point on any one side is a distance at most δ from some point on one of the other two sides [**GH**, Proposition 21 of chapter 2]. The validity of this condition is independent of the choice of generators although the value of δ is not. Hyperbolic groups are finitely presented [**Sa**, Proposition 17]. Small cancellation groups satisfying the hypothesis $C'(1/6)$ or the hypotheses $C'(1/4)$ and $T(4)$ are hyperbolic [**Str**], but it is easy to see that $Z \times Z$ is not.

THEOREM C. *If G is hyperbolic, then either*

(1) *G is virtually free, and for any set of generators $\Delta(n)$ is bounded; or*
(2) *G is not virtually free, and for any set of generators there are constants M, P, Q, R such that $M \log n + P \leq \Delta(n) \leq Q \log n + R$.*

Our results show that roughly speaking $\Delta(n)$ is either linear, logarithmic, or bounded, and that the logarithmic case characterizes hyperbolic groups which are not virtually free.

2. Proof of Theorem A and Theorem B

Throughout this section $w = a_1 \ldots a_n$, $n \geq 1$, will stand for a word in the generators of G representing the identity. For a fixed set of relators $\mathcal{R}$, define $\alpha(w)$ to be the least integer such that w is freely equivalent to a product of $\alpha(w)$ conjugates of relators in $\mathcal{R}$. If there is no such product, $\alpha(w) = \infty$. For $n \geq 1$ define $\beta(n) = \max_{1 \leq |w| \leq n} \alpha(w)$.

PROOF OF THEOREM A. To prove the first assertion of Theorem A construct a triangulation of w by picking distinguished points $p_1, \ldots, p_n$ on a circle in the plane, labeling them with the group elements represented by the successive prefixes of w, and drawing chords

(1) From p_n to p_i for all i with $2 \leq i \leq \lceil n/3 \rceil$; and
(2) From $p_{\lceil n/3 \rceil}$ to p_i for all i with $\lceil n/3 \rceil + 2 \leq i \leq 2\lceil n/3 \rceil$; and
(3) From $p_{2\lceil n/3 \rceil}$ to p_i for all i with $2\lceil n/3 \rceil + 2 \leq i \leq n$.

For the second part of Theorem A assume $\Delta(n) < \lceil n/3 \rceil$ for all $n \geq N \geq 4$, and let $\mathcal{R}$ be the set of all relators of length at most N. We will show $\beta(n) \leq 2^n$. Clearly $\beta(n) \leq 1$ if $n \leq N$; we may assume $n > N$. By induction on n, $\beta(n-1) \leq 2^{n-1}$, so we need only show $\alpha(w) \leq 2^n$. By hypothesis w has a triangulation with all chords of length less than $\lceil n/3 \rceil$. Because chord length is an integer, all chords have length less than $n/3$. Any chord C divides the circle into two arcs with labels $w_3 w_1$ and w_2 where $w = w_1 w_2 w_3$. Choosing a label v for C in the appropriate direction, we see that $w_1 v w_3$ and $w_2 v^{-1}$ are both words representing the identity in G. It follows that $\alpha(w) \leq \alpha(w_1 v w_3) + \alpha(w_2 v^{-1})$. Since labels of chords have minimum length, $|v| \leq |w_1 w_3|$ and $|v| \leq |w_2|$. If both these inequalities are strict, then by induction $\alpha(w) \leq \alpha(w_1 v w_3) + \alpha(w_2 v^{-1}) \leq 2^{n-1} + 2^{n-1} \leq 2^n$.

It remains to find a chord C which divides the circle into two arcs both longer than C. Since $|w| > N \geq 4$, the triangulation does have chords. Any chord divides the circle into two arcs of length, say, d and e. Pick C with d as large as possible subject to $d \leq e$. We claim $d \geq n/3$. To see this observe that C is one side of a triangle T which has its third vertex on the arc of length e. This third vertex divides that arc into two shorter arcs of length d' and d'' with $e = d' + d''$, and our claim will follow from $d' \leq d$ and $d'' \leq d$. If $d' = 1$, then clearly $d' \leq d$. Otherwise the corresponding side of T is not a subarc of the circle but a chord C' which divides the circle into arcs of length d' and $d + d''$. Since $d + d'' > d$, our choice of C implies first that $d + d'' > d'$ and consequently that $d' \leq d$. Thus $d' \leq d$ in all cases; and as $d'' \leq d$ by symmetry, our claim is valid. Since C has length less than $n/3$, it is the desired chord. $\square$

PROOF OF THEOREM B. The proof that (3) implies (1) is straightforward and is omitted. Assume (1) holds; that is, $\Delta(n) \leq n/6 + K$. By [Ol] or [Pa] conclusion (2) holds once we know that G has a subquadratic isoperimetric inequality, i.e., $\lim_{n \to \infty} \beta(n)/n^2 = 0$. Take N to be an integer larger than $\max\{3, 1000K\}$. We will show that $\beta(n) \leq \beta(N)n^{1.9}$.

If $n \leq N$, there is nothing to prove, so assume $n > N$; by induction on n it suffices to show $\alpha(w) \leq \beta(N)n^{1.9}$. As in the proof of Theorem A find a chord C which divides the circle into two arcs of lengths $e \geq d \geq n/3$. Note that $e = n - d \leq 2n/3$ and $d \leq n/2$. Since C has length at most $\Delta(n)$, $\alpha(w) \leq \beta(d + \Delta(n)) + \beta(e + \Delta(n)) \leq \beta(d + n/6 + K) + \beta(e + n/6 + K)$. Because

$d + n/6 + K \le e + n/6 + K \le 2n/3 + n/6 + K < n$, induction on n yields

$$\alpha(w) \le \beta(d + n/6 + K) + \beta(n - d + n/6 + K)$$
$$\le \beta(N)n^{1.9}\big((d/n + 1/6 + .001)^{1.9} + (1 - d/n + 1/6 + .001)^{1.9}\big)$$
$$\le \beta(N)n^{1.9}$$

where $(d/n + 1/6 + .001)^{1.9} + (1 - d/n + 1/6 + .001)^{1.9} \le 1$ follows from $1/3 \le d/n \le 1/2$.

To complete the proof of Theorem B we will show that (2) implies (3) by proving that if G is hyperbolic, then $\Delta(n) \le C \log(n)$ for some constant C. By assumption G is $\delta/4$ hyperbolic for some δ (this odd choice of δ is made to correspond to the hypothesis of [**CDP**, Lemma 1.6 of Chapter 3], which will be employed later). Since it does no harm to increase δ, assume $\delta \ge 1$, and take $C > 10\delta$.

Let $\gamma = g_0 \ldots g_n$ be a cycle in Γ. As $\Delta(n) = 0$ for $n = 1, 2, 3$, we may assume $n \ge 4$. Choose distinguished points $p_1, \ldots, p_n$ on a circle as in Definition 1, and give each p_i the label g_i as in Definition 2. Start constructing a triangulation by adding a chord from p_n to p_2. As this chord has length at most 2, we are done if $n = 4$. Otherwise it suffices to show that whenever a chord of length at most $C \log(n)$ has endpoints p_i, p_j with $3 \le j - i$, then we can add a chord from p_i to p_{j-1} or from p_{i+1} to p_j or we can add chords from p_i and p_j to p_k for some k with $i + 2 \le k \le j - 2$. In other words

(1) $d(g_i, g_{j-1}) \le C \log(n)$; or

(2) $d(g_{i+1}, g_j) \le C \log(n)$; or

(3) $d(g_i, g_k), d(g_j, g_k) \le C \log(n)$ for some k with $i + 2 \le k \le j - 2$.

Suppose (1) and (2) do not hold. Thus $d(g_i, g_j) > C \log(n) - 1$. Let γ' be a geodesic in Γ from g_i to g_j and consider a ball of radius $r = (C/2) \log(n) - 2$ around the midpoint x of γ'. If the part of γ from g_i to g_j intersects the ball, then there is a vertex g_k on γ with $i \le k \le j$ and with g_k a distance at most $(C/2) \log(n) + r + 1 \le C \log(n) - 1$ from each endpoint. Since $d(g_i, g_j) > C \log(n) - 1$, we have $i < k < j$; and one of the conditions above must hold.

If γ does not intersect the ball, then by [**CDP**, Lemma 1.6 of Chapter 3], $j - i \ge \delta(2^{(r/\delta)-1} - 2)$. As $\delta \ge 1$ and $i \ge 2$, we have $n \ge j \ge j - i + 2 \ge 2^{(r/\delta)-1} = 2^{C \log(n)/2\delta - 2/\delta - 1} \ge n^{C/2\delta} 2^{-3} \ge n^5/8$. But $n > n^5/8$ is impossible as $n \ge 2$. $\square$

3. Proof of Theorem C

LEMMA 3.1. *If S and S' are generating sets for G, there is a constant K such that $\Delta_S(n) \le K \Delta_{S'}(Kn) + K$.*

PROOF. First note that given a triangulation of the circle in the sense of Definition 1, with distinguished points $p_1, \ldots, p_n$, $n > 3$, we can generate a new triangulation with $n - 1$ distinguished points by allowing a point p_i to move counterclockwise on the circle until it becomes identified with p_{i+1}. Here $i+1, i-1$ etc. are understood modulo n. Let p_j be the third vertex of the triangle whose other two vertices are p_i, p_{i+1}.

(1) If $j \notin \{i+2, i-1\}$, then the chord $\overline{p_i p_j}$ is identified with the chord $\overline{p_{i+1} p_j}$, and the triangle $\overline{p_i p_{i+1} p_j}$ disappears. All other triangles with p_i as a vertex have that vertex replaced by p_{i+1}.

(2) If $j = i+2$, then the chord $\overline{p_i p_{i+2}}$ is identified with the edge $\overline{p_{i+1} p_{i+2}}$, and the triangle $\overline{p_i p_{i+1} p_{i+2}}$ disappears. All other triangles with p_i as a vertex have that vertex replaced by p_{i+1}.

(3) If $j = i-1$, then the original triangulation has the triangle $\overline{p_{i-1} p_i p_{i+1}}$; and the new triangulation is obtained by removing the chord $\overline{p_{i-1} p_{i+1}}$.

Now choose K_1 so that each generator in S can be expressed as a word of length at most K_1 in S' and vice-versa. Let Γ and Γ' be the Cayley diagrams of G with respect to S and S' respectively, and take d and d' be the corresponding metrics. We have $(1/K_1)d'(g,h) \leq d(g,h) \leq K_1 d'(g,h)$. Suppose $\gamma = g_0 \ldots g_n$ is a cycle of length n in Γ. If $n \leq 3$, then there is nothing to prove as $\Delta_S(n) = 0$. Thus we may assume $n > 3$. Since $d(g_{i-1}, g_i) \leq 1$ implies $d'(g_{i-1}, g_i) \leq K_1$, γ can be expanded to a cycle γ' of length $K_1 n$ or less in Γ' by interpolating at most $K_1 - 1$ additional vertices between each g_{i-1} and g_i.

Consider a $\Delta_{S'}(K_1 n)$-triangulation of γ' with distinguished points $p_1, \ldots, p_n$ corresponding to the original cycle γ and additional points $q_{i,1} \ldots q_{i,j(i)}$, $j(i) < K_1$, corresponding to the additional vertices between g_i and g_{i+1}. Modify this triangulation as above so that the points $q_{i,j}$ are all identified with p_{i+1} to obtain a triangulation with distinguished points $p_1, \ldots, p_n$. As each point $q_{i,j}$ moves a distance at most K_1 on the circle, each chord lengthens by at most $2K_1$ in the metric d'. We obtain a $\Delta_{S'}(K_1 n) + 2K_1$-triangulation in terms of d' and consequently a $K_1 \Delta_S(K_1 n) + 2K_1^2$-triangulation of γ with respect to d. $\square$

LEMMA 3.2. *Suppose G is a free product of H and K with finite subgroups of H and K amalgamated or G is an HNN extension with base H and two finite subgroups of H associated. There is a set of generators for H which extends to a set of generators for G in such a way that the subgroup H is isometrically embedded in G with respect to the corresponding metrics d_H, d_G.*

PROOF. We treat the free product case first. Choose a set of generators S_H for H which includes every element of its amalgamated subgroup (and is closed under inverse). Choose S_K likewise for K. $S_G = S_H \cup S_K$ is a set of generators for G. Clearly $d_H(h_1, h_2) \geq d_G(h_1, h_2)$. To prove the reverse inequality suppose u and w are words of minimum length in S_H and S_G respectively both representing $h = h_1^{-1} h_2$. The word w factors uniquely as $w = w_1 \ldots w_n$ in such a way that w_i and w_{i+1} are words in different alphabets from $\{S_H, S_K\}$. Among all words of minimum length representing h we choose w with n minimum. It suffices to show that

(3.1) $|u| \leq |w|$.

If $n > 1$, then no subword w_i represents an element of an amalgamated subgroup. Otherwise w_i could be replaced by a single generator from the other alphabet to obtain a new word which represented the same element of h, was no longer than w, and whose factorization as a product of words in different alphabets had fewer than n terms. The word $u^{-1}w_1 \ldots w_n$ represents the identity, and the first term of its factorization into a product of words from different alphabets is either u^{-1} or $u^{-1}w_1$ depending on whether w_1 is a word in S_K or S_H. By the normal form theorem for free products with amalgamation [LS, Chapter 4] one of the terms in the factorization must represent an element of an amalgamated subgroup, and the only possibility is u^{-1} or $u^{-1}w_1$ respectively. In the first case $|u| \leq 1$ because it represents a word in the amalgamated subgroup of H, and (3.1) follows directly. Likewise in the second case the group element represented by $u^{-1}w_1$ is also represented by a word of length at most 1 whence $|u| \leq |w_1|+1$. Consequently (3.1) holds unless $w = w_1$; but then as w_1 is a word in S_H, (3.1) holds by choice of u.

The HNN case is similar to the free product case. Suppose that G is an HNN extension with base H and stable letter t and that A and B are the associated subgroups of H with $t^{-1}At = B$. Choose generators S_H for H with $A \cup B \subset S_H$, and let $S_G = S_H \cup \{t, t^{-1}\}$. It suffices to prove (3.1) when u and w are chosen as before. More precisely any word w in S_G factors uniquely as $w = w_0 t^{\epsilon_1} \ldots t^{\epsilon_n} w_n$ where $\epsilon_i = \pm 1$, the w_i's are words (possibly empty words) in S_H; and w is chosen with n minimum among words of minimum length representing h. It is straightforward to check that this factorization does not include any subsequences of the form $t^{-1}w_i t$ with w_i representing an element of A or $tw_i t^{-1}$ with w_i representing an element of B. Applying Britton's Lemma [LS, Chapter 4] to $(u^{-1}w_0)t^{\epsilon_1} \ldots t^{\epsilon_n} w_n$, we conclude that $n = 0$ whence $|u| \leq |w_0| = |w|$ by choice of u. $\square$

PROOF OF THEOREM C. If G is virtually free, then conclusion (1) holds by [MS1]; the upper bound of (2) comes from Theorem B. Thus it suffices to show

$$(3.2) \qquad\qquad M \log n + P \leq \Delta(n)$$

when G is hyperbolic but not virtually free. First we show that although the constants M and P may change from one generating set to another, the validity of (3.2) is independent of the choice of generators for G.

Suppose that (3.2) holds for a set of generators S. If S' is another set, then $M \log n + P \leq \Delta_S(n) \leq K\Delta_{S'}(Kn)+K$ by Lemma 3.1. We claim $M' \log m + P' \leq \Delta_{S'}(m)$ for some constants M', P'. If this inequality holds for all but finitely many m, then with a change of P' it holds for all m. Thus we may assume $m \geq 2K$ and choose $n \geq 2$ so that $nK \leq m < (n+1)K$. We obtain $m \leq Kn+K$ and $M \log n + P \leq K\Delta_{S'}(Kn) + K \leq K\Delta_{S'}(m) + K$. The desired inequality follows in a straightforward way.

By the preceding argument it suffices to show that (3.2) holds for one set of generators. Since hyperbolic groups are finitely presented, we may use induction on the accessibility length of G [**Dun**]. The accessibility length of G is the length of the longest series $G = G_0 \supset G_1 \supset \ldots \supset G_n$ such that each G_i has a decomposition as a nontrivial free product with amalgamation or as an HNN extension where one of the factors or the base is G_{i+1} and the amalgamated or associated subgroups are finite. A free product with amalgamation is nontrivial if the amalgamated subgroups are both proper. As G is not virtually free, the results of [**Sta**] imply that the number of ends of G is either 1 or ∞, and in the latter case G has a decomposition of the type mentioned.

Suppose G is a nontrivial free product $G = H *_F K$ with F finite. Choose the generators of Lemma 3.2. It is an immediate consequence of that lemma that H and K are hyperbolic. If H and K are virtually free, then by [**Gre**] or [**KPS**] so is G. Thus we may assume H is not virtually free. As H has shorter accessibility length than G, the induction assumption yields $\Delta_H(n) \geq M \log n + P$, and $\Delta_G(n) \geq M \log n + P$ follows from Lemma 3.2. Likewise if G is an HNN extension with base H and finite associated subgroups, then as before H is hyperbolic but not virtually free by [**Gre**] or [**KPS**]. The induction hypothesis yields $\Delta_H(n) \geq M \log n + P$, and the desired inequality follows.

It remains to deal with the case in which G has one end. Let Γ be the Cayley diagram of G with respect to some set of generators S, and let d be the corresponding word metric. Since G is hyperbolic, geodesic triangles in Γ are δ-thin for some δ [**GH**, Definition 16]. The meaning of δ-thin is that there is a map from the perimeter of the triangle to three lines in the Euclidean plane with a common endpoint such that

(1) The vertices of the triangle are mapped onto the other endpoints of the lines;

(2) The restriction of f to each side of the triangle is an isometry;

(3) Points with the same image under f are a distance at most δ apart.

Since increasing δ does no harm, take δ to be a positive integer. Pick a geodesic segment γ in Γ of length $2n$ with ends g, g' and midpoint 1. As G has one end, there is a path $\gamma' = g_1, g_2, \ldots g_N$ in Γ such that $g_1 = g$, $g_N = g'$, and $d(1, g_i) \geq n$ for all i.

We will find a new path γ'' from g to g'. For each i, $1 < i < N$, pick a geodesic segment from 1 to g_i. Let h_i be the vertex on this segment with $d(1, h_i) = n$ and γ_i the subsegment from 1 to h_i. Define $h_1 = g_1$ and $h_N = g_N$ and take γ_1 and γ_N to be the subsegments of γ from 1 to g_1 and g_N respectively. For each i, $1 \leq i < N$, consider the geodesic triangle with vertices $1, g_i, g_{i+1}$, and whose edges are the geodesic segments from 1 to g_i and g_{i+1} previously chosen together with the edge in γ from g_i to g_{i+1}. As this triangle is δ-thin and the side opposite vertex 1 has length 1, it follows in a straightforward way that $d(h_i, h_{i+1}) \leq 2\delta + 1$. Construct a path from h_1 to h_N by joining each h_i to h_{i+1} with a geodesic segment of length at most $2\delta + 1$. Clearly the distance from 1 to any point on this path is at least $n - (2\delta + 1)$. As the labels of the γ_i's are words of length n in S, there are at most $|S|^n$ distinct γ_i's and hence at most that many distinct h_i's. Thus by deleting loops from the path just constructed, we obtain a path γ'' from h_1 to h_N of length at most $(2\delta + 1)|S|^n$. Further the distance from 1 to any point on γ'' is at least $n - (2\delta + 1)$.

Consider any triangulation of the cycle formed by γ and γ''. This cycle has length at most

$$(3.3) \qquad\qquad L(n) = 2n + (2\delta + 1)|S|^n$$

By [**MS**, Lemma 5] any diagonal triangulation has the property that if the circle is divided into three arcs each beginning and ending at distinguished points, then there is a triangle with vertices on each arc. Thus there is a triangle with vertex h on γ'', and vertices on γ_1 and γ_N. It follows that $d(1, h)$ is at most equal to the sum of the lengths of two sides of this triangle whence $n - 2\delta - 1 \leq d(h, 1) \leq 2\Delta(L(n))$.

Pick any $m \geq 2 + (2\delta + 1)|S|$ and choose n with $L(n) \leq m \leq L(n+1)$. From the preceding paragraph $n/2 - \delta - 1/2 \leq \Delta(L(n)) \leq \Delta(m)$. On the other hand as $m \leq L(n+1)$, (3.3) yields $M \log m + P \leq n/2 - \delta - 1/2 \leq \Delta(m)$ for some constants M, P. $\square$

Acknowledgement

I thank Panagiotis Papasoglu for showing me how to improve an earlier version of Theorem C.

REFERENCES

[CDP] M. Coornaert, T. Delzant, and A. Papadopoulos, *Géométrie et théorie des groupes*, Lecture Notes in Math., vol. 1441, Springer Verlag, Berlin, 1990.

[Dun] M. J. Dunwoody, *The accessibility of finitely presented groups*, Invent. Math. (1985), 449–457.

[Eps] D. B. A. Epstein, J. W. Cannon, D. F. Holt, S. V. F. Levy, M. S. Paterson, and W. P. Thurston, *Word Processing in Groups*, Jones and Bartlett, Boston, 1992.

[Ger] S. M. Gersten, *Dehn functions and ℓ_1 norms of finite presentations*, Algorithms and Classification in Combinatorial Group Theory (G. Baumslag and C. F. Miller III, eds.), Math. Sci. Res. Inst. Publs., vol. 23, Springer Verlag, New York, 1992, pp. 195-224.

[GH] E. Ghys and P. de la Harpe, *Espaces métriques hyperboliques*, Sur les Groupes Hyperboliques d'apres Mikhael Gromov (E. Ghys and P. de la Harpe, eds.), Birkhäuser, Boston, 1990, pp. 27–45.

[Gre] R. Gregorac, *On generalized free products of finite extensions of free groups*, J. London Math. Soc. **41** (1966), 662–666.

[Gro] M. Gromov, *Hyperbolic groups*, Essays in Group Theory (S. M. Gersten, ed.), Springer Verlag, New York, 1987, pp. 75–263.

[KPS] A. Karrass, A. Pietrowski, and D. Solitar, *Finite and infinite cyclic extensions of free groups*, J. Austral. Math. Soc. **16** (1973), 458-466.

[MS1] D. E. Muller and P. E. Schupp, *Groups, the theory of ends and context-free languages*, J. Computer and System Sciences **26** (1983), 295–310.

[MS2] _____, *The theory of ends, pushdown automata, and second-order logic*, Theoretical Computer Science **37** (1985), 51–75.

[Ol] A. Yu. Ol'shanskii, *Hyperbolicity of groups with subquadratic isoperimetric inequality*, International J. Algebra and Computation **1** (1991), 281–289.

[Pa] P. Papasoglu, *On the sub-quadratic inequality for groups*, to appear.

[Sa]   E. Salem, *Premières propriétés des groupes hyperboliques*, Sur les Groupes Hyper-
        boliques d'apres Mikhael Gromov (E. Ghys and P. de la Harpe, eds.), Birkhäuser,
        Boston, 1990, pp. 67–77.
[Sta]   J. Stallings, *Group Theory and Three Dimensional Manifolds*, Yale University Press,
        New Haven and London, 1971.
[Str]   R. Strebel, *Small cancellation groups*, Sur les Groupes Hyperboliques d'apres Mikhael
        Gromov (E. Ghys and P. de la Harpe, eds.), Birkhäuser, Boston, 1990, pp. 227–273.

DEPARTMENT OF MATHEMATICS, STEVENS INSTITUTE OF TECHNOLOGY, HOBOKEN, NEW
JERSEY 07030

E-mail address: rgilman@vaxc.stevens-tech.edu

Contemporary Mathematics
Volume 169, 1994

Braids, Riemann surfaces and Moduli

W. J. HARVEY

ABSTRACT. This is an expository essay on the subject of topological ideas
in complex analysis.

We discuss several contemporary developments in discrete group theory
relating directly to braids and to Riemann surfaces and attempt to trace at
least part of their venerable origins. Each concerns a symbiotic interaction
of complex function theory and geometric group theory through the notion
of monodromy, which figured greatly in nineteenth century mathematics
and which has returned to prominence recently.

Introduction

As a brief historical prelude, with no pretensions to completeness, we discuss
a few salient features of early developments in the theory of discrete groups
and complex analysis. The hypergeometric equation as studied by Euler, Gauss
and Riemann provides an appropriate starting point. In the landmark paper by
H.A. Schwartz of 1873 [18], the analytic continuation of solutions to this equation
with controlled polar behaviour at three singular points was used to introduce the
famous triangle groups and functions. The particular case of the ideal hyperbolic
triangle, where all three angles are zero, leads to Legendre's modular function
$\lambda(\tau)$, which turns out to be automorphic for the congruence subgroup $\Gamma(2)$ of
the classical modular group $SL_2(\mathbb{Z})$, acting on the upper half plane $\{\text{Im } \tau > 0\}$
by fractional linear transformations. We observe (with hindsight) that the group
$\Gamma(2)$ appears here as avatar for the notion of *braid monodromy*, a certain natural
geometric representation of the braid group $B_4(\mathbb{P}^1)$, which acts as the director
for a crucial interplay between the differential equation, the topology of the
punctured plane and the metric structure of hyperbolic geometry.

A further ingredient in this tangled tale is the notion of covering surface.
The study of algebraic plane curves as ramified coverings of the sphere was

1991 *Mathematics Subject Classification.* Primary 32G15, 14H15; Secondary 11F55, 30F35.
This paper is in final form and no version of it will be submitted for publication elsewhere

developed by Clebsch, Lüroth and others; equally, these structures had formed the basis for Riemann's conception of the complex analytic surface defined by an algebraic function and for subsequent work of many authors on complex ordinary differential equations, including that of Schwarz noted above. Later, A.Hurwitz introduced a systematic method for the study of ramified coverings, using what amounted to permutation monodromy representations of braid groups some 30 years before the latter received an official definition. This viewpoint, which is vital for our considerations, was reactivated by Wilhelm Magnus in the late 1960s and employed by his student, J.S. Birman in her doctoral thesis; since then it has formed part of an important synthesis, bringing the topological (and complex analytic) theory of moduli of Riemann surfaces into close formation with combinatorial group theory via the representation of braid groups as subgroups of surface mapping-class groups. We describe this relationship in more detail in sections 1-3.

With the exception of Dehn's combinatorial study of discrete hyperbolic plane groups and the work of Picard and his students (to be discussed in section 5) on certain examples of complex surfaces which are the base spaces for families of algebraic curves, the pace of developments in geometric topology of surfaces then slackened during the uniformisation era, perhaps because of the considerable efforts needed to absorb and clarify that dramatic advance on the complex analytic flank. Further diversions of topological interest were in any case plentiful at that time. The dormant period lasted until J. Nielsen's pioneering work on surface mapping class groups and the definition of braid group by E. Artin in 1925, which was then employed by Dehn, Mangler, Markoff, Magnus and others during the 1930's in the study of surfaces and 3-manifolds.

The revival and success of these ideas in low dimensional topology some thirty years later owes much to interim advances in algebraic techniques and in particular to the development of combinatorial group theoretic methods by Magnus and his school. For instance, the well known algebraic structure of braid groups led to similar explicit descriptions of other algebraic automorphism groups, putting them into a digestible, and even presentable, form for computations. More recently, with the assimilation of further techniques from algebraic topology of manifolds, it has been possible to apply results on the cohomology of braid and modular groups coupled with important progress in complex analysis through the theory of Teichmüller spaces to develop a better understanding of the complex geometric structures underlying the deformations (moduli) of Riemann surfaces; these matters will be discussed briefly in §4.

At present, aside from the natural place that braids occupy within contemporary geometry and topology, one encounters them in a bewildering variety of settings, reflecting the influence which the concept exerts in areas such as algebraic geometry and number theory as well as the structure of von Neumann algebras, knot invariants, quantum groups and theoretical physics; the catalytic power of the simple braid relation

$$aba = bab$$

to generate richer structures and to trigger fresh insights in unsuspected corners
of mathematics is quite remarkable. No account will be given of these exciting
wider developments – the interested reader may consult the excellent survey article by Cartier [6], and the extensive bibliography provided there. My purpose
in this essay is a restricted one: to survey the gradual development of a simple
topological method within complex analysis, leading from the early pioneers of
function theory to the present day. It seems appropriate, however, to record it
here as a measure of their power and flexibility that braid groups are deeply
involved in current attempts both to bridge the (predominantly three dimensional) gaps in our understanding of the structure of manifolds and to construct
a viable theory of quantum gravitation, two steps in the continuing struggle to
comprehend the interface between mathematics and the real world. Few topics
from the treasury could better represent that unity of purpose within the growing diversity of human interests which Wilhelm Magnus embodied throughout
his life.

1. Braids and Riemann surfaces.

An observation going back at least to Hurwitz [13], and presumably with
origins in Riemann's memoir on Abelian functions [Crelle J., 1857] connects the
variational theory of Riemann surfaces with braids; here we follow in part the
formulation of Magnus [15].

Because every compact Riemann surface has meromorphic functions, defining
ramified covering mappings $f : X \rightarrow \mathbb{CP}^1$, a natural way to study complex
analytic families of surfaces is to concentrate on the functions, aiming to preserve
the basic topology of the map f but to vary the *location* of the ramification points
in the sphere: this is easily done by moving the points holomorphically within
the space of distinct tuples in $\mathbb{CP}^1$. Filling in the complex structure at each of the
branch points, a special version of the familiar removal of isolated singularities,
then provides a holomorphically varying family of compact surfaces.

Using $\{P_1, \ldots, P_r\}$ as ramification points in $\mathbb{CP}^1$, an n-sheeted covering f is
tantamount to a *permutation representation*

$$\gamma_j \mapsto \pi_j \ , \qquad j = 1, \ldots, r$$

of $\pi_1(\mathbb{P}^*)$ into the symmetric group S_n, where $\mathbb{P}^*$ denotes the sphere with the
points P_j deleted: a loop in $\mathbb{P}^*$ encircling one such point P_j from a base point
P_0 determines (by the path lifting property) a permutation of the n points lying
over P_0, depending only on the homotopy class of the loop, and this rule is seen
(by uniqueness of lifting) to extend to a group homomorphism. Conversely, such
combinatorial data enables one to construct a topological model of the covering
space X over the punctured sphere, with equivalent coverings corresponding to

conjugate permutation homomorphisms; transitivity of the permutation group action on the n sheets is equivalent to connectedness of the covering surface.

The data furnished by this representation is sometimes called the *permutation monodromy* of the pair $\{X, f\}$, usually viewed up to equivalence. Note that for complete precision in setting up the appropriate topological framework for ramified coverings one needs (c.f. [15]) to choose generic f-related base points on X and on $\mathbb{CP}^1$ for paths and homotopy.

We write C_r for the *configuration space* of unordered r-tuples $\{P_1, \dots, P_r\}$ of distinct points P_j in the sphere $\mathbb{CP}^1$; the local structure of the sphere gives a natural topological (even holomorphic) structure to the totality of such pairs. At the same time we introduce the notation $\overline{C}_r$ for the space of *ordered* r-tuples of distinct points, which is a ramified Galois covering of C_r with cover group the symmetric group Σ_r acting by permutation of the points. Hurwitz defined his *monodromy groups* A_r, B_r as the fundamental groups of C_r, $\overline{C}_r$ respectively. Of course one has $A_r \rhd B_r$ with quotient Σ_r.

The action of the group A_r on the set of all Riemann surfaces represented as such a covering over the sphere with n sheets and r branch points produces a finite index stabiliser for a given X. Magnus proved in [15] that this subgroup of A_r has a representation inside the group of outer automorphisms of $\pi_1(X)$, the *mapping class* or *modular group* of the surface X, by studying the effect of generating elements on the structure of $\pi_1(X)$. This involves some careful footwork on the ramified covering $\{X, f\}$; X is first punctured as above over the P_j to make the covering smooth and then the effect of filling-in points is considered, with a concomitant study of the combinatorial group theory bringing out the relationship between the various homotopy groups.

In fact, this approach provided part of the motivation for my own study (and that of co-workers including C. Maclachlan and G. Gonzalez-Diez) of moduli theory for certain Galois coverings of $\mathbb{CP}^1$. Some of our results will be described in §3 and §4.

2. The hypergeometric function.

The hypergeometric equation (HGE) is an O.D.E. with three regular singular points (at 0, 1, and ∞ say) which takes the form

$$\frac{d^2u}{dx^2} + \left(\frac{1-\lambda}{x} + \frac{1-\mu}{x-1}\right)\frac{du}{dx} + \frac{(1-\mu-\lambda)^2 - \nu^2}{4x(x-1)}u = 0.$$

The notation here is perhaps slightly unusual: it represents the differential equation for the hypergeometric function $F(\alpha, \beta, \gamma; x)$, where

$$\alpha = \frac{1}{2}(1 - \lambda - \mu + \nu), \qquad \beta = \frac{1}{2}(1 - \lambda - \mu - \nu), \qquad \gamma = 1 - \lambda,$$

and

$$F(\alpha, \beta, \gamma; x) = \sum \frac{(\alpha, n)(\beta, n)}{(\gamma, n)} \frac{x^n}{n!} \qquad (|x| < 1).$$

A complete discussion of early work on this seminal equation may be found in the famous Göttingen lectures of Felix Klein (which appeared as a 1933 volume of Springer Verlag's Grundlehren series) running to several hundred pages; we begin pre-emptively from the claim (known in essence to Euler) that

$$\frac{\Gamma(\beta)\Gamma(\gamma - \beta)}{\Gamma(\gamma)} F(\alpha, \beta, \gamma; x) = \int_0^1 u^{\beta-1}(1 - u)^{\gamma-\beta-1}(1 - xu)^{-\alpha} \, du$$
$$= \int_1^{\infty} \phi \tag{2.1}$$

where ϕ may be regarded as a multivalued holomorphic 1-form on $\mathbb{P}^1 \backslash \{0, 1, x, \infty\}$.

For elementary reasons, the space of all solutions to the HGE has a basis of functions $F_1(x) = \int_0^1 \phi$, $F_2 = \int_1^x \phi$. The quotient $f = F_1/F_2$ is a meromorphic function, which is at first only definable on the simply connected $\mathbb{C} - \mathbb{R}_+$, say. However the form ϕ extends inside $\mathbb{C} - \{0, 1, x\}$ by analytic continuation, a process which determines an action of the Hurwitz monodromy group A_4, also nowadays known as the *braid group* of the sphere $B_4(\mathbb{P}^1)$, on the space of solutions: sending x on some closed path in $\mathbb{C} - \{0, 1\}$, for instance, results in a linear change in $\{F_1, F_2\}$, and by working out the effect on the quotient a representation of the braid group into the group $PGL_2(\mathbb{C})$ ensues which is known as the *monodromy group* of the HG equation.

Schwarz proved several important results about this in his paper [18], centering on the elementary geometry of the developing map and monodromy image. The behaviour of the form ϕ in (2.1) provides the key to understanding how f transforms. Under each motion of one ramification point $s \in S = \{0, 1, x, \infty\}$ around another, ϕ is multiplied by a nonzero constant and any path of integration between points of S is correspondingly modified, the combined effect on f being expressible simply in terms of the corresponding pair of exponents in the HGE. We state the two main results of Schwarz below.

THEOREM 1. *If the monodromy image is a finite subgroup of $PGL_2(\mathbb{C})$, then the fundamental quotient f of solutions to the HG equation is algebraic as a function of x, and conversely.*

Let $\lambda_s = 1 - \mu_s - \alpha$ for $s = 0, 1, \infty$, where $\mu_0 = \gamma - \alpha$, $\mu_1 = 1 + \beta - \gamma$ and $\mu_\infty = 1 - \beta$.

THEOREM 2. *Writing $f = F_1/F_2$, with F_1, F_2 as above, f is a locally meromorphic multivalued function in $\mathbb{C} - \{0, 1\}$ mapping the upper half x plane onto a triangular region bounded by circular arcs or lines with angles $\pi\lambda_0$, $\pi\lambda_1$, $\pi\lambda_\infty$ at the points $f(s), s = 0, 1, \infty$.*

When λ_0^{-1}, λ_1^{-1}, $\lambda_\infty^{-1} \in \mathbb{N} \cup \infty$ and $\lambda_0 + \lambda_1 + \lambda_\infty < 1$, the circles or lines containing sides of the triangle are orthogonal to a common circle, which forms the boundary of a hyperbolic plane on which the monodromy image acts as a discrete triangle group of hyperbolic isometries.

Some classic illustrations arising from this, including the notable cases where $\lambda_0^{-1} = 2$, $\lambda_1^{-1} = 3$, $\lambda_\infty^{-1} = 7, 8$ or ∞, may be found together with an account of the underlying group theory and geometry in the book by Wilhelm Magnus [*Non-Euclidean Tesselations and their groups* Academic Press, New York (1974)]. In Figure 1 we have given only a schematic view of the mapping f.

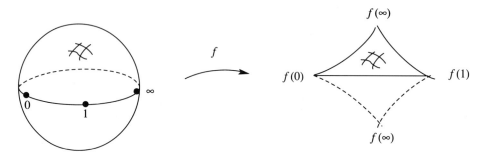

FIGURE 1. Developing image under f of the x plane.

Reflecting the developing image triangle in the edges gives rise in the situation of theorem 2 to a tesselation of the hyperbolic plane. One obtains a *triangle relation* in the image of the (pure) braid group $\overline{B}_4(\mathbb{P}^1)$.

$$A_{\infty x} \cdot A_{0x} \cdot A_{1x} = Id,$$

where $A_{ij}, i, j \in S$ are certain (generating) elements in $\overline{B}_4(\mathbb{P}^1)$ depicted below in Figure 2 (the braid with label j passes *over* those between it and i, circling behind i and returning to base); the monodromy images of elements A_{ix} are hyperbolic rotations through angle $2\pi\lambda_i$ centred at $f(i)$.

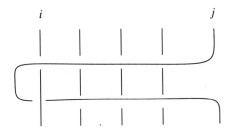

FIGURE 2. The braids A_{ij}, with $i < j$.

Picard [17] and his student Alezais [1] extended this hypergeometric theory to integrals $\int_s^t \phi$; $s, t \in S = \{0, 1, x, y, \infty\}$ analogous to (2.1) with multivalued holomorphic 1-form

$$\phi = u^{\mu_0}(1-u)^{\mu_1}(1-xu)^{\mu_x}(1-yu)^{\mu_y}\,du \qquad (2.2)$$

defined on $\mathbb{C} - \{0,1,x,y\}$, whose (rational) exponents μ_s dictate the behaviour of a corresponding developing map from the space of positions $\{(x,y); x \neq y\}$ in $\mathbb{C} - \{0,1\}$, viewed as a quotient of the configuration space C_5, into a complex projective 2-space. In this way they discovered many interesting monodromy images of $B_5(\mathbb{P}^1)$ in the automorphism group of the complex unit 2-ball. More recently, renewed interest in that work has yielded a wealth of significant extensions of their results with contributions by Deligne and Mostow, Holzapfel, Terada and Shiga. We comment further on this in section 5 after describing the general relationship between orbifold surfaces and *Fuchsian groups*, which are discrete groups of orientation-preserving isometries of the hyperbolic plane; the monodromy triangle groups appearing in theorem 2 are of course examples of Fuchsian groups.

3. Fuchsian groups and moduli of surfaces.

By the uniformisation theorem, due to Poincaré, Klein and Koebe, each simply connected Riemann surface is furnished with a *geometric structure*, a representation as a homogeneous space carrying one of the three standard two-dimensional geometries. It then follows easily, by regarding a Riemann surface as a quotient of its universal covering by a group of cover transformations, that every complex analytic surface has such a local geometric structure. This determines in particular a way to pass from any Riemann surface with negative Euler characteristic to a discrete group of transformations of the hyperbolic plane, unique up to conjugacy.

In general, however, it is not easy to make this passage explicit, even between the equivalent categories of compact Riemann surfaces (or complete algebraic curves) and the co-compact Fuchsian groups which uniformise them. A class for which this *is* possible is the family of curves $\mathcal{X} = \{X_\alpha\}$ arising from the affine equation

$$w^n = a_0 + a_1 z + \ldots + a_d z^d = a_d \prod_{1}^{r}(z - \alpha_j)^{m_j}, \qquad (3.1)$$

where $\alpha = \{\alpha_j\} \in C_r$. Here one has a powerful algebraic lever, the automorphism $\sigma = \{\sigma_\alpha\}$ of $\mathcal{X}$ given by

$$w \mapsto \zeta w, \qquad z \mapsto z, \qquad \text{where} \qquad \zeta = e^{2\pi i/n}.$$

These are often called *Kummer curves* because their function fields are Kummer extensions of the rational field $\mathbb{C}(z)$ with cyclic Galois group generated by σ. The rule $(w,z) \mapsto z$ defines the function f and fixes the topological type of the covering over the z−sphere.

In terms of uniformisation, an equivalent description comes from the class of Fuchsian groups with torsion generators

$$\langle x_1, \ldots, x_r; \quad x_1^{n_1} = 1, \ldots, x_r^{n_r} = 1, \quad \prod x_i = 1 \rangle \tag{3.2}$$

where the periods are $n_j = n/(n, m_j)$ and the x_j correspond to homotopy classes of loops in the punctured sphere $\mathbb{P}^*$ around the ramification points α_j. The prescription of an epimorphism ρ from such a Fuchsian group Γ to $\mathbb{Z}_n$ whose kernel is a torsion-free subgroup K defines a ramified covering surface $U/K = X$ having an algebraic equation of the given type. The choice of homomorphism is tantamount to prescribing multiplicities for the roots of the polynomial in the corresponding equation (3.1); note that the point at ∞ in this representation must be counted as ramified if and only if $n \nmid d$.

Now as observed in §1, there is a close relationship between braids and the structure of the family $\mathcal{X}$ viewed as a subset of the moduli space of curves: the genus of each surface X_α is determined via the Hurwitz branching formula as $g = 1 + (1/2)\Sigma(n - n/m_j)$, and we can summarise results of [9], [10] and [11] in the following statement.

THEOREM 3. *Inside the space of moduli $\mathcal{M}_g$, the locus of all curves with a fixed representation (3.1) as a cyclic covering of $\mathbb{CP}^1$ forms an irreducible subvariety $V = V(H)$ that is unirational and quasi-projective but not (in general) normal. The normalisation $\tilde{V}$ is a complex analytic orbifold isomorphic to a certain (explicitly given) finite covering of $\mathcal{C}_r$.*

The proofs involve considerations from Teichmüller space theory which will not be addressed in detail here. An important part is played by the notion of *relative modular group $Mod_g(H)$* associated to the cyclic symmetry group H of $\mathcal{X}$. It transpires that this group is closely linked with the group of H−symmetric mapping classes of the compact surface X, obtained by Magnus in [15] as a homomorphic image of an appropriate stabiliser subgroup in the Hurwitz monodromy group A_r. An alternative construction in [11] shows that $Mod_g(H)$ is an extension by the group H of a subgroup of finite index in the modular group of the r-times punctured sphere which contains a homomorph of the pure Artin braid group $\overline{B}_r(\mathbb{C})$. With the help of the presentation (3.2) for the uniformising Fuchsian group Γ of the quotient Riemann sphere and the representative epimorphism ρ from Γ to $\mathbb{Z}_n$, the modular subgroup concerned is rendered into algebraic form as a group of braid automorphisms of the free group on r generators (up to inner automorphism), which induce automorphisms of the quotient Γ compatible with the epimorphism. The topological effect of this is to single out those mapping-classes on the reference surface X which project to the sphere, inducing permutations of the periodic orbits for the group H that preserve the local data; for the case of prime n this means permutations of the fixed points of the generating automorphism of H that preserve the rotation number at each point.

This relationship between braids and symmetric mapping classes has origins in a result proved by V. I. Arno'ld [2] in the hyperelliptic case of (3.1), where the automorphism α has order $n = 2$ and the subgroup concerned is in fact the full modular group of the sphere with $r = 2g + 2$ cone points of order 2. The special properties of symmetric mapping classes were later studied by various authors including Magnus & Peluso [14], Birman and Hilden [5] and the author with C. Maclachlan [11].

Unirationality for the modular subvarieties $\widetilde{V}$ amounts to nothing more than the observation [9] that each one is finitely covered by the affine algebraic set

$$\Omega_{r-3} = \mathbb{C}^{r-3} \setminus \Delta_*$$

where $\Delta_* = \{\alpha_j = 0 \text{ or } 1, \text{ for some } j\} \cup \{\alpha_j = \alpha_k, j \neq k\}$ is the *normalised diagonal set*.

The general lack of normality for $V(H)$ reflects the fact that a Riemann surface X may possess two automorphisms which generate groups H, H' that are topologically conjugate while not actually being conjugate in the automorphism group $Aut(X)$; an example of this (see [9]) is the class of Fermat curves F_{2p} with equation $\{y^{2p} = 1 - x^{2p}\}$. This implies for instance that the corresponding modular subvariety $V(H)$ is not a complex orbifold; however the normalisation is always an orbifold, isomorphic to the quotient of the Teichmüller space $T_{0,r}$ by the group of mapping classes that are compatible with the epimorphism ρ as discussed above.

4. Further Comments on Kummer Families.

1.   Because the cohomology of the braid groups has been extensively worked out, it is possible to infer certain facts about the topological type of the relative modular subvarieties discussed in §3, at least in the case where all exponents m_j are equal. This happens in the hyperelliptic case and for all families of curves of non-singular Kummer type $\{y^n = \prod_1^{nk}(x - \alpha_j)\}$ with genus $g = (n-1)(nk-2)/2$.

THEOREM 4. *The subspace $V(H)$ of moduli space $\mathcal{M}_g$ defined by a relative modular family of Galois coverings of the Riemann sphere has cohomology ring generated by elements of degree 1. If the Galois group H is cyclic, with equation (3.1) having all exponents equal to 1, then this modular subspace is simply connected and hence contractible.*

As I know no precise reference for these facts, an outline of the proof will be sketched. The first statement (pointed out in conversation by Peter Zograf) is a consequence of a theorem of Arnol'd [3] which gives the structure of the cohomology ring for Artin's braid group $B_r(\mathbb{C})$ in terms of the de Rham algebra of forms on the configuration space $C_r(\mathbb{C})$ of unordered tuples $\{\lambda_j\}$ of points in the plane; the algebra is generated by the 1-forms $\omega_{ij} = d\log(\lambda_i - \lambda_j)$. This space is a $K(\pi, 1)$ with fundamental group $B_r(\mathbb{C})$ and the obvious inclusion into C_r (see for instance [4] or [15] for the induced effect on the braid groups) leads to the first conclusion for the quotient modular subvariety, which is biholomorphically isomorphic to a finite quotient of C_r.

Now the action of the relative modular group on the Teichmüller space of the r-pointed Riemann sphere permutes the ramification points transitively under the second assumption on the covering, and the quotient is then isomorphic to the moduli space $\mathcal{M}_{0,r}$ of the r-pointed sphere. Therefore the 1-forms ω_{ij}, which generate, are all equivalent in the quotient space. To reach the second conclusion one must prove the triviality of the first homotopy group; this is a consequence of the fact (a result of C. Maclachlan) that the pointed moduli spaces $\mathcal{M}_{0,r}$ are, like $\mathcal{M}_g$, simply connected because they arise as quotients of simply connected (Teichmüller) spaces by discrete groups generated by mapping classes with fixed points.

It may be noted that the second part of the theorem follows trivially for the case $n = 2$ from a result of Katsylov and Bogomolov in invariant theory which shows that the hyperelliptic subvariety is rational. It would be interesting to know which modular subvarieties $\tilde{\mathcal{V}}$ are rational.

2. In work of G. González-Díez [8], the theory of Riemann theta functions is deployed to define a higher λ-function, an analogue of Legendre's elliptic modular invariant, for families of surfaces $\mathcal{X}$ with equation as in (3.1) when the integer n is prime. This function identifies the (normalised) locations of the r branch points as products of quotients of theta functions (for the surfaces of the family $\mathcal{X}$) with $\frac{1}{n}$-characteristics; the result extends the famous formula of Jacobi for the Legendre elliptic modular function, which relates to the case $n = 2, r = 4$.

3. Recently D. Piponi (Ph.D. thesis at King's College London, 1993) has computed the multiplier system of the Hodge line bundle over C_r for families of Kummer surfaces with n prime; he has also proved a kind of "bosonisation formula" on these families (a generalisation of the classical formula of Thomae for hyperelliptic surfaces) which relates an expression involving theta constants to products of cross ratios of the branch points. A further ingredient in this work provides explicitly the corresponding symplectic representations of the braid group $B_r(\mathbb{P}^1)$, amounting to an independent calculation of the homological part of the monodromy for these families; this occurs, at least implicitly, in the results of Deligne-Mostow to be described in the next section.

5. Monodromy for higher dimensional families.

Deligne and Mostow [7] studied the extension of the monodromy problems solved by Schwarz (and discussed in §2) to an arbitrary finite number of punctures, including in particular the case of 5 ramification points examined earlier by Picard and his students. Their method is based on the analytic continuation of analogues for the multivalued 1-forms ϕ written down in (2.2); however they pass to an equivalent monodromy problem for sections of complex line bundles on the family of punctured spheres $\mathbb{P}_\alpha, \alpha \in C_r$, the bundles being defined by the multiplier system created from continuation of ϕ around the omitted ramification points. The process may also be regarded as the study of solutions to a many variable linear system of partial differential equations (a natural generalisation of the hypergeometric equation) first studied by P. Appell, again with prescribed singular behaviour at the omitted points.

Their results apply to the families (3.1) of compact Riemann surfaces considered in §3 and 4, including, for instance, the so-called *Picard family* of genus 3 curves given by

$$y^3 = x(x - 1)(x - \alpha_1)(x - \alpha_2). \tag{5.1}$$

Here ∞ is also a ramification point. The multiplicities m_j in the equation (3.1) under consideration lead to a specific choice of multiplier system involving n-th roots of unity; this can be worked out in terms of the formulation given in §4. Necessary and sufficient conditions are derived on the exponents $\{\mu_j\}$ of the forms ϕ in (2.2) for the monodromy action on the projective space of sections to be discrete; the dimension of this space is $d = r - 3$. As Mostow explains in his survey [16], their work leads to the construction of many new lattices as monodromy image groups, which are discrete subgroups of finite volume in symmetric spaces of dimension d and type fixed by the exponents; these include the Picard - Le Vavasseur catalogue of lattice subgroups in the complex hyperbolic group $SU(2, 1)$. As in the classical case, the lattices are homomorphic images of braid groups which accompany a developing mapping of the universal covering of the configuration quotient space Ω_{r-3}. Mostow exhibits several close analogies in this regard with the situation discussed in §2; for instance the image is generated by complex reflections, the natural version in complex hyperbolic space of the elliptic generators for the Schwarz triangle groups, and defining relations occur which are in the same spirit as the triangle relation mentioned after theorem 2, though more numerous.

Independently, H. Shiga [19] has studied the case (5.1) in great detail, using the transformation properties of theta functions and a careful derivation of the precise period relations satisfied by these curves. He verified directly the striking original assertion of Picard that the monodromy image group is the Eisenstein lattice in $SU(2, 1; \mathcal{O})$, where $\mathcal{O}$ is the ring of integers in the number field $Q(\sqrt{-3})$. He has also derived criteria on algebraic points of (a suitable model for) the developing image symmetric space for the values of a certain modified inverse function to be algebraic, thus extending the classical results of Kronecker and Weber on the relationship between special values of the modular j-function and the class fields of elliptic curves with complex multiplication.

A completely different approach to the family of Picard curves (5.1) was developed by Feustel and Holzapfel within the framework of algebraic geometry, using the theory of variations of Hodge structure. These deep methods are not easily summarised, and the interested reader is referred to the very clear account in the text [12].

As a final comment on this line of research, it seems important to mention that the Picard family is one of just ten cases in complex dimension $d = 2$ for which the monodromy image group is *arithmetic*; independently, these were determined by Terada [20] as verification that a relationship exists between the generalised hypergeometric equation and a list of arithmetic curve families given twenty years earlier by Shimura. The methods of automorphic function theory on the complex hyperbolic unit ball then provide a relatively explicit inverse mapping from the ball quotient to the base space of the modular family.

REFERENCES

1. R. Alezais, *Sur une classe de fonctions hyperfuchsiennes*, Ann. Ecole Norm. Sup. 19 (1902), 261–323.
2. V. I. Arnol'd, *Remark on the branching of hyperelliptic integrals as functions of the parameters*, Functional Anal. Appl. 2 (1968), 187–189.
3. _____, *The cohomology ring of the colored braid group*, Math. Notes of the Academy of Sci. of the USSR 5 (1969), 138–140.
4. J. S. Birman, *Braids, Links, and Mapping Class Groups*, Annals of Mathematics Studies, Number 82, Princeton University Press, Princeton, 1975.
5. J. S. Birman and H. M. Hilden, *On isotopies of homeomorphisms of Riemann surfaces*, Ann. of Math. (2) 97 (1973), 424–439.
6. P. Cartier, *Développements récents sur les groupes de tresses*, Séminaire Bourbaki 716 (1989), 1–42.
7. P. Deligne and G. D. Mostow, *Monodromy of hypergeometric functions and non-lattice integral monodromy*, I.H.E.S. Publications Mathémtiques 63 (1986), 5–106.
8. G. González-Díez, *Loci of curves which are prime Galois coverings of* $\mathbb{P}^1$, Proc. London Math. Soc. 62 (1991), 469–489.
9. G. González-Díez and W. J. Harvey, *Moduli of Riemann surfaces with symmetry*, Discrete Groups and Geometry (W. J. Harvey and C. Maclachlan, eds.), London Math. Soc. Lecture Note Series 173, Cambridge University Press, 1992, pp. 75–93.
10. W. J. Harvey, *On branch loci in Teichmüller space*, Trans. Amer. Math. Soc. 153 (1971), 387–399.
11. W. J. Harvey and C. Maclachlan, *On mapping class groups and Teichmüller spaces*, Proc. London Math. Soc. (3) 30 (1975), 496–512.
12. R.-P. Holzapfel, *Geometry and Arithmetic around Euler Partial Differential Equations*, D. Reidel Publishing Company, Dordrecht/Boston/Lancaster, 1986.
13. A. Hurwitz, *Über Riemann'sche Flächen mit gegebenen Verzweigungspunkten*, Math. Ann. 39 (1891), 1–61.
14. W. Magnus and A. Peluso, *On a theorem by V. I. Arnold*, Comm. Pure and App. Math 22 (1969), 683–692.
15. W. Magnus, *Braids and Riemann surfaces*, Comm. Pure and App. Math 25 (1972), 151–161.
16. G. D. Mostow, *Braids, hypergeometric functions, and lattices*, Bull. Amer. Math. Soc. 16 (1987), 225–246.
17. E. Picard, *Sur les fonctions de deux variables indépendentes analogues aux fonctions modulaires*, Acta Math. 2 (1883), 114-135.
18. H. A Schwarz, *Uber diejenige Falle, in welche die Gaussche hypergeometriche Funktion algebraische Funktion ihres vierten Elementes darstellt*, Crelle J. 54 (1873), 292-335.
19. H. Shiga, *On the representation of the Picard modular function by* θ *constants I-II*, Publ. R.I.M.S. Kyoto University 24 (1988), 311–360.
20. T. Terada, *Fonctions hypergéométriques* F_1 *et fonctions automorphes II*, J. Math. Soc. Japan 37 (1985), 173–186.

DEPARTMENT OF MATHEMATICS, KING'S COLLEGE LONDON, STRAND, LONDON WC2R 2LS
E-mail address: w.harvey@uk.ac.kcl.cc.oak

Contemporary Mathematics
Volume **169**, 1994

SOME REMARKS ON J REPLACEMENT
IN DIRECT PRODUCTS

R. HIRSHON

ABSTRACT. The concept of J replacement in direct products has been useful in several previous studies. The object of this paper is to gain some further insight into this concept. Some results are obtained for finitely generated residually finite groups with a finitely generated center and for general finitely generated groups.

1. Introduction

Let J represent the infinite cyclic group. The group M is said to be J replaceable if

(1.1) $$G = M \times C = N \times D, \quad M \approx N$$

implies $J \times C \approx J \times D$. The concept of J replacement has been useful in [2], [3], [4] and [5] and has been studied somewhat more extensively in [6]. J replacement studies led to a study related to direct products and produced unexpected results on residually finite groups [8]. The purpose of this paper is to gain some further insight into the concept of J replacement. While some of our results hold for finitely generated M with a finitely generated center $Z(M)$, we also investigate the case when M is a finitely generated residually finite group with a finitely generated center. For example, we may state a weak form of Theorem 3.1 as follows: Finitely generated residually finite groups with finitely generated centers are J replaceable if and only if finitely generated residually finite centerless groups are cancellable in finitely generated residually finite centerless direct products. A finitely generated residually finite group with a finitely generated center is J replaceable provided that its centerless finitely generated residually finite homomorphic images are cancellable in finitely generated residually finite centerless direct products. Also, see Corollary to Theorem 2.2, Theorem 3.1, Theorem 3.2, and Corollary 3.1. Our investigation leads naturally to a study of (1.1) when C and D are perfect and centerless. We develop some properties of such a decomposition which are interesting in their own right.

1991 *Mathematics Subject Classification.* Primary 20E26, 20E34, 20D30, 20F22.
This paper is in final form and no version of it will be submitted for publication elsewhere.

2. Some Results on Tame Groups

2.1 Preliminaries. We say that an abelian group G is *tolerable* if G has no subgroup which is expressible as a direct product of an infinite number of nontrivial cyclic groups.

Suppose G is a group having a sequence of direct decompositions of the form

$$(2.1) \qquad G = G_1 \times G_2 \times \cdots \times G_n \times F_n, \; F_n = G_{n+1} \times F_{n+1}$$

If the intersection of the descending chain of subgroups F_i in (2.1) is the identity, then G is isomorphic to a subgroup of the unrestricted direct product of the G_i. If g is in G and if g_i is the component of g in G_i, we may identify g with the infinite sequence $g_1, g_2, g_3, \ldots$ in the unrestricted direct product. In this case, we write $g = g_1 g_2 g_3 \cdots$ or $g = [g_i]$, $i \geq 1$, or simply $g = [g_i]$ if the index range is clear.

A major tool in the study of J replacement is Theorem 2.4 in [6] which we state here as

Lemma 2.1. *Let M/M' be finitely generated, and let $Z(M)$ be tolerable in (1.1). Suppose that $C \times J \not\approx D \times J$. Then we can find $\tilde{C}, \tilde{D}$ isomorphic to centerless perfect direct factors of M and a group $\tilde{M}$ with $\tilde{M}/\tilde{M}'$ finitely generated, $Z(\tilde{M})$ tolerable and $\tilde{M} \times \tilde{C} \approx \tilde{M} \times \tilde{D}$ and $\tilde{C} \times J \not\approx \tilde{D} \times J$.*

With the aid of [6, Lemma 3.2, p. 360], we may show

Lemma 2.2. *Suppose that in (1.1) either C and D are both perfect and centerless or G is centerless. If R is any direct factor of G, then $R = (M \cap R) \times (C \cap R) = (N \cap R) \times (D \cap R)$.*

We define a *prefactor* of group K as a subgroup H of K such that H is a homomorphic image of K and $K = H H_1$ where H_1 is a subgroup of K such that H and H_1 commute elementwise.

Let $J(R)$ be the class of groups $\tilde{R}$ which may be obtained from the group R by a finite number of applications of the process of taking prefactors or taking direct products with a finitely generated abelian group. That is, $\tilde{R}$ is in $J(R)$ if and only if there exist a sequence of groups $R_1, R_2, \ldots R_n$, $n \geq 1$, with $R_1 \approx R$, $R_n \approx \tilde{R}$ and if $i > 1$ either R_i is a prefactor of R_{i-1} or $R_i \approx R_{i-1} \times F_i$ for some finitely generated abelian group F_i.

Lemma 2.3. *The group $\tilde{M}$ of Lemma 2.1 may be taken to be in $J(M)$.*

For a proof of the above, see [6, Corollary of Theorem 2.4].

Lemma 2.4. *If $\tilde{R}$ is in $J(R)$ and $\tilde{R}/K$ is a centerless homomorphic image of $\tilde{R}$, then $\tilde{R}/K$ is a homomorphic image of R.*

Since the class $J(R)$ arises quite naturally in studying the J replacement of R, it seems appropriate to call $J(R)$ the J closure of R, and we do so in the sequel. Lemma 2.1 shows that a careful investigation of (1.1) is in order when

$$(2.2) \qquad C \text{ and } D \text{ are perfect and centerless.}$$

2.2 C and D Perfect and Centerless. Now suppose (2.2) holds. Set $N_{-1} = G, N_0 = N, D_0 = D, L_{-1} = M \cap D$. As in [6, Section 4], we construct the groups N_i, P_i, L_i and S_i and we have for $i \geq 0$

$$(2.3) \qquad L_i = M \cap N_i \cap D_{i+1}, \quad S_i = N_i \cap C \cap D_{i+1}$$

$$(2.4) \qquad N_i \cap C = (N_{i+1} \cap C) \times S_i, \quad M \cap N_i = (M \cap N_{i+1}) \times L_i$$

$$(2.5) \qquad L_i \theta = L_{i+1} \times S_{i+1}, \quad i \geq -1$$

$$(2.6) \qquad D_{i+1} = (C \cap D_i) \times (M \cap N_{i-1} \cap D_i)\theta, \quad (M \cap N_i)\theta = N_{i+1}$$

$$(2.7) \qquad (M \cap N_i) \times C = N_{i+1} \times D_{i+1}, \quad D_i \approx D$$

Instead of working with the isomorphism θ of M onto N we may work with the isomorphism θ^{-1} of N onto M and repeat the above ideas to construct groups M_i, C_i, P_i and T_i which are as in [6, Section 4]. We have $P_{-1} = N \cap C, M_{-1} = G, M_0 = M, C_0 = C$ and, corresponding to (2.3), (2.4), (2.5), (2.6), (2.7), we have for $i \geq 0$

$$(2.8) \qquad P_i = N \cap M_i \cap C_{i+1}, \quad T_i = M_i \cap D \cap C_{i+1}$$

$$(2.9) \qquad M_i \cap D = (M_{i+1} \cap D) \times T_i, \quad N \cap M_i = (N \cap M_{i+1}) \times P_i$$

$$(2.10) \qquad P_i \theta^{-1} = P_{i+1} \times T_{i+1}, \quad i \geq -1$$

$$(2.11) \qquad C_{i+1} = (D \cap C_i) \times (N \cap M_{i-1} \cap C_i)\theta^{-1}, \quad (N \cap M_i)\theta^{-1} = M_{i+1}$$

$$(2.12) \qquad (N \cap M_i) \times D = M_{i+1} \times C_{i+1}, \quad C_i \approx C$$

We depict below the interaction among θ, the T_i, S_i, and the $L_i \cap P_j = E_{ij}$.

$$(2.13)$$

	$T_0 \to S_0$	S_1	S_2	$S_3 \ldots$
		$\uparrow$	$\uparrow$	$\uparrow$
	$T_1 \quad \to$	E_{00}	E_{10}	$E_{20} \ldots$
			$\nearrow$	$\nearrow$
	$T_2 \quad \to$	E_{01}	E_{11}	$E_{21} \ldots$
			$\nearrow$	$\nearrow$
	$T_3 \quad \to$	E_{02}	E_{12}	$E_{22} \ldots$

In summary, we have $E_{j0}\theta = S_{j+1}, T_j\theta = E_{0,j-1}, j \geq 1, T_0\theta = S_0$, and $E_{ij}\theta = E_{i+1,j-1}, j > 0$. This interaction may be verified by using (2.5) and (2.10). We note that the N_i and M_i are a descending sequence of normal subgroups of N and M respectively with N_i and M_i being characterized for $i \geq 1$ as

$$N_i = \{n | (n \in N, n\theta^{-j} \in N, 1 \leq j \leq i\},$$
$$M_i = \{m | (m \in M, m\theta^{j} \in M, 1 \leq j \leq i\}$$

Note by Lemma 2.2

(2.14) $$M = (M \cap N) \times (M \cap D), N = (M \cap N) \times (N \cap C),$$
$$C = (C \cap N) \times (C \cap D), D = (D \cap M) \times (D \cap C)$$

Since

(2.15)
$$N \cap M = P_0 \times P_1 \times \cdots \times P_i \times (N \cap M_{i+1})$$
$$= L_0 \times L_1 \times \cdots \times L_i \times (M \cap N_{i+1})$$

by using Lemma 2.2, we may deduce that, for a suitable sequence of groups F_n, $N \cap M$ has a sequence of decompositions of the form

(2.16) $$N \cap M = E_1 \times E_2 \times \cdots \times E_n \times F_n$$

where $E_1 = E_{00}$ and E_i is the direct product of the groups E_{uv} with $u + v = i - 1, u \geq 0, v \geq 0$ if $i > 2$ and where $F_n = E_{n+1} \times F_{n+1}$. From (2.13) we deduce that E_i is isomorphic to the direct product of i copies of T_i, that is,

(2.17) $$E_i \approx \times^i T_i$$

2.3 Tame Groups. In studying the case when R is not J replaceable, we may pass by means of Lemma 2.3 to a decomposition (1.1) where M is in $J(R)$, (2.2) is satisfied and C and D are isomorphic to direct factors of R. For such a decomposition, the sequence of groups T_i come up in an interesting and natural manner, and it seems appropriate to study this sequence. For lack of a better name, we call this sequence the T *sequence* of 1.1 (which of course depends on θ, M, N, C, D). If $T_i = 1$ for all i sufficiently large, we say that the T *sequence is finite*. These remarks motivate the following

Definition: We say that R is *tame* provided that the T sequence is finite for all decompositions (1.1) obeying (2.2) with C and D isomorphic to direct factors of R and with M in $J(R)$.

A better understanding of the above definition might be obtained via the next paragraph and Section 2.4 below.

We recall from [6] that a group G is said to obey the *W.M.C.O.D.* (weak minimal condition on direct factors) if, for every sequence of decompositions (2.1) with G_{i+1} isomorphic to a direct factor of G_i, $G_i = 1$ ultimately. A finitely generated group with a tolerable center and the *W.M.C.O.D.* is J replaceable [6]. If R obeys the *W.M.C.O.D.*, it is easy to show that every member of $J(R)$ does also. Thus, by using (2.5),(2.14) and (2.15), it is clear that a group with the *W.M.C.O.D.* is tame. For the sequel, it will be useful to note that if G is a finitely generated group and $d(G)$ designates the minimal number of generators of G, then for a finite group, F, $F \neq 1$, from [17]

(2.18) $$\lim_{n \to \infty} d(\times^n F) = \infty$$

2.4 Untamed Groups. It may be useful to point out some properties of untamed groups. First note that from (2.9)

$$(2.19) \qquad M \cap D = T_0 \times T_1 \times \cdots \times T_{r-1} \times (M_r \cap D)$$

$$(2.20) \qquad N \cap C = S_0 \times S_1 \times \cdots \times S_{r-1} \times (N_r \cap C)$$

If M is a finitely residually finite group and infinitely many of the T_i are distinct from 1, then by using (2.18) we see that the isomorphism classes of finite simple homomorphic images of the T_i are infinite in number. Consequently, by using Lemma 2.4, we see that if R is a finitely generated residually finite group which is not tame, then R has infinitely many isomorphism classes of finite simple homomorphic images. In fact, we can find a centerless homomorphic image G of R such that G has a sequence of decompositions of the form (2.1) such that there exist finite simple groups $A_i, i \geq 1$ representing infinitely many isomorphism classes with $\times^i A_i \approx G_i$. Also, it is not difficult to show that if R does not satisfy the *W.M.C.O.D.*, then the automorphism group of R contains a subgroup isomorphic to the restricted direct product of all symmetric groups. (A modification of this idea is used in [8] to obtain some results on residually finite groups.) Consequently, if R is not tame, the automorphism group of R contains a subgroup isomorphic to the restricted direct product of all symmetric groups.

2.5 Easy Decompositions. Let (2.2) hold. Let $\tilde{N}$ be the intersection of the descending chain of subgroups N_i, and let $\tilde{M}$ be the intersection of the descending chain of subgroups M_i. We say (1.1) is an *easy decomposition* if $\bar{N} = \bar{M} = 1$. In this case we call θ an *easy isomorphism*. In studying J replacement of finitely generated residually finite groups, we will show that it suffices to study easy decompositions. As we show below, this in turn reduces the study of J replacement in finitely generated residually finite groups with a tolerable center to the study of centerless finitely generated residually finite groups.

In an easy decomposition, we see from (2.4) that $M \cap N$ and $N \cap C$ are subgroups of the unrestricted direct products of the L_i and the S_i $i \geq 0$ respectively. From (2.9), we see that $M \cap N$ and $M \cap D$ are subgroups of the unrestricted direct products of the P_i and the $T_i, i \geq 0$ respectively. With the aid of Lemma 2.2, we may conclude that each L_i is a subgroup of the unrestricted direct product of the $L_i \cap P_j, j \geq 0$ and each P_i is a subgroup of the unrestricted direct product of the $P_i \cap L_j, j \geq 0$. In turn this gives $M \cap N$ as a subgroup of the unrestricted direct product of the groups $E_{ij}, i \geq 0, j \geq 0$. Since D is centerless and each T_k is a direct factor of $M \cap D$, we see from (2.14) that if (1.1) is an easy decomposition, then M is centerless. In addition, if D is residually finite, then so is M.

Remark. When $\bar{N} = \bar{M} = 1$, one can visualize to some extent the relation among $\theta, M \cap D, N \cap C, M \cap N$ and the T_i, S_i, E_{ij} via (2.13). In this case, $M \cap D, N \cap C, M \cap N$ are given by certain sequences $[t_i], [s_i]$ and $[e_{ij}]$ with $t_i \in T_i, s_i \in S_i, e_{ij} \in E_{ij}$. One problem that arises is that, for $t = [t_i]$ in $M \cap D$, the element $[t_i \theta^{i+1}]$ is in the unrestricted direct product of the S_i but not necessarily in $N \cap C$. Similarly, for $s = [s_i]$ in $N \cap C$, the element $[s_i \theta^{-i-1}]$ is in the unrestricted direct product of the T_i but not necessarily in $M \cap D$. We return to this observation in Section 3.2.

Theorem 2.1. *If (1.1) is an easy decomposition with a finite T sequence, then $C \approx D$.*

Proof. Let θ be an easy isomorphism of M onto N. Construct the groups E_{ij} as in Section 2.2. By hypothesis there is an integer r with $T_i = 1$ for $i \geq r$. Then, from (2.9), $M_i \cap D = M_{i+1} \cap D, i \geq r$ so that $M_r \cap D \subset \bar{M}$. Similarly, $N_r \cap C \subset \bar{N}$. Therefore, $M_r \cap D = N_r \cap C = 1$. Consequently, from (2.19) and (2.20), $M \cap D = T_0 \times T_1 \times \ldots \times T_{r-1}$ and $N \cap C = S_0 \times S_1 \times \ldots \times S_{r-1}$ so that $M \cap D \approx N \cap C$. From (2.14) we deduce that $C \approx D$.

As in [6], we say a group L is *terrible* if there exists a group $K, K \neq 1$ such that $L \approx L \times K$. We say that K is a *terrible kernel* for L. We remark that a method for constructing finitely generated groups L which have terrible finitely presented nontrivial kernels K is given via [7, Theorems 1 and 2]. Also, see [9, 10, 13].

THEOREM 2.2. *Let R be a tame finitely generated group with a tolerable center. Suppose that there is no direct factor K of R which is a terrible kernel for any finitely generated group L. Then R is J replaceable.*

Proof. Deny the assertion. By means of Lemma 2.1, we may pass to a counterexample (1.1) with $J \times C \not\approx J \times D$ where C, D are perfect centerless direct factors of R and where $M \subset J(R)$. Construct the subgroups T_i, S_i relative to (1.1). By hypothesis, there is an integer r such that $T_i = S_i = 1, i \geq r$. Let $S = S_0 \times S_1 \times S_2 \times \cdots \times S_{r-1}, T = T_0 \times T_1 \times \cdots \times T_{r-1}$.

By [6, 4.2], $(N \cap C)/S$ is either 1 or a terrible kernel for a finitely generated group. Similarly $(M \cap D)/T$ is either 1 or a terrible kernel for a finitely generated group. Since $(N \cap C)/S$ and $(M \cap D)/T$ are direct factors of $N \cap C$ and $M \cap D$ which in turn are direct factors of C and D, we conclude that $N \cap C = S$ and $M \cap D = T$ so that $N \cap C \approx M \cap D$ and thus $C \approx D$, a contradiction. By using (2.18), we see that the hypothesis on direct factors in Theorem 2.2 always holds for a finitely generated residually finite group R. Thus, we have the

Corollary. *Let R be a tame finitely generated residually finite group with a tolerable center. Then R is J replaceable.*

If R is a finitely generated residually finite group, then so is any member of $J(R)$. This remark together with (2.16), (2.17), (2.18), Section 2.4 and the above corollary prompt the following

Question: Is a finitely generated residually finite group with a tolerable center tame?

3. Residually Finite Groups and Easy Decompositions

Let (1.1) be a decomposition with C and D perfect, centerless, and residually finite and M, N, C, D finitely generated. We will now show how to pass to an easy decomposition. If G is a group, $I(G)$ will designate the set of finite homomorphic images of G.

Lemma 3.1 [1]. *If $I(A) = I(A_1)$ and $I(A \times B) = I(A_1 \times B_1)$ where A, A_1, B, B_1 are finitely generated, then $I(B) = I(B_1)$.*

Lemma 3.2. *If $I(A) = I(A_1)$ and $I(B) = I(B_I)$, then $I(A \times B) = I(A_1 \times B_1)$.*

Lemma 3.3 [14]. *If H is a finitely generated residually finite group and $\bar{H} \neq 1$, then $I(H) \neq I(A/\bar{H})$.*

Now we construct the groups N_i and $\bar{N}$ as in Sections 2.2 and 2.5. Set $B = \bar{N}$ and $A = M \cap B$. Then $A\theta = \cap(M \cap N_i)\theta = \cap N_{i+1}$. Since the N_i form a descending chain, $A\theta = B$. However, $N_i = (N_i \cap M) \times (N_i \cap C)$. Therefore, $B = A \times \bar{C}$ where $\bar{C} = \cap(N_i \cap C), i \geq 0$. Hence $(M \times C)/A\theta = (N \times D)/A\theta = (M \times C)/(A \times \bar{C})$ and $(M/A) \times (C/\bar{C}) \approx (N/A\Theta) \times D$. Now we may apply Lemma 3.1 to see that $C/\bar{C}$ and D have the same finite images. Again Lemma 3.1 applied to (1.1) implies C and D have the same finite images. Since C is residually finite, $\bar{C} = 1$ and $A\theta = A = B$. Similarly, if $E = \cap M_i$, then $E\theta^{-1} = E = E \cap N$. By using the description of the M_i and N_i before (2.14), we see that if R is any subgroup of $M \cap N$ with $R\theta = R$, then R is in $\bar{M}$ and R is in $\bar{N}$. Consequently, $\bar{N} = \bar{M} = \bar{N}\theta = \bar{M}\theta$.

Now write $M^1 = M/A, C^1 = (CA)/A, N^1 = N/A, D^1 = (DA)/A$ so that

$$(3.1) \qquad M^1 \times C^1 = N^1 \times D^1, C^1 \approx C, D^1 \approx D$$

Note that θ induces an isomorphism θ^1 of M^1 onto N^1 in (3.1) and that one may easily check that (3.1) is an easy decomposition with respect to θ^1. Consequently, M^1 is residually finite. Hence if a finitely generated residually finite group M is not J replaceable in (1.1), we can pass to an easy decomposition with C and D perfect and centerless by using Lemma 2.1 and our comments above. By Lemma 2.4, the result is a finitely generated centerless residually finite homomorphic image of M which is not J replaceable.

We recall that a group M is *cancellable* in direct products if (1.1) always implies $C \approx D$. If $Z(C) = 1$ in (1.1) and $M \approx J$, then $C \approx D$. With these considerations, we may summarize our previous comments as

Theorem 3.1. *Finitely generated residually finite groups with tolerable centers are J replaceable if and only if finitely generated residually finite centerless groups are cancellable in easy decompositions. A finitely generated residually finite group with a tolerable center is J replaceable provided that its centerless finitely generated residually finite homomorphic images are cancellable in easy decompositions.*

3.1 Some Final Remarks. This last section is motivated by our remarks immediately preceding Theorem 2.1 of Section 2.5. In particular, this section retrieves some information in easy decompositions by embedding an easy decomposition (1.1) into decompositions formed by adjoining appropriate elements in the unrestricted direct product of the T_i, S_i and E_{ij}. An application to residually finite groups is given in Theorem 3.2 and Corollary 3.1.

To begin, we would like to point out some of the properties of the finite homomorphic images of certain groups which arise naturally. If G is finitely generated in (1.1) and (2.2) holds and if

$$T = T_0 \times T_1 \times \cdots, S = S_0 \times S_1 \times \ldots, S^r = S_0 \times S_1 \times \cdots \times S_r$$

are the restricted direct product of the T_i, the S_i and the $S_i, i \leq r$ respectively then

$$I(N \cap C) = I(S) = I(T) = I(M \cap D)$$

Any member of $I(S)$ is a member of $I(S^r)$ for a suitable r. Hence $I(S) = \cup I(S^r)$. Since S^r is a direct factor of $N \cap C$, $I(S^r)$ is contained in $I(N \cap C)$ for all r. On the other hand, $(N \cap C)/S$ is a terrible kernel for a finitely generated group, so by (2.18) $N \cap C/S$ has no nontrivial finite homomorphic image. This implies that if R is a normal subgroup of finite index in $N \cap C$, $RS = N \cap C$ and $(N \cap C)/R \approx S/(R \cap C)$. Hence $I(N \cap C) = I(S)$. Similarly, $I(M \cap D) = I(T)$. By (2.14) and Lemma 3.2, this gives another independent observation of the fact that $I(C) = I(D)$ when C and D are perfect and centerless and M is finitely generated.

Let (1.1) be an easy decomposition with C and D finitely generated, and let θ be an easy isomorphism of M onto N. Let $t = [t_i], i \geq 0$ be an element of $M \cap D$. We form elements u_i^t in the unrestricted direct product of the E_{ij} of Section 2.2 by setting for $i \geq 1$

$$(3.2) \qquad u_i^t = t_i \theta t_{i+1} \theta^2 \ldots = [t_{i+k-1} \theta^k], k \geq 1$$

Let $\tilde{E}$ be the unrestricted direct product of the E_{ij}, and let $\tilde{S}$ be the unrestricted direct product of the $S_i, i \geq 1$. From (2.13) we see that the isomorphism θ may be extended to act on $\tilde{E}$ so that θ may be thought of as mapping $\tilde{E}$ into $\tilde{E} \times \tilde{S}$ and

$$u_1^t \theta = (t_1 \theta^2)(t_2 \theta^3) \ldots = [t_i \theta^{i+1}], i \geq 1$$

Note that if $i \geq 2$, then

$$(3.3) \qquad u_i^t \theta = (t_{i-1}^{-1} \theta) u_{i-1}^t$$

Let M_* be the subgroup generated by the $u_i^t, i \geq 1, t \in M \cap D$. Let C_* be the subgroup generated by the $u_1^t \theta, t \in M_1 \cap D$.

In the sequel, if H and K are subgroups of a group, HK designates the subgroup generated by H and K. In view of (3.3), θ induces an isomorphism of MM_* onto NM_*C_*, and we have

$$(3.4) \qquad G_* = MM_* \times C_*C = NM_*C_* \times D, \quad MM_* \approx NM_*C_*$$

The group CC_* has some interesting properties. We will argue below that (a) CC_* is centerless, (b) if C is a finitely generated residually finite group, then so is CC_*, (c) CC_* is perfect, (d) $I(CC_*) = I(C)$. Note that G_* may not be finitely generated so that we cannot invoke previous comments in (3.4) to conclude $I(C_*C) = I(D)$.

Since $C_*C = C_*(N \cap C) \times (C \cap D)$, in view of our previous remarks, (2.14) and Lemma 3.2, to show $I(C_*C) = I(D)$, it suffices to show $I[C_*(N \cap C)] = I(S)$. The map β defined by $t\beta = u_1^t \theta, t \in M_1 \cap D$ is an isomorphism of $M_1 \cap D$ onto C_*. Note that if we write

$$\bar{T} = T_1 \times T_2 \times T_3 \times \cdots, \quad \bar{S} = S_1 \times S_2 \times S_3 \times \cdots,$$

then $T = T_0 \times \bar{T}$, $S = S_0 \times \bar{S}$ and $\bar{T}\beta = \bar{S}$, $(M_1 \cap D)\beta = C_*$ so that $(M \cap D)/T \approx (M_1 \cap D)/\bar{T} \approx C_*/\bar{S}$. As in previous arguments, this implies $I(\bar{S}) = I(C_*)$. Let L be a normal subgroup of finite index in $C_*(N \cap C)$. Then, as in previous arguments, $(L \cap C_*)\bar{S} = C_*$, $(L \cap N \cap C)S = N \cap C$ and consequently $LS = C_*(N \cap C)$.

Hence $I[C_*(N \cap C)] \subset I(S)$. However, S^r is a direct factor of $C_*(N \cap C)$, so $I(S^r) \subset I[C_*(N \cap C)]$ and consequently $I(S) = I[C_*(N \cap C)]$. We remark that the map α defined by $t\alpha = t_0 \theta u_1^t \theta$ defines an isomorphism of $M \cap D$ into $C_* C$ and the subgroup of $C_* C$ generated by $(M \cap D)\alpha(C \cap D)$ is isomorphic to D. Also, by using (2.9), we can see that C_* is isomorphic to a direct factor of D. To verify that $C_* C$ is centerless, it suffices to verify that $Z[C_*(N \cap C)] = 1$. Since $C_*(N \cap C)$ is a subgroup of the centerless group $\tilde{S}$ and since $C_*(N \cap C)$ contains S, it is clear that $C_*(N \cap C)$ is centerless. Since $N \cap C/S$ and $C_*/\bar{S}$ have no nontrivial finite homomorphic images, it follows that if $R = C_*(N \cap C)$, then R/S has no nontrivial finite homomorphic images. However $S \subset S'$ so that R/R' is a finitely generated abelian group with $I(R/R') = 1$. Hence $R = R'$. Hence $(C_* C)' = C_* C$. Also, $C_* C$ is a subgroup of $\tilde{S}$ so that $C_* C$ is residually finite when C is residually finite.

Now, in analogy to (3.4), we reverse the roles of M and N, C and D and θ and θ^{-1}. The result is an extension $\tilde{G}_*$ with

$$\tilde{G}_* = NN_* \times D_* D = MN_* D_* \times C, \quad NN_* \approx MN_* D_*$$

Here, D_* is defined by

$$D_* = \left\{ s_1 \theta^{-2} s_2 \theta^{-3} \ldots = [s_i \theta^{-i-1}], i \geq 1 \right\}$$

where $[s_i], i \geq 1$ is an arbitrary element of $N_1 \cap C$. Let γ be an isomorphism which maps the unrestricted direct product of the T_i onto the unrestricted direct product of the S_i where

$$[t_i]\gamma = [t_i \theta^{i+1}], \quad [s_i]\gamma^{-1} = [s_i \theta^{-i-1}], i \geq 0$$

Then $CC_* = (C \cap N_1)[(M_1 \cap D)\gamma] \times S_0 \times (C \cap D)$, $\quad DD_* = (M_1 \cap D)[(C \cap N_1)\gamma^{-1}] \times T_0 \times C \cap D$. Clearly, γ maps $(M_1 \cap D)(C \cap N_1)\gamma^{-1}$ onto $(C \cap N_1)(M_1 \cap D)\gamma$ so that $CC_* \approx DD_*$.

Theorme 3.2. *Let R be a finitely generated residually finite group with a tolerable center. Then either*

(a) we may embed R in a J replaceable finitely generated residually finite group R^1, $Z(R^1)$ tolerable and $I(R) = I(R^1)$ or

(b) we may embed R in a group G_* containing a chain of subgroups R^k such that, for each $k \geq 1$, R^k is properly contained in R^{k+1}, R^k is a finitely generated residually finite group with a tolerable center, $I(R^k) = I(R^{k+1})$, R^k is not J replaceable and $R^1 \approx R$.

Proof. If R is J replaceable, (a) is obviously true with $R^1 = R$. Suppose R is not J replaceable and that (a) is false. It suffices to show that we may properly embed R in a finitely generated residually finite group R^2 with $I(R^2) = R$ and $Z(R^2)$ tolerable. For then since (a) is false, R^2 is not J replaceable. We will then be able to proceed inductively to construct a sequence of groups R^i where R^i is embedded properly in R^{i+1} and where R^{i+1} is constructed from R^i precisely as R^2 was constructed from R. An application of the construction in [12, Exercise 18, p. 33] then yields G_*.

If R is not J replaceable, we may invoke Lemma 2.1 and pass to a decomposition (1.1) with C and D perfect and centerless and each isomorphic to direct factors of R and $C \not\approx D$. As in Section 3.1, we may then pass to an easy decomposition (1.1) with $C \not\approx D$. We then construct the groups C_* and D_* as in Section 3.1. Since $CC_* \approx DD_*$ and $C \not\approx D$, either C is a proper subgroup of CC_* or D is a proper subgroup of DD_*. Say C is a proper subgroup of CC_*. Since C is a direct factor of R, we may write $R \approx C \times R_*$ for suitable R_*. We then set $R^2 \approx CC_* \times R_*$. Clearly, R^2 has the desired properties.

If G is a polycyclic group, then the polycyclic groups H for which $I(G) = I(H)$ fall into finitely many isomorphism classes (see [15,16]). With respect to this type of phenomenon, we may state

Corollary 3.1. *Let R be a finitely generated residually finite group with a tolerable center. Suppose that (up to isomorphism classes) there are only finitely many finitely generated residually finite groups*

$$R_1, R_2, \ldots R_k$$

with tolerable centers and with the same finite homomorphic images as R, and suppose further that there is no R_i which is isomorphic to a proper subgroup of itself. Then alternative (a) of Theorem 3.2 holds.

Corollary 3.2. *If (1.1) is an easy decomposition and C and D are finitely generated, then there exists a finitely generated centerless perfect group K that has two subgroups $\bar{C}, \bar{D}$ isomorphic to C and D respectively and $I(K) = I(C) = I(D)$ and $K = \bar{C}\bar{D}$.*

Proof. Take $K = CC_*, \bar{C} = C, \bar{D} = (M_1 \cap D)\gamma \times S_0 \times (C \cap D)$.

References

1. G. Baumslag, *Some metacyclic groups with the same finite images*, Composito Math **29** (1974), 241–252.
2. R. Hirshon, *Some cancellation theorems with applications to nilpotent groups*, Aust. J. Math. **23** (1977), 147–166.
3. R. Hirshon, *The equivalence of $x^t C \approx x^t D$ and $J \times C \approx J \times D$*, Trans. Amer. Math. Soc. **249** (1979), 331–340.
4. R. Hirshon, *The number of indecomposable terms in direct decompositions of groups with the maximal condition*, Jour. Algebra **75** (1982), 75–82.
5. R. Hirshon, *On uniqueness of direct decompositions of groups into directly indecomposable factors*, J. Pure Appl. Algebra **63** (1990), 155–160.
6. R. Hirshon, *Some properties of direct products*, Jour. Algebra **115** (1988), 352–365.
7. R. Hirshon and D. Meier, *Groups with a quotient that contains the original groups as a direct factor*, Bull. Aust. Math. Soc. **45** (1992), 513–520.
8. R. Hirshon, *Residually finite extensions of periodic groups*, Jour. Algebra **154** (1993), 1–11.
9. R. Hirshon, *Some properties of groups which allow homomorphisms onto their direct square*, Jour. Algebra. (to appear).
10. J.M. Tyrer Jones, *Direct products and the Hopf property*, J. Aust. Math. Soc. **17** (1974), 174–196.
11. W. Magnus, *Residually finite groups*, Bull. Amer. Math. Soc. **75** (1969), 305–316.
12. W. Magnus, A. Karrass, D. Solitar, *Combinatorial Group Theory*, Interscience Publishers, 1966.
13. D. Meier, *Non-hopfian groups*, J. London Math. Soc. (2) **26** (1982), 265–270.

14. P. Pickel, *A property of finitely generated residually finite groups*, Bull. Aust. Math. Soc. **15** (1976), 347–350.

15. P. Pickel, *Finitely generated nilpotent groups with isomorphic finite quotients*, Trans. Amer. Math. Soc. **160** (1971), 327–341.

16. P. Pickel, F.J. Grunewald and D. Segal, *Polycyclic groups with isomorphic finite quotients*, Annals of Math. **111** (1980), 155–195.

17. J. Wiegold, *Growth sequences of finite groups*, J. Aust. Math. Soc. **27** (1974), 133–143.

DEPARTMENT OF MATHEMATICS, POLYTECHNIC UNIVERSITY, BROOKLYN, NY 11201

Contemporary Mathematics
Volume **169**, 1994

WILHELM MAGNUS, APPLIED MATHEMATICIAN

Harry Hochstadt

Wilhelm Magnus is well known for his many significant contributions to group theory. But he is also known for his expertise in special functions. In conjunction with Fritz Oberhettinger he produced the famous handbook *Formulas and Theorems for the Functions of Mathematical Physics* [1]. A first edition of this book appeared in Germany in 1943 and a second edition in 1948. Shortly thereafter the English edition appeared in America. Many applied mathematicians have good reason to be grateful for these marvelous compendia. Magnus was also one of the chief collaborators in the preparation of *Higher Transcendental Functions* [2] produced under the auspices of the Bateman Manuscript Project.

Magnus's contributions to research on problems of wave propagation are less well known and certainly deserve wider recognition. Aside from studying these contributions one might well ask how he came to these problems. From the time he received his Ph.D. until the late 1930's he worked on group theory. Between 1939 and 1949 he made numerous contributions to applied mathematics and then he returned to his first love, group theory. Nevertheless even after that he inspired a number of students who completed doctoral dissertations under his supervision, working on special functions or applied mathematics. He had not been able to obtain a position as professor of mathematics, because he was not a member of the Nazi Party. He was a docent at the University of Frankfurt and in 1939 obtained a position as a research mathematician at the large industrial company Telefunken. It was there that he did most of his applied research. The great mathematical physicist Arnold Sommerfeld was a consultant at Telefunken and it is more than probable that some of Magnus's work was directly or indirectly inspired by him. In fact, in one publication helpful discussions with Sommerfeld are explicitly acknowledged.

As indicated above, Magnus turned to applied mathematics as a matter of necessity, but that still does not explain why hebecame so successful with these problems. Naturally his basic mathematical talents served him well. But these were enhanced by the broad education that he received at the University ofFrankfurt. There are two moving essays describing the atmosphere at Frankfurt during his student days, one by Magnus [14] and another by Carl Ludwig Siegel [15]. Frankfurt had been unusually fortunate in attracting an outstanding faculty among whom were Max Dehn, Ernst Hellinger, Paul Epstein and Otto Szasz. While each was a specialist in some area of mathematics, they did notoverlap and they each also had broad cultural interests. Asdescribed by Siegel, Dehn conducted a seminar on the history of mathematics which was attended by both faculty and students. The atmosphere was informal and the purpose was not to lead to publications but to learning. They

1991 *Mathematics Subject Classification*. Primary 01A70 Secondary 78-03

were even criticized by professors at other universities who said "Those gentlemen haveno dignity." In such an atmosphere Magnus must have flourished. Sadly all this ended with the coming of the Nazis. By then Siegel had left Frankfurt and some of the others fled Germany. In [15] Siegel tells how the Magnus family assisted Dehn in his flight from Germany.

I now want to describe some of his most important and fundamental contributions to applied mathematics. In [3], entitled About aBoundary Value Problem of the Wave Equation for the Parabolic Cylinder, Magnus solves a problem where a function $u(x,y)$ is sought which should satisfy the reduced wave equation with prescribed boundary values on the inner surface of a parabolic cylinder. The choice of a parabolic surface is plausible because of the well-known reflecting and focusing properties of a parabola. With the great importance of the then new developments in radio communications and radar these represented a new frontier in such research. The coordinates $x = \dfrac{\eta\xi}{k}$ and $y = \dfrac{\xi^2 - \eta^2}{2k}$ are introduced. The reduced wave equation now takes the form

$$\frac{\partial^2 u}{\partial \xi^2} + \frac{\partial^2 u}{\partial \eta^2} + (\xi^2 + \eta^2)u = 0$$

Here k is the wave number. The surfaces $\xi =$ constant and $\eta=$constant form a system of confocal parabolas with a common focal point at the origin. The solution u is prescribed on the parabola $\eta = \eta_0$ as

$$u(\xi,\eta_0) = e^{\dfrac{-i}{2}(\xi^2 - \eta_0^2)}\nu(\xi)$$

where $\nu(\xi)$ is a prescribed function.But to obtain a unique solution one also needs a "boundary condition" at infinity. In 1912 Sommerfeld had formulated this so-called radiation condition as

$$\lim_{r \to \infty} r \left| \frac{\partial u}{\partial r} + iku \right| = 0$$

for outgoing waves in a three dimensional problem. To do so he also required that if a surface scatters such a wave the surface had to be sufficiently smooth so that it had a well defined normal at every point. We shall see in the article under discussion and in later articles that Magnus played a major rolein replacing Sommerfeld's condition by weaker ones. In the problem under discussion Magnus shows that an appropriate radiation condition is

$$\left| \xi^{-1}\frac{\partial u}{\partial \xi} + iu \right| \leq \xi^{-\epsilon}|u|$$

for sufficiently large ξ . Here ϵ is an arbitrary positive number. Later on he replaces it by an even weaker form

$$\left| \int_{\eta_0}^{-\eta_0} \bar{u}\left(\xi^{-1}\frac{\partial u}{\partial \xi} + iu \right) d\eta \right| \leq \xi^{-\epsilon} \int_{\eta_0}^{-\eta_0} |u|^2 \, d\eta$$

Finally he provides the solution in the form of a line integral

$$u(\xi, \eta) = e^{\frac{-i}{2}(\xi^2 - \eta^2)} \int_{\nu - i\infty}^{\nu + i\infty} \frac{\bar{p}(\eta, \bar{n})}{\bar{p}(\eta_0, \bar{n})} p(\xi, n) h(n) \, dn \qquad \left(\frac{-1}{2} < \nu < 0\right)$$

where the function $h(n)$ depends on the given boundary condition. The function $p(\xi, \eta)$ is an even solution of the differential equation that arises from the separation of variables in the reduced wave equation, namely

$$\frac{d^2 p}{d\xi^2} + 2i(-\xi \frac{dp}{d\xi}) + np = 0, \qquad p(0, n) = 2^{-n-1} e^{n\pi i/4} \Gamma(-n/2)$$

Here $h(n)$ is given by

$$h(n) = \frac{-i}{\pi^2} \int_0^\infty \bar{p}(\sigma, -\bar{n} - 1) \nu(\sigma) d\sigma$$

All this seems rather simple and straightforward, but to justify the many operations such as the convergence of the integral and all the manipulations with these integrals, the properties of the special functions used must be proved. The functions in question are all confluent hypergeometric functions. The initial value problem for the corresponding equations were understood since at least the time of Gauss. The solutions at infinity could also be determined by asymptotic methods, but how the solutions at infinity could be related to those at the origin is known as the connection problem. It is necesary to know the connection in this type of problem, but these were not yet available in the extant literature.

In a later publication [4], entitled On the Theory of the Cylindrical Parabolic Mirror, he turns to a related problem. Here a point source is placed at the focal point of the parabola and the solution sought represents the wave reflected from the surface. Let $H_0^{(2)}$ be a Hankel function of the second kind, which represents the source. Then let

$$E = -u(\xi, \eta) + H_0^{(2)} \left(\frac{\xi^2 + \eta^2}{2}\right)$$

and u has to be a solution of the reduced wave equation so that E vanishes on $|\eta| = \eta_0$. Now let $\delta(\xi, \eta)$ satisfy the same differential equation as $p(\xi, \eta)$ in the previous problem, but with the initial conditions

$$\delta(0, n) = \frac{\Gamma(1/2) 2^{n/2}}{\Gamma(\frac{1-n}{2})} \qquad \delta'(0, n) = (1 + i)\frac{\Gamma(-1/2) 2^{(n-1)/2}}{\Gamma(\frac{-n}{2})}$$

He then develops the following integral representation for the Hankel function.

$$e^{i(\xi^2 + \eta^2)/2} H_0^{(2)}\left(\frac{\xi^2 + \eta^2}{2}\right) = \frac{1}{\sqrt{2\pi^2}} \int_{\nu - i\infty}^{\nu + i\infty} \delta(\xi, n) \delta(\eta, -n - 1) \Gamma(\frac{-n}{2}) \Gamma(\frac{n+1}{2}) dn$$

Using that he can find an integral representation for the solution.

$$E(\xi, \eta) = \frac{-1}{\sqrt{2\pi^2}} e^{-i(\xi^2 + \eta^2)/2}$$

$$\times \int_{\nu - i\infty}^{\nu + i\infty} \delta(\xi, n) \left\{ \frac{p(\eta, -n - 1)}{p(\eta_0, -n - 1)} \delta(\eta_0, -n - 1) - \delta(|\eta|, -n - 1) \right\}$$

$$\times \Gamma(\frac{-n}{2}) \Gamma(\frac{n+1}{2}) dn$$

with $n = \nu + i\mu$, $-1/2 < \nu < 0$. Next he shows, using asymptotic estimates, that one can close the contour of the integral and that this results in a residue integration over the zeros of $p(\eta_0, -n-1)$ as functions of the parameter $-n - 1 = -\frac{1}{2} + i\mu$. Finally he is able to localize these zeros thereby showing that the resulting series converges. The smaller the focal length the faster the series will converge. Let $\{\mu_1, \mu_2, \cdots\}$ denote these zeros. He shows that these lie in the intervals

$$\left(\pi \frac{2l-1}{2\eta_0}\right)^2 > 2\mu_l > \left(\pi \frac{2l-1}{2\eta_0}\right)^2 - \eta_0^2$$

so that all but a finite number must be positive. In fact if $\eta_0 = 2.006$ then $\mu_1 = 0$ and all the rest of the μ_l are all positive. The more zeros are positive the faster the series will converge. Finally he is able to present the solution in the form

$$e^{i\xi^2} 2E(\xi, \eta) = \frac{-e^{3\pi i/4} 2^{1/4}}{\sqrt{\pi}} \sum_{l=1}^{\infty} \delta(\xi, i\mu_l - \frac{1}{2}) A_{\mu_l}(\eta) 2^{-i\mu_l/2} \Gamma(\frac{1}{4} - \frac{i}{2}\mu_l)$$

$$\times \left\{ \int_0^{\eta_0} A_{\mu_l}^2(s) ds \right\}^{-1}$$

where the $A_\mu(s)$ are suitable solutions of the differential equation

$$A_\mu'' + A_\mu(2\mu + \eta^2) = 0$$

In these two articles we see that Magnus used well–known tools of applied mathematics, but one must also realize that while the general techniques are familiar, the details had to be worked out ab initio. The special functions used had to be studied, their asymptotic structures had to be determined to provide convergence estimates and also to be able to localize the eigenvalues. He shows a skill that one would hardly expect from someone so new to this type of research.

While the above two problems are the only two on which he himself worked that involved parabolic coordinates, he was still responsible for suggesting related problems to doctoral students, namely Epstein [5] and Hochstadt [6]. Epstein's problem was to determine the addition theorems under rotations of the coordinate system for the solutions of the reduced wave equation in parabolic cylinder coordinates. Hochstadt found addition theorems under rotations of the coordinate system in paraboloidal coordinates. Hochstadt also investigated the solution of diffraction of a plane wave coming in from infinity parallel to the axis for both parabolic cylindrical surfaces as well as paraboloidal surfaces. The writing down of solutions in terms of integral representations was not too difficult, but the challenge was to find asymptotic representations for large wave numbers [7].

Another problem solved by Magnus is entitled [8] About Diffraction of an Electromagnetic Wave by a Half Plane. An incoming plane wave is represented by

$$E_0 = Ae^{-ik(x\cos\alpha + y\sin\alpha)}, \quad A \; constant$$

and the solution sought is given by

$$E = E_0 + u$$

where u is a solution of the reduced wave equation

$$\frac{\partial^2 u}{\partial x^2} + \frac{\partial^2 u}{\partial y^2} + k^2 u = 0$$

and u is supposed to satisfy the radiation condition, and on the half plane $y = 0$, $x \geq 0$, it has to satisfy the boundary condition

$$u = -e^{-ik \sin \alpha}$$

This problem had already been solved by Sommerfeld in a very ingenious way [9]. Sommerfeld wanted to use the method of images. To do so he introduced a second sheet à la Riemann surfaces with a cut along the positive x axis and a wave of period 4π instead of 2π with respect to the radial variable. Poincaré had also proposed a method of using a dipole distribution on the half plane, but this leads to an integral equation, and methods for solving such equations were not truly understood at that time. Magnus proposed to carry out Poincaré's program. For convenience he lets $k = 1$ and seeks a solution in the form

$$u = A \int_0^\infty f(\eta) H_0^{(2)}(\sqrt{x^2 + (y - \eta)^2}\,d\eta$$

where $f(\eta)$ is the density of the dipole distribution. The boundary condition now leads to the integral equation

$$-e^{-iy \sin \alpha} = \int_0^\infty f(\eta) H_0^{(2)}(|y - \eta|)\, d\eta$$

Magnus then proposes to solve the more general equation

$$g(y) = \int_0^\infty f(\eta) H_0^{(2)}(|y - \eta|)\, d\eta$$

He then represents the density function in the form

$$f(\eta) = \frac{\pi}{2i} e^{i\pi/4} \sum_{m=0}^\infty i^m (2m + 1) c_m \eta^{-1} J_{(2m+1)/2}(\eta)$$

using Bessel functions, where the coefficients c_m are to be determined. Then, if $g(y)$ is represented by

$$g(y) = \sum_{n=0}^\infty i^n a_n J_n(y)$$

one obtains an infinite system of linear equations.

$$a_n = \sum_{m=0}^\infty c_m \epsilon_n \left(\frac{1}{\frac{2m+1}{2} - n} + \frac{1}{\frac{2m+1}{2} + n} \right) ; \epsilon_0 = 1, \epsilon_n = 2 \text{ for } n \geq 1$$

At this point he returns to the diffraction problem for which the coefficients are given by

$$a_n = \epsilon_n \cos(\alpha + \frac{\pi}{2})$$

By very clever manipulations he then inverts this system to obtain

$$c_m = \frac{1}{\pi} \sin\left(\frac{2m+1}{2}(\alpha + \frac{\pi}{2})\right), \frac{3\pi}{2} > \alpha > -\frac{\pi}{2}$$

Finally he sums the series for the distribution function and obtains

$$f(\eta) = i \sum_{m=0}^{\infty} e^{i(2m+1)\pi/4} \frac{2m+1}{2} \sin\left\{\frac{2m+1}{2}(\alpha + \frac{\pi}{2})\right\} J_{(2m+1)/2}(\eta)$$

From this he can then find the representation found much earlier by Sommerfeld. In reviewing this method one sees that he not only solved the diffraction problem under investigation, but in fact solved a more general integral equation, by reducing it to an algebraic system.

There are two more basic contributions that I wish to discuss. In [10] Magnus presents a generalization of Sommerfeld's radiation condition. First he presents a simple proof under Sommerfeld's assumptions, but then he considers more general surfaces. In general these surfaces are smooth with tangent planes at almost every point. But they may have edges where two such surface meet and they may also have free boundaries so that the surfaces need not be closed. To prove the result he again follows Poincaré's suggestion to use integral equations. He has to consider equations of the form

$$E(x, y) = \int \int_F \frac{e^{-ikR_0}}{R_0} g(x_0, y_0) dx_0 dy_0$$

I shall not say much about the proof other than to comment that he needs to use the expansion of plane waves in terms of Bessel functions. At that time this article represented the greatest advance in this field. Naturally since then more work has been done by numerous researchers.

One due to W.L. Miranker [11] deals with a variable index of refraction. Miranker treats the reduced wave equation in the form

$$\Delta u + h^2(x)u = 0$$

where $|h^2(x) - k^2| \leq 0(|x|^{-\mu}), k^2 > 0, \mu > 2$ and

$$\lim_{r \to \infty} \int \int |\frac{\partial u}{\partial r} - ih(x)u|^2 ds = 0$$

If u is bounded everywhere then

$$\lim_{r \to \infty} \int \int |u|^2 ds \neq 0$$

The latter result is known as the Rellich growth estimate, but Rellich only proved it for a constant index of refraction.

A second result is due to L.M. Levine [12]. Levine defines a regular surface G to be one which is composed of a finite number of $C^{2+\lambda}$ pieces, such that wherever two such pieces are adjacent they don't form a zero exterior angle. On each such surface element, u satisfies a boundary condition which may be a Dirichlet condition $u = 0$ or $\frac{\partial u}{\partial n} + \beta u = 0$, $\beta > 0$. If u also satisfies a radiation condition in the form

$$\lim_{r\to\infty} \int |\frac{\partial u}{\partial r} - iku|^2 ds = 0$$

then $u \equiv 0$ outside G.

The last result I wish to discuss is an article entitled [13] Questions of Uniqueness and the Behavior at Infinity of Solutions of $\Delta u + k^2 u = 0$. In this case he works in a p dimensional space. He goes on to define a generalized hemisphere as follows. Consider a sphere of unit radius centered at the origin and on it a system of closed smooth curves. Suppose that if any ray ξ through the origin that passes through the interior of one of the regions bounded by one of these curves then $-\xi$ does not pass through the interior of one of these curves. For example for $p = 2$, using polar coordinates, consider

$$0 \leq \theta \leq \frac{\pi}{3}, \qquad \frac{2\pi}{3} \leq \theta \leq \pi, \qquad \frac{4\pi}{3} \leq \theta \leq \frac{5\pi}{3}$$

These form a generalized hemisphere. Magnus then proves the following startling result. If u is a solution of the reduced wave equation such that it satisfies the boundedness condition

$$|r^{(p-1)/2}u| \leq M$$

everywhere and furthermore

$$\lim_{r\to\infty} r^{(p-1)/2}u = 0$$

on every ray passing through a generalized hemisphere then u must vanish identically. The proof is then accomplished by using an expansion of a solution of the form

$$u = \sum_{n=0}^{\infty} c_n r^{-(p-2)/2} J_{n+(p-2)/2}(kr) Y_n^{(p-1)}(\xi)$$

where the $Y_n^{(p-1)}$ are p dimensional spherical harmonics.

There is an important method called the Watson Transform, first developed by G.N. Watson in 1918 in a study of radio waves. It is important in studying how electromagnetic waves move around obstacles such as the earth's surface or steep mountain ranges. The same technique has also become important in modern physics under the heading of Regge poles. While Magnus did not directly contribute to such methods, he proposed several such problems and supervised several dissertations, but I shall not dwell on these.

The results discussed above represent those I consider most significant and most fundamental. There are other excellent papers of an applied character, as well as a number of dissertations dealing with problems of diffraction. Magnus's contributions to applied mathematics were not as extensive as his contributions to group theory but were of a basic and fundamental character, as I hope I have shown, and deserve wider recognition in the mathematical world.

REFERENCES

[1] W. Magnus and F. Oberhettinger, Formulas and Theorems for the Special Functions of Mathematical Physics, Chelsea Publishing Company, New York, 1949.

[2] A. Erdélyi, W. Magnus, F. Oberhettinger, F.G. Tricomi, Higher Transcendental Functions, Vol.s 1-3, McGrawHill Book Company Inc., New York 1953.

[3] W. Magnus, Ueber eine Randwertaufgabe der Wellengleichung fuer den parabolischen Zylinder, Jahresbericht der Deutschen Mathematiker-Vereinigung 50, 140-161, 1940.

[4] W. Magnus, Zur Theorie des zylindrischparabolischen Spiegels, Zeitschrift fuer Physik 118, 343-356, 1941.

[5] D. Epstein, Diffraction Problems for the Parabolic Cylinder, NYU Dissertation 1956.

[6] H. Hochstadt, Addition Theorems and Applications of Functions of the Paraboloid of Revolution, NYU Dissertation 1956, also Pac. Jour. Math., 7, 1365, 1957.

[7] H. Hochstadt, Asymptotic Formulas for Diffraction by Parabolic Surfaces, Comm. Pure and Appl. Math., 10, 311, 1957.

[8] W. Magnus, Ueber die Beugung elektromagnetischer Wellen an einer Halbebene, Zeitschrift fuer Physik 117, 168-179, 1941.

[9] A. Sommerfeld, Optics, Academic Press Inc., Pub., New York, 1954.

[10] W. Magnus, Ueber Eindeutigkeitsfragen bei einer Randwertaufgabe von $\Delta u + k^2 = 0$, Jahresbericht der Deutschen MathematikerVereinigung 52, 177-188, 1943

[11] W.L. Miranker, The Reduced Wave Equation with a Variable Index of Refraction, Comm. Pure and Appl. Math., 10, 491-502, 1957.

[12] L.M. Levine, A Uniqueness Theorem for the Reduced Wave Equation, Comm. Pure and Appl. Math., 17, 147-176, 1964.

[13] W. Magnus. Fragen der Eindeutigkeit und des Verhaltens im Unendlichen fuer Loesungen von $\Delta u + k^2 = 0$, Abhandlungen aus dem Mathematischen Seminar der Universitaet Hamburg 16, 77-94, 1949.

[14] W. Magnus, Vignette of a Cultural Episode, Wilhelm Magnus, Collected Papers, 623-629, Springer Verlag, 1984.

[15] C.L. Siegel, Zur Geschichte des Frankfurter Mathematischen Seminars, Collected Papers, 3, 462-474. Springer Verlag, 1966.

POLYTECHNIC UNIVERSITY, SIX METROTECH CENTER, BROOKLYN, NEW YORK 11201

Contemporary Mathematics
Volume **169**, 1994

ON THE COMBINATORIAL CURVATURE OF GROUPS OF F-TYPE AND OTHER ONE-RELATOR FREE PRODUCTS

A. JUHÁSZ AND G. ROSENBERGER

Dedicated to the memory of Wilhelm Magnus

Abstract. We consider certain one-relator free products with respect to the combinatorial curvature and hyperbolicity.

Introduction

Recall [3, p. 134] that a presentation $P = \langle X | R \rangle$ has nonpositive (negative) combinatorial curvature if the hyperbolic area of every inner face of every van Kampen diagram over R has nonnegative (positive) hyperbolic area. It is known that groups with a presentation having negative curvature are hyperbolic in the sense of Gromov [3. p. 134]. In this paper we consider the combinatorial curvature of a class of one-relator products which generalize groups of F-type (see [2] and 1.6 below). Our main result is the following

THEOREM A. *Let* $G^{(1)} = \underset{\alpha \in \mathcal{T}_1}{*} G_\alpha^{(1)}$, $\quad G^{(2)} = \underset{\beta \in \mathcal{T}_2}{*} G_\beta^{(2)}$ *and*

let $G^{(0)} = G^{(1)} * G^{(2)}$. *Let* $U \in G^{(1)}$, $V \in G^{(2)}$ *be cyclically*

1991 Mathematics subject classification. Primary 20F06. Secondary 20F10.

The research of the first author was supported by the Fund for the Promotion of Research at the Technion.

This paper is in final form and no version of it will be submitted for publication elsewhere.

reduced words of infinite order in $G^{(0)}$. Assume that if $U \in G_\alpha^{(1)}$, then $G_\alpha^{(1)}$ is free and U is cyclically reduced in $G_\alpha^{(1)}$. Assume the analogue for V. Let G be the quotient of $G^{(0)}$ by the normal closure of UV in $G^{(0)}$. Then

(a) G has nonpositive combinatorial curvature as a quotient of $G^{(0)}$;

(b) G is hyperbolic in the sense of Gromov if and only if $G^{(i)}$ hyperbolic, $i = 1,2$, and

(*)   At least one of U and V is neither a proper power nor a product of two elements of order 2.

Since groups of F-type are special cases of the groups mentioned in the theorem and cyclic groups are hyperbolic, we get

COROLLARY. Groups of F-type are hyperbolic unless U is a proper power or a product of two elements of order 2 and V also is a proper power or a product of two elements of order 2.   In the last case they have nonpositive combinatorial curvature.   In   particular,   they   satisfy   a   quadratic isoperimetric inequality.

We consider van Kampen diagrams.   All the unexplained terms concerning them can be found in [4].

1. Notation and preliminary results

1. Let M be a diagram over a free group or free product F. We shall denote by $|\partial M|$ the length of a cyclically reduced label of M over F.

2. Assume A is given by a presentation $\langle X|R \rangle$.   Then we shall always assume that R is cyclically reduced.   Also we shall assume that our van Kampen diagram contains a minimal number of regions for a simply connected given boundary label.

3.   Recall   from   [3]   that   a   presentation $\langle X|R \rangle$ has nonpositive (negative) combinatorial curvature if every inner region D of every van Kampen diagram over R satisfies that the excess $\kappa(D) = 2\pi + \sum_{i=1}^{n}(\theta_i - \pi)$ is nonpositive (negative).   Here $\partial D = v_1 e_1 v_2 e_2 \ldots v_n e_n v_1$, v_i vertices on ∂D, e_i edges on ∂D,

and θ_i are the inner angles of the polygon D at v . In general, one takes $\theta_i = \dfrac{2\pi}{d(v_i)}$ where $d(v_i)$ is the valency of v_i, i.e., distribute the curvature equally among the regions containing v_i. However, from the point of view of the theory of groups with nonpositive curvature [1] and the theory of hyperbolic groups [3], it is immaterial how the angles around v_i are distributed, as long as they sum up to 2π and $\theta_i - \dfrac{2\pi}{d(v_i)} < \dfrac{2\pi}{6}$. In the present work we shall make use of this remark.

4. Recall from [2] that a group G is of F-type if it has a presentation

$$G = \langle a_1,\ldots,a_n; \quad a_1^{\ell_1} = a_n^{\ell_n} = 1, \quad U(a_1,\ldots,a_p)V(a_{p+1},\ldots,a_n) = 1\rangle$$

where $n \geq 2$, $\ell_i = 0$ or $\ell_i \geq 2$, $1 \leq p \leq n-1$, $U(a_1,\ldots,a_p)$ is a cyclically reduced word in the free product on $a_1,\ldots,a_p$ which is of infinite order and $V(a_{p+1},\ldots,a_n)$ is a cyclically reduced word in the free product on $a_{p+1},\ldots,a_n$ which is of infinite order.

5. The following lemma is an immediate consequence of the construction of diagrams over free products (see [4]).   We omit its proof.

LEMMA. *Let* $A = \langle X|R\rangle$, $B = \langle Y|S\rangle$ *be finitely generated, and let* $G = A*B/D$, *where* D *is the normal closure of* $T \leq A*B$. *Let* $Q = R\cup S\cup T$. *Thus* $Q \leq F\ (X\cup Y)$. *Assume that there are constants* C *and* a *such that the following isoperimetric inequalities hold for R-diagrams* M, *S-diagrams* N *and T-diagrams* H *respectively:*

$\mathrm{Vol}(M) \leq C|\partial M|^a$, $\mathrm{Vol}(N) \leq C|\partial N|^a$ *and* $\mathrm{Vol}(H) \leq C|\partial H|^a$ *where for a P-diagram* Z, $\mathrm{Vol}(Z)$ *is the number of regions of* Z.

Then there is a constant $C' \geq C$ *depending on* Q *such that for every word* $W \in F(X\cup Y)$ *which represents* 1 *in* G *there is a Q-diagram* U *with boundary label* W *such that* $\mathrm{Vol}(U) \leq C'|W|^a$.

COROLLARY. *(a) If* A *and* B *are hyperbolic and the presentation of* G *as a quotient of the free product* $A*B$ *satisfies a linear isoperimetric inequality, then* G *is hyperbolic.*

(b) If A and B satisfy a polynomial isoperimetric inequality of degree k and G has nonpositive combinatorial curvature as a quotient of A*B, then G satisfies a polynomial isoperimetric inequality of degree max(k,2).

2. The structure of van Kampen diagrams

Let R be the set of all the cyclically reduced cyclic conjugates of UV and $(UV)^{-1}$.

Let M be a reduced R-diagram, and let D be an inner region of M. Then D has a boundary cycle $v_0 \mu v_1 v v_0$ such that the label of μ is U and the label of v is V. From now on we shall not distinguish between a path in M and its label in $G^{(0)}$.

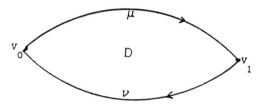

Fig. 1

Since the vertices v_0 and v_1 separate μ from v we shall call them *separating vertices*.

PROPOSITION 1.   *Separating vertices of D have valency at least 4.*

Proof. Assume v_0 has valency 3.   Then there are regions D_1 and D_2 which contain v_0 on their boundary.

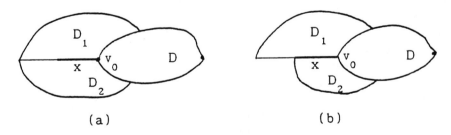

(a)   (b)

Fig. 2

Let x be the first letter on the edge common to D_1 and D_2 which contains v_0. If $x \in G^{(1)}$, then v_0 is a separating vertex of D_2 (Fig. 2(a)). Consequently a non-trivial tail of v is a common edge of D and D_2. But then since U is cyclically reduced and $U \neq U^{-1}$, D completely cancels D_2:

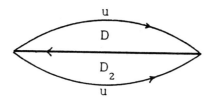

Fig. 3

This violates the assumption that M is reduced.

Similarly, if $x \in G^{(2)}$, then v_0 is a separating vertex of D^1 (Fig 2(b)) leading to the same contradiction. Thus v_0 has valency $\neq 3$. Finally assume v_0 has valency 2. Then, since U and V are cyclically reduced, v_0 is a separating vertex of D_1 and D_1 cancels D completely as in Fig. 3.

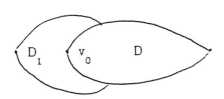

Fig. 4

COROLLARY 1. *If M satisfies $C(5)$, then M has a nonpositive combinatorial curvature.*

For D always has two vertices with valency ≥ 4 (namely v_0 and v_1).

REMARK. Neither U nor V can be a piece, for then as we have seen, the diagram is not reduced. Consequently U and V are each the product of at least two pieces. Since by the

Proposition no region D_1 can be a neighbour of D with a common edge (see Fig. 4) which contains v_0 as an inner vertex, D has at least 4 neighbours. This proves

COROLLARY 2. (a) M satisfies C(4).

(b) An inner region D has 4 neighbours if and only if its boundary cycle is subdivided into 4 edges in such a way that the label on the first two edges is U and the label on the second two edges is V. In particular, if such a region exists, then U and V are each the product of two pieces.

Thus, in what follows, we shall study the situation when U is the product of two pieces. Denote by $\mathcal{D}_0$ the set of all the inner regions in M which have a boundary path with label in $\{U^{\pm 1}, V^{\pm 1}\}$ and has a decomposition to two pieces by a vertex which is not a separating vertex.

Denote the set of all the inner vertices of M which are not separating vertices and divide a boundary path of a region D with label $U^{\pm 1}$ or label $V^{\pm 1}$ to two parts by V_0. For $i \geq 3$ denote by V_i the set of all the vertices of M not in V_0 which have valency i. Denote by $\kappa(U)$ the contribution of all the vertices of an inner region D which are inner vertices of the boundary path μ with label U to the excess of D, and let $\kappa(V)$ be defined analogously. Then

$$(1) \qquad \kappa(D) = 2\pi + \kappa(U) + \kappa(V) + (\theta_0 - \pi) + (\theta_t - \pi) = \kappa(U) + \kappa(v) + \theta_0 + \theta_t,$$

where θ_0 and θ_t denote the angles of the two separating vertices v_0 and v_t (see Fig. 1).

The following definition and the exposition of the next proposition and its proof are due to the referee. It greatly improves the original version, and we are grateful to him.

DEFINITIONS.   (a) A corner consists of a pair (e, e^1) of edges which have the same initial vertex and are such that the path $e^{-1}e^1$ is a subpath of a boundary cycle of a region associated with the corner.

(b) Given a vertex $u \in V_0$, a "bad" corner at u is a corner (e, e^1) such that the path $e^{-1}e^1$ has label one of $U^{\pm 1}, V^{\pm 1}$. A "good" corner is a corner which is not bad.

(c) Call two corners "adjacent" if they have an edge in common.

PROPOSITION 2.   Let the edge pair (α^{-1},β) define a bad corner at the vertex v of the region D.   Then
(a)   The two corners at v which are adjacent to the given corner are good corners;
(b)   The vertex v is preceded or followed (in the boundary cycle of D) by a vertex of valency four which is a separating vertex of D.

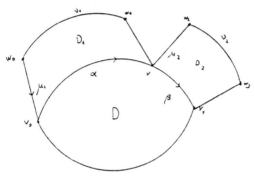

Fig. 5

Proof.   Let w_0 and w_1 be the separating vertices of D_1 (see Fig. 5).   Then $w_1 \neq v$ since v is not a separating vertex by assumption and $w_0 \neq v_0$, otherwise either D_1 cancels D (if μ_1 has label U) or U is not cyclically reduced (if μ_1 has label U^{-1}).

The lemma is proved.

Now define the angles in M.   If $v \in V_i$, $i \geq 3$, then assign the value $\dfrac{2\pi}{i}$ to every angle having v as its vertex.   If $v \in V_0$ and $d = d(v) \geq 5$, then assign to each angle having v as its vertex the value $\dfrac{2\pi}{d}$.

Let $\varepsilon = \dfrac{2\pi}{8m^2}$, $m = 2(|U|+|V|)$.   If $v \in V_0$ and $d(v) = 4$, then by Proposition 2 there are at most two regions which contain v on their boundary and belong to $\mathcal{D}_0$ (D_1 and D_3 in Fig. 6).

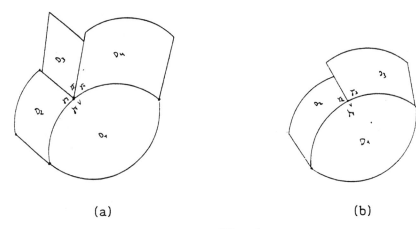

(a)   (b)

Fig. 6

If D_1 and D_3 are in $\mathcal{D}_0$, then assign $\frac{2\pi}{4} - \varepsilon$ to γ_1 and γ_3 and assign $\frac{2\pi}{4} + \varepsilon$ to γ_2 and γ_4.

If $D_1 \in \mathcal{D}_0$ and $D_3 \notin \mathcal{D}_0$, then assign $\frac{2\pi}{4} - \varepsilon$ to γ_1, $\frac{2\pi}{4} + \frac{\varepsilon}{2}$ to γ_2 and γ_4 and $\frac{2\pi}{4}$ to γ_3.

Finally, if $d(v) = 3$, then by Proposition 2 exactly one of the regions containing v, say D_1, belongs to $\mathcal{D}_0$ (see Fig. 6(b)). Assign $\frac{2\pi}{4} - \varepsilon$ to γ_1 and assign $\frac{2\pi}{3} + \frac{2\pi}{24} + \frac{\varepsilon}{2}$ to γ_2 and to γ_3.

PROPOSITION 3. *Let D be an inner region of M. If D has a vertex $v \in V_0$ on its boundary, then $\kappa(D) < 0$.*

Proof. Denote by a_i the number of inner vertices of μ (see Fig. 1) in V_i, $i \geq 3$, and let $a_4^* = \sum_{i \geq 4} a_i$. For $i \geq 5$, denote by b_i the number of inner vertices of μ in V_0 with valency i.

Define numbers b_4', b_4'', c_4 and c_3 as follows:

b_4' is the number of inner vertices of μ in V_0 which are of degree 4 and such that
(a)   the corner associated with D is a good corner;
(b)   exactly one of the two corners adjacent to the corner associated with D is a bad corner;

b_4'' is the number of inner vertices of μ in V_0 which are of degree 4 and such that

(a)   the corner associated with D is a good corner;

(b)   both the two corners adjacent to the corner associated with D are bad corners;

c_4 is the number of inner vertices of μ in V_0 which are of degree 4 and such that

(a)   the corner associated with D is a good corner;

(b)   neither of the two corners adjacent to the corner associated with D are bad corners (so that the remaining corner which is "opposite" to the corner associated with D must be a bad corner);

c_3 is the number of inner vertices of μ in V_0 of degree 3 such that the corner associated with D is a good corner (in which case exactly one of the other two corners adjacent to the corner associated with D is a bad corner).

Assume $D \notin \mathcal{D}_0$ and evaluate $\kappa(U)$. Thus,

$$\kappa(U) = \sum_{i \geq 3} a_i \left(\frac{2\pi}{i} - \pi\right) + \sum_{i \geq 5} b_i \left(\frac{2\pi}{i} - \pi\right) + b_4' \left(\frac{2\pi}{4} + \varepsilon - \pi\right)$$

$$+ b_4'' \left(\frac{2\pi}{4} + \frac{\varepsilon}{2} - \pi\right) + c_4 \left(\frac{2\pi}{4} - \pi\right) + c_3 \left(\frac{2\pi}{3} + \frac{2\pi}{24} - \pi + \frac{\varepsilon}{2}\right)$$

$$\leq - \frac{\pi}{4}c_3 - \frac{\pi}{2}a_4^* - \frac{\pi}{2}\left(b_4' + b_4''\right) + \frac{\varepsilon}{2}\left(2b_4' + b_4' + c_3\right)$$

$$\leq - \frac{\pi}{4}c_3 - \frac{\pi}{2}a_4^* - \frac{\pi}{2}\left(b_4' + b_4''\right) + \frac{\pi}{8m^2}(2m).$$

Assume $c_3 > 0$. Then, by Proposition 2, $a_4^* \geq \frac{1}{2}c_3$, hence

$$- \frac{\pi}{2}\left(a_4^* + \frac{1}{2}c_3\right) \leq - \frac{\pi}{2} - \frac{\pi}{4}, \text{ and}$$

$$\kappa(U) \leq - \left(\frac{\pi}{4} + \frac{\pi}{2}\right) + \frac{\pi}{8m^2}(2m) = \frac{3}{4}\pi + \frac{\pi}{4m} < - \frac{3}{4}\pi + \frac{\pi}{4} = - \frac{\pi}{2},$$

as $m > 1$. Thus

(2)   If $D \notin \mathcal{D}_0$ and $c_3 > 0$, then $\kappa(U) < - \frac{\pi}{2}$.

Assume $c_3 = 0$. Then $b_4' + b_4'' \neq 0$, hence, by Proposition 2, $a_4 > 0$. Consequently

$$\kappa(U) \leq - \frac{\pi}{2}a_4 - \frac{\pi}{2}\left(b_4' + b_4''\right) + \frac{\pi}{4m} \leq - \frac{\pi}{2} - \frac{\pi}{2} + \frac{\pi}{4m}$$

$$< - \frac{\pi}{2} - \frac{\pi}{2} + \frac{\pi}{4} < \frac{\pi}{2} .$$

(3)   If $D \notin \mathcal{D}_0$ and $c_3 = 0$, then $\kappa(U) < -\frac{\pi}{2}$.

Assume now that $D \in \mathcal{D}_0$. Then by the choice of γ_1, $\kappa(U) = -\frac{\pi}{2} - \varepsilon$.

(4)   If $D \in \mathcal{D}_0$, then $\kappa(U) < -\frac{\pi}{2}$.

Evaluate now $\kappa(D)$. By (1) $\kappa(D) = \kappa(U) + \kappa(V) + \frac{\pi}{2} + \frac{\pi}{2}$
$< -\frac{\pi}{2} - \frac{\pi}{2} + \frac{\pi}{2} + \frac{\pi}{2} = 0$ by (2), (3) and (4).

The proposition is proved.

COROLLARY 3. D *has nonpositive combinatorial curvature.   D*
has zero curvature if and only if D has four neighbours and
every boundary vertex is a separating vertex with valency 4.

We prove that in this case the condition * of Theorem A is
satisfied.
Let the notation be as in Fig. 7.   Then either v is a
separating vertex for D_1 or v is a separating vertex for D_2
(see Fig. 7).

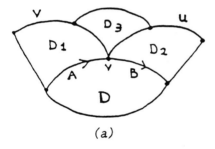 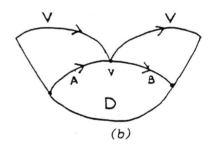

(a)   (b)

Fig. 7

If v is a separating vertex for only one of D_1 and D_2, then it
is a separating vertex for D_3 and hence for two neighbouring
regions of D (D_1 and D_3 or D_2 and D_3).   But the boundary
labels of such neighbours cancel each other, contradicting our
assumption that M contains a minimal number of regions with
the given boundary label. Consequently v is a separating vertex
for both D_1 and D_2 (see Fig. 7(b)).

Let A be the label of the piece common to D and D_1, and let B be the label of the piece common to D and D_2. Then one of the following equations holds: $(AB)(AB) = U(AB)V$ (see Fig. 8(a)) or $A = \overline{A}$ and $B = \overline{B}$ (see Fig. 8(b)). In the first case $A = D^\alpha$ and $B = D^\beta$ and in the second case A and B have order 2, as required.

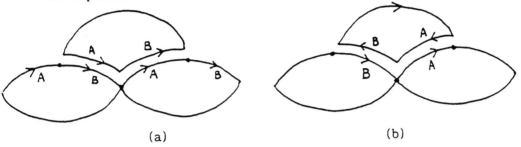

(a)   (b)

Fig. 8

3. The Proof of Theorem A

Assume (*) holds. Then by Proposition 3(e) the combinatorial curvature at each inner region of a derived van Kampen diagram over R is strictly negative. Therefore G is hyperbolic.

Assume now that (*) does not hold. Then R has one of the following forms.

(a) $A^n B^m$ $n,m \geq 2$;

(b) ABCD, $A = \overline{A}$, $B = \overline{B}$, $C = \overline{C}$, $D = \overline{D}$;

(c) ABC^n, $n \geq 2$, $A = \overline{A}$, $B = \overline{B}$.

In case (a) let $H = \langle A,B \rangle$. Then it follows from the fact that every van Kampen diagram over the symmetric closure of UV has a nonpositive combinatorial curvature that $H \cong \langle a,b \mid a^n = b^{-m} \rangle$. Consequently G is not hyperbolic.

A similar argument shows that the subgroup of G generated by A, B, C and D in Case (b) is isomorphic to $\langle a,b,c,d \mid a^2,b^2,c^2,d^2,abcd \rangle$ which contains a normal free abelian subgroup generated by ab and bc.

Finally in Case (c) let $H = \langle B,BC^n \rangle$. Then, as above, $H \cong \langle a,b \mid a^2,(ab^n)^2 \rangle$ in which $K = \langle b^n \rangle$ is an infinite cyclic normal subgroup such that $H/K = \mathbb{Z}_2 * \mathbb{Z}_n$, $n \geq 2$. Thus in all cases, G is not hyperbolic.

REFERENCES

1. M.R. Bridson, Geodesics and curvature in metric simplicial complexes, in "Group Theory from a Geometrical Viewpoint," Proc. ICTP Trieste, World Scientific, 1991.

2. B. Fine and G. Rosenberger, Generalizing algebraic properties of Fuchsian groups, *Groups St. Andrews,* 1989, Vol. 1, 124–147; *London Math. Soc. Lecture Note Series* 159, (1991).

3. M. Gromov: "Hyperbolic Groups" in Essays in Group Theory, S.M. Gersten, editor, Springer Verlag M.S.R.I. Lecture Note Series No. 8, 1987.

4. R.C. Lyndon and P.E. Schupp: *Combinatorial Group Theory,* Springer, Berlin, 1977.

Acknowledgement. We are grateful to the referee for his criticism and valuable remarks.

A. Juhász
Department of Mathematics
Technion— Israel Institute of Technology
32000 Haifa, Israel

G. Rosenberger
Fachbereich Mathematik
Universität Dortmund
4600 Dortmund 50, Germany

Contemporary Mathematics
Volume **169**, 1994

BRANCHED DIHEDRAL STRUCTURES
ON RIEMANN SURFACES

KATHRYN KUIKEN AND JOHN T. MASTERSON

ABSTRACT. In this paper, we construct and classify new families of branched structures, called *dihedral structures*, which are subordinate to certain branched projective structures on Riemann surfaces M. It is shown that there exist at most three dihedral structures subordinate to a given projective structure provided that the monodromy group of the projective structure is of order greater than two. For low branch divisor degree, these structures are shown to depend uniquely on their branch data and equivalence classes of monodromy homomorphisms. Finally, these structures are applied to construct families of ordinary differential equations on M having monodromy groups which are Z_2-extensions of Abelian groups.

1. Introduction

Unbranched projective and affine structures on compact Riemann surfaces were first constructed by Gunning [4,5] and were developed further by Kra [6,7,8,9]. Branched projective and affine structures were classified by Mandelbaum [14,15,16] and have recently been developed further in [17,18,19]. Theorem 4 of [6] establishes that any unbranched projective structure having a subordinate affine structure must have only one or two subordinate affine structures; this result also applies in the branched case. A more detailed classification of certain subordinate affine structures on surfaces of genus $g = 1$ has been carried out in [18]. The reader is referred to [14] and [17] for detailed definitions of branched affine and projective structures and their divisors and connections.

Recall that [4,5] each projective structure on a Riemann surface M can be identified with a projective equivalence class $\{A \circ f, A \in PL(2, \mathbb{C})\}$ containing its locally meromorphic geometric realizations. Each (possibly multivalued) nonconstant representative f of this class satisfies a functional equation

$$f([\lambda]z) = \psi([\lambda]) \circ f(z)$$

where $[\lambda] \in \pi_1(M)$, $\psi : \pi_1(M) \to PL(2, \mathbb{C})$ is the monodromy homomorphism and $f([\lambda]z)$ represents the result of analytic continuation of $f(z)$ along the arbitrary closed loop λ on M. Note that $\psi(\pi_1(M))$ is the monodromy group of the structure. The branch divisor of the structure is $B = \sum_{i=1}^{n}(r_i - 1)p_i$ where the p_i are the points of ramification order r_i of f. The given projective structure admits a subordinate affine structure provided that its projective equivalence class contains

---

1991 *Mathematics Subject Classification.* Primary 30F99.
This paper is in final form and no version of it will be submitted for publication elsewhere.

an affine equivalence subclass $\{A \circ f, A \in GA(1, \mathbb{C})\}$ with representative f having monodromy homomorphism $\psi : \pi_1(M) \to GA(1, \mathbb{C})$, where $GA(1, \mathbb{C})$ is the general affine group of dimension one. Each projective structure may be identified with a unique meromorphic quadratic differential $\theta_2 f$ and each affine structure with a unique affine connection $\theta_1 f(= f''/f')$, where $\theta_2 f(= (\theta_1 f)' - \frac{1}{2}(\theta_1 f)^2)$ is the classical Schwarzian derivative.

In this paper, we develop a parallel to this theory of affine substructures by constructing and classifying a new category of structures subordinate to certain branched projective structures. These new structures will be called *branched dihedral structures* since their associated monodromy homomorphisms ψ map $\pi_1(M)$ into the Lie group $D = \{T_1(z) = c_1 z, \ T_2(z) = c_2/z \ \forall c_1, c_2 \in \mathbb{C}^*\}$ and since they are each associated with a dihedral equivalence subclass $\{A \circ f, A \in D\}$ of the projective equivalence class for the corresponding projective structure on M. We omit a formal development of these structures from branched dihedral coordinate coverings since this is analogous to the corresponding development of branched projective and affine structures found in [14]. Our structures are not required to be regular, i.e., the image of a geometric realization may be $P^1(= \mathbb{C} \cup \{\infty\})$ and not merely some subset of $\mathbb{C}$.

The plan of this paper is as follows: First, we determine some invariants of dihedral structures and properties of these invariants. Next, it is shown that modules of these structures exist on certain compact γ-hyperelliptic [3] Riemann surfaces M of genus $g \geq 0$. Then, it is demonstrated that these structures are, within certain constraints, uniquely determined by their branch data and equivalence classes of their monodromy homomorphisms. Furthermore, we prove that if $o(\psi(\pi_1(M))) > 2$, then there exist at most three dihedral structures subordinate to a given projective structure. Finally, we apply these structures to construct families of second order linear ordinary differential equations on M having monodromy groups which are subgroups of the group D.

2. Invariants of Dihedral Structures

It is henceforth assumed, unless we restrict further, that M is an arbitrary compact Riemann surface of genus $g \geq 1$. Let $\{A \circ f, A \in D\}$ be the dihedral equivalence class of an arbitrary fixed dihedral structure on M. This structure inherits a unique branch divisor $B = \sum_{i=1}^m (r_i - 1)p_i$ determined by the associated projective structure on M. Since all elements $A \in D$ preserve the nonzero-nonpolar set of f and also preserve the zero-polar set, the divisor B may be expressed in terms of dihedral-structure subinvariants as follows:

$$B = B_{NPZ} + B_{PZ}$$

where $B_{NPZ} = \sum_{i=1}^l (r_i - 1)p_i$ is the divisor of nonpolar-nonzero branch points of f and $B_{PZ} = \sum_{i=l+1}^m (r_i - 1)p_i$ is the divisor of zero-polar branch points of f. Furthermore, the divisor $B_{SZP} = \sum_{k=1}^n s_k$ of simple poles or zeros of f is also an additional invariant for the given dihedral structure. We can now state

Lemma 1. *For any dihedral structure on M, $deg(B) \geq 2g - 2$. In fact, $deg(B) = 2g - 2 + sum(positive\ ramification\ orders\ of\ poles/zeros\ of\ f)$.*

Proof. Since f has monodromy homomorphism into D, $\frac{f'}{f}$ is a Prym differential on M. Hence, $deg\left(\frac{f'}{f}\right) = 2g - 2$. Also, the divisor

$$\left(\frac{f'}{f}\right) = B_{NPZ} - \sum_{i=\ell+1}^{m} 1 \cdot p_i - B_{SZP}$$

(1)

$$= B - \sum_{i=\ell+1}^{m} r_i p_i - B_{SZP}$$

Thus,

$$deg(B) = deg\left(\frac{f'}{f}\right) + \sum_{i=\ell+1}^{m} r_i + deg(B_{SZP})$$

and the conclusion follows. ‖

We now introduce a differential invariant which, unlike B, uniquely determines a given dihedral structure. For any locally meromorphic function t, define $\theta_d t = \left(\frac{t'}{t}\right)^2$ and call this operator the *dihedral derivative*. It is easily seen that $\theta_d f \equiv \theta_d g$ in some neighborhood if and only if $f \equiv A \circ g$ for some $A \in D$. Hence, we easily obtain the following

Lemma 2. *There is a natural one-one correspondence between {dihedral structures on M} and {quadratic differentials ϕ of the form $\theta_d f$, where f is a representative geometric realization of any dihedral structure on M}.*

Since dihedral structures are subordinate to projective structures (and dihedral equivalence is finer than projective equivalence), one would expect that $\theta_2 f$ can be expressed as a function of $\theta_d f$. In fact, by a direct computation we obtain

Lemma 3. $\theta_2 f = \frac{1}{2}\frac{(\theta_d f)''}{\theta_d f} - \frac{5}{8}\left[\frac{(\theta_d f)'}{\theta_d f}\right]^2 - \frac{1}{2}\theta_d f$

Equation (1) implies that

$$(\theta_d f) = \left(\left(\frac{f'}{f}\right)^2\right) = \sum_{i=1}^{\ell} 2(r_i - 1)p_i - \sum_{i=\ell+1}^{m} 2p_i - 2B_{SZP}$$

Thus, $\phi = \theta_d f$ is a quadratic differential having even order zeros at the points of B_{NZP} and having poles of order two at all zeros/poles of f. In fact, $\theta_d f$ has quadratic residue (i.e., coefficient of $\frac{1}{(z-z(p))^2}$) of the form $k^2, k \in Z^*$, at any zero/pole p of f of ramification order k. Furthermore, any quadratic differential ϕ on M having only even order zeros and double poles with quadratic residue $k^2, k \in Z$, determines a dihedral structure on M with geometric realization $f = e^{\int \sqrt{\phi}}$. Thus, we are led to the following extension of Lemma 2.

Lemma 4. *There is a natural one-one correspondence between {dihedral structures on M with branch divisors B_{NZP} and B_{ZP} and simple zero/polar divisor B_{SZP}} and {quadratic differentials ϕ on M having zero divisor $2B_{NZP}$ and having double* *poles at the points of B_{ZP} and B_{SZP} with residues k^2, where k is the order of the zero/pole of any corresponding geometric realization}.*

One might surmise that Lemma 4, together with the Riemann-Roch theorem, would provide an effective method to show the existence of dihedral structures with prescribed branching. Unfortunately, this does not appear to be the case. Let $N(= \text{card}(B_{ZP}) + \text{card}(B_{SZP}))$ be the prescribed number of double poles of the quadratic differentials in Lemma 4. The Riemann-Roch theorem implies that the quadratic differentials with second order poles (and square residues $k^2, k \in Z^*$) form an affine space L of dimension $3g - 3 + N$ consisting of the differentials $\{\phi_0 + \sum_{j=1}^{3g-3+N} t_j \phi_j\}$, where ϕ_0 is a differential of this form and the differentials ϕ_j have at most simple poles at the points of B_{PZ} and B_{SPZ}. The additional requirement, from Lemma 4, of even order zeros at the points of B_{NPZ} forces a restriction to the affine subspace of L for which the following nonhomogeneous linear system of $\sum_{i=1}^{\ell} 2(r_i - 1)$ equations holds:

$$\phi_0^{(k)}(p_i) + \sum_{j=1}^{3g-3+N} t_j \phi_j^{(k)}(p_i) = 0$$

$$\forall k \in \{0, \dots, 2(r_i - 1) - 1\}, \forall i \in \{1, \dots, \ell\}$$

However, Eq. (1) implies that $\sum_{i=1}^{\ell} 2(r_i - 1) = 4g - 4 + 2N$. Thus, this nonhomogeneous system is generically overdetermined since it has $4g - 4 + 2N$ equations in the $3g - 3 + N$ unknowns t_j. Thus, this nonconstructive approach is unwieldy since it ignores moduli information on M. In the next section, we shall directly construct dihedral structures in a moduli-dependent way using function-field elements on certain classes of algebraic curves; these curves represent large classes of possible surfaces M.

3. Existence of Dihedral Structures

In Theorem 1, we restrict M to be a γ-hyperelliptic surface with corresponding algebraic curve of the form

(2) $w^{2n} = \Pi_{\nu=1}^{2\ell}(t - e_\nu)$, e_ν distinct, $n|\ell$

or of the form

(3) $w^{2n} = \Pi_{\nu=1}^{2\ell-1}(t - e_\nu)$, e_ν distinct, $(n, 2\ell - 1) = 1$

An elementary calculation using the genus formula [1, p.144] shows that the Riemann surfaces for the curves (2) and (3) have genus $g = 2n\ell - \ell - 2n + 1$.

Theorem 1. *(i) Each map*

$$f_{(i,j,k,m)}(p) = \left(\frac{\sqrt{e_j - e_k}}{\sqrt{e_j - e_i}} \frac{\sqrt{t(p) - e_i}}{\sqrt{t(p) - e_k}} \right.$$

$$\left. + \frac{\sqrt{e_i - e_k}}{\sqrt{e_j - e_i}} \frac{\sqrt{t(p) - e_j}}{\sqrt{t(p) - e_k}} \right)^{1/m}, \; m|n, \forall p \in M$$

determines a dihedral structure on a surface M, associated with an arbitrary curve of form (2), having monodromy group D_{2m} and having branch divisor

$$B = \sum_{\nu \in \{1,\dots,2\ell\}-\{i,j,k\}} (2n-1)p_\nu + (n-1)p_i + (n-1)p_j + \left(\frac{n}{m}-1\right)p_k$$

where $t(p_\nu) = e_\nu \ \forall \nu = 1,\dots,2\ell$ and p_k is the only zero/polar branch point.

(ii) Each map

$$f_{(i,j,m)}(p) = \left(\frac{\sqrt{t(p)-e_i} - \sqrt{e_j-e_i}}{\sqrt{t(p)-e_i} + \sqrt{e_j-e_i}} \right)^{1/m} , \quad m|2n, \ \forall p \in M$$

determines a dihedral structure on a surface M associated with an arbitrary curve of form (3) having monodromy group D_m and having branch divisor

$$B = \sum_{\substack{\nu=1 \\ \nu \neq i,j}}^{2\ell-1} (2n-1)p_\nu + (n-1)p_i + (n-1)p_\infty + \left(\frac{2n}{m}-1\right)p_j$$

where $t(p_\nu) = e_\nu \ \forall i = 1,\dots,2\ell$, $t(p_\infty) = \infty$ and p_j is the only zero/polar branch point.

Proof. (i) The Riemann surface M for (2) may be represented as a multi-sheeted surface consisting of $2n$ t-planes with branch lines as follows:

SHEET $\mu, \mu \in \{1,\dots,2n\}$

A curve through the branch lines is closed on M if and only if the sum of its winding numbers with respect to the points $t(p_\nu)$, $\nu = 1,\dots,2\ell$, is a multiple of $2n$. It is easily verified that $z = f_{(i,j,k,m)}$ is locally single-valued (and locally meromorphic) on M. Furthermore, the analytic continuation of z gives

(a) z if $t(p_i)$, $t(p_j)$, $t(p_k)$ are exterior to the closed curve (i.e., the curve has winding number zero with respect to these points),

(b) $e^{\pm 2\pi ir/m}z$ if the curve has winding number $r(\in Z^*)$ with respect to $t(p_i)$, $t(p_j)$ and $t(p_k)$,

(c) $e^{\pm \pi ir/m}z$ if the curve has winding number r with respect to (resp. winding number zero w.r.t.) $t(p_i)$ and $t(p_j)$ while it also has winding number zero with respect to (resp. winding number r w.r.t.) $t(p_k)$,

(d) $\frac{e^{\pm 2\pi ir/m}}{z}$ if the curve has winding number r with respect to (resp. winding number zero w.r.t.) $t(p_j)$ while it also has winding number zero with respect to (resp. winding number r w.r.t.) $t(p_i)$ and $t(p_k)$,

(e) $\frac{e^{\pm \pi ir/m}}{z}$ if the curve has winding number r with respect to (resp. winding number zero w.r.t.) $t(p_i)$ and winding number zero with respect to (resp. winding number r w.r.t.) $t(p_j)$ and $t(p_k)$.

Clearly, cases (a)-(e) include a generating set of closed curves Λ for $\pi_1(M)$ as $r \in Z^*$ is varied. Hence, $z = f_{(i,j,k,m)}$ represents a projective structure on M. Also, the above set of elements generates monodromy group D_{2m}. Therefore, $z = f_{(i,j,k,m)}$ determines a dihedral structure subordinate to the corresponding projective structure.

An elementary calculation shows that $z = f_{(i,j,k,m)} = 0$ or ∞ iff $p = p_k$. In fact, $p = p_k$ is a pole or zero of ramification order n/m. Furthermore,

$$f'_{(i,j,k,m)}(p) = \frac{1}{m}(f_{(i,j,k,l)})^{\frac{1-m}{m}} \cdot \frac{1}{2} \frac{t'(p)}{(\sqrt{t(p)} - e_k)^3} \times$$

$$\left[\frac{\sqrt{e_j - e_k}}{\sqrt{e_j - e_i}} \frac{\sqrt{e_i - e_k}}{\sqrt{t(p)} - e_i} + \frac{\sqrt{e_i - e_k}}{\sqrt{e_j - e_i}} \frac{\sqrt{e_j - e_k}}{\sqrt{t(p)} - e_j} \right]$$

Thus, $f'_{(i,j,k,m)}(p)$ has zeros of order $2n - 1$ at p_ν, $\nu \in \{1, \dots, 2\ell\} - \{i, j, k\}$ and has zeros of order $n - 1$ at p_i and p_j. Consequently, the branch divisor is as stated in the theorem.

(ii) The proof of this part is very similar to the proof of (i); so, we omit almost all details. Note that here p_∞ is a branch point of $t(p)$ of ramification order $2n$ and a point of local single-valuedness (and, in fact, branching if $n > 1$) of $f_{(i,j,m)}(p)$. ‖

Remark 1. There undoubtedly exist on many compact Riemann surfaces constructions of dihedral structures which are different from those given in Theorem 1. However, one interesting feature of both (i) and (ii) is that the differential equations

$$(4) \qquad\qquad y''(t) + \frac{1}{2}\theta_2(h(t))y(t) = 0,$$

where $h = f_{(i,j,k,m)}(p) \circ t^{-1}$ or $f_{(i,j,m)}(p) \circ t^{-1}$, are hypergeometric equations [2] on $\widehat{\mathbb{C}}$ having dihedral monodromy groups. Hence, the dihedral structures obtained in (i) and (ii) have associated projective structures which can be viewed as lifts [10,11,12] under $t = t(p)$ of the hypergeometric equations (4).

We now develop a theorem which permits us to combine certain dihedral structures to obtain other dihedral structures; the resulting classes of structures form a module over the integers. We say that the monodromy homomorphisms ψ_1 and ψ_2 from $\pi_1(M)$ to D are *compatible* if $\psi_1([\Lambda]) = (T_{[\Lambda]}(z) = cz)$ iff $\psi_2([\Lambda]) = (T^*_{[\Lambda]}(z) = c^*z)$ and $\psi_1([\Lambda]) = (T_{[\Lambda]}(z) = c/z)$ iff $\psi_2([\Lambda]) = (T^*_{[\Lambda]}(z) = c^*/z)$. Similarly, the dihedral structures determined by the geometric realizations f_1 and f_2 on M are *compatible* if they have compatible monodromy homomorphisms ψ_1 and ψ_2. Clearly, compatibility is an equivalence relation for monodromy homomorphisms and also for dihedral structures. We are now in a position to state

Theorem 2. *Let* $f_1, f_2, \dots, f_n$ *be geometric realizations for compatible dihedral structures on* M *where no power of any* f_i *can be expressed as a product of powers of the remaining* f_j, $j \neq i$. *Then, the products* $\Pi_{i=1}^n f_i^{k_i}$, $k_i \in Z^*$, *determine a set of dihedral structures on* M *which are in natural one-one correspondence with the module* $Z^n/\langle v \to -v, \ v \in Z^n \rangle$.

Proof. The products of the form $\Pi_{i=1}^n f_i^{k_i}$ have monodromy homomorphisms compatible with the homomorphisms of the f_i. Hence, these products determine dihedral structures on M compatible with the structures for the f_i. The non-constancy of the f_i and the assumption that no power of any f_i can be expressed as a product of powers of the remaining f_j together imply that $\Pi_{i=1}^n f_i^{k_i} \equiv \Pi_{i=1}^n f_i^{\ell_i}$ iff $k_i = \ell_i \, \forall i \in \{1, \dots, n\}$. Furthermore, $\Pi_{i=1}^n f_i^{k_i}, c\Pi_{i=1}^n f_i^{k_i}$ and $c/\Pi_{i=1}^n f_i^{k_i}$ determine the same dihedral structure on M. Therefore, the dihedral structures determined by products of the form $\Pi_{i=1}^n f_i^{k_i}$ are in a natural one-one correspondence with the equivalence classes of vectors $\{\pm v, v = (k_1, k_2, \dots, k_n)\}$. Hence, the result follows. ||

Illustrations of this theorem are readily constructed. The functions $f_{(i,j,k,m)}$ in Theorem 1(i) with i and j fixed and $k \in \{1, \dots, 2\ell\} - \{i, j\}$ determine compatible structures each with distinct unique zero/pole p_k. Therefore, the assumption on powers in Theorem 2 is satisfied. Also, $f_{(i,j,k,m)} = (f_{(i,j,k,n)})^{n/m}$. Thus, the products $\Pi_{k=1, k \neq i,j}^{2\ell} f_{(i,j,k,n)}^{\mu_k}$ determine a family of dihedral structures in one-one correspondence with $\hat{v} = Z^{2\ell-2}/\langle v \to -v, v \in Z^{2\ell-2}\rangle$. Similarly, the products $\Pi_{j=1, j \neq i}^{2\ell-1} f_{(i,j,2n)}^{\mu_j} f_{(i,j,2n)}$ in Theorem 1(ii) determine a family of dihedral structures which is also in one-one correspondence with $\hat{v}$.

4. Uniqueness of Dihedral Structures

We are now in a position to state

Theorem 3. *Assume that $f_j, j = 1, 2$, are geometric realizations for two dihedral structures on M having compatible monodromy homomorphisms ψ_1 and ψ_2 and having the same branch divisors B_{NPZ} and B_{PZ}. Assume further that the branch divisors satisfy $2g - 1 \leq deg(B_j) \leq 3g - 2 + \sum_{i=\ell+1}^n r_i, j = 1, 2$, and that at least one of the divisors $B_{SPZ,j}, j = 1, 2$, is of minimal index of specialty. Then, $f_j, j = 1, 2$, determine the same dihedral structure on M.*

Proof.. The Prym differentials $f'_j/f_j, j = 1, 2$, have divisors given by Eq. (1) as follows:

$$(5) \qquad \left(\frac{f'_j}{f_j}\right) = B - \sum_{i=\ell+1}^m r_i p_i - B_{SPZ,j}, \, j = 1, 2$$

where B and the r_i and p_i are the same for the two dihedral structures. Eq. (5) and the assumed inequalities $deg(B_j) \leq 3g - 2 + \sum_{i=\ell+1}^m r_i, j = 1, 2$, together imply the inequalities

$$(6) \qquad deg(B_{SPZ,j}) \leq g, \, j = 1, 2.$$

The minimality of $i(B_{SPZ,j}), j = 1$ or 2, and inequality (6), together with the Riemann-Roch theorem and the Noether gap theorem [3], imply that $i(B_{SPZ,j}) = g - deg(B_{SPZ,j}), j = 1$ or 2, and that $B_{SPZ,j}$ and all its nontrivial proper subdivisors are Noether gaps.

Define $h = (f_2'/f_2)/(f_1'/f_1)$. Then, h has divisor of zeros and poles given by $(h) = B_{SPZ,1} - B_{SPZ,2}$. Furthermore, the compatibility of the monodromy homomorphisms $\psi_j, j = 1, 2$, implies that the Prym differentials $f_j'/f_j, j = 1, 2$, belong to the same multiplicative character homomorphisms $\tilde{\psi}_j : \pi_1(M) \to C^*$, i.e., $\tilde{\psi}_1 \equiv \tilde{\psi}_2$, where $\tilde{\psi}_j([\Lambda]) = 1$ if $\psi_j([\Lambda]) = kz$ and $\tilde{\psi}_j([\Lambda]) = -1$ if $\psi_j([\Lambda]) = k/z$. Hence, h is a single-valued meromorphic function on M. This observation and the previous conclusion concerning Noether gaps together imply that $B_{SPZ,1} = B_{SPZ,2}$. Thus, $h \equiv c \in C^*$. Consequently, $f_1^c \equiv f_2$. Since $deg(B_j) \geq 2g - 1, j = 1, 2$, Lemma 1 implies that the dihedral structures associated with $f_j, j = 1, 2$, must have zero(s)/pole(s). Hence, $c = \pm 1$; otherwise, either f_1 or f_2 would have common zero/polar points of different ramification orders or would have points of local multivaluedness. Therefore, $f_j, j = 1, 2$, determine the same dihedral strucutre. ∥

Remark 2. In view of the inequality $deg(B_{SPZ,j}) \leq g, j = 1, 2$, and the fact that minimality of the index of specialty holds for generic divisors of degree at most g on M, Theorem 3 can be paraphrased as an assertion of the non-existence on M of generic low-degree simple zero/polar variations of dihedral structures having fixed branch data and fixed compatibility class $\{\psi\}$ of monodromy homomorphisms.

We now state

Lemma 5. *If two dihedral structures on M are compatible and are both subordinate to a given projective structure with $o(\psi(\pi_1(M))) > 2$, then these dihedral structures are the same.*

Proof. A given dihedral structure with $o(\psi(\pi_1(M))) > 2$ must have monodromy group $\psi(\pi_1(M))(\subset D)$ containing an element $T(z) = az, a \neq 1, -1$ or containing two distinct order 2 elements $T_1(z) = \frac{b}{z}$ and $T_2(z) = \frac{c}{z}$. In the second case, $T(z) = \frac{b}{c}z(\neq id) \in \psi(\pi_1(M))$. Hence, in either case, $\psi(\pi_1(M))$ contains a nontrivial multiplicative element $T_\Lambda(z) = az, a \neq 1$, associated with some closed curve Λ on M and a geometric realization f of the particular dihedral structure. A possibly different dihedral structure subordinate to the same projective structure would have geometric realization $C \circ f, C \in PL(2, C)$, with monodromy element $C \circ T_\Lambda \circ C^{-1} \in D$ associated with the curve Λ. Furthermore, $C \circ T_\Lambda \circ C^{-1}(z) = dz, d \in C^*$, since this second dihedral structure is compatible with the first. In fact, the conjugating transformation C must map the fixed point set $\{0, \infty\}$ of T_Λ to the same fixed point set of $C \circ T_\Lambda \circ C^{-1}$. Hence, $C(w) = kw$ or $C(w) = k/w$ and the two dihedral structures are the same. ∥

Theorem 4. *A given projective structure on M with $o(\psi(\pi_1(M))) > 2$ and with some subordinate dihedral structure must have either one or three subordinate dihedral structures. There exist three subordinate dihedral structures iff $\psi(\pi_1(M)) = D_2$, the Klein 4-group.*

Proof. There are two cases.

Case 1. If the monodromy group $\psi(\pi_1(M))$ contains any element of order greater than two, then some subordinate dihedral structure with realization f contains an element $T_\Lambda(z) = az, a \neq 1, -1$, associated with some closed curve Λ on M. Any other subordinate dihedral structure has element $\tilde{T}_\Lambda = C \circ T_\Lambda \circ C^{-1}$ associated with Λ. $\tilde{T}_\Lambda$ cannot assume the form $\tilde{T}_\Lambda(z) \neq b/z$, an order 2 element. Hence,

$\tilde{T}_\Lambda(z) = bz$, $b \neq 1, -1$. The argument in the proof of Lemma 5 establishes that $C(w) = kw$ or $C(w) = k/w$. Thus, f and $C \circ f$ determine just one dihedral structure in this case.

Case 2. The monodromy group $\psi(\pi_1(M))$ contains only (at least two distinct) order 2 elements and the identity. Thus, $\psi(\pi_1(M)) = D_2$, the Klein 4-group. Therefore, there is a subordinate dihedral structure on M with realization f_1 whose specific monodromy homomorphism $\psi_1 : \pi_1(M) \to D$ may be expressed as

$$\psi_1 : \quad \{\Lambda_i\} \to (T(z) = \frac{a_1}{z})$$
$$\{\Gamma_j\} \to (T(z) = -\frac{a_1}{z}) \, , \, a_1 \in \mathbb{C}^*$$
$$\{\Omega_k\} \to (T(z) = -z)$$

where Λ_i, Γ_j and Ω_k are all closed curves on M which do not map to the trivial monodromy element. Since Lemma 5 assures us that other compatible subordinate dihedral structures are the same, we need only determine the compatibility classes of any other incompatible subordinate dihedral structures. In fact, these structures can have only the following two compatibility classes of homomorphisms:

$$\psi_2 : \quad \{\Lambda_i\} \to (T(z) = -z)$$
$$\{\Gamma_j\} \to (T(z) = \frac{a_2}{z})$$
$$\{\Omega_k\} \to (T(z) = -\frac{a_2}{z}) \, , \, a_2 \in \mathbb{C}^*$$

and

$$\psi_3 : \quad \{\Lambda_i\} \to (T(z) = \frac{a_3}{z})$$
$$\{\Gamma_j\} \to (T(z) = -z)$$
$$\{\Omega_k\} \to (T(z) = -\frac{a_3}{z}) \, , \, a_3 \in \mathbb{C}^*$$

Observe that ψ_2 is realized by $f_2 = C_2 \circ f_1$ where $C_2(w) = \frac{w+1}{w-1}$ and ψ_3 is realized by $f_3 = C_3 \circ C_2 \circ f_1$ where $C_3(w) = \frac{iw+1}{iw-1}$. Hence, f_1, f_2 and f_3 represent three distinct dihedral structures subordinate to the same projective structure. ||

Remark 3. The functions $f_{(i,j,k,1)}$ in Theorem 1(i) and $f_{(i,j,2)}$ in Theorem 1(ii) determine dihedral structures with monodromy group D_2. Hence, the associated projective structures have three subordinate dihedral structures.

For the trivial cases satisfying $o(\psi(\pi_1(M))) \leq 2$, there may exist infinitely many dihedral structures subordinate to the projective structure determined by a given realization f. For example, any dihedral structure with trivial monodromy group $\psi(\pi_1(M)) = id$ and with only nonzero-nonpolar branch points may have its simple poles/zeros moved by a 3-parameter family of Möbius transformations A. Thus, the realizations $A \circ f$, with A varied in this way, determine inequivalent dihedral structures but the same projective structure.

5. An Application to Monodromy of Differential Equations

In Theorem 5, we restrict M to be a γ-hyperelliptic surface as in Theorem 1. We now use the mappings $f_{(i,j,k,m)}$ and $f_{(i,j,m)}$ in Theorem 1 to construct classes of differential equations on M having monodromy groups given by Z_2-extensions of Abelian groups.

To construct the monodromy homomorphism and monodromy group of a differential equation

$$(7) \qquad\qquad y''(p) + J(p)y(p) = 0$$

on M, we analytically continue any ratio $f(p)$ of linearly independent solutions to Eq. (7) along a canonical set of cross cuts Λ_i, $i \in I$, for the homotopy group $\pi_1(M - \{\text{singular points of Eq. (7)}\})$ to get new ratios $\psi([\Lambda_i]) \circ f(p)$. Here, the monodromy homomorphism of Eq. (7) is $\psi : \pi_1(M-\{\text{singular points of Eq. (7)}\}) \to PL(2,\mathbb{C})$ and the monodromy group is $\psi(\pi_1(M - \{\text{singular points of Eq. (7)}\}))$. Note that $f(p)$ satisfies the equation $\theta_2 f(p) \equiv 2J(p)$.

We are now in a position to state

Theorem 5. *The differential equations (7) with*

$$(8) \qquad\qquad f(p) = \prod_{\substack{k=1 \\ k \neq i,j}}^{2\ell} f_{(i,j,k,1)}^{\alpha_k}, \; \alpha_k \in \mathbb{C}$$

or

$$(9) \qquad\qquad f(p) = \prod_{\substack{j=1 \\ j \neq i}}^{2\ell-1} f_{(i,j,1)}^{\alpha_j}(p), \; \alpha_j \in \mathbb{C}$$

can be viewed as Fuchsian equations on M. The monodromy groups of Eqs. (8) and (9) are given by

$$(10) \qquad \begin{aligned} & \langle A, B_i(i = 1, \dots, m); A^2 = 1, \\ & \quad B_j B_k = B_k B_j \text{ for all } j, k = 1, \dots, m, \; j < k, \\ & \quad AB_i = B_i^{-1}A \text{ for all } i = 1, \dots, m \rangle \end{aligned}$$

or

$$(11) \qquad \begin{aligned} & \langle A, B_i(i = 1, \dots, m); A^2 = 1, B_m^\ell = 1, \\ & \quad B_j B_k = B_k B_j \text{ for all } j, k = 1, \dots, m, \; j < k, \\ & \quad AB_i = B_i^{-1}A \text{ for all } i = 1, \dots, m \rangle \end{aligned}$$

where $m \leq 2\ell - 2$. As the α_k in Eq. (8) and the α_j in Eq. (9) are varied, each of these groups, $\forall m \in \{0, \dots, 2\ell - 2\}$, are realized as a monodromy group for some equation (8) and some equation (9) on each M.

Proof. The Eqs. (7) with $f(p)$ of form (9) can be regarded as lifts by the single-valued meromorphic map $z = t(P) - e_i$ (determined by Eq. (3)) of Eqs. (8) on $\widehat{\mathbb{C}}$ in [13], where the α_j, r_j and n in Eq. (8) are given respectively by $\sqrt{e_j - e_i}, \alpha_j$ and $2\ell - 2$. Hence, as in [10], the monodromy group G' of each equation (7) is a subgroup of the monodromy group G of the corresponding equation (8) in [13]. By Theorem 6 of [13], the Eqs. (8) have monodromy groups G which include all of the groups given in (10) and (11) as the α_j are varied. Also, it is easily seen that G' must have some generator of the form $T(z) = \frac{k}{z}$. Consequently, G' must be isomorphic to one of the groups listed in (10) or (11). Thus, the proof will be complete provided we can choose constants α_j for which G' can belong to an arbitrary isomorphism class of groups in (10) or (11).

The surface M can be viewed as consisting of $2n$ t-planes with branch lines as follows:

$$\times\!\!-\!\!-\!\!-\!\!\times \qquad \times\!\!-\!\!-\!\!-\!\!\times \qquad \cdots \qquad \times\!\!-\!\!-\!\!-\!\!\times$$
$$t(p_1) \quad t(p_2) \qquad t(p_3) \quad t(p_4) \qquad\qquad t(p_{2\ell-1}) \quad t(p_\infty) = \infty$$

Choose m of the numbers $\alpha_j, j = 1, \ldots, 2\ell - 1, j \neq i$, so that the numbers $2\pi i \alpha_j$ are algebraic and linearly independent over Q, and choose the remaining $2\ell - 2 - m$ numbers $\alpha_j = 1$. A closed curve Λ on M determines a multiplicative monodromy element $t(z) = k_1 z$ of G' iff the sum of the winding numbers of Λ with respect to $t(p_\infty)$ and $t(p_i)$ is even. In fact, the most general element of this form has $k_1 = e^{2\pi i \sum_{\substack{j=1 \\ j \neq i}}^{2\ell-1} m_j \alpha_j}$, where $\sum_{j=1}^{2\ell-1} m_j, j \neq i$, is even. Furthermore, any nonmultiplicative element $t(z) = \frac{k_2}{z}$ has $k_2 = e^{2\pi i \sum_{\substack{j=1 \\ j \neq i}}^{2\ell-1} m_j \alpha_j}$. An argument using the Lindemann-Weierstrass theorem, as in the proof of Theorem 1 of [13], establishes that the group G' with these elements is an arbitrary group of form (10). Similarly, choosing $m - 1$ of the numbers $\alpha_j, j = 1, \ldots, 2\ell - 1, j \neq i$, so that the numbers $2\pi i \alpha_j$ are algebraic and linearly independent over Q, choosing one of the $\alpha_j = \frac{1}{\ell}$ and choosing the remaining $2\ell - 2 - m$ numbers $\alpha_j = 1$, we obtain a group G' which is an arbitrary group of form (11).

The proof with $f(p)$ of form (8) is similar. Here, a closed curve Λ on M determines a multiplicative monodromy element of G' iff the sum of the winding numbers with respect to $t(p_i)$ and $t(p_j)$ is even. ‖

REFERENCES

1. H. Cohn, *Conformal Mappings on Riemann Surfaces*, Dover, New York, 1980.
2. A. Erdélyi, et.al., *Higher Transcendental Functions*, Bateman Manuscript Project, Vol. 1, McGraw-Hill, New York, 1953.
3. H. Farkas and I. Kra, *Riemann Surfaces*, Springer-Verlag, New York, 1980.
4. R.C. Gunning, *Special Coordinate Coverings of Riemann Surfaces*, Math. Ann. **170** (1966), 67-86.
5. R.C. Gunning, *Lectures on Riemann Surfaces*, Princeton University Press, Princeton, 1966.
6. I. Kra, *On Affine and Projective Structures on Riemann Surfaces*, J. Analyse Math. **22** (1969), 285-298.

7. I. Kra, *Deformations of Fuchsian Groups*, Duke Math. J. **36** (1969), 537-546.

8. I. Kra, *Deformations of Fuschsian Groups, II*, Duke Math. J. **38** (1971), 499-508.

9. I. Kra, *A Generalization of a Theorem of Poincaré*, Proc. Am. Math. Soc. **27** (1971), 299-302.

10. K. Kuiken, *The Effect of Rational Substitutions on the Monodromy Group of the General nth Order Equation of the Fuchsian Class*, Math. Z. **163** (1978), 111-119.

11. K. Kuiken and J.T. Masterson, *On the Monodromy Groups of Lifted Euler Equations*, Glasgow Math. Jour. **25** (1984), 47-57.

12. K. Kuiken and J.T. Masterson, *Infinite Groups and Hill's Equation*, Arkiv för Matematik **24** (1986), 119-130.

13. K. Kuiken and J.T. Masterson, *On Global Solutions and Monodromy Groups of Differential Equations Having at Least Two Singularities on the Riemann Sphere*, Indag. Math. **90** (1987), 301-312.

14. R. Mandelbaum, *Branched Structures on Riemann Surfaces*, Trans. Am. Math. Soc. **163** (1972), 261-275.

15. R. Mandelbaum, *Branched Structures and Affine and Projective Bundles on Riemann Surfaces*, Trans. Am. Math. Soc. **183** (1973), 37-58.

16. R. Mandelbaum, *Unstable Bundles and Branched Structures on Riemann Surfaces*, Math. Annalen **214** (1975), 49-59.

17. J. T. Masterson, *Branched Affine and Projective Structures on Compact Riemann Surfaces*, Indag. Math. **91** (1988), 309-319.

18. J. T. Masterson, *Branched Structures Associated with Lamé's Equation*, Arkiv för Matematik **28** (1990), 131-137.

19. J.T. Masterson, *Uniqueness of Affine Structures on Riemann Surfaces*, Tamkang Jour. Math. **23** (1992), 87-94.

Department of Mathematics, Polytechnic University, Six Metrotech Center, Brooklyn, New York 11201

E-mail address: kkuiken@photon.poly.edu

Department of Mathematics, Seton Hall University, 400 South Orange Avenue, South Orange, New Jersey 07079

E-mail address: fmasters@setonvm.bitnet; f105@setonmus.bitnet

Contemporary Mathematics
Volume **169**, 1994

Groups and Lie Algebras: The Magnus Theory

JOHN P. LABUTE

ABSTRACT. In this article we give an exposition of a connection between groups and Lie algebras that was discovered by Magnus. We also outline results of J. Labute that answer many of the questions Magnus posed concerning the determination of the Lie ring associated to the lower central series of a finitely presented group.

We dedicate this article to the
memory of Wilhelm Magnus

1. Introduction. There is a very strong analogy between the theory of groups and the theory of Lie algebras—many group theoretical results have Lie algebra counterparts that are also true but sometimes with completely different proofs. The connections between the two theories are deep but not completely understood. The most well-known and best understood connection is the one between Lie groups and Lie algebras; every finite dimensional (complex) Lie algebra is the Lie algebra of some (complex) Lie group which is locally determined by the Lie algebra.

Another connection, which was discovered by Magnus [**12,13**] and developed by Lazard [**11**], associates to the lower central series of a group a Lie algebra which, under certain conditions, also determines the group 'locally'. However, we can compute this Lie algebra for relatively few groups. Magnus [**13**] and Witt [**17**] did it for the case of a free group, and Magnus was very much interested in the case of finitely presented groups. He asked a number of questions, cf. [**15, Sec. 5.11**], which only recently have we been able to answer [**8,9**]. In fact we are now able to compute the Lie algebra associated to the lower central series of a large class of finitely presented groups, an important example being the fundamental group of a compact Riemann surface.

1991 *Mathematics Subject Classification.* Primary 20F14, 20F40; Secondary 20F05, 20F12, 17B70.

Partially supported by Quebec FCAC Grant 88-EQ-2489

This paper is in final form and no version of it will be submitted for publication elsewhere.

In this article we will outline the work of Magnus on this connection and show that perhaps his contribution has not received the recognition it deserves. Many interesting questions remain, and we hope that this article will stimulate further research in this area.

2. The Magnus Connection. The seeds of this connection lie in the paper of Philip Hall [6] which introduces the commutator calculus. There Magnus found the following commutator identities

(2.1) $$(x, yz) = (x, z)(x, y)((x, y), z),$$

(2.2) $$(xy, z) = (x, z)((x, z), y)(y, z),$$

where x, y, z are elements of a group and

$$(x, y) = x^{-1}y^{-1}xy = (y, x)^{-1}.$$

Perhaps it was his attempt to 'linearize' these identities in a way that was analogous to the construction of the Lie algebra of a linear Lie group that led him to his representation of the elements of a free group by formal power series in non-commuting indeterminates with integral coefficients. We now describe this representation, cf. [12].

Let F be the free group on the letters x_i $(i \in I)$, and let M be the ring (associative $\mathbb{Z}$-algebra) of formal power series in the non-commuting indeterminates ξ_i $(i \in I)$ with integral coefficients. This ring is now commonly known as the *Magnus algebra* although Magnus was not the first to use it. Any element u of M can be uniquely written as a formal sum

$$u_0 + u_1 + ... + u_n + ...,$$

where each term u_n is an integral linear combination of monomials

$$\xi_{i_1}\xi_{i_2}...\xi_{i_n}.$$

The natural number n is called the *degree* of u_n. We let M_n denote the submodule of M generated by the monomials of degree n. The sum of these submodules is a subring of M isomorphic to the free associative ring on the letters ξ_i $(i \in I)$. Let I be the ideal of formal power series having constant term 0. If $u \in I$, then $1 + u$ is invertible and

$$(1 + u)^{-1} = 1 - u + u^2 - u^3 + ... + (-1)^n u^n + ...$$

The invertible elements $1 + \xi_i$ generate a subgroup of the group of units of M. This group, called the *Magnus group*, is a homomorphic image of F under the homomorphism which takes x_i to $1 + \xi_i$. Magnus proved that this mapping is injective; in fact, it is even injective taking coefficients in $\mathbb{Z}/p\mathbb{Z}$ for any prime p. To prove this it suffices to remark that, for $n \in \mathbb{Z}$ prime to p, q a power of p,

$$(1 + \xi_i)^{nq} \equiv 1 + n\xi_i^q \quad (\mathrm{mod}\ (p, I^{q+1})).$$

Identifying F with its image in M, we have $x_i = 1 + \xi_i$. Moreover, if $x \in F$, $x \neq 1$, we have $x = 1 + u$ with $u \in I^n$, $u \notin I^{n+1}$, for some $n \geq 1$. Magnus called n the *dimension* of x. If $y = 1 + v \in F$ is of dimension m, then

$$xy = 1 + u + v + uv$$
$$(x, y) = 1 + uv - vu \pmod{I^{m+n+1}}.$$

If D_n is the set of elements of F of dimension $\geq n$ (the dimension of the identity element of F being defined to be ∞), we see that D_n is a subgroup of F (the so-called n-th *dimension subgroup* of F). These subgroups have the following properties:

$$D_1 = F, \quad D_{n+1} \subseteq D_n, \quad (D_m, D_n) \subseteq D_{m+n},$$

where, for subgroups H, K of a group G, (H, K) denotes the subgroup of G generated by the commutators (h, k) with $h \in H$, $k \in K$. Such a sequence of subgroups is called a *central filtration* of G. If f is the mapping of F into I defined by $f(x) = x - 1$, then f maps D_n into I^n. If $x \in D_m$, $y \in D_n$, we have

$$f(xy) \equiv f(x) + f(y) \pmod{I^{m+n}},$$
$$f((x, y)) \equiv uv - vu \pmod{I^{m+n+1}}.$$

We therefore obtain, for each $n \geq 1$, a mapping $\phi_n : D_n \to I^n/I^{n+1}$ such that $\phi_n(x) = f(x) + I^{n+1}$. If we use the canonical identification of M_n with I^n/I^{n+1}, then ϕ_n has another description; namely, if $x = 1 + u_n + u_{n+1} + \ldots$ with $u_i \in M_i$, then $\phi_n(x) = u_n$. If $x \notin D_{n+1}$, then $u_n \neq 0$ and is called the *initial form* of x. For example, $\phi_1(x_i) = \xi_i$ and

$$\phi_2((x_i, x_j)) = \xi_i\xi_j - \xi_j\xi_i, \quad \phi_2((x_1, x_2)(x_3, x_4)) = \xi_1\xi_2 - \xi_2\xi_1 + \xi_3\xi_4 - \xi_4\xi_3.$$

The direct sum A of the I^n/I^{n+1} ($n \geq 0$), where $I^0 = M$, has a natural graded ring structure; if $u \in I^m$, $v \in I^n$, then $(u + I^{m+1})(v + I^{n+1}) = uv + I^{m+n}$. Under the identification of I^n/I^{n+1} with M_n, we recover the multiplication in M, and so the graded ring A can be identified with the free associative ring on the ξ_i ($i \in I$).

The mapping ϕ_n has the following properties:

$$\phi_n(xy) = \phi_n(x) + \phi_n(y) \quad \text{if} \ x, y \in D_n;$$
$$\phi_n(x) = 0 \iff x \in D_{n+1};$$
$$\phi_{m+n}((x, y)) = \phi_m(x)\phi_n(y) - \phi_n(y)\phi_m(x) \quad \text{if} \ x \in D_m, \ y \in D_n.$$

This is the linearization that Magnus introduced in [12]. It transforms multiplication in D_n into addition in M_n and the commutator of two elements $x \in D_m, y \in D_n$ into $[\xi, \eta] = \xi\eta - \eta\xi$, where ξ, η are the images of x, y respectively. We call $[\xi, \eta]$ the (Lie) *bracket* of the elements ξ, η in the associative ring A. This bracket defines a non-associative operation on A. More precisely, we have the Jacobi identity

$$[u, [v, w]] + [w, [u, v]] + [v, [w, u]] = 0$$

along with the identities

$$[u, u] = 0,$$
$$[u + v, w] = [u, w] + [v, w],$$
$$[u, v + w] = [u, v] + [u, w].$$

The last two identities are the linearizations of (2.1) and (2.2). The identities $[u, u] = 0$ and $[u, v] = -[v, u]$ are the linear forms of $(x, x) = 1$ and $(x, y) = (y, x)^{-1}$ in F. Thus A becomes a Lie ring if we replace multiplication by the bracket operation; we denote this Lie ring by A_L. The axioms for a Lie ring are the same as that for a ring except that the associative law for multiplication is replaced by the Jacobi identity.

Let $L_D(F)$ be the direct sum of the abelian groups $L_{D,n}(F) = D_n/D_{n+1}$, where the abelian groups $L_{D,n}(F)$ are denoted additively. The above shows that the commutator operation defines a Lie ring structure on $L_D(F)$ and that the mappings ϕ_n induce an isomorphism ϕ of this Lie ring with a Lie subring of the Lie ring A_L associated to the associative algebra A. It should be noted that to prove that $L_D(F)$ is a Lie ring the only axiom that needs ϕ is the Jacobi identity; the other axioms need only (2.1) and (2.2) along with $(x, x) = 1$.

The lower central series $C_n = C_n(F)$ $(n \geq 1)$ of F is defined by

$$C_1 = F, \quad C_{n+1} = (F, C_n).$$

In [6] Philip Hall proves the identity

(2.3) $(H, (K, L)) \subseteq (L, (H, K))(K, (L, H))$

with which one proves by induction that C_n $(n \geq 1)$ is a central filtration. We have $C_n \subseteq D_n$, and Magnus showed in [12] that the image of C_n in $L_{D,n}(F)$ is of finite index. He also conjectured that $C_n = D_n$. If $L_C(F)$ is the direct sum of the abelian groups $L_{C,n}(F) = C_n/C_{n+1}$ (again denoted additively), the natural mappings of $L_{C,n}(F)$ into $L_{D,n}(F)$ define a homomorphism α of $L_C(F)$ into $L_D(F)$. We have $C_n = D_n$ for all n if and only if this mapping is injective. In fact, if this mapping is injective, suppose that we have already shown that $C_n = D_n$ for some n (this is true for $n = 1$). Then any element $x \in D_{n+1}$ lies in $D_n = C_n$, and so the image of x in $D_n(F)$ is zero, which means that $x \in C_{n+1}$.

To prove the injectivity of $\alpha : L_C(F) \to L_D(F)$ Magnus needed to know that the commutator operation also induced a Lie ring structure on $L_C(F)$, and for this he needed to show that for $x \in C_m$, $y \in C_n$, $z \in C_k$

(2.4) $(x, (y, z))(z, (x, y))(y, (z, x)) \in C_{m+n+k}.$

In [13] he states that this is easily done using results of [6], but later in a note added to this paper in his Collected Works [14] he states that there is gap in his proof due to the fact that the above identity is not as obvious as claimed. While not obvious it is not difficult to prove using Hall's identities (2.1) and (2.2). In fact,

$$
\begin{aligned}
(x, y), z) &= z^{-1}(y, x)z(x, y) \\
&= (z^{-1}yz, z^{-1}xz)(x, y) \\
&= (y(y, z), x(x, z))(x, y) \\
&= (y(y, z), (x, z))(y(y, z), x)((y(y, z), x), (x, z))(x, y) \\
&\equiv (y, (x, z))((y, z), x) \pmod{C_{m+n+k}}
\end{aligned}
$$

from which (2.4) follows immediately. This also yields a proof of (2.3).

The above procedure yields a graded Lie ring $L(G)$ for any central filtration on a group G. If $f : G \to G'$ is a surjective homomorphism of groups, then $G'_n = f(G_n)$ $(n \geq 1)$ is a central filtration on G' and f induces a Lie ring homomorphism $f^* : L(G) \to G'$. If H is the kernel of f, then $H_n = H \cap G_n$ $(n \geq 1)$ is a central filtration on H and $L(H)$ is the kernel of f^*. For a systematic account of all this, see [11].

3. The Magnus-Witt Theorems. Let L be the free Lie ring on the letters ξ_i $(i \in I)$, and let β be the unique homomorphism of L into $L_C(F)$ which sends ξ_i to the class of x_i in $L_{C,1}(F)$. Since $L_C(F)$ is generated by the elements of degree 1, the homomorphism β is surjective. The mapping $\gamma = \phi\alpha\beta$ is a homomorphism of L into A_L sending ξ_i to ξ_i. The injectivity of α would follow from the injectivity of γ which would also show that β is bijective and hence that $L_C(F)$ and L were isomorphic. To prove the injectivity of γ one easily reduces to the case $I = \{1, 2, .., q\}$.

In a talk given at Bad Salzbrunn in September 1936 Magnus conjectured, cf. [17], that in this case γ is injective and that L_n is a free $\mathbb{Z}$-module of rank

$$
\frac{1}{n} \sum_{d|n} \mu(n/d)q^d,
$$

where μ is the Möbius function. Witt was in attendance at this lecture and within a month he had submitted his proof of these facts in a greatly generalized form. Two weeks later Magnus submitted his proof of the injectivity of γ. In this paper [13] he also proved that L was a free $\mathbb{Z}$-module and gave an iterative procedure for finding a basis. Curiously, he did not prove his conjectured formula for the rank even though, as we shall see, it follows easily from his proof.

To describe this procedure, which Magnus calls *elimination*, we let $\mathfrak{h}$ be a Lie ring generated by the elements η_i $(i \in I)$ and let $\mathfrak{h}'$ be the ideal of $\mathfrak{h}$ generated by the elements η_j $(j \neq i)$ for some fixed i. Then $\mathfrak{h}'$ is generated, as an ideal of $\mathfrak{h}$, by the elements

$$
\eta_{j,k} = \mathrm{ad}^k(\eta_j)(\eta_i) \quad (j \neq i, k \geq 1),
$$

where $\mathrm{ad}(x)(y) = [x, y]$. Let A be the free associative ring on the letters ξ_i $(i \in I)$, and suppose that we are given a homomorphism of $\mathfrak{h}$ into A such that

$\psi(\eta_i) = \xi_i$. Let A' be the subalgebra of A generated by the elements

$$\xi_{j,k} = ad^k(\xi_j)(\xi_i) \quad (j \neq i, k \geq 1).$$

If ψ' is the restriction of ψ to $\mathfrak{h}'$, then ψ' is a homomorphism of $\mathfrak{h}'$ into A' such that $\psi'(\eta_{j,k}) = \xi_{j,k}$. In [**12**] Magnus proves the following two Lemmas.

LEMMA 1. A' is a free associative ring on the $\eta_{j,k}$.

LEMMA 2. As an abelian group, A is the direct sum of the subgroups $\xi_i^k A'$ $(k \geq 0)$.

COROLLARY 1. As abelian groups, $\mathfrak{h} = \mathfrak{h}' \oplus \mathbb{Z}\xi_i$.

COROLLARY 2. As abelian groups, $A \cong A' \otimes \mathbb{Z}[\xi_i]$.

We now suppose that $\mathfrak{h}$ is a graded Lie ring and that the η_i $(i \in I)$ are homogeneous of integral degree $d_i \geq 1$. We also assume the d_i $(i \in I)$ satisfy the following finiteness property:

(F)   For $n \geq 1$, $\{i \in I : d_i \leq n\}$ is a finite set.

The elements $\eta_{j,k}$ $(j \neq i, k \geq 1)$ also satisfy property (F). If we grade A by decreeing d_i to be the degree of ξ_i, then ψ and ψ' are homomorphisms of graded Lie rings. By repeatedly splitting off as above a rank 1 component of lowest degree, we find homogeneous non-zero elements $\zeta_1, ..., \zeta_n$ of $\mathfrak{h}$ of non-decreasing degrees, a homogeneous Lie subring $\mathfrak{h}^{(n)}$ of $\mathfrak{h}$ which is generated by homogeneous elements $\eta_j^{(n)}$ $(j \geq 1)$, whose degrees satisfy the property (F), such that

$$\mathfrak{h} = \mathfrak{h}^{(n)} \oplus \mathbb{Z}\zeta_1 \oplus \mathbb{Z}\zeta_2 \oplus ... \oplus \mathbb{Z}\zeta_n$$

and such that the subring $A^{(n)}$ of A generated by the $\psi(\eta_j^{(n)})$ is a free associative ring on these elements. Moreover, since each component of $\mathfrak{h}$ is finitely generated because of (F), we can for any given $N \geq 1$ find n so that the homogeneous components of $\mathfrak{h}^{(n)}$ are zero in degrees $< N$. Hence the elements ζ_n $(n \geq 1)$ are a basis for $\mathfrak{h}$ and ψ is injective. Applying this to the case where A is the free associative ring on $\xi_1, ..., \xi_q$, $\mathfrak{h}$ is the free Lie ring on $\xi_1, ..., \xi_q$ and $\psi = \gamma$, we get the injectivity of γ. The proof also yields the fact that the elements

$$\zeta_{i_1} ... \zeta_{i_n} \quad (n \geq 0, i_1 \leq ... \leq i_n)$$

are a $\mathbb{Z}$-basis for A. It can be shown, cf [**4**, §**2.11**], that the basis ζ_i $(i \in I)$ is in fact what is now commonly called a Hall basis of the free Lie ring L. We also obtain that $\mathfrak{h}'$ is a free Lie ring on the $\xi_{j,k}$; this result is called the *elimination theorem*.

To get the rank formula we define the Poincaré series $P_M(t)$ of a graded K-module $M = \oplus_{n \geq 0} M_n$ (K any commutative ring), where the K-modules M_n are free of finite rank a_n, by

$$P_M(t) = \sum_{n \geq 0} a_n t^n.$$

If M and N are two such modules, we have

$$P_{M \oplus N} = P_M(t) + P_N(t), \quad P_{M \otimes_K N} = P_M(t) P_N(t)$$

since $M \oplus N$ and $M \otimes_K N$ are graded with

$$(M \oplus N)_n = M_n \oplus N_n, \quad (M \otimes N)_n = \oplus_{i+j=n} M_i \otimes N_j.$$

Using Corollary 1 , we obtain

$$P_A(t) = \prod_{n \geq 1} \frac{1}{(1 - t^n)^{a_n}},$$

where a_n is the rank of L_n. Since $P_A(t) = (1 - qt)^{-1}$, we have

$$\prod_{n \geq 1} \frac{1}{(1 - t^n)^{a_n}} = \frac{1}{1 - qt}$$

from which the rank formula follows easily.

The pair (γ, A) has the following universal property with $K = \mathbb{Z}$:

If B is an associative K-algebra and λ_0 is a Lie algebra homomorphism of L into A_L, there is a unique algebra homomorphism $\lambda : A \to B$ such that $\lambda_0 = \lambda\gamma$.

Given any Lie algebra L over a commutative ring K, there exists an associative K-algebra A and a Lie algebra homomorphism $\gamma : L \to A_L$ having the above universal property. This algebra is called the *enveloping algebra* of L. It is the analogue of the group algebra of a group.

Witt [17] and Birkhoff [3] proved that, for any Lie algebra L over a commutative ring K which has a K-basis ζ_i $(i \in I)$ with I ordered, the images of the elements

$$\zeta_{i_1} \dots \zeta_{i_n} \quad (n \geq 0, i_1 \leq \dots \leq i_n)$$

in its enveloping algebra A are a K-basis for A. This result is now commonly known as the Birkhoff-Witt theorem although Bourbaki also attributes it to Poincaré. Witt in [18] refers to it, in the case of a free Lie algebra, as the Birkhoff-Magnus-Witt Theorem. The rank formula is now known as Witt's formula—it should be more appropriately called the Magnus-Witt formula.

4. Lie Rings of Finitely Presented Groups. Let F be the free group on $x_1, \dots, x_q$, and let R be the normal subgroup of F generated by the elements $r_1, \dots, r_\ell$. If L, $\mathfrak{g}$ are respectively the Lie rings associated to the lower central series F_n, G_n $(n \geq 1)$ of F and $G = F/R$, we have $\mathfrak{g} = L/\mathfrak{R}$, where $\mathfrak{R}$ is the Lie ring associated to the central filtration $R_n = F_n \cap R$ $(n \geq 1)$ of R. Since R_n/R_{n+1} is isomorphic to $R_n F_{n+1}/F_{n+1}$, the homogeneous elements of $\mathfrak{R}$ are the initial forms of elements of R. In particular, the initial forms $\rho_1, \dots, \rho_\ell$ of the relators $r_1, \dots, r_\ell$ are elements of $\mathfrak{R}$. If $\mathfrak{r} = (\rho_1, \dots, \rho_\ell)$ is the ideal of L generated by $\rho_1, .., \rho_\ell$, we have $\mathfrak{r} \subseteq \mathfrak{R}$.

An immediate question to ask is: *When do we have* $\mathfrak{r} = \mathfrak{R}$? To see that $\mathfrak{r} \neq \mathfrak{R}$ in general, consider the case $q = 2$, $s = 1$, $r_1 = r = x_1^2$. Then the initial form ρ of r is $2\xi_1$. However, we have

$$(x_2, x_1^2) = (x_2, x_1)^2((x_2, x_1), x_1)$$
$$((x_2, x_1), ((x_2, x_1^2)) = ((x_2, x_1), ((x_2, x_1), x_1),$$

which shows that $[[\xi_1, \xi_2], [\xi_1, [\xi_1, \xi_2]]$ is an element of $\mathfrak{R}$ which is not in $\mathfrak{r} = (2\xi_1)$. The same thing is true if $r = x^2s$ with $s \in F_3$.

Magnus [15] called a relator $r \in R$, $r \neq 1$, of dimension m a *power relator* if for some $n > 1$

$$r \equiv s^n \pmod{F_{m+1}}$$

with $s \notin R$. Otherwise, r was called a *commutator relator*. If ρ is the initial form of r, then r is a commutator relator iff $\rho \neq m\sigma$ with $m > 1, \sigma \notin \mathfrak{R}$. There are no power relators in R if and only if $\mathfrak{g}$ is torsion free. Magnus asked whether G was torsion free in the case the defining relators $r_1, ..., r_\ell$ were commutator relators of dimension m that were linearly independent modulo F_{m+1}.

In the case $\ell = 1$ the question has a positive answer and we even have $\mathfrak{R} = \mathfrak{r}$, cf [2]. There is also a formula for the rank a_n of $\mathfrak{g}_n$. In fact, since the Poincaré series of the enveloping algebra of $\mathfrak{g}$ is

$$\frac{1}{1 - qt + \ell t^m} = \frac{1}{(1 - \alpha_1 t)(1 - \alpha_2 t) \cdots (1 - \alpha_m t)},$$

we have

$$a_n = \frac{1}{n} \sum_{d|n} \mu(n/d)(\alpha_1^d + \alpha_2^d + ... + \alpha_m^d).$$

As a result we see that the rank of the n-th homogeneous component of $\mathfrak{g}_n$ depends only on n, q and m and not on the particular form of the initial form ρ of r.

For example, if G is the fundamental group of a compact Riemann surface of genus g, we have $G = F/R$ with F free on $x_1, ..., x_{2g}$ and $R = (r)$ with

$$r = (x_1, x_2)(x_3, x_4) \cdots (x_{2g-1}, x_{2g}).$$

The relator r is a commutator relator of dimension 2 with initial form

$$\rho = [\xi_1, \xi_2] + [\xi_3, \xi_4] + ... + [\xi_{2g-1}, \xi_{2g}].$$

The Lie ring associated to the lower central series G_n of G therefore has a single defining relation $\rho = 0$, and the abelian group G_n/G_{n+1} is torsion-free of rank

$$a_n = \frac{1}{n} \sum_{d|n} \mu(n/d)(\alpha^d + \beta^d),$$

where α, β are the roots of the polynomial $X^2 - qX + 1$.

The essential ingredient of the proof of this result is the following theorem, cf. [7], which is the analog of the identity theorem for a one-relator group:

THEOREM. *If $\mathfrak{g} = \mathcal{L}/\mathfrak{r}$ is a one-relator Lie ring, where the defining relator ρ is homogeneous and not a proper multiple of an element of L, then $\mathfrak{g}$ is a free $\mathbb{Z}$-module and $\mathfrak{r}/[\mathfrak{r}, \mathfrak{r}]$ is a free module of rank 1 over the enveloping algebra of $\mathfrak{g}$, with basis the class of ρ.*

The module structure on $\mathfrak{r}/[\mathfrak{r}, \mathfrak{r}]$ is induced by the adjoint representation

$$ad : L \to \mathrm{Der}(L),$$

where $ad(\xi)(\eta) = [\xi, \eta]$ and $\mathrm{Der}(L)$ consists of the derivations of L ($D : L \to L$ is a derivation of L if $D(x+y) = D(x) + D(y)$ and $D([x, y]) = [D(x), y] + [x, D(y)]$). $\mathrm{Der}(L)$ is a Lie algebra under the bracket $[D, D'] = DD' - D'D$. The proof of the theorem uses the Birkhoff-Witt theorem to prove that, for any prime p, we have an exact sequence

$$0 \to \mathfrak{r}(p)/[\mathfrak{r}(p), \mathfrak{r}(p)] \to U(p)^q \to U(p) \to \mathbb{Z}/p\mathbb{Z} \to 0,$$

where U is the enveloping algebra of $\mathfrak{g}$, $U(p) = U/pU$ and $\mathfrak{r}(p) = \mathfrak{r}/p\mathfrak{r}$. From this we get a formula for the Poincaré series of $U(p)$ which depends only on q and ℓ. This yields the freeness of U and $\mathfrak{g}$ as $\mathbb{Z}$-modules as well as the formula for the Poincaré series of U. The assertion about $\mathfrak{r}/[,]$ follows easily.

In the case of several commutator relators, the simple linear independence of their initial forms is not sufficient in general to guarantee that $\mathfrak{g}$ is torsion free, cf [**9, p. 55**]. A sufficient (but not necessary) condition is the requirement that, for every prime p, $\mathfrak{r}(p)/[\mathfrak{r}(p), \mathfrak{r}(p)]$ be a free module over the enveloping algebra of $\mathfrak{g}(p)$, cf. [**9**]; from this we also get the equality of $\mathfrak{r}$ and $\mathfrak{R}$. However, this independence condition is not easy to verify; in fact Anick [**1**] has shown that there is no algorithm for it. For certain groups, e.g., link groups, the initial forms of the defining relators can be computed from local data, cf. [**2,5,10**]. A pretty application of this is to the case where G is the fundamental group of a tame link in S^3 with q components. Using a Milnor presentation, cf. [**10**], we can get a presentation with q generators (corresponding to the components of the link) and $\ell = q - 1$ defining relators whose initial forms are

$$\rho_i = \sum_{j \neq i} \ell_{ij} [\xi_i, \xi_j]$$

with ℓ_{ij} the linking number between the i-th and j-th components. The verification of our independence condition can, in this case, be done using the linking diagram of the link which is a graph with vertices v_i corresponding to the components and an edge of weight ℓ_{ij} joining v_i, v_j; there is no edge if $\ell_{ij} = 0$. The independence condition is equivalent to this graph being connected modulo p for any prime p which, in turn, is equivalent to the relators being linearly independent and generating a subgroup which is a direct summand of L_2. This is a strengthened form of the independence condition of Magnus.

It would be interesting to find weaker conditions on $\mathfrak{r}$ that would enable us to compute $\mathfrak{g}$. Perhaps the existence of a certain type of free resolution of U

would furnish such a condition. This would have interesting applications for link groups.

If G is the fundamental group of a compact Riemann surface, then the quotients $G/C_n(G)$ are completely determined by the Lie ring associated to the lower central series of G. **Problem**: *Find other interesting examples of such groups.* Taking a lead from the work of Morgan [16], one should perhaps look at fundamental groups of smooth complex analytic varieties.

REFERENCES

1. D. Anick, *Diophantine equations, Hilbert series, and undecidable spaces*, Annals of Math. **122** (1985), 87–112.
2. D. Anick, *Inert sets and the Lie algebra associated to a group*, J. Algebra **111** (1987), 154–165.
3. G. Birkhoff, *Representability of Lie Algebras and Lie Groups by Matrices*, Annals of Math. **38** (1937), 526–532.
4. N. Bourbaki, *Groupes et Algèbres de Lie, Ch.2*, Hermann, Paris, 1972.
5. R. Hain, *Iterated integrals, intersection theory and link groups*, Topology **24** (1985), 45–66.
6. P. Hall, *A Contribution to the Theory of Groups of Prime-Power Order*, Proc. London Math. Soc., Ser. 2 **36** (1933), 29-95.
7. J.P. Labute, *Algèbres de Lie et pro-p-groupes définis par une seule relation*, Invent. Math **4** (1967), 142–158.
8. J.P. Labute, *On the Descending Central Series of Groups with a Single Defining Relator*, J. Algebra **14** (1970), 16–23.
9. J.P. Labute, *The determination of the Lie algebra associated to the lower central series of a group*, Trans. Amer. Math. Soc. **288** (1985), 51–57.
10. J.P. Labute, *The Lie Algebra Associated to the Lower Central Series of a Link Group and Murasugi's Conjecture*, Proc. AMS **109** (1990), 951–956.
11. M. Lazard, *Sur les groupes nilpotents et les anneaux de Lie*, Ann. École Normale Supér. **71** (1954), 101–190.
12. W. Magnus, *Beziehungen zwischen gruppen und Idealen in einem speziellen Ring*, Math. Ann. **111** (1935), 259-280.
13. W. Magnus, *Über Bezeihungen Zwischen hoheren Kommutatoren*, Jour. Crelle **177** (1937), 105–115.
14. W. Magnus, (G. Baumslag and B. Chandler, Editors), *Wilhelm Magnus Collected Papers*, Springer-Verlag, New York, 1984.
15. W. Magnus, A. Karrass and D. Solitar, *Combinatorial Group Theory*, Dover Publications, New York, 1976.
16. J. Morgan, *The algebraic topology of smooth algebraic varieties*, Publ. Math. IHES **48** (1978), 137-204.
17. E. Witt, *Treue Darstellungen Lieschen Ringe*, J. reine angew. Math. **177** (1937), 152-160.
18. E. Witt, *Die Unterringe der freien Lieschen Ringe*, Math. Z. **64** (1956), 152-160.

DEPARTMENT OF MATHEMATICS, McGILL UNIVERSITY,
CENTRE INTERUNIVERSITAIRE EN CALCUL MATHÉMATIQUE ALGÉBRIQUE,
805 SHERBROOKE STREET WEST, MONTREAL, QUEBEC H3A 2K6

E-mail address: labute@math.mcgill.ca

Contemporary Mathematics
Volume **169**, 1994

Semiregular continued fractions
whose partial denominators are 1 or 2

JOSEPH LEHNER

TO THE MEMORY OF WILHELM MAGNUS

1. Within the set of all semiregular continued fractions we define a subset F by imposing a very restrictive condition on the partial denominators. We then show that F is still large enough to represent all real numbers modulo 1.

We shall first consider *infinite* continued fractions (CF):

$$(1) \qquad b_0 + \cfrac{a_1}{b_1 + \cdots} = \left[b_0, \frac{a_1}{b_1}, \ldots \right],$$

where

$$(1a) \qquad (b_i, a_{i+1}) = (1,1) \quad \text{or} \quad (2,-1), \ i \geq 0$$

These continued fractions satisfy the conditions of Tietze's Convergence Theorem [P, 135], namely

$$b_i \geq 1, \quad b_i + a_{i+1} \geq 1,$$

so we can assert:

LEMMA. *The continued fraction (1), (1a) converges. Moreover,*

$$(2) \qquad q_n \geq 1, \ n \geq 0; \ \lim_{n \to \infty} q_n = +\infty$$

where

$$\frac{p_n}{q_n} = \left[b_0, \frac{a_1}{b_1}, \ldots, \frac{a_n}{b_n} \right]$$

are the convergents of (1). Finally, if α_0 is the value of (1), we have

$$(3) \qquad \alpha_0 \in (b_0, b_0 + a_1]$$

1991 *Mathematics Subject Classification.* Primary 11A55.

Next suppose CF is finite. We assume it is of the form

(4) $[1]$ or $\left[b_0, \dfrac{a_1}{b_1}, \ldots, \dfrac{a_{n-1}}{1}, \dfrac{1}{2}\right], \quad i \le n - 2$

the (b_i, a_{i+1}) satisfying (1a).

Define F to be the set of continued fractions satisfying (1), (1a), and (4). We are going to prove:

THEOREM. *Every real number is the sum of an integer and an element of F. Given a real number α, we can write*

$$\alpha = [\alpha] - 1 + \alpha_0,$$

where $\alpha_0 = [b_0, a_1/b_1, \ldots] \in F$. There is an algorithm ("farthest integer" algorithm) for determining a_i, b_i. Moreover, if α is irrational, we have

(5) $\left| \alpha_0 - \dfrac{p_{n-1}}{q_{n-1}} \right| \le \dfrac{1}{q_{n-1}}, \quad n \ge 1$

If α is rational, α_0 has one of the forms (4).

2. For the continued fraction (1) we have the following well-known formula:

(6) $p_n = b_n p_{n-1} + a_n p_{n-2}, \quad q_n = b_n q_{n-1} + a_n q_{n-2}, \quad n \ge 2$

$$p_0 = b_0, \quad p_1 = b_0 b_1 + a_1, \quad q_0 = 1, \quad q_1 = b_1$$

$$p_n q_{n-1} - q_n p_{n-1} = \pm 1, \quad n \ge 1$$

We also have (2), (3) and the following estimates [P, 136]:

(7) $q_n \ge n \qquad$ if $\qquad a_n = 1$

$\qquad\qquad\qquad\qquad q_{n-1} \ge n \qquad$ if $\qquad a_n = -1$

Given a real number α, we define α_0 by

$$\alpha = [\alpha] - 1 + \alpha_0,$$

hence $1 \le \alpha_0 < 2$. First, let α_0 be irrational, so that

$$1 < \alpha_0 < 2.$$

We shall show how α_0 can be expanded as an element of F.

Define α_n by

$$\alpha_n = \left[b_n, \frac{a_{n+1}}{b_{n+1}}, \ldots\right], \quad n \ge 0.$$

Suppose $\alpha_1, \ldots, \alpha_{n-1}, b_0, \ldots, b_{n-2}, a_1, \ldots, a_{n-1}$ have been chosen. Then for $n \ge 1$,

(8)   if $1 < \alpha_{n-1} < 3/2$, set $b_{n-1} = 2$, $a_n = -1$, and $\alpha_n = 1/(2 - \alpha_{n-1})$

$\qquad$ if $3/2 < \alpha_{n-1} < 2$, set $b_{n-1} = 1, a_n = 1$, and $\alpha_n = 1/(\alpha_{n-1} - 1)$.

Thus all α_n, a_n, b_n are determined and

$$\alpha_{n-1} = b_{n-1} + \frac{a_n}{\alpha_n}.$$

Moreover,

(9) $$1 < \alpha_n < 2$$

The continued fraction $[b_0, a_1/b_1, \ldots]$ is a member of F.

We must show that $\alpha_0 = [b_0, a_1/b_1, \ldots]$ i.e., that the continued fraction defined in (8) represents α_0. This we shall do by proving (5).

From (6) we have

(10)
$$\alpha_0 - \frac{p_{n-1}}{q_{n-1}} = \left[b_0, \ldots, \frac{a_{n-1}}{b_{n-1}}, \frac{a_n}{\alpha_n} \right] - \frac{p_{n-1}}{q_{n-1}} = \frac{\alpha_n p_{n-1} + a_n p_{n-2}}{\alpha_n q_{n-1} + a_n q_{n-2}} - \frac{p_{n-1}}{q_{n-1}},$$

$$\left| \alpha_0 - \frac{p_{n-1}}{q_{n-1}} \right| = \frac{1}{q_{n-1}^2 (\alpha_n + a_n q_{n-2}/q_{n-1})}$$

Hence if $a_n = 1$ we get from (9)

(11) $$\left| \alpha_0 - \frac{p_{n-1}}{q_{n-1}} \right| < \frac{1}{q_{n-1}^2}, \quad a_n = 1$$

We next consider $a_n = -1$. It suffices to show that

(12) $$\alpha_n q_{n-1} + a_n q_{n-2} \geq 1,$$

as we see from (10). We proceed by induction. When $n = 2$ we have

$$\alpha_2 q_1 + a_2 q_0 = \alpha_2 b_1 + a_2 = 2\alpha_2 - 1 > 1,$$

by (9). Now assume

(13) $$\alpha_{n-1} q_{n-2} + a_n q_{n-3} \geq 1, \quad n > 2.$$

By (8) we have $\alpha_n = 1/(2 - \alpha_{n-1})$ and $b_{n-1} = 2$. Hence using (6) and (9) we get

$$\alpha_n q_{n-1} + a_n q_{n-2} = \frac{1}{2 - \alpha_{n-1}} (b_{n-1} q_{n-2} + a_{n-1} q_{n-3}) - q_{n-2}$$

$$= \frac{1}{2 - \alpha_{n-1}} (\alpha_{n-1} q_{n-2} + a_{n-1} q_{n-3} + (b_{n-1} - 2) q_{n-2})$$

$$\geq \frac{1}{2 - \alpha_{n-1}} > 1,$$

completing the induction.

Now assume α_0 is rational. We may suppose as before that $1 \leq \alpha_0 < 2$. If $\alpha_0 = 1$ we are finished: $\alpha_0 = [1]$. Suppose $1 < \alpha_0 < 2$; write $\alpha_n = h_n/k_n$, $k_n \geq 1$, $(h_n, k_n) = 1$, $n \geq 1$. We divide the interval $(1, 2)$ into two subintervals $(1, 3/2)$ and $[3/2, 2)$ and make the following definitions:

If $1 < \alpha_{n-1} = h_{n-1}/k_{n-1} < 3/2$, set $b_{n-1} = 2$, $a_n = -1$, and

$$\alpha_n = \frac{h_n}{k_n} = \frac{1}{2 - h_{n-1}/k_{n-1}} = \frac{k_{n-1}}{2k_{n-1} - h_{n-1}},$$

so $k_n = 2k_{n-1} - h_{n-1} < 2k_{n-1} - k_{n-1} = k_{n-1}$. Likewise, if $3/2 \leq \alpha_{n-1} = h_{n-1}/k_{n-1} < 2$, set $b_{n-1} = 1$, $a_n = 1$, and

$$\alpha_n = \frac{h_n}{k_n} = \frac{1}{h_{n-1}/k_{n-1} - 1} = \frac{k_{n-1}}{h_{n-1} - k_{n-1}},$$

so $k_n = h_{n-1} - k_{n-1} < 2k_{n-1} - k_{n-1} = k_{n-1}$. The k_n decrease and eventually reach 2, at which point the continued fraction terminates as $\left[\ldots, \frac{1}{2}\right]$.

Thus α_0 is expanded in a finite continued fraction that belongs to F:

$$\alpha_0 = [1] \quad \text{or} \quad \alpha_0 = \left[b_0, \frac{a_1}{b_1}, \ldots, \frac{a_{n-1}}{b_{n-1}}, \frac{1}{2}\right] \quad \text{for some } n \geq 1.$$

The Theorem is now completely proved.

References

[P] O. Perron, *Die Lehre von den Kettenbrüche*, vol. 1, Stuttgart, Teubner, 1929.

NOTE ADDED IN PROOF. The above expansion of a real number is not unique. E.g., $1 = [1] = [2, -1/2, -1/2 \ldots]$. To achieve uniqueness, we introduce a further condition on F, namely: If $CF\alpha = [b_0, a_1/b_1, \ldots]$ is infinite, then $b_i + a_{i+1} \geq 2$ for infinitely many indices i. If $CF\alpha$ is finite and has at least 2 terms, the last term is $1/2$. Let $F' \subset F$ be the set of CF in F satisfying this additional condition. *Then the expansion of a number as an element of F' is unique.*

Consider the case of CFs $[b_0, a_1/b_1, \ldots]$, $[b_0', a_1'/b_1', \ldots]$, both infinite, which agree through index $n - 1$ but no further. Then $\alpha_j = \alpha_j'$, $0 \leq j < n$, but $\alpha_n \neq \alpha_n'$. Hence $1 + 1/\alpha_n = \alpha_{n-1} = \alpha_{n-1}' = 2 - 1/\alpha_n'$ say, so

$$(14) \qquad\qquad \frac{1}{\alpha_n} + \frac{1}{\alpha_n'} = 1.$$

By (3), $\alpha_n \in (b_n, b_n + a_{n+1}]$. There are three possibilities: (i) $\alpha_n, \alpha_n' \in (1, 2]$; (ii) $\alpha_n, \alpha_n' \in (2, 1]$; (iii) $\alpha_n \in (1, 2]$, $\alpha_n' \in (2, 1]$. Now (ii) and (iii) contradict (14). As for (i), note that $\alpha_n = \alpha_n' = 2$, otherwise (14) is contradicted. Hence $b_n = b_n' = 1$ and the CFs agree through index n, against the assumption.

The remaining cases are similar.

314-N SHARON WAY, JAMESBURG, NJ 08831

Contemporary Mathematics
Volume **169**, 1994

Testing for the center of a one-relator group

FRANK LEVIN

ABSTRACT. We describe a simple algorithm which shortens
an algorithm to determine the center of a one-relator group
given by Baumslag and Taylor in [1].

1. Introduction

In [1] G.Baumslag and T.Taylor give an algorithm which determines
the center of a one-relator group. In Section 5 we will describe the algorithm in some detail. For the present it suffices to observe that the
relator of a one-relator group G with center determines an integral
matrix M, and G will have a nontrivial center only if M has finite order r,
say. It was observed in [1] that r could be determined in finitely many
steps, but the suggested approximation to the number of steps in [1] is
an exponential function of the degree, d, of M. The purpose of this note
is to describe a simple algorithm which will accomplish this part of the
Baumslag-Taylor algorithm in $O(\log_p d)$ time. In general, d can be large,
so this can be significant. The final part of the Baumslag-Taylor algorithm
requires determining that the r-th power of an automorphism of a free
group of rank d is inner, is, at best, tedious and will not be considered
here.

2. Notation and preliminary remarks

For the sequel, p will denote a fixed prime, ω a primitive p-th root
of unity and $P_n(x)$ the n-th cyclotomic polynomial of degree $\varphi(n)$, the
Euler φ-function. Next we list a few elementary facts regarding roots of
unity and cyclotomic polynomials.

For any $\alpha \varepsilon \mathbb{C}$, $j \varepsilon \mathbb{N}$, $x^P - \alpha^P$ has a factor $x \omega^j - \alpha$. In particular,

$$\prod_{j=1}^{P} (x \omega^j - \alpha) = x^P - \alpha^P.$$

More generally,

1991 Mathematics Subject Classification Primary 20E99, 20F05.
This paper is in final form and no version of it will be submitted
for publication elsewhere.

(2.1) *For any monic polynomial $f(x) \varepsilon C[x]$ of degree d with zeros α_j, $j = 1, \cdots, d$,*

$$\text{if } f_p(x) = \prod_{j=1}^{d}(x - \alpha_j^P), \text{ then } f_p(x^P) = \prod_{j=1}^{P} f(x \omega^j).$$

For instance, for $p = 2$, $f_p(x^2) = f(x) \cdot f(-x)$.

In the sequel, for any monic polynomial $f(x) \varepsilon C[x]$, $f_p(x)$ will denote the polynomial defined in (2.1). The following is then an immediate corollary to (2.1).

(2.2) (i) *If $f(x) = g(x)h(x)$, then $f_p(x) = g_p(x) h_p(x)$.*

(ii) *If either all zeros of f(x) are roots of unity or some zero of f(x) is not a root of unity, the same will be true of $f_p(x)$.*

If $\omega_1, \cdots, \omega_k$, $k = \varphi(n)$, are the primitive nth roots of unity for some $n > 0$ and p does not divide n, the map $\omega_j \to \omega_j^P$ merely permutes the ω_j. If $n = pm$, ω_j^P is a primitive mth root of unity. If $(m, p) = 1$, each ω_j will be repeated p-1 times, while if $(p, m) = p$, each ω_j will be repeated p times. This gives the following.

(2.3) *Let $f(x) = P_n(x)$.*

(i) *If $(p, n) = 1$, then $f_p(x) = f(x)$.*

(ii) *If $n = pm$, then $f_p(x) = P_m(x)^{P-1}$ if $(p, m) = 1$, and $f_p(x) = P_m(x^P)$ if $(p, m) = p$.*

Finally, we need one more fact, routinely verified.

(2.4) *If n is divisible by p^2, then $P_n(x) \varepsilon Q[x^P]$.*

3. The algorithm.

Let $f(x) \varepsilon Z[x]$ and p be a fixed prime. The above remarks lead to the following easy algorithm for determining if any zero of f(x) is not a root of unity. The first step is to construct $f_p(x)$, as defined in (2.1). If $f_p(x) = f(x)$, all zeros are roots of unity. Otherwise, continue with the sequence

(3.1) $f^{(1)}(x) = f_p(x)$, $f^{(2)}(x) = f_p^{(1)}$, $\cdots$, $f^{(n)}(x) = f^{(n-1)}(x)$, $\cdots$.

It follows from the remarks in §2 that all zeros of f(x) are roots of unity or, equivalently, f(x) is a product of cyclotomic polynomials if and only if (3.1) stabilizes for some $n \geq 1$, that is, for some $n \geq 1$,

$$f^{(n+1)}(x) = f^{(n)}(x).$$

Further, if it stabilizes, then this occurs in at most $O(\log_p d)$ steps, where $d = \deg(f)$.

If (3.1) stabilizes there are some tests to apply to determine the cyclotomic factors involved. Firstly, one of the various factorization methods can be tried. Also, with $p=2$, for instance, if (3.1) stabilizes at $n=2$, there will be a $2m$-th root of unity, m odd, and $f_p(x)$ and $f(x)$ will have a common factor. Otherwise, if (3.1) stabilizes at $n \geq 3$, then $f(x)$ has a factor in $\mathbb{Z}[x^2]$, which is directly found by computing $\gcd(f(x)+f(-x), f(x)-f(-x))/2$.

4. The Baumslag—Taylor algorithm

The following is a brief description of the algorithm by G. Baumslag and T. Taylor [1] for determining the center of a one-relator group.

Let G be a nonabelian one-relator group with nontrivial center $Z(G)$. It follows from Murasugi [2] that G is a 2-generator group, and using standard techniques, we may assume G has a presentation $G = \langle x, y; R=1 \rangle$, where R has zero exponent sum in y, and, for some $n \geq 1$, $R = R(x_0, \cdots, x_n)$, where $x_i = y^i x y^{-i}$, $i \geq 0$. By [1], the x_i generate a free group K, which is the kernel of the epimorphism $G \to \langle x, y; R=1, y=1 \rangle$. Further, x_0 and x_n will occur once each in R with exponents 1 or -1, and since $R=1$, we can solve for x_n: $x_n = w$, say, where $w = w(x_0, \cdots, x_{n-1})$. In addition, y will induce an automorphism ψ of K defined by $x_i \psi = x_{i+1}$, $i=0, \cdots, n-2$, $x_{n-1}\psi = w$, and for some r, ψ^r will be inner. In particular, the automorphism ψ_{ab} which ψ induces on K abelianized will have order r, as will the $n \times n$-matrix M which represents ψ_{ab}. The characteristic polynomial $F(\lambda)$ of M is easily obtained as follows. First, write w additively. Next, replace each x_i in w by λ^i. $F(\lambda)$ is then given by $F(\lambda) = \lambda^n - w$. Since $M^r = I$ only if all zeros of $F(\lambda)$ are roots of unity, the algorithm in Section 3 is directly applicable for determining if r is finite and, with a bit of luck, the precise value of r, if finite.

If M has infinite order, $Z(G)$ will be trivial. However, if r is finite, then the further steps to verify that the r-th power of ψ is inner can become unwieldy, even for small values of r.

Finally, as an exercise to illustrate the difficulties involved the reader is invited to show that the center of the group

$$G = \langle x, y; xyxy^3xy^3xyxy^{-8} \rangle$$

is trivial given that $r=60$.

REFERENCES

1. Gilbert Baumslag and Tekla Taylor, The centre of groups with one defining relator, Math. Ann. **175** (1968), 315-319.

2. Kunio Murasugi, The center of a group with a single defining relation, Math. Ann. **155** (1964), 246-251.

MATHEMATICS INSTITUTE, RUHR UNIVERSITY, D-44780 BOCHUM, FEDERAL REPUBLIC OF GERMANY

Contemporary Mathematics
Volume **169**, 1994

Generalizing the Baer-Stallings Pregroup

Seymour Lipschutz

ABSTRACT. One of the axioms of the Baer-Stallings pregroup is replaced by a set of weaker axioms and the resultant structure P is still embeddable in a group.

1. Introduction

Let P be a _pree_, that is, let P be a nonempty set with a partial operation $m:D \longrightarrow P$ where $D \subseteq P \times P$. We say pq is _defined_ if $(p,q) \in D$ and we will usually denote $m(p,q)$ by pq. (Baer [1] used the term "add" for a pree and denoted $m(p,q)$ by $p + q$.) The _universal group_ $G(P)$ of P is the group with the following presentation:

$G(P) = gp[P; z = xy$, where $z = m(x,y)]$

In other words, the generators of $G(P)$ are the elements of P and the defining relations of $G(P)$ come from the partial operation m on P. A pree P is said to be _group-embeddable_ or, simply, _embeddable_, if P can be embedded in its universal group $G(P)$ (see Rimlinger [6]).

Next follow two classical examples of embeddable prees.

Example 1.1. Let H and K be groups which intersect in a subgroup A. Schreier [7] proved in 1927 that the amalgam $P = H \cup_A K$ is an embeddable pree. Here $G(P)$ is the free product of H and K with A amalgamated.

1991 AMS Subject Classification. Primary 20E06.
This paper is in final form and no version of it will be submitted for publication elsewhere.

Example 1.2. Let $T_n = (H_i; A_{st})$ be a tree
graph of groups with vertex groups H_i, with edge
groups A_{st}, and with diameter n. (Here A_{st} is a
subgroup of groups H_s and H_t.) The amalgam
$P = \cup_i H_i$ of the groups in T_n (called a tree
pree) is known to be embeddable (cf. Serre [8]).
Here G(P) is the tree product of the groups H_i
with the subgroups A_{st} amalgamated.

Baer [1] generalized Schreier's result
(Example 1.1). Specifically, Baer defined a
pregroup P as a pree satisfying certain axioms
and showed that P is embeddable. Stallings, who
invented the name pregroup, proved additional
properties of pregroups in [9] and [10]. These
results were extended by Rimlinger in [6].
For reference, we list Stallings' five
axioms in [9] which define a pregroup P:

[P_1] There exists $1 \in P$ such that, for all $a \in P$,
la and al are defined and la = al = a.
[P_2] For each $a \in P$, there is $a^{-1} \in P$ such that aa^{-1}
and $a^{-1}a$ are defined and $aa^{-1} = a^{-1}a = 1$.
[P_3] If pq is defined, then $q^{-1}p^{-1}$ is defined
and $(pq)^{-1} = q^{-1}p^{-1}$.
[P_4] Supposing ab and bc are defined, then a(bc)
is defined if and only if (ab)c is defined,
in which case the two are equal. (We then
say abc = a(bc) = (ab)c is defined.)
[P_5] If ab, bc and cd are defined, then either
(ab)c or (bc)d is defined.

We emphasize that a pregroup P generalizes
the pree in Example 1.1, but it does not
generalize the tree pree in Example 1.2.
Specifically, the tree pree satisfies axioms
[P_1]-[P_4] but need not satisfy axiom [P_5] as
illustrated by the following example.

Example 1.3. Consider a tree pree
$T = H\cup_A K\cup_B L$ where A and B are nontrivial proper
subgroups of the vertex groups H and L,
respectively, and where K = A + B. Suppose
$a \neq 1$, $b \neq 1$ and
$x \in (H \backslash A)$, $a \in A$, $b \in B$, $y \in (L \backslash B)$
Then xa, ab and by are defined, but neither
x(ab) nor (ab)y are defined.

This example leads us to the following:

Problem 1.1: Find an axiom or set of axioms (to replace Stallings' axiom $[P_5]$) such that: (i) The prees P which satisfy the new axioms and $[P_1]-[P_4]$ are embeddable. (ii) These prees P include some or all of the tree prees in Example 1.2.

This paper considers the following axioms (where reduced, fully reduced, and equivalent will be defined later):

$[T_n]$ If a_1a_2, a_2a_3, ..., $a_{n+2}a_{n+3}$ are defined, then $(a_1a_2)a_3$, $(a_2a_3)a_4$, ..., or $(a_{n+1}a_{n+2})a_{n+3}$ is defined.

$[K]$ If ab, bc, cd and (ab)(cd) are defined, then (ab)c or (bc)d is defined.

$[L]$ If ab, bc and cd are defined but $[ab,cd]$ and $[a,bc,d]$ are reduced, and if (ab)z and $z^{-1}(cd)$ are defined, then bz and $z^{-1}c$ are defined.

$[B]$ Equivalent fully reduced words have the same length.

Observe first that $[P_5] = [T_1]$ and that axiom $[T_n]$ only holds for a tree pree P when the diameter of the tree does not exceed n. On the other hand, we prove in Section 6 that axioms $[K]$, $[L]$ and $[B]$ hold for any tree pree P, (they are independent of the diameter). Axiom $[B]$, which we call Baer's axiom, is analagous to his axiom [1, p.648] which states:

"Similar irreducible vectors have the same length."

(See also [10].)

The following results have already been proved.

Theorem A (Kushner and Lipschutz [3]). Suppose P satisfies $[P_1]-[P_4]$ and $[T_2]$ and $[K]$. Then P is embeddable.

Theorem B (Kushner [5], Hoare [2]): <u>Suppose P
satisfies $[P_1]$-$[P_4]$ and $[T_2]$. Then P is
embeddable.</u>

Theorem C (Kushner and Lipschutz [4]): <u>Suppose P
satisfies $[P_1]$-$[P_4]$ and $[T_3]$ and [K]. Then P is
embeddable.</u>

Theorem B was proved independently by
Kushner and Hoare. However, by using three
cases instead of five to define the permutation
F_a and by using a localized type argument, Hoare
proved Kushner's original result with a
considerably shorter and less involved proof.

One purpose of this paper is to introduce
the axioms [L] and [B] and their consequences.
In particular, we have the following main
result:

Theorem 1.1. <u>Suppose P satisfies $[P_1]$-$[P_4]$ and
[K], [L] and [B]. Then P is embeddable.</u>

ACKNOWLEDGEMENT: The author thanks Dr.
Hoare for invaluable discussions and the proof
of Proposition 2.1.

DEDICATION: The author dedicates this paper
to his teacher, advisor and friend Wilhelm
Magnus (1907-1990); he was a shining example of
a kind, considerate and concerned human being.

2. <u>Preliminaries</u>

Suppose P satisfies $[P_1]$-$[P_4]$. The
notation and definitions in [9] and [3] are used
below. We also use, without explicit mention,
the immediate consequences of axioms $[P_1]$-$[P_4]$
appearing in both [9] and [3]. However, for
clarity, we also repeat some basic definitions
below.

Suppose $X = [x_1, x_2, \ldots, x_m]$ is an m-tuple of
elements of P. We say X is a <u>word of length m</u>.
Suppose $A = [a_1, \ldots, a_{m-1}]$ such that each triple
$a_{i-1}^{-1} x_i a_i$ is defined (where $a_0 = a_m = 1$).
Then the <u>interleaving</u> of X by A is said to be
defined and we set

$$X \ast A = [x_1 a_1, \; a_1^{-1} x_2 a_2, \; \ldots, \; a_{m-1}^{-1} x_m]$$

We write $X*A*B$ for $(X*A)*B$ and we write $X \longrightarrow Y$
if $Y = X*A_1*\ldots*A_k$, i.e., if Y can be obtained
from X by a sequence of interleavings. (We note
that the relation $X \longrightarrow Y$ is symmetric.)

A word $E = [1,\ldots,a,\ldots,1]$, that is, where
E contains 1 in every position except possibly
one, is called an elementary word, and $X*E$,
when defined, is called an elementary
interleaving. Note that $X*A$ can be obtained
from X using a sequence of elementary
interleavings. Accordingly, if $X \longrightarrow Y$, then Y
can also be obtained from X by a sequence of
elementary interleavings. We also write
$I_k = [1,1,\ldots,1]$ for the elementary word of
length k with all 1's.

A word $X = [x_1,x_2,\ldots,x_m]$ is said to be
reduced if no pair $x_{i-1}x_i$ is defined. The word X
is fully reduced if X is reduced and
$X*A_1*A_2*\ldots*A_m$ [or $X*E_1*\ldots*E_r$] is reduced
whenever defined. We note that any reduced word
X of length $m = 2$ is fully reduced, and every
word X of length $m = 1$ is automatically reduced
and fully reduced. Although, in a pregroup,
every reduced word is also fully reduced, this
may not be true for tree prees. Consider
$X = [x,ab,y]$ and $A = [a,1]$ in Example 1.3. Then
X is reduced but $X*A = [xa,b,y]$ is not reduced.

The notion of a splitting, which follows,
is new and basic to this paper. Consider a word
$X = [x_1,\ldots,x_m]$ where $m > 2$. Suppose $x_k = a_k b_k$,
for $k = 2, \ldots,m-1$, and suppose $x_1 a_2$, $b_j a_{j+1}$
($j=2,\ldots,m-2$), and $b_{m-1}x_m$ are defined. We then
say X is reducible to Y, written $X \Rightarrow Y$, where
$$Y = [x_1 a_2, b_2 a_3, \ldots, b_{m-2}a_{m-1}, b_{m-1}x_m]$$
For example, $X = [x,ab,y]$ in Example 1.3. Then
$X \Rightarrow Y$ where $Y = [xa,by]$.

If X is reducible to Y, as above, then X is
said to split or to be splitable, and the
factorization $x_k = a_k b_k$ ($k=2,\ldots,m-1$) is called
a splitting of X. For $m = 2$, we say that
$X = [x_1,x_2]$ is splitable if $x_1 x_2$ is defined and,
in such a case, $X \Rightarrow Y$ where $Y = [x_1 x_2]$. The
term nonsplitable will be used for the negation
of splitable. If $m = 1$, then X is automatically
nonsplitable.

Observe that if $X \Rightarrow Y$, as above, then
$[Y,1] \longrightarrow X$. Specifically, X can be obtained

from [Y,1] by elementary interleavings using
x_m^{-1}, a_{m-1}^{-1}, ..., a_2^{-1}, respectively. Thus, in
particular, X is not fully reduced. Also
observe that a splitting $x_k = a_k b_k$, $i < k < j$,
of a subword $[x_i,...,x_j]$ of X can be extended to
a splitting of X by setting $x_k = 1x_k$, for $k \leq i$,
and $x_k = x_k 1$, for $k \geq j$. Lastly, observe that
every word X may be repeatedly reduced to yield
a nonsplitable word Z since each such reduction
decreases the length of the word by one. In
particular, we say X <u>splits n times</u> if
$X \Rightarrow Y_1 \Rightarrow ... \Rightarrow Y_n$; and, in such a case,
$X \rightarrow [Y_n, I_n]$. We also note that $[X, I_k]$ splits
at least k times.

Throughout the remainder of this paper, we
assume P satisfies $[P_1]$-$[P_4]$ and [K], [L] and
[B]. The following proposition, proved in
Section 4, describes the fully reduced words
in P.

Proposition 2.1 (Hoare): <u>X is fully reduced if
and only if X is nonsplitable.</u>

Corollary 2.2: <u>Every word X may be reduced to a
fully reduced word by applying the above
reductions.</u>

The notion of fully reduced words, and the
notion of $x_k = ab$ splitting in $X = [x_1,...,x_m]$,
when $x_{k-1}a$ and bx_{k+1} are defined, already
appeared in [3] and [4]. However, the above
definition of a general splitting and the
following definition of equivalence are new.

<u>Definition</u>: Suppose X and X' are fully reduced
            words. Define $X \approx X'$ if there exist
            k and k' such that $[X, I_k] \rightarrow [X', I_{k'}]$.

Clearly, $X \approx X'$ is an equivalence relation.
Since we are using axiom [B], X and X' must have
the same length. Hence, we can assume $k = k'$.

Corollary 2.3: <u>Suppose $X \rightarrow X'$. Then X splits
n times if and only if X' splits n times.</u>

Corollary 2.4: <u>Suppose X splits n times and
$[X, I_k] \rightarrow [X', I_k]$. Then X' splits n times.</u>

Observe that if Corollary 2.3 or 2.4 were not true, then fully reduced words of different lengths would be equivalent. This will then contradict axiom [B].

3.   The Mapping F_a

Suppose X is a fully reduced word. For each $a \in P$, we define a mapping $F_a(X)$ on the collection of fully reduced words as follows:

(1) If [a,X] is fully reduced, then
    $F_a(X)$ = [a,X].

(2) If [a,X] $\Rightarrow$ Y where Y is fully reduced,
    then $F_a(X)$ = Y.

(3) If [a,X] $\Rightarrow$ Y $\Rightarrow$ Z, then $F_a(X)$ = Z.

Remark. Since a splitting may occur in more than one way, the map $F_a(X)$ may be multivalued. However, all such values are equivalent. Specifically, suppose [a,X] $\Rightarrow$ Y where Y is fully reduced and suppose [a,X] $\Rightarrow$ Y'. Then [Y,1] $\longrightarrow$ X $\longrightarrow$ [Y',1]. Thus, by Corollary 2.4, Y is also fully reduced, and hence Y $\approx$ Y'. Similarly, suppose [a,X] $\Rightarrow$ Y $\Rightarrow$ Z and [a,X] $\Rightarrow$ Y' $\Rightarrow$ Z' where Z and Z' are fully reduced. Then [Z,1,1] $\longrightarrow$ [a,X] $\longrightarrow$ [Z',1,1] and, therefore, Z $\approx$ Z'.

The main properties of the map $F_a(X)$ follow:

Theorem 3.1: $\underline{F_a(X) \text{ is fully reduced.}}$

Theorem 3.2: $\underline{\text{If } X \approx X', \text{ then } F_a(X) \approx F_a(X').}$

Theorem 3.3: $\underline{\text{If ab is defined, then}}$
$\underline{F_{ab}(X) \approx F_a(F_b(X)).}$

The proof of the above theorems will occupy Section 5. Each proof consists of looking at the different cases that can occur. Theorem 1.1 follows from these theorems using standard techniques (see [3, 6, 9]).

4.   Proof of Proposition 2.1

If X splits, then X is not fully reduced; hence we need only prove the converse which follows from the following result.

Lemma 4.1: **Suppose X is reduced and X∗E splits. Then X splits.**

Let $X = [x_1, x_2, \ldots, x_m]$ and let
$X*E = [x_1, \ldots, x_k Z, Z^{-1} x_{k+1}, \ldots, x_m]$. Suppose the following defines a splitting of X∗E:

$$X*E = [x_1, A_2 B_2, \ldots, A_k B_k,$$
$$A_{k+1} B_{k+1}, \ldots, A_{m-1} B_{m-1}, x_m]$$

Then $B_k A_{k+1}$ is defined. Also, since $x_k Z = A_k B_k$ and $Z^{-1} x_{k+1} = A_{k+1} B_{k+1}$, we have $x_k = (A_k B_k) Z^{-1}$ and $x_{k+1} = Z(A_{k+1} B_{k+1})$ are defined.

Suppose X∗E is not reduced. Then, since X is reduced, we either have $x_{k-1}(x_k Z)$ is defined, in which case $x_k = (x_k Z) Z^{-1}$ is a splitting of $[x_{k-1}, x_k, x_{k+1}]$, or we have $(Z^{-1} x_{k+1}) x_{k+2}$ is defined, in which case $x_{k+1} = Z(Z^{-1} x_{k+1})$ is a splitting of $[x_k, x_{k+1}, x_{k+2}]$. In either case, X splits. Thus we can now consider the case that X∗E is reduced.

Case 1. $[A_k, B_k A_{k+1}, B_{k+1}]$ is reduced.

Since X∗E is reduced, we have $[A_k B_k, A_{k+1} B_{k+1}]$ is reduced. Also,
$x_k = (A_k B_k) Z^{-1}$ and $x_{k+1} = Z(A_{k+1} B_{k+1})$ are defined. Thus, by axiom [L], we have $B_k Z^{-1}$ and $Z A_{k+1}$ are defined. Hence the following defines a splitting of X:
$$X = [x_1, A_2 B_2, \ldots, A_k (B_k Z^{-1}),$$
$$(Z A_{k+1}) B_{k+1}, \ldots, A_{m-1} B_{m-1}, x_m]$$

Case 2. $A_k B_k A_{k+1}$ is defined.

Since $x_{k+1} = Z(A_{k+1} B_{k+1})$ is defined, we can apply axiom [K] to
$$[x_k^{-1}, x_k Z, A_{k+1}, B_{k+1}] = [x_k^{-1}, A_k B_k, A_{k+1}, B_{k+1}]$$
The second triple is not defined since X∗E is reduced; hence $Z A_{k+1}$ is defined. Therefore, $x_{k+1} = (Z A_{k+1}) B_{k+1}$ and so $x_k (Z A_{k+1}) = A_k B_k A_{k+1}$ is defined. Thus the following defines a splitting of $[x_k, \ldots, x_m]$:
$$[x_k, (Z A_{k+1}) B_{k+1}, A_{k+2} B_{k+2}, \ldots, A_{m-1} B_{m-1}, x_m]$$
Thus X also splits.

Case 3. $B_k A_{k+1} B_{k+1}$ is defined.

Similarly to Case 2, we obtain that $B_k Z^{-1}$ is defined and that the following is a splitting of $[x_1, \ldots, x_{k+1}]$:

$$[x_1, A_2 B_2, \ldots A_{k-1} B_{k-1}, A_k (B_k Z^{-1}), x_{k+1}]$$

Thus X splits in this case.

Accordingly, Lemma 4.1 is proved, and hence Proposition 2.1 is proved.

5. Proofs of Theorems

The theorems use the following lemmas.

Lemma 5.1: <u>Suppose X is fully reduced and a $\in$ P.</u>
(a) <u>If $[a,X] \Rightarrow Y$, then $X \longrightarrow [x_1', x_2', \ldots, x_m']$ where ax_1' is defined.</u>
(b) <u>If $[a,X] \Rightarrow Y \Rightarrow Z$, then $X \longrightarrow [a^{-1}, Z]$.</u>

Suppose $[a,X] \Rightarrow Y$. Say,
$$X = [x_1, \ldots, x_m],$$
$$Y = [a\bar{A}_1, B_1 A_2, B_2 A_3, \ldots, B_{m-1} x_m],$$
where $x_j = A_j B_j$, $j = 1, \ldots, m-1$, is the splitting of $[a,X]$, Then
$$X = [A_1 B_1, \ldots, A_{m-1} B_{m-1}, x_m]$$
$$\longrightarrow [\bar{A}_1, B_1 A_2, B_2 A_3, \ldots, B_{m-1} x_m],$$
where aA_1 is defined. Hence (a) is true.

Now suppose $[a,X] \Rightarrow Y \Rightarrow Z$. Say
$$X = [x_1, \ldots, x_m],$$
$$Y = [a\bar{A}_1, B_1 A_2, B_2 A_3, \ldots, B_{m-1} x_m],$$
$$Z = [(a\bar{A}_1) \bar{E}_1, F_1 \bar{E}_2, \ldots, F_{m-2} (B_{m-1} x_m)]$$
where $x_j = A_j B_j$, $j = 1, \ldots, m-1$, is the splitting of $[a,X]$, and where $B_j A_{j+1} = E_j F_j$, $j = 1, \ldots, m-2$, is the spliting of Y. We have:
$$X = [A_1 B_1, A_2 B_2, \ldots, A_{m-1} B_{m-1}, x_m]$$
$$\longrightarrow [A_1, B_1 A_2, \ldots, B_{m-2} A_{m-1}, B_{m-1} x_m]$$
$$= [a^{-1}(aA_1), E_1 F_1, \ldots, E_{m-2} F_{m-2}, B_{m-1} x_m]$$
$$\longrightarrow [a^{-1}, (aA_1) E_1, F_1 E_2, \ldots, F_{m-2} (B_{m-1} x_m)]$$
$$= [a^{-1}, Z]$$

Thus (b) is also true.

Lemma 5.2: <u>Suppose X is fully reduced and a $\in$ P.Then $[a,X]$ splits at most twice.</u>

We need only show that if $[a,X] \Rightarrow Y \Rightarrow Z$, then Z is fully reduced. By Lemma 5.1, $X \longrightarrow [a^{-1}, Z]$. Since X is fully reduced, so is $[a^{-1}, Z]$ fully reduced; hence Z is fully reduced.

<u>Remark</u>:   Lemma 4.1, Proposition 2.1, and
           Lemmas 5.1 and 5.2 require only axioms
           $[P_1]$-$[P_4]$ and $[K]$ and $[L]$, that is, do
           not require Baer's axiom $[B]$.

Proof of Theorem 3.1:   <u>$F_a(X)$ is fully reduced.</u>

   Follows directly from Lemma 5.2.

Proof of Theorem 3.2: <u>If $X \approx X'$, then
                       $F_a(X) \approx F_a(X')$.</u>

   Since $X \approx X'$, we have $[X,I_k] \rightarrow [X',I_k]$.
The proof is reduced to the three ways that
$F_a(X)$ is defined.
<u>Case 1.</u> $F_a(X) = [a,X]$.
   We have $[a,X,I_k] \rightarrow [a,X',I_k]$.   Note $[a,X]$
is fully reduced; hence, by Corollary 2.4, so is
$[a,X']$.   Thus $F_a(X') = [a,X']$.   Also,
$$[F_a(X'),I_k] = [a,X',I_k] \rightarrow [a,X,I_k]$$
$$= [F_a(X),I_k]$$
Hence $F_a(X) \approx F_a(X')$.
<u>Case 2.</u> $F_a(X) = Y$ where $[a,X] \Rightarrow Y$.
   Then $[Y,1] \rightarrow [a,X]$.   Hence
$$[Y,1,I_k] \rightarrow [a,X,I_k] \rightarrow [a,X',I_k]$$
where, by Corollary 2.4, $[a,X']$ splits only
once.   Say $[a,X'] \Rightarrow Y'$.   Then $F_a(X') = Y'$ and
also $[Y',1] \rightarrow [a,X']$.   Thus
$$[F_a(X'),1,I_k] = [Y',1,I_k]$$
$$\rightarrow [a,X',I_k] \rightarrow [Y,1,I_k]$$
$$= [F_a(X),1,I_k].$$
Hence $F_a(X) \approx F_a(X')$.
<u>Case 3.</u> $F_a(X) = Z$ where $[a,X] \Rightarrow Y \Rightarrow Z$.
   Then $[Y,1] \rightarrow [a,X]$ and, furthermore,
$[Z,1,1] \rightarrow [Y,1] \rightarrow [a,X]$.   Hence
$$[Z,1,1,I_k] \rightarrow [Y,1,I_k] \rightarrow [a,X,I_k]$$
$$\rightarrow [a,X',I_k]$$
where now $[a,X']$ splits twice.   Say
$[a,X'] \Rightarrow Y' \Rightarrow Z'$.   Then $F_a(X') = Z'$ and also
$[Z',1,1] \rightarrow [Y',1] \rightarrow [a,X']$.   Thus
$$[F_a(X'),1,1,I_k] = [Z',1,1,I_k] \rightarrow [a,X',I_k]$$
$$\rightarrow [a,X,I_k] \rightarrow [Z,1,1,I_k]$$
$$= [F_a(X),1,1,I_k]$$
Hence $F_a(X) \approx F_a(X')$.
   Therefore, Theorem 3.2 is proved.

Proof of Theorem 3.3: <u>If ab is defined, then</u>
$$\underline{F_{ab}(X) \approx F_a(F_b(X)).}$$

Here X is fully reduced. The proof is
divided into the three different ways that $F_b(X)$
can be defined. Each of the cases will then
have three subcases, the three ways that
$F_a(F_b(X))$ can be defined.

<u>Case (A)</u>: $F_b(X) = [b,X]$.

There are three subcases.
<u>Case (A.1)</u>: $F_a(F_b(X)) = [a,b,X]$.
    This case is impossible since ab is defined.
<u>Case (A.2)</u>: $F_a(F_b(X)) = Y$ where $[a,b,X] \Rightarrow Y$.
    Since ab is defined, we can assume
Y = [ab,X].   Since Y is fully reduced,
$F_{ab}(X) = [ab,X] = F_a(F_b(X))$
<u>Case (A.3)</u>: $F_a(F_b(X)) = Z$ where $[a,b,X] \Rightarrow Y \Rightarrow Z$.
    Again we can assume Y = [ab,X].   Then
$F_{ab}(X) = Z$ since $[ab,X] \Rightarrow Z$ where Z is fully
reduced.   Thus $F_{ab}(X) = F_a(F_b(X))$.

<u>Case (B)</u>: $F_b(X) = V$ where $[b,X] \Rightarrow V$.

By Lemma 5.1 and Theorem 3.2, we can assume
that X = [x,y,...,z] and V = [bx,y,...,z].
Since ab is defined,
$$[a,V] = [a,bx,y,...,z] \longrightarrow [ab,x,...,z]$$
$$= [ab,X]$$
There are three subcases.
<u>Case (B.1)</u>: $F_a(F_b(X)) = [a,V]$
    Since $[a,V] \longrightarrow [ab,X]$ and   [a,V] is fully
reduced, we have [ab,X] is fully reduced.   Hence
$$F_{ab}(X) = [ab,X] \longrightarrow [a,V] = F_a(F_b(X)).$$
Thus $F_{ab}(X) \approx F_a(F_b(X))$.
<u>Case (B.2)</u>: $F_a(F_b(X)) = Y$ where $[a,V] \Rightarrow Y$.
    Then $[Y,1] \longrightarrow [a,V] \longrightarrow [ab,X]$.   Since [Y,1]
splits once, so does [ab,X]; say $[ab,X] \Rightarrow Y'$.
Then $F_{ab}(X) = Y'$ and $[Y',1] \longrightarrow [ab,X]$. Thus
$$[F_{ab}(X),1] = [Y',1] \longrightarrow [ab,X] \longrightarrow [Y,1]$$
$$= [F_a(F_b(X)),1]$$
Thus $F_{ab}(X) \approx F_a(F_b(X))$.
<u>Case (B.3)</u>: $F_a(F_b(X)) = Z$ where $[a,V] \Rightarrow Y \Rightarrow Z$.
    Then $[Z,1,1] \longrightarrow [a,V] \longrightarrow [ab,X]$.   Since
[Z,1,1] splits twice, so does [ab,X]; say
$[ab,X] \Rightarrow Y' \Rightarrow Z'$.   Then $F_{ab}(X) = Z'$ and

[Z',1,1] $\rightarrow$ [ab,X]. Thus
    [F_{ab}(X),1,1] = [Z',1,1] $\rightarrow$ [ab,X]
                $\rightarrow$ [Z,1,1] = [F_a(F_b(X)),1,1]
Thus F_{ab}(X) $\approx$ F_a(F_b(X)).

<u>Case (C)</u>: F_b(X) = W where [b,X] $\Rightarrow$ V $\Rightarrow$ W.

    By Lemma 5,1 and Theorem 3.2, we can assume
X = [b^{-1},y,...,z] and W = [y,...,z]. Since ab
is defined, we have
        [ab,X] = [ab,b^{-1},y,...,z] $\Rightarrow$ [a,y,...,z]
              = [a,W]
There are three subcases.
<u>Case (C.1)</u>: F_a(F_b(X)) = [a,W].
    We obtain F_{ab}(X) by reducing [ab,X]. We
have [ab,X] $\Rightarrow$ [a,W] where [a,W] is fully
reduced. Hence F_{ab}(X) = [a,W] = F_a(F_b(X)).
<u>Case (C.2)</u>: F_a(F_b(X)) = Y where [a,W] $\Rightarrow$ Y.
    We obtain F_{ab}(X) by reducing [ab,X]. We
have [ab,X] $\Rightarrow$ [a,W] $\Rightarrow$ Y where Y is fully
reduced. Hence F_{ab}(X) = Y = F_a(F_b(X)).
<u>Case (C.3)</u>: F_a(F_b(X)) = Z where [a,W] $\Rightarrow$ Y $\Rightarrow$ Z.
    We obtain F_{ab}(X) by reducing [ab,X]. Thus
        [ab,X] $\Rightarrow$ [a,W] $\Rightarrow$ Y $\Rightarrow$ Z.
This case is impossible since [ab,X] cannot
split three times.

    Consequently, Theorem 3.3 is proved.

6. Application to Tree Products

    Let T = (H_i; A_{st}) be a tree graph of groups,
let P = $\cup_i H_i$, the amalgam of the vertex groups,
and let G = G(P). Thus G is the tree product of
the vertex groups H_i with the edge groups A_{st}
amalgamated.
    Note first that P is a pree which satisfies
axioms [P_1]-[P_4] and that ab (resp. abc) is
defined in P iff a and b (resp. a, b and c)
belong to the same vertex group H_i. We claim
that P also satisfies axioms [K], [L] and [B].

Proposition 6.1: <u>P satisfies axiom [K]</u>.

    Suppose (ab)c is not defined. Then a and c
belong to distinct vertex groups, say H_i and H_j,
where b belongs to all edge groups between H_i

and H_j but neither a nor c belong to any edge
group between H_i and H_j. Then ab belongs to H_i
but to no edge group between H_i and H_j.
Moreover, since (ab)(cd) is defined, cd also
belongs to H_i. Thus b(cd) = (bc)d is defined.

Proposition 6.2: **P satisfies axiom [L].**

Since [a,bc,d] is reduced, and ab and cd
are defined, there exist distinct vertex groups,
say H, K, L, with the following properties:
   (1) K lies between H and L in the graph.
   (2) H, K, and L contain a, bc and d,
       respectively.
   (3) b belongs to all edge groups between K
       and H; and c belongs to all edge groups
       between K and L.
   (4) ab belongs to H but no edge group
       between K and H; and cd belongs to L
       but no edge group between K and L.
Since (ab)z and z^{-1}(cd) are defined, z and z^{-1}
belong to all edge groups between H and L. Thus
bz and z^{-1}c are defined.

   Remark. Since P satisfies axioms $[P_1]$-$[P_4]$
and [K] and [L], we note that P satisfies
Proposition 2.1, that nonsplitable words are
fully reduced, and that P satisfies Lemma 5.2,
that [a,X] splits at most twice when X is fully
reduced.

Proposition 6.3: **P satisfies Baer's axiom [B].**

   Suppose P does not satisfy [B]. Let
X = $[x_1,...,x_m]$ be a fully reduced word of
minimal length contradicting [B]; hence there
exists a fully reduced word Y = $[y_1,...,y_n]$ such
that XY = $x_1...x_m y_1...y_n$ = 1 in G and m < n.
   We first show that X ≠ 1. Suppose X = 1
and hence Y = 1. Let T' be the minimal subgraph
of T containing Y, and let G' be the tree
product of the groups in T'. Some y_j in Y
belongs to a vertex group H' of T' such that H'
contains only a single edge group, say A'. Then
G' = $K *_A$ H', where K is the tree product of the
groups in T' excluding H'. Then Y = 1 in G'.
This contradicts the Normal Form Theorem for the

free product of two groups with a single
amalgamation. Accordingly, $X \neq 1$ and so $m > 0$.

Now let $X' = [x_1, \ldots, x_{m-1}]$. Since Y is
fully reduced, we can only reduce $[x_m, Y]$ at most
twice to obtain a fully reduced word Y'. Then
the length of Y' must be greater than the length
of X' and $X'Y' = 1$ in G. This contradicts the
minimality of X. Thus P satisfies Baer's axiom
[B], that is, Proposition 6.3 is proved.

The above give rise to a normal form
theorem for elements in a tree product of
groups. First, however, we state a normal form
theorem for a generalized free product.

Normal Form Theorem A: <u>Let $G = G(P) = H *_A K$,
where $P = A \cup B$. Suppose $X = [x_1, x_2, \ldots, x_r]$ and
$Y = [y_1, y_2, \ldots, y_s]$ are reduced words in P and
$X = Y$ in G. Then:</u>
    <u>(1) $r = s$.</u>
    <u>(2) There exits $Z = [z_1, \ldots, z_{r-1}]$ such
        that $Y = X * Z$.</u>

The analagous result for tree products
follows:

Normal Form Theorem B: <u>Let $T = (H_i; A_{st})$ be a
tree graph of groups, let $P = \cup_i H_i$, and let
$G = G(P)$. Suppose $X = [x_1, x_2, \ldots, x_r]$ and
$Y = [y_1, y_2, \ldots, y_s]$ are fully reduced words in P
and $X = Y$ in G. Then:</u>
    <u>(1) $r = s$.</u>
    <u>(2) There exists k, with $k \leq 2r$, such that
        $[X, I_k] \rightarrow [Y, I_k]$.</u>

Part (1) follows from Proposition 6.3.
Part (2) follows from the fact that $X^{-1}Y = 1$ so
   $[X, I_k] \rightarrow [X, X^{-1}, Y] \rightarrow [I_k, Y] \rightarrow [Y, I_k]$

7. <u>Problems</u>

An argument similar to the one in the proof
of Proposition 6.3 shows that Baer's axiom [B]
in Theorem 1.1 can be replaced by the following
weaker axiom:

[B'] If X is a fully reduced word with positive length, then X is not equivalent to 1.

Problem 7.1: Replace axiom [B] in Theorem 1.1 by an even weaker axiom (or simply eliminate axiom [B]), and prove that the pree P is still embeddable.

Problem 7.2: Find a universal bound, if it exists, for the k in our definition of equivalent fully reduced words. In particular, determine if a bound on k exists for the tree pree P in Example 1.2. For example, k may depend on the diameter n of the tree graph of groups T_n. We do know (cf. [3, 4]) that k = 0 for n $\leq$ 4.

REFERENCES

1. R. Baer, Free sums of groups and their generalizations II, III, Amer. J. of Math. 72 (1950), 625-670.

2. A.H.M. Hoare, On generalizing Stallings' pregroup, preprint.

3. H. Kushner and S. Lipschutz, A generalization of Stallings' pregroup, Jour. Algebra 119 (1988), 170-184.

4. H. Kushner and S. Lipschutz, On embeddable prees, Jour. Algebra, to appear.

5. H. Kushner, On pree-stars and their universal groups, Ph.D. thesis, Temple University (1978).

6. F.S. Rimlinger, Pregroups and Bass-Serre Theory, AMS Memoirs Series 361 (1987).

7. O. Schreier, Die Untergruppen der freier Gruppen, Abh. Math. Sem. Univ. Hamburg 5 (1927), 161-183.

8. J.-P. Serre (with H. Bass), Trees, Springer-Verlag, New York,1980.

9. J.R. Stallings, Group theory and three-dimensional manifolds, Yale Monographs 4, Yale Univ. Press, 1971.

10. J.R. Stallings, Adian groups and pregroups, Essays in group theory, Math. Sci. Res. Inst. Publ. 8, Springer-Verlag New York (1987), 321-342.

Department of Mathematics, Temple University, Philadelphia, PA, 19122

Contemporary Mathematics
Volume **169**, 1994

ON BINARY σ-INVARIANT WORDS IN A GROUP

O. MACEDONSKA AND DONALD M. SOLITAR

To Wilhelm Magnus, who led the way for many to realms of depth and beauty.

ABSTRACT. We investigate the fixed-points of σ, an automorphism of a 2-generator group G which switches the generators. For the special case where G is a free 2-generator metabelian group or a free 2-generator 3-soluble group, we show that the σ-fixed-point subgroup $\mathfrak{S}(G)$ is the smallest possible, namely, $gp(cc^\sigma)$, where c ranges over a set of generators for G' in the metabelian case and c ranges over a set of generators for G'' in the 3-soluble case. As an application to groups with more than 2 generators, we compute the fixed points of an extended permutation automorphism of order 2 which maps each generator into the inverse of a generator, for a free metabelian group of finite or countably infinite rank, by using the notion of a σ-retract. The problem as to which automorphisms of a reduced free group G can be induced by regarding G as the reduced free group of a σ-invariant subgroup of $F = \langle x, y \rangle$ is briefly considered, especially in connection with σ-retracts.

1. INTRODUCTION

Definition 1. We denote the mapping $w(x,y) \to w(y,x)$ which switches x, y by σ (called **switching**).

We define $w^{-\sigma}$ by $(w^\sigma)^{-1} = (w^{-1})^\sigma$.

Let $G = \langle x, y | R_i \rangle$. A word $w(x,y)$ is σ-**invariant** in G if, as elements of G, $w = w^\sigma$, i.e., w is a fixed point of σ.

As is customary, if the identity element of G is the only fixed point of an automorphism α of G, we say that α **has no fixed points** in G. $\square$

Remark 1. If G is a 2-generator reduced free group, then $w(x,y)$ is σ-invariant iff $w(g,h) = w(h,g)$ for all g, $h \in G$. The latter condition is used in [5, p. 95] to define a **symmetric binary word**. $\square$

Definition 2. The elements of G determined by the σ-invariant words in G are called the σ-**invariant elements** of G and form a maximum σ-invariant subgroup

1991 *Mathematics Subject Classification.* Primary 20F14, 20E36; Secondary 20F18, 06C05, 12K05.

Key words and phrases. fixed points of automorphisms, σ-invariant words, binary symmetric words, σ-invariant presentations, metabelian groups, 3-soluble groups, 2-generator groups, σ-retracts, switchable automorphisms.

The first author wishes to thank the Department of Mathematics and Statistics of York University for its support, and was also supported by her University's Grant PB/2 P301 02905/93.

The second author was previously supported by NSERC Grant OGP0007909.

of G, which we denote by $\mathfrak{S}(G)$. More generally, if $H < G$, then $\mathfrak{S}(H)$ denotes the maximum σ-invariant subgroup of H. $\square$

Remark 2. If $w(x,y)$ is σ-invariant in G then so are $w(y,x)$ and $w^n(x,y)$; also, $w(x,y)$ is σ-invariant in G iff its freely reduced form is σ-invariant in G. $\square$

Example 1. If G is the free group on x,y, then $\mathfrak{S}(G) = 1$. If G is the free abelian group on x,y, then $\mathfrak{S}(G) = gp(xy)$. If G is free 2-nilpotent, then $\mathfrak{S}(G) = gp(x^2y^2[y,x]^2)$ (see [5]). For results for other two-generator free nilpotent groups, see [5, 2, 1]. $\square$

The subgroup of σ-invariant words in a group $G = \langle x,y|R_i \rangle$ depends on the presentation for G. Indeed, the infinite cyclic group $Z = \langle x \rangle$ can be presented by $\langle x,y\,|\,xy^{-1} \rangle$ and by $\langle x,y\,|\,xy \rangle$. In the first presentation every element of Z is σ-invariant since $x = y$. In the second presentation only the identity element is σ-invariant; for if $w(x,y)$ is σ-invariant and a,b is the exponent sum of w on x,y respectively, then since $y = x^{-1}$ we have $w(x,y) = x^{a-b}$ while $w(y,x) = x^{b-a}$, so that $a - b = 0$ and $w = 1$.

Definition 3. The presentation $G = \langle x,y|R_i \rangle$ is σ**-invariant** if the mapping $\sigma : x \mapsto y,\ y \mapsto x$ is an automorphism of G. $\square$

Remark 3. G has a σ-invariant presentation iff there are two elements u, v of G such that $G = gp(u, v)$ and an automorphism α of G such that $u^\alpha = v$, $v^\alpha = u$. This is equivalent to either of the following conditions, where $F = \langle x,y \rangle$ and σ is the switching automorphism of $F = \langle x,\ y \rangle$:

(1) there exists a normal subgroup N of F such that $N^\sigma = N$ and $G = F/N$;

(2) there exists a presentation $G = \langle x,y\,|\,R_i \rangle$ for which the set $\{R_i\}$ is σ-invariant .

More generally, if $G = \langle x,y\,|\,R_i \rangle$ is a σ-invariant presentation and $\{T_j(x,y)\}$ is a σ-invariant set of words in G, then

(3) $G = \langle x,y\,|\,R_i\,,T_j \rangle$ is a σ-invariant presentation.

Moreover, if $G = \langle x,y|R_i \rangle$ is a σ-invariant presentation for G and N is a σ-invariant normal subgroup of G with $N < H < G$, then

(4) $\mathfrak{S}(H)/N < \mathfrak{S}(H/N)$.

In particular,

(5) every 2-generator free group in a variety has a σ-invariant presentation.

Finally, let G have a σ-invariant presentation, and let A, B be σ-invariant subgroups of G.

(6) If $G = A * B$ or $A \times B$, where A, B are σ-invariant subgroups of G, then $\mathfrak{S}(G) = \mathfrak{S}(A) * \mathfrak{S}(B)$ or $\mathfrak{S}(A) \times \mathfrak{S}(B)$, respectively. More generally, let G be an **unique alternating product group with factors** $A_i\,, i \in I$, i.e., every element $g \in G$ has some expression as a product of non-identity elements, $g = g_1 \cdots g_n$, alternating from A_i, and for all such products $g = h_1 \cdots h_n$ with h_j in the same factor as g_j, we have $h_j = g_j$. If the subgroups A_i are σ-invariant , then $\mathfrak{S}(G) = gp(\mathfrak{S}(A), \mathfrak{S}(B))$. $\square$

Example 2. Every abelian 2-generator group G has a σ-invariant presentation.

Indeed, it is known that G has a presentation

$$G = \langle x, y \,|\, [x, y], x^r, y^s; s | r, 2 < r \le \infty \rangle.$$

We take $u = x$ and $v = xy$. Then $G = gp(u, v)$, and the map $x \to xy$, $y \to y^{-1}$ defines an automorphism interchanging u and v. The σ-invariant presentation is

$$G = \langle u, v \,|\, [u, v], u^r, v^r, u^s = v^s \rangle. \quad \square$$

We consider which free products of two cyclic groups have a σ-invariant presentation.

Theorem 1. *A free product $G = \langle x, y \,|\, x^r, y^s, \; 0 \le s \le r \rangle$ has a σ-invariant presentation iff $r = s$ or $s = 0$ or at least one of r and s is odd.*

Proof. If $r = s$, then the given presentation for G is σ-invariant.

If $s = 0$, i.e., $G = \langle x, y \,|\, x^r \rangle$, we take $u = yx$, $v = y$. Clearly $G = gp(u, v)$, and the map $x \to x^{-1}$, $y \to yx$ defines an automorphism interchanging u and v. The σ-invariant presentation is

$$G = \langle u, v \,|\, (v^{-1}u)^r \rangle.$$

If at least one of the exponents is odd, say $s = 2k + 1$, we take $u = xy$, $v = xy^{-1}$. We have $G = gp(u, v)$ because $v^{-1}u = y^2$ generates y, and y, u generate x. The required automorphism interchanging u and v is defined by the map $x \to x$, $y \to y^{-1}$. The σ-invariant presentation is

$$G = \langle u, v \,|\, (v^{-1}u)^{ks}, (u(v^{-1}u)^k)^r \rangle.$$

Finally, suppose $r \ne s$, $s \ne 0$, and both r and s are even. We consider G as a factor group F/N where $F = \langle x, y \rangle$ and $\bar{x}, \bar{y}$ are the N-cosets of x, y respectively.

Suppose that G has a σ-invariant presentation $G = gp(\bar{u}, \bar{v})$, where the map $\sigma : \bar{u} \to \bar{v}$, $\bar{v} \to \bar{u}$ defines an automorphism in G. Now $\bar{x}$ has the largest order of any element of G, and $\bar{y}$ is not conjugate in G to a power of $\bar{x}$. Since each element of order r in G is conjugate to $\bar{x}^\epsilon$, ϵ odd, and each element of G of order s which is not conjugate to a power of $\bar{x}$ is conjugate to an odd power of $\bar{y}$, it is not difficult to check that each automorphism in G maps $\bar{x}$ into a conjugate of an odd power of itself, and that $\bar{y}$ is mapped into that same conjugate of an odd power of itself. Hence, for some $\bar{w} \in G$ and ϵ and δ odd, we have $\sigma : \bar{x} \to (\bar{w})^{-1}(\bar{x})^\epsilon(\bar{w})$, $\bar{y} \to (\bar{w})^{-1}(\bar{y}^\delta)(\bar{w})$. Hence, for any word $\bar{z}(\bar{x}, \bar{y})$ in G, we have $\bar{z}^\sigma(\bar{x}, \bar{y}) = \bar{w}^{-1}\bar{z}(\bar{x}^\epsilon, \bar{y}^\delta)\bar{w}$. Thus if $\bar{u} = \bar{u}(\bar{x}, \bar{y}^\delta)$, then $\bar{v} = \bar{u}^\sigma = (\bar{w})^{-1}\bar{u}(\bar{x}^\epsilon, \bar{y}^\delta)\bar{w}$. To simplify computations, we map G homomorphically onto $H = \langle a, b \,|\, a^2, b^2 \rangle$ by mapping $\bar{x} \to a$, $\bar{y} \to b$. Then $\bar{u}, \bar{v}$ go into conjugate elements which must generate H. Now the distinct conjugate classes of H are $\{1\}$, $\{$conjugates of $a\}$, $\{$conjugates of $b\}$, $\{(ab)^k, (ba)^k, k > 0\}$. But if H is divided by the normal subgroup generated by any one of these classes, the identity subgroup does not result; therefore, two generators of H cannot lie in a single conjugate class. $\square$

2. Free 2-generator metabelian groups

Definition 4. Let G be a 2-generator group with a σ-invariant presentation. If H is a σ-invariant subgroup of G, then we define the **symmetrizer** of H, denoted by $\mathfrak{Z}(H)$, to be $gp(hh^\sigma \; ; \forall h \in H)$.

More generally, if α is an automorphism of G and H is an α-invariant subgroup of G, then we define the **α-symmetrizer** of H, denoted by $\mathfrak{Z}(\alpha, H)$, to be $gp(hh^\alpha \; ; \forall h \in H)$. $\square$

Remark 4. Clearly $\mathfrak{Z}(\alpha, H)$ is α-invariant.

Moreover, if G is a 2-generator group with an automorphism α of order 2 and $w \in G$, ww^α is α-invariant iff w commutes with w^α. If H is an α-invariant (e.g., verbal) abelian subgroup of G, then each element of $\mathfrak{Z}(\alpha, H)$ is α-invariant. In particular, if G has a σ-presentation, then $\mathfrak{Z}(H) < \mathfrak{S}(H)$. $\square$

Definition 5. Let G have a σ-presentation, and let $A < H$ where A, H are σ-invariant subgroups of G. We call A a **σ-retract** of H if A is a retract of H and the kernel of the retraction is σ-invariant. $\square$

Lemma 1. *Let G have a σ-presentation, and let A, B be σ-invariant subgroups of G. If $H = gp(A, B)$ and $A \cap B^H = 1$, then A is a σ-retract of H.*

Proof. Since $H = gp(A, B)$, we have that H is σ-invariant. Moreover, if $N = B^H$, then N is σ-invariant, $H = AN$, and for every element $h \in H$, $h = an$, where $a \in A$, $n \in N$; moreover, the elements a, n are unique. The mapping $\eta: h \mapsto a$ is a σ-retraction of H onto A with kernel N. $\square$

Lemma 2. *Let G have a σ-presentation, and let H be a σ-invariant subgroup of G for which*

(1) $$\mathfrak{S}(H) < \mathfrak{Z}(V(H))U(H),$$

where $V = V(H)$, $U = U(H)$ are verbal subgroups of H. If A is a σ-retract of H then

(2) $$\mathfrak{S}(A) < \mathfrak{Z}(V(A))U(A).$$

Proof. Let η be the σ-retraction of H onto A with kernel N. First of all, if $h \in \mathfrak{S}(H)$ and $an = h = h^\sigma = a^\sigma n^\sigma$, then $a^\sigma = a$. Therefore, $\mathfrak{S}(H)^\eta = \mathfrak{S}(A)$. Also, since $U = U(H)$ is a verbal subgroup of H, $U^\eta = U(A)$. If $v(h_1, \ldots, h_r) \in V$ then $v = v(a_1, \ldots, a_r) \cdot n = v^\eta \cdot n$ for some $n \in N$, so $v^\sigma = (v^\eta)^\sigma \cdot n^\sigma$, and $vv^\sigma = v^\eta (v^\eta)^\sigma \cdot m$ where $m \in N$. Hence, $(\mathfrak{Z}(V))^\eta = \mathfrak{Z}(V^\eta) = \mathfrak{Z}(V(A))$. Thus, applying η to the LHS and RHS of (1) results in (2). $\square$

It is important to further classify the generators c_{rs} of F'.

Definition 6. We say that a pair (r, s) is **non-trivial** if $r \neq 0 \neq s$. We say that a non-trivial pair (r, s) is **pure** if $r = s$, and **regular** if $r \leq s$. If $i = (r, s)$ then we call $i' = (s, r)$ the **reverse** of i. $\square$

Theorem 2. *The subgroup of σ-invariant words in a free metabelian group G on two generators x, y is the free abelian group, freely generated by the words $c_{rs} c_{rs}^\sigma$ where $c_{rs} = x^r y^s x^{-r} y^{-s} = [x^r, y^s]$ and (r, s) is impure and regular.*

Proof. It is known that the commutator subgroup F' in the free group F generated by x, y is freely generated by the elements $c_{rs} = [x^r, y^s] \neq 1$, $r, s \in Z$, and that $c_{rs} \neq 1$ iff $r \neq 0 \neq s$ (see [3, p. 229]). These c_{rs} are therefore free abelian generators for F'/F''.

If w is σ-invariant mod F'', then it is also σ-invariant mod $[F', F]$, and by [5] has the form

$$w = x^p y^p \prod c_{rs}^{a_{rs}} \text{ for distinct pairs } (r, s) \text{ and } a_{rs} \neq 0.$$

Suppose $p \neq 0$. Since

$$(3) \qquad\qquad c_{rs}^{\sigma} = [y^r, x^s] = c_{sr}^{-1},$$

the σ-invariance of w implies $x^p y^p \prod c_{rs}^{a_{rs}} \equiv y^p x^p \prod c_{sr}^{-a_{rs}} \mod F''$, which can be written as

$$(4) \qquad\qquad c_{pp} \prod c_{rs}^{a_{rs}} \prod c_{sr}^{a_{rs}} \equiv 1 \mod F''.$$

Now, c_{pp} is non-trivial, and if the integer ξ is its exponent sum in the LHS of equation (4), then $1 + 2\xi = 0$, which is impossible. So $p = 0$ and

$$w = \prod c_{rs}^{a_{rs}} \in F'.$$

Now let ξ_{rs} be the exponent sum of c_{rs} for a fixed r, s, in the LHS of equation (4). Then $\xi_{rs} = a_{rs} + a_{sr} = 0$, i.e., $a_{rs} = -a_{sr}$. Clearly, a_{rr} would have to be 0, and therefore, for all pairs (r, s) in equation (4), $r \neq s$. Since all c_{rs} commute modulo F'', we conclude that w is a product of elements $c_{rs} c_{sr}^{-1}$ where $r < s$, which are of the required form $c_{rs} c_{rs}^{\sigma}$. Moreover, these elements freely generate the abelian group $\mathfrak{S}(F/F'') < F'/F''$. $\square$

Corollary to Theorem 2.1. *Let* $G = gp(x, y)$ *be a free 2-generator metabelian group, and let* $Q = gp(c_{rs})$ *for all impure and regular* (r, s). *Then* $\mathfrak{S}(G) = \mathfrak{Z}(Q)$.

Proof. This result is immediate. $\square$

Corollary to Theorem 2.2. *If* F *is the free group on* x, y *and* $F^{(n)}$ *is the* n-*th term of its derived series for* $n \geq 2$, *then* $\mathfrak{S}(F/F^{(n)}) < F'/F^{(n)}$.

Proof. If $u \in F$ and $u^{\sigma} \equiv u \mod F^{(n)}$, then $u^{\sigma} \equiv u \mod F''$, and so $u \in F'$. $\square$

It will be useful to introduce an equivalence relation $\cong$ on the set of non-trivial pairs (r, s), and a linear order $\prec$ on these, in which each pair has only finitely many pairs before it.

Remark 5. A linear order such that each element has only finitely many elements before it, is a well-order of ω-type, which we shall refer to as an ω-*order*.

Moreover, given a relation $\prec$ (or any binary relation) between the elements of a set B, we can induce a relation between elements $b \in B$ and subsets $P \subseteq B$ by defining $b \prec P$ if $b \prec p$ for every $p \in P$; similarly, we can induce a relation between subsets P, Q of B by defining $P \prec Q$ if for every $p \in P$, $q \in Q$ we have $p \prec q$. $\square$

Definition 7. We define the relation $\cong$ to mean **is equal to or the reverse of**.

We define the relation $\prec$, called **precedes**, using lexicographic order on the quintuple

$$(\min(|r|,|s|),\ \max(|r|,|s|),\ r+s,\ |r-s|,\ r).$$

We use $\preceq$ to denote **precedes or equals** and $\precdot$ to denote **immediately precedes**. $\square$

Lemma 3. *The relation $\cong$ on non-trivial pairs is an equivalence relation with at most two elements in each equivalence class, and $i \cong i'$. The relation $\prec$ on non-trivial pairs is an ω-order. Moreover, if i is a regular pair, then $i = i'$ or $i \precdot i'$.*

Proof. Clearly, $\cong$ is an equivalence relation, and $j \cong i$ iff $j = i$ or $j = i'$. The equivalence class $\{i\}_\cong$ has one element if i is pure and two otherwise.

Since the first two integers of the quintuples whose lexicographic order determine $\prec$ are $|r|,|s|$ in some order, there are only finitely many pairs which precede a given one, and so $\prec$ is an ω-order.

Moreover, if $i = (r,s)$ is regular then $r \le s$. If $r = s$ then $i = i'$; otherwise $i \prec i'$. If $j = (p,q)$ and $i \prec j \prec i'$, then the quintuples for (r,s), (s,r), and (p,q) have the same first four terms. But coincidence of the first four terms implies $\{r,s\} = \{p,q\}$, and so $j = i$ or $j = i'$, and thus $i \precdot i'$. $\square$

Remark 6. An $\cong$-equivalence class has one element or one of its elements immediately precedes the other. Hence, if $j \prec i$ then $\{j\}_\cong = \{i\}_\cong$ or $\{j\}_\cong \prec \{i\}_\cong$. Thus $\prec$ induces an ω-order on the equivalence classes of $\cong$. $\square$

Corollary to Theorem 2.3. *If $w \in \mathfrak{S}(F/F'')$, then $w = \prod_j w_j$, where $w_j = c_j^{a_j} c_{j'}^{-a_j}$, $a_j \in Z$, $a_j \ne 0$, and j is impure and regular. Moreover, if the pairs j have increasing precedence then this representation is unique.*

Proof. Since $c_j^g = c_{j'}^{-1}$ by equation (3), and $\mathfrak{S}(F/F'')$ is a free abelian group freely generated by $c_{rs}c_{rs}^g$ where (r,s) is impure and regular by Theorem 2, the result follows easily. $\square$

3. Free 2-Generator 3-Soluble Groups

We shall obtain free generators for F'', using a Kurosh rewriting process, with the non-trivial free generator c_{rs} of type (r,s) for F', and an extended Schreier right coset representative system for F'/F'' (see [3, pp. 229–237]). The following notation is useful:

Definition 8. Let w be an element of F' which is a product of different syllable types. We call the largest pair m for which m occurs in w, the **maximum syllable type** or **mxsltp** of w, and call the regular pair of m, m' the **maximum regular syllable type** or **mxrgsltp** of w; if m is impure and both m, m' occur in w, we say that w is **full**. If every syllable type occurring in w is impure, and for each j occurring in w also j' occurs in w, we say the w is **completely full**. $\square$

Remark 7. Clearly,

$$(5) \qquad\qquad (c_i^a)^\sigma = c_{i'}^{-a} .$$

Hence, if $w \in F'$ is a product of different syllable types, m is the mxrgsltp of w, and β is any representative type then we have the following:
m is the mxrgsltp of ${}^\beta w$ and $({}^\beta w)^\sigma$; either m or m' is the mxsltp of ${}^\beta w$ and hence occurs in ${}^\beta w$; w is full iff $({}^\beta w)^\sigma$ is full and, in that case, m' is the mxsltp for both ${}^\beta w$, $({}^\beta w)^\sigma$; if m alone occurs in w then m, m' is the mxsltp for ${}^\beta w$, $({}^\beta w)^\sigma$ respectively; if m' alone occurs in w then m', m is the mxsltp for ${}^\beta w$, $({}^\beta w)^\sigma$ respectively.

If w is a representative mod F'', then w is completely full iff ${}^\beta w$ is completely full iff $({}^\beta w)^\sigma$ is completely full.

It follows from Corollary to Theorem 2.3 that if w is σ-invariant mod F'', then ${}^\beta w$ is completely full. □

Lemma 4. *Let the free generator c_i be of type i in F', where i is not necessarily regular. Then the following representatives form an extended Schreier system for F'/F'':*
for $w \in F'/F''$, the neutral representative of w is

$$(6) \qquad\qquad {}^*w = \prod_{j \in J} w_j ,$$

and $w_j = c_j^{a_j}$, where the pairs j are in ascending order;
moreover, the i-th representative of w is

$$(7) \qquad\qquad {}^i w = \prod_{\substack{j \prec i \\ j \in J}} w_j \cdot c_{i'}^{a_{i'}} \cdot \prod_{\substack{i' \prec h \\ h \in J}} w_h \cdot c_i^{a_i}$$

where the pairs j, h are in ascending order, and $a_{i'} = 0$ if i is pure or $c_{i'}$ does not occur in w, and $a_i = 0$ if c_i does not occur in w.
For the i-type representatives, we have

$$(8) \qquad\qquad {}^i(wc_i) \approx ({}^i w)c_i .$$

Finally, ${}^i w \approx {}^* w$ iff either i does not occur in w or i is the mxsltp of w.

Proof. That the given representatives form an extended Schreier system with the given generator types, and that equation (8) holds for the i-th representatives, follows immediately from equations (6) and (7), and the definition of an extended Schreier system (see [3, p. 234]). □

Remark 8. Extended Schreier representatives must be freely reduced, so that two representatives can be freely equal iff they are equal. Let β, γ be any representative types (including neutral type $*$). Clearly, ${}^\beta w = {}^\beta({}^\gamma w)$. Moreover, since F'' is σ-invariant, ${}^\beta(w^\sigma) = {}^\beta(({}^\gamma w)^\sigma)$.
If w consists of a single syllable then ${}^\beta w = {}^* w$. □

Lemma 5. *The free group F'' is freely generated by the non-identity t-symbols $t_{i_w} = {}^i w \cdot ({}^* w)^{-1}$ which, if ${}^i w$ is as in equation (7), are described by:*

(9)
$$t_{i_w} = \prod_{j \prec i} w_j \cdot c_{i'}^{a_{i'}} \cdot \prod_{i' \prec h} w_h \cdot c_i^{a_i} \cdot$$
$$\left(\prod_{j \prec i} w_j \cdot c_k^{a_k} c_{k'}^{a_{k'}} \cdot \prod_{i' \prec h} w_h \right)^{-1}, \text{ where } k \text{ is the smallest of } i, i'.$$

Suppose now that w is the product of different types of syllables.
Then $t_{i_w} = 1$ iff ${}^i w = {}^* w$ iff either i does not occur in w or i is the mxsltp of w.
If n is the mxsltp of w, then

(10)
$$t_{n_w} = 1 = t_{n(w^\sigma)}.$$

If m is the mxrgsltp of w, then

(11)
$$t_{m_w} = 1 \text{ iff } w \text{ or } w^\sigma \text{ is not full } \text{ iff } t_{m(w^\sigma)} = 1.$$

If w is a single syllable then

(12)
$$t_{i_w} = 1 = t_{i(w^\sigma)} \text{ for every } i.$$

There is a natural mapping between $t_{i_w} \neq 1$ and the couples $(i, {}^* w)$ with ${}^i w \neq {}^* w$, and this mapping is bijective.

Proof. By [3, Theorem 4.7 and Corollary 4.8], F'' is freely generated by its t-symbols described in equation (9); its s-symbols define the identity because of equation (8).

The conditions under which $t_{i_w} = 1$ are equivalent to those under which $t_{i_w} \approx 1$, by Remark 8, and are given at the end of Lemma 4.

We suppose now that w is a product of different type syllables.

Let n be the mxsltp of w, so that n occurs in w. Clearly, $t_{n_w} = 1$. If n is pure, then n is the mxsltp of w^σ and $t_{n(w^\sigma)} = 1$. On the other hand, if n is impure but n' does not occur in w, then n does not occur in w^σ, and again, $t_{n(w^\sigma)} = 1$. If w is full, then n is the mxsltp of w^σ, and again, $t_{n(w^\sigma)} = 1$.

Let m be the mxrgsltp of w; hence, m is also the mxrgsltp of w^σ. Clearly w is full iff w^σ is full. Suppose w is not full. If m is pure then m is the mxsltp of both w and w^σ, and by equation (10) with $n = m$, we have equation (11). If m is impure then m' does not occur in w, and m does not occur in w^σ. Hence, $t_{m(w^\sigma)} = 1$; but m is the mxsltp of w, and so $t_{m_w} = 1$. Suppose next that w or w^σ is full. Then m is impure, and both m and m' occur. Since m', and not m, is the mxsltp of both w and w^σ, $t_{m_w} \neq 1 \neq t_{n(m^\sigma)}$.

If w is a single syllable then by equation (5) so is w^σ. Hence, by the final statement in Remark 8, equation (12) holds.

Finally, if $t_{i_w} \neq 1$ then ${}^i w \neq {}^* ({}^i w) = {}^* w$. In particular, $w \neq 1$ so i is the syllable type of the last syllable of ${}^i w$. Thus the couple $(i, {}^* w)$ with ${}^i w \neq {}^* w$ is uniquely determined by $t_{i_w} \neq 1$. On the other hand such a couple uniquely determines $t_{i_w} \neq 1$. Thus, the mapping is bijective. $\square$

Definition 9. An **initial segment** of a word is an initial word segment made up of whole syllables. $\square$

Remark 9. Since F' is the free product of the infinite cyclic groups $gp(c_{rs})$ where (r, s) is non-trivial, every word $w \in F'$ has a reduced syllable length $\lambda(w)$ in these factors. Suppose now that

$$(13) \qquad w = \prod_{j \in J} c_j^{a_j}, \ c_j^{a_j} \neq 1 \text{ and } J \text{ has distinct pairs.}$$

Clearly if J has p elements then

$$p = \lambda(w) = \lambda(^i w) = \lambda(^* w) \text{ for any non-trivial pair } i.$$

If u is a proper initial segment of w, then $\lambda(u) < p$.
   Moreover,

$$(14) \ w^\sigma = \prod_{j \in J} c_{j'}^{-a_j}, \ c_{j'}^{-a_j} \neq 1, \text{ with } j' \text{ running over distinct non-trivial pairs,}$$

so that $\lambda(w^\sigma) = \lambda(w)$.
   The element $t_{i_w} = {}^i w(^* w)^{-1}$ of F'', if non-identity, determines $^i w$ and therefore $\lambda(^i w)$. The fact that $\lambda(w) = \lambda(w^\sigma)$ allows us to use the following definition to find $\mathfrak{S}(F/F''')$. $\square$

Definition 10. We define $\mathcal{L}_k$, $\mathcal{E}_k$ to be the subgroups of F'' generated by all t-symbols t_{j_w} with $\lambda(^j w) < k$, $\lambda(^j w) = k$ respectively. $\square$

Remark 10. Clearly, $\mathcal{L}_{k+1} = gp(\mathcal{L}_k, \mathcal{E}_k)$. The normal subgroup of $\mathcal{L}_{k+1}$ generated by $\mathcal{L}_k$ is just $\mathcal{L}_k^{\mathcal{E}_k} = gp(\mathcal{L}_k^u)$, where u ranges over the elements of $\mathcal{E}_k$, and is contained in $\mathcal{L}_k F'''$. $\square$

Remark 11. If $k \leq 2$ then $\mathcal{L}_k = 1$; also $\mathcal{E}_1 = 1$. For, if $^i w$ has one syllable then its t-symbol is 1 by equation (12). $\square$

Lemma 6. *If $\{t_{w_i}\}_{i \in I}$, $\{t_{v_j}\}_{j \in J}$ are sets of non-identity t-symbols of F'', if a_i, b_j are non-zero integers, and if $k \leq \lambda(w_i)$, $k \leq \lambda(v_j)$, then*

$$(15) \qquad \prod_{i \in I} t_{w_i}^{a_i} \equiv \prod_{j \in J} t_{v_j}^{b_j} \bmod \mathcal{L}_k F'''$$

iff $\{t_{w_i}\}_{i \in I} = \{t_{v_j}\}_{j \in J}$ and $(a_i)_{i \in I} = (b_j)_{j \in J}$ in the corresponding order.

Proof. We first note that since F''/F''' is the free abelian group freely generated by the non-identity t-symbols, $F'''/\mathcal{L}_k F'''$ is the free abelian group freely generated by the non-identity t-symbols t_{h_u} with $\lambda(^h u) \geq k$. Hence, $\{t_{w_i}\}_{i \in I} = \{t_{v_j}\}_{j \in J}$ and $(a_i)_{i \in I} = (b_j)_{j \in J}$ in the corresponding order. $\square$

Remark 12. We may compute $\tau(w)$ using the Kuros rewriting process arising from the given extended Schreier system for F' mod F''. Since the s-symbols are 1, a word $w \in F''$ can be expressed in terms of t-symbols, replacing each symbol x in w as follows:
if $w = LxR^{-1}$ with L, R words in c_{rs} with non-trivial (r, s), and $x = c_i^\epsilon$, $c_i \neq 1$, $\epsilon = \pm 1$, then

$$(16) \qquad t_{i_L}^{-1} \cdot t_{i_{Lx}} = t_{i_L}^{-1} \cdot t_{i_R}.$$

Hence, if $w = DBE^{-1}$ where B is a block of consecutive symbols of type i in w then the intermediate t-symbols from B cancel in $\tau(w)$, and one can repace B with

$$(17) \qquad t_{i_D}^{-1} \cdot t_{i_{(DB)}} = t_{i_D}^{-1} \cdot t_{i_{(E)}} = t_{i_{(EB^{-1})}}^{-1} \cdot t_{i_{(E)}}.$$

In particular, whole syllables can be replaced in computing $\tau(w)$. $\square$

Lemma 7. *The subgroups $\mathcal{L}_k$ and $\mathcal{L}_k^{\mathcal{E}_k}$ are σ-invariant.*

Proof. Suppose $\lambda(^i w) < k$. Then

$$(18) \qquad (t_{i_w})^\sigma = (^i w (^* w)^{-1})^\sigma = (^i w)^\sigma (^* w)^{-\sigma}.$$

In expressing the RHS of equation (18) in terms of t-symbols by computing τ of it, using equation (17), all the D and DB entering from syllables in $(^i w)^\sigma$ have syllable length $< k$, while all the EB^{-1} and E entering from syllables in $((^* w)^\sigma)^{-1}$ also have syllable length $< k$. Hence, $\mathcal{L}_k$ is σ-invariant.

Now $N = \mathcal{L}_k^{\mathcal{E}_k}$ is the normal subgroup of $\mathcal{L}_{k+1}$ generated by $\mathcal{L}_k$, both of which are σ-invariant; hence, N is also σ-invariant. $\square$

Remark 13. We shall now compute the σ-image of t-symbols $t_{j_w} \neq 1$ for a fixed word $w \in F'$, where w is a representative mod F''. (In fact, the results we obtain will hold, with slight modification, even for $t_{j_w} = 1$. We state these results, for completeness, as corollaries.) $\square$

Lemma 8. *Let $w \in F'$ be the product of different type syllables; let m be the mxrgsltp of w and $k = \lambda(^* w)$. Then*

$$(19) \qquad t_{m_w}^\sigma \equiv t_{m_{(w^\sigma)}}^{-1} \mod \mathcal{L}_k^{\mathcal{E}_k}.$$

Proof. In the following, we use Remark 7 and Lemma 5.

We may compute $t_{m_w}^\sigma$ using equation (9) with $i = m$, the mxrgsltp of w. Since $t_{m_w} \neq 1$, by equation (11) w is full, and so $^* w$ is full. In $^m w$ the last two syllables correspond to m', m; in $^* w$ the last two syllables correspond to m, m'; in $(^m w)^\sigma$ the last two syllables correspond to m, m'; in $(^* w)^\sigma$ the last two syllables correspond to m', m. Now in rewriting $(^m w)^\sigma \cdot ((^* w)^\sigma)^{-1}$, the symbols occurring will be those arising from the syllables of $(^m w)^\sigma$ and the inverse of those arising from the syllables of $(^* w)^\sigma$. Since both of these words have syllable length k, all the t-symbols arising will belong to $\mathcal{L}_k$ except possibly for $t_{m'((^m w)^\sigma)}$ and $t_{m((^* w)^\sigma)}^{-1} = t_{m_{(w^\sigma)}}^{-1}$, which come from the last syllable of each word. But m' is the mxsltp of $(^m w)^\sigma$ so the former t-symbol is 1.

Thus we have established that

$$(20) \qquad t_{m_w}^\sigma = P t_{m_{(w^\sigma)}}^{-1} Q^{-1} \text{ where } P, Q \in \mathcal{L}_k.$$

Clearly,

$$(21) \qquad P t_{m_{(w^\sigma)}}^{-1} Q^{-1} = t_{m_{(w^\sigma)}}^{-1} \cdot t_{m_{(w^\sigma)}} P t_{m_{(w^\sigma)}}^{-1} Q^{-1} \equiv t_{m_{(w^\sigma)}}^{-1} \mod \mathcal{L}_k^{\mathcal{E}_k}. \square$$

Corollary to Lemma 8.1. *Let $w \in F'$ be the product of different type syllables; let m, n be the mxrgsltp, mxsltp of w respectively; let $k = \lambda(^*w)$ and $t_{m\,w} = 1$. Then*

(22)
$$t^\sigma_{m\,w} = t^{-1}_{m(w^\sigma)},$$
$$t^\sigma_{n\,w} = t^{-1}_{n(w^\sigma)}.$$

Proof. These equations follow immediately from equations (11) and (10). $\square$

Corollary to Lemma 8.2. *Let $w \in F'$ be a representative mod F''; let m be the mxrgsltp of w, $k = \lambda(^*w)$, and suppose that $t_{m\,w} \neq 1$. Then*

$$t^\sigma_{m\,w} = P t^{-1}_{m(w^\sigma)} Q^{-1} \text{ where } P, Q \in \mathcal{L}_k.$$

Proof. This is just equation (20). $\square$

Lemma 9. *Let $w \in F'$ be the product of different type syllables; let m be the mxrgsltp of w, $i \prec m$, $k = \lambda(^*w)$, and $t_{iw} \neq 1$. Then*

(23)
$$t^\sigma_{i\,w} \equiv t_{i'(w^\sigma)} t^{-1}_{m(w^\sigma)} \text{ mod } \mathcal{L}_k{}^{\mathcal{E}_k}.$$

Proof. The proof is similar to that of Lemma 8.

Since $t_{iw} \neq 1$, i occurs in w, i' occurs in w^σ, and both i, i' precede m. The last syllable of ^iw has type i, and the last syllable of w^σ has type i'.

We first assume that i is regular. Now

(24)
$$(^i w)^\sigma = (\prod_{j \prec i} w_i \cdot c^e_{i'} \cdot \prod_{i' \prec j \prec m} w_j \cdot c^a_m c^b_{m'} c^d_i)^\sigma$$
$$= \prod_{i \prec j} w^\sigma_j \cdot c^{-e}_i \cdot \prod_{i' \prec j \prec m} w^\sigma_j \cdot c^{-a}_{m'} c^{-b}_m c^{-d}_{i'},$$

and

(25)
$$(^* w)^\sigma = (\prod_{j \prec i} w_j \cdot c^d_i c^e_{i'} \cdot \prod_{i' \prec j \prec m} w_j \cdot c^a_m c^b_{m'})^\sigma$$
$$= \prod_{j \prec i} w^\sigma_j \cdot c^{-d}_{i'} c^{-e}_i \cdot \prod_{i' \prec j \prec m} w^\sigma_j \cdot c^{-a}_{m'} c^{-b}_m,$$

where $a = 0$ or not, and $b = 0$ or not, depending on which of m, m' occurs in w, and $d \neq 0$ but $e = 0$ or not, depending on whether i' occurs in w or not. Now all the t-symbols that arise in computing $\tau((^iw)^\sigma (^*w)^{-\sigma})$ belong to $\mathcal{L}_k$, except possibly for the last t-symbol $t_0 = t_{i'(w^\sigma)}$ from $(^iw)^\sigma$ and the last t-symbol t_2 from $(^*w)^\sigma$. If m occurs in w^σ, i.e., $b \neq 0$, then $t2 = t^{-1}_{m(w^\sigma)}$, and equation (23) holds. If m does not occur in w^σ, then $a \neq 0$, and $t2 = t^{-1}_{m'(w^\sigma)} = 1 = t^{-1}_{m(w^\sigma)}$ since m' is the mxsltp of w^σ, while m does not occur in w^σ. Again, equation (23) holds.

If i is not regular then i' is, and i' occurs in w^σ, and since m is the mxrgsltp for w^σ, the hypothesis of the first part of our proof is satisfied, with i' replacing i and w^σ replacing w. Hence

(26)
$$(t_{i'(w^\sigma)})^\sigma \equiv t_{i\,w} t^{-1}_{m\,w} \text{ mod } \mathcal{L}_k{}^{\mathcal{E}_k}.$$

Applying σ to both sides of equation (26) and using equation (19), we obtain

$$(27) \qquad t_{i'(w^\sigma)} \equiv (t_{i_w})^\sigma (t^\sigma_{m_w})^{-1} \equiv (t_{i_w})^\sigma t_{m(w^\sigma)} \mod \mathcal{L}_k^{\varepsilon_k}.$$

Solving this last equation using its first and third terms yields equation (23). $\square$

Corollary to Lemma 9.1. *Let $w \in F'$ be a representative mod F''; let m be the mxrgsltp of w, $i \prec m$, $k = \lambda(^*w)$, and $t_{i_w} = 1$. Then*

$$(28) \qquad (t_{i_w})^\sigma = t_{i'(w^\sigma)}.$$

Proof. Since $i \prec m$, i is not the mxsltp of w. Hence, $t_{i_w} = 1$ implies i does not occur in w. But then i' does not occur in w^σ, and $t_{i'(w^\sigma)} = 1$. $\square$

Definition 11. Let $w \in F'$ be a representative $\neq 1$, mod F'', let $k = \lambda(w)$, and let m be the mxrgsltp of w. Then we define the *t*-symbols **connected to** w to be the set w^κ which is the union of the sets

$$(29) \qquad w^\kappa(j) = \{t_{j_w}, t_{j'(w^\sigma)}\},$$

$$(30) \qquad w^\kappa(m) = \{t_{m_w}, t_{m(w^\sigma)}\},$$

for all j, m occurring in w, with $j \prec m$ and m different from the mxsltp of w.

We say that the *t*-symbols in w^κ are **connected by** w and use $\sim$ to denote this relation among non-identity *t*-symbols.

We say that the *t*-symbols in $w^\kappa(i)$ are **strongly connected by** w where $i = j$ or $i = m$, and use $\simeq$ to denote this relation among non-identity *t*-symbols.

We define the mapping χ, called the **strong connection**, on the set of all non-identity *t*-symbols by

$$(31) \qquad \chi : t_{j_w} \mapsto t_{j'(w^\sigma)}, \quad t_{j'(w^\sigma)} \mapsto t_{j_w}, \quad \text{if } j \prec m$$

and

$$(32) \qquad \chi : t_{m_w} \mapsto t_{m(w^\sigma)}, \quad t_{m(w^\sigma)} \mapsto t_{m_w}.$$

We call $gp(w^\kappa)$ the **subgroup connected by** w.

We define $\mathfrak{I}(w)$, called the **includer of the *t*-symbols connected by** w, to be the subgroup $\mathcal{L}_{k+1}$. We define $\mathfrak{X}(w)$, called the **excluder of the *t*-symbols connected by** w, to be the normal subgroup of $\mathcal{L}_{k+1}$ generated by the *t*-symbols not in w^κ. $\square$

Remark 14. It is not difficult to show that $w^\kappa = (^*w)^\kappa = (w^\sigma)^\kappa$; moreover, $(^jw)^\kappa = (^iu)^\kappa$ iff $\{^*w, ^*(w^\sigma)\} = \{^*u, ^*(u^\sigma)\}$. For, if $\lambda(w) = 1$ then $w^\kappa = \emptyset$, while if $\lambda(w) \geq 2$ then $w^\kappa \neq \emptyset$, and any element of w^κ determines $\{^*w, ^*(w^\sigma)\}$, and hence, w^κ. Specifically, let $u = {}^*w$ or $u = {}^*(w^\sigma)$; if $j \prec m$ occurs in u, then $\{t_{j_u}, t_{j'(u^\sigma)}\} < w^\kappa$; if m occurs in ${}^*w$ and ${}^*(w^\sigma)$, then $\{t_{m_u}, t_{m(u^\sigma)}\} < w^\kappa$.

Clearly every $t_{i_w} \neq 1$ is connected to itself by ${}^*w$. Thus:

   (1) The relation $\sim$ is an equivalence relation among the non-identity *t*-symbols of F''.
   (2) The relation $\simeq$ is an equivalence relation among the non-identity *t*-symbols of F''.

(3) If $k = \lambda(w)$, then $w^\kappa < \mathcal{E}_k$ while $\mathcal{L}_k^{\mathcal{E}_k} < \mathfrak{X}(w)$.

(4) Equations (19), (23), (22), and (28) imply that $gp(w^\kappa)$ is σ-invariant mod $\mathcal{L}_k{}^{\mathcal{E}_k}$, where $k = \lambda(w)$. Hence, $gp(w^\kappa, \mathcal{L}_k{}^{\mathcal{E}_k})$ and $gp(u^\kappa, \mathcal{L}_k{}^{\mathcal{E}_k})$ are σ-invariant where $\lambda(u) = k$, $u^\kappa \neq w^\kappa$). Hence $\mathfrak{X}(w^\kappa)$ is σ-invariant.

(5) The t-symbols connected by w are $\neq 1$ mod F'''. Indeed, if j occurs in w but is not its mxsltp, $t_{jw} \neq 1$. Hence, if m is not the mxsltp of w, it follows from equation (22) that $t_{m(w^\sigma)} \neq 1$. Moreover, if $j \prec m$, it follows from equation (28) that $t_{j'(w^\sigma)} \neq 1$.

(6) If m is pure then m is the mxsltp of w, and so $w^\kappa(m)$ is not a subset of w^κ.

(7) If $j \preceq m$ and j occurs in w then, $t_{jw} = t_{j(w^\sigma)}$ iff $^j w$ is σ-invariant mod F'' iff $^j w$ is as in Corollary to Theorem 2.3.

In each of the following, we assume that j and m occur in w.

(8) It follows immediately from (7) that $w^\kappa(m)$ consists of a single element iff w is σ-invariant mod F''.

(9) If $j \prec m$ then $w^\kappa(j)$ has two elements. For if j is impure, this is immediate from equation (29). On the other hand, if j is pure and has exponent a in w, then $j' = j$ has exponent $-a$ in w^σ, and so again $w^\kappa(j)$ consists of two elements.

(10) The mapping χ is bijective and maps each w^κ onto itself. Moreover, χ has order two on w^κ unless w is σ-invariant mod F'' and $\lambda(w) = 2$, in which case w^κ has a single element. This follows immediately from (8) and (9).

(11) If $j \prec m$,

$$(33) \qquad\qquad t_{jw} \simeq t_{j(w^\sigma)}$$

iff j is pure or w is σ-invariant mod F''. Indeed, if j is pure or w is σ-invariant mod F'', the equivalence holds. On the other hand, if $j \neq j'$ and w is not σ-invariant mod F'', then it is immediate that $t_{j(w^\sigma)} \neq t_{j'(w^\sigma)}$, while from (7) $t_{jw} \neq t_{j(w^\sigma)}$. $\square$

Lemma 10. *If $w \in F'$ is a representative mod F'' and $k = \lambda(w)$, then w^κ is a retract of $\mathfrak{I}(w)$ with kernel $\mathfrak{X}(w)$. Moreover, $\mathfrak{I}(w)F'''/\mathfrak{X}(w)F'''$ is the abelian group freely generated by w^κ.*

Proof. Let η be the endomorphism of $\mathfrak{I}(w)$ which maps the t-symbols of w^κ into themselves and maps the other t-symbols of $\mathfrak{I}(w)$ into 1. Clearly w^κ is a retract of $\mathfrak{I}(w)$, and $\mathfrak{X}(w)$ is the kernel of η.

Since w^κ is a free factor of $\mathfrak{I}(w)$ which is itself a free factor of F'', we have $\mathfrak{I}(w)F'''/\mathfrak{X}(w)F'''$ is the abelian group freely generated by w^κ. $\square$

Lemma 11. *If $t_{jw} \neq 1$ then t_{jw} is not σ-invariant mod $\mathcal{L}_k F'''$, where $k = \lambda(^j w)$.*

Proof. We may assume $w = {}^* w$. Since $t_{jw} \neq 1$, if m is the mxrgsltp of w then $j \preceq m$. Also, we have $k = \lambda(^j w) = \lambda(w) = \lambda(^m w) = \lambda(^m(w^\sigma)) = \lambda(^{j'}(w^\sigma))$. Moreover, both F''' and $\mathcal{L}_k^{\mathcal{E}_k}$ are subgroups of $\mathcal{L}_k F'''$. Hence, the results in equations (19) and (23) together with Lemma 6 hold mod $\mathcal{L}_k F'''$. This implies that t_{mw} is not fixed by σ; if $j \prec m$ then (9) of Remark 14 implies that t_{jw} is not fixed by σ. $\square$

Corollary to Lemma 11.1. *If $t_{jw} \neq 1$ then t_{jw} is not σ-invariant mod F'''.*

Proof. Since F''' is a subgroup of $\mathcal{L}_k F'''$, where $k = \lambda(^j w)$, this result follows immediately from Lemma 11. $\square$

Definition 12. We shall use $\mathcal{Z}$ to denote $\mathfrak{Z}(F'')$, the symmetrizer of F'' (see Definition 4). $\square$

We already know from Corollary to Theorem 2.2 that $\mathfrak{S}(F/F''') < F'/F'''$. We shall now proceed to establish that $\mathfrak{S}(F/F''')$ is contained in F''/F''' and, in fact, is precisely $\mathcal{Z} F'''/F'''$.

Lemma 12. *If w is σ-invariant mod F''' then $w \in F''$.*

Proof. It follows from Corollary to Theorem 2.3 that if w is σ-invariant mod F''', then $w \in F'$ and w is the product of elements of the form $c_j^a c_{j'}^{-a}$, $a \neq 0$, mod F''. Clearly then any representative of w mod F'' has the same form; in particular, if m is the mxrgsltp of $^* w$ then $^m w$ has that form and is completely full. Therefore the last syllable type in $(^m w)^\sigma$ is m'. We wish to show that $^m w = 1$, and so $w \in F''$. Suppose $^m w \neq 1$.

Let $k = \lambda(^m w)$. Now for some $v \in F''$, $w = v \cdot {}^m w$, and so

$$v \cdot {}^m w \equiv w \equiv w^\sigma \equiv (v \cdot {}^m w)^\sigma \equiv v^\sigma \cdot {}^m w^\sigma \mod F'''$$

and $^m w \cdot (^m w)^{-\sigma} \equiv v^{-1} v^\sigma \mod F'''$. Since $^m w$ is completely full, so is $^m(w^\sigma)$.

We now compute in $\mathfrak{I}(w)F''' \mod \mathfrak{X}(w)F'''$, which modulus includes both F''' and $\mathcal{L}_k^{\mathcal{E}_k}$. In computing $\tau(^m w \cdot (^m w)^{-\sigma}) \mod \mathcal{L}_k^{\mathcal{E}_k}$, we have, from the last syllables in $^m w$ and $(^m w)^\sigma$,

$$(34) \qquad {}^m w(^m w)^{-\sigma} \equiv t_{mw} t_{m'((^m w)^\sigma)}^{-1} = t_{mw} \mod \mathfrak{X}(w).$$

Since $w \equiv w^\sigma \mod F''' < F''$, for any representative type j, we have that $t_{jw} = t_{j(w^\sigma)}$. Hence, in computing $v^{-1} v^\sigma \mod \mathfrak{X}(w)F'''$, we may assume that

$$(35) \qquad v \equiv \prod_{\substack{j \in J \\ j \prec m}} t_{jw}^{a_j} t_{j'w}^{a_{j'}} \cdot t_{mw}^{a_m} \mod \mathfrak{X}(w)F''',$$

where J is the set of impure regular syllable types occurring in $^m w$.

To compute $v^\sigma \mod \mathfrak{X}(w)F'''$, we may use equations (19) and (23). We obtain

$$(36) \qquad v^\sigma \equiv \prod_{\substack{j \in J \\ j \prec m}} t_{j'w}^{a_j} t_{jw}^{a_{j'}} \cdot t_{mw}^{b_m} \mod \mathfrak{X}(w)F''',$$

where

$$b_m = -a_m - \sum_{\substack{j \in J \\ j \prec m}} (a_j + a_{j'}).$$

Hence,

$$(37) \qquad v^{-1} v^\sigma \equiv \prod_{\substack{j \in J \\ j \prec m}} t_{jw}^{a_{j'} - a_j} t_{j'w}^{a_j - a_{j'}} \cdot t_{mw}^{b_m - a_m} \mod \mathfrak{X}(w)F'''.$$

By combining equations (34) and (37), we obtain

(38) $$t_{mw} \equiv \prod_{\substack{j \in J \\ j \prec m}} t_{jw}^{a_{j'} - a_j} t_{j'w}^{a_j - a_{j'}} \cdot t_{mw}^{b_m - a_m} \mod \mathfrak{X}(w)F'''.$$

From Lemma 10, w^κ freely generates the abelian group $\mathfrak{I}(w)F'''/\mathfrak{X}(w)F'''$. Hence $a_{j'} = a_j$ and $1 = b_m - a_m = -2a$, where

$$a = a_m + \sum_{\substack{j \in J \\ j \prec m}} a_j ,$$

an obvious contradiction. □

Lemma 13. *Let* $w = zv$, *where* $z \in \mathcal{L}_k^{\mathcal{E}_k}$, $v \in \mathcal{E}_k$. *If* $w^\sigma \equiv w \mod \mathcal{L}_k^{\mathcal{E}_k}$ *then* $v^\sigma \equiv v \mod \mathcal{L}_k^{\mathcal{E}_k}$.

Proof. $w^\sigma = z^\sigma v^\sigma \equiv w = zu \mod \mathcal{L}_k^{\mathcal{E}_k}$. It follows from equations (19) and (23) that $v^\sigma = ru$, where $r \in \mathcal{L}_k^{\mathcal{E}_k}$ and $u \in \mathcal{E}_k$. Therefore, $z \cdot v \equiv z^\sigma r \cdot u \mod \mathcal{L}_k^{\mathcal{E}_k}$, and by Lemma 6, $v = u$. Thus, $v^\sigma \equiv u = v \mod \mathcal{L}_k^{\mathcal{E}_k}$. □

Lemma 14. *Let* $w \in F''$ *and*

(39) $$w = \prod_{q \in Q} t_{\rho_q}^{a_q},$$

where $t_{\rho_q}^{a_q} \neq 1$, ρ_q *are different representatives for* F' *mod* F'' *with* $\lambda(\rho_q) = 2$ *and* $a_q \in Z$. *If* w *is* σ-*invariant mod* F''' *then* $w \in \mathfrak{I}(\mathcal{E}_2)$.

Proof. Since we are working mod F''', we may assume all elements of F'' commute and hence that connected, and also, strongly connected t-symbols are contiguous in w. Since $\mathcal{L}_2 = 1$, it follows from (4) of Remark 14, that the subgroup connected by ρ_q is σ-invariant. Moreover, mod F''' for distinct $(\rho_q)^\kappa$, these subgroups have trivial intersection. Hence, if w is σ-invariant mod F''', then the same is true for each segment of w that is in a connected subgroup. Consider such a segment v in the subgroup connected by ρ_q, where we can assume $\rho_q = n^e j^d$, where n, m is the mxsltp, mxrgsltp of ρ_q respectively, and $j \preceq m \preceq n$. We consider now three cases. If $n = m$ or $j \prec m$ then the t-symbols connected by ρ_q are

  (1) $g = t_{j(n^e j^d)}$ and $h = t_{j'(n' - e j'^{-d})}$.

Let $n \neq m = j$ so that $n' = m$, $\rho_q = n^e m^d$.
If $e \neq -d$ then the t-symbols connected by ρ_q are

  (2) $g = t_{m(n^e m^d)}$ and $h = t_{m(m - e n - d)}$.

If $e = -d$ then the t-symbol connected by ρ_q is

  (3) $g = t_{m(n - {}^d m^d)}$.

The segment v of t-symbols connected by ρ_q is $v = g^a h^b$ in cases (1) and (2), and g^a in case (3), where $a \neq 0$.

  In case (1), m does not occur in ρ_q or is its mxsltp, so that equation (23) reduces to $g^\sigma = h$. Thus, $v^\sigma = h^a g^b = v = g^a h^b \mod F'''$, and so $a = b$. But in this case, $v = (gh)^a = (gg^\sigma)^a \in \mathcal{Z}$.

In case (2), equation (19) reduces to $g^\sigma = h^{-1}$. In this case, $v^\sigma = h^{-a}g^{-b} \equiv v = g^a h^b \bmod F'''$, and so $a = -b$. Now, $v = (gh^{-1})^a = (gg^\sigma)^a \in Z$.

In the final case (3), $v = g^a$ with $v^\sigma = (g^\sigma)^a = g^{-a}$, and so $a = 0$. This contradiction shows that the case (3) cannot occur for v which is σ-invariant mod F'''. $\square$

Lemma 15. *Let $w \in F''$ and*

$$(40) \qquad\qquad w = \prod_{q \in Q} t^{a_q}_{\rho_q},$$

where $t^{a_q}_{\rho_q} \neq 1$, $\rho_q \in F'$ are different representatives mod F'', with $\lambda(\rho_q) \leq k$ and a_q an integer. If w is σ-invariant mod F''' then

$$(41) \qquad\qquad w \equiv us \bmod F''',$$

where $s \in 3(\mathcal{E}_k)$, $u \in \mathcal{L}_k$, and u is σ-invariant mod F'''.

Proof. Since we are working mod F''', $\mathcal{L}_k = \mathcal{L}_k^{\mathcal{E}_k}$, and all elements of F'' commute. Hence, $w = z v_1 \cdots v_r$, where $z \in \mathcal{L}_k$, $v_d \neq 1$ is a maximal segment of w connected by ρ_d with $\lambda(\rho_d) = k$, and strongly connected t-symbols in v_d are contiguous. It follows from equations (19), (23), (31) and (32) that $v_d^\sigma \in gp(\rho_d^\kappa) \bmod \mathcal{L}_k$. Hence, from Lemma 13, $(v_1 \cdots v_r)^\sigma \equiv v_1^\sigma \cdots v_r^\sigma \bmod \mathcal{L}_k$. But $\{gp(\rho_d^\kappa)\}$ generate their direct product mod $\mathcal{L}_k$. Thus, $v_d^\sigma \equiv v_d \bmod \mathcal{L}_k$.

Suppose v_r is connected by ρ_r, which has mxrgsltp m. Let there be ℓ_r strongly connected classes of t symbols $t_{j\rho_q}$ in v_r, $j \prec m$. We use a double induction on r, ℓ_r to establish equation (40).

If $r = 0$ then equation (40) is immediate. Let $r > 0$, $k = 0$.

If ρ_r is not σ-invariant, then

$$(42) \qquad\qquad v_r = t^a_{m\,\rho_r} t^b_{m\,(\rho_r^\sigma)}$$

with $a \neq 0$. Hence, from equation (19),

$$(43) \qquad\qquad v_r^\sigma \equiv t^{-a}_{m\,(\rho_r^\sigma)} t^{-b}_{m\,\rho_r} \bmod \mathcal{L}_k.$$

Thus, from Lemma 6, $b = -a$,

$$(44) \qquad\qquad v_r \equiv s_r = (t_{m\,\rho_r}(t_{m\,\rho_r})^\sigma)^a \bmod \mathcal{L}_k,$$

and $v_r = h s_r$, $h \in \mathcal{L}_k$, $s_r \in 3(\mathcal{E}_k)$.

The case ρ_r is σ-invariant leads to a contradiction. For then

$$(45) \qquad\qquad v_r = t^a_{m\,\rho_r}$$

with $a \neq 0$, and so

$$(46) \qquad\qquad v_r^\sigma \equiv t^{-a}_{m\,\rho_r} \bmod \mathcal{L}_k.$$

But then from Lemma 6, $a = -a$, and this contradicts $a \neq 0$.

Thus $w \equiv z v_1 \cdots v_{r-1} v_r \equiv h z v_1 \cdots v_{r-1} s_r$ is σ-invariant mod F''', and since s_r is σ-invariant mod F''', so is $h z v_1 \cdots v_{r-1}$. We may use our inductive hypothesis since $r - 1 < r$, and obtain equation (40).

Assume then that $r > 0$, $k > 0$. In this case we may assume that $j \prec m$ and

$$(47) \qquad v_r = u_r \cdot t^a_{j\rho_r} t^b_{j'(\rho^\sigma_r)}$$

with $a \neq 0$. Hence, from equations (19) and (23),

$$(48) \qquad v^\sigma_r \equiv (u_r)^\sigma t^{e_1}_{m\rho_r} t^{e_2}_{m(\rho^\sigma_r)} \cdot t^a_{j'(\rho^\sigma_r)} t^b_{j\rho_r} \mod \mathcal{L}_k.$$

From Lemma 6 it follows that $b = a$. But from equation (23), we have

$$(49) \qquad t_{j'(\rho^\sigma_r)} = (t_{j\rho_r})^\sigma t_{m(\rho^\sigma_r)} h,$$

where $h \in \mathcal{L}_k$. Hence,

$$(50) \qquad v_r = u_r h^a t^a_{m(\rho^\sigma_r)} s_r,$$

where $s_r = t_{m\rho_r} (t_{m\rho_r})^\sigma \in \mathfrak{Z}(\mathcal{E}_k)$.

Thus $w \equiv z v_1 \cdots v_{r-1} \cdot v_r \equiv h^a z v_1 \cdots v_{r-1} \cdot u_r t^a_{m(\rho^\sigma_r)} s_r$ is σ-invariant mod F''', and since s_r is σ-invariant mod F''', so is $h^a z v_1 \cdots v_{r-1} \cdot u_r t^a_{m(\rho^\sigma_r)}$. We may use our inductive hypothesis since $\ell_r - 1 < \ell_r$, and obtain equation (40). $\square$

We may now establish our principal result about the σ-invariant elements of the free 2-generator 3-soluble group.

Theorem 3. $\mathfrak{S}(F/F''') = \mathfrak{Z}(F''/F''')$.

Proof. We use induction on k to show that $\mathfrak{S}(\mathcal{L}_k F'''/F''') < \mathfrak{Z}(\mathcal{L}_k F'''/F''')$.

For $k \leq 2$, $\mathcal{L}_k = 1$ and the result is immediate. For $k = 3$, the result is a restatement of Lemma 14 since $\mathcal{L}_3 = \mathcal{E}_2$.

The result for $k+1$ is immediate from Lemma 15, using the inductive hypothesis for k. $\square$

Corollary to Theorem 3.1. *Let Q be a subcollection of commutators c_{pq} such that $A = gp(Q)$ is σ-invariant. If $B = gp(c_{rs}; c_{rs} \notin Q)$, then B is σ-invariant, $F' = gp(A, B)$, and*

$$(51) \qquad \mathfrak{S}(AF'''/F''') = \mathfrak{Z}(A'F'''/F''').$$

Proof. From Lemma 1, A is a σ-retract of F', and so Lemma 2 implies equation (51). $\square$

4. Induced Automorphisms of Order 2

Remark 15. Our results, so far, were stated in terms of the fixed points of the switching automorphism of certain 2-generator groups. However, we can apply them to some automorphisms of some groups with two or more generators.

For example, we can immediately obtain the fixed points of any automorphism α of $F = \langle x, y \rangle$, F/F'', or F/F''' which switches any two free generators of its reduced free group. Clearly, α must be of order 2 and have no fixed points if the reduced free group is F. $\square$

Example 3. If $F = \langle x, y \rangle$, then the automorphism $\alpha \colon x \mapsto xy$, $y \mapsto y^{-1}$ is the automorphism of F which switches the two free generators x, xy. Hence, the maximum α-invariant subgroup of F/F'' is $\mathfrak{Z}(\alpha, F'/F'')$, and the maximum α-invariant subgroup of F/F''' is $\mathfrak{Z}(\alpha, F''/F''')$. $\square$

Remark 16. To apply our results to groups with more than two generators, we note that σ induces an automorphism of order 2 on each σ-invariant subgroup H of F. Moreover, since $w \in F$ is σ-invariant iff $w = 1$, such induced automorphisms are fixed point free. However, if V is a verbal (or even characteristic) subgroup of H, then σ will also induce an automorphism of H/V, and this may have σ-invariant elements, e.g., if $H = F'$, $V = F'''$. Thus, in some cases, our previous results can be applied to finding fixed points of automorphisms α of reduced free groups with more than 2-generators, and to 2-generator reduced free groups where α is not just a renaming of σ. $\square$

Definition 13. Let G be a reduced free group, where $G = K/V$, K is an absolutely free n-generator group, n finite or countably infinite, V is a verbal subgroup of K, and let α be an automorphism of G of order 2. We say that α is **strongly switchable via** H if K is a σ-retract of a σ-invariant subgroup H of $F = \langle x, y \rangle$ and α is induced by σ; if K is merely a subgroup of H and α is induced by σ then we say α is **weakly switchable via** H. $\square$

Remark 17. If α is strongly switchable via H and η is the retraction of H onto K then $\mathfrak{S}(K) = (\mathfrak{S}(H))^\eta$.

It should be noted that if K were a retract of H, not σ-invariant, then $\mathfrak{S}(K) = \mathfrak{S}(H) \cap K$, but this may be smaller than $(\mathfrak{S}(H))^\eta$. An example is provided by choosing the automorphism $\alpha: x \mapsto xy$, $y \mapsto y^{-1}$ in Example 3 which switches x, xy and setting $H = gp(x^3, y, (xy)^3)$. These three generators are Nielsen reduced and are free generators for H (see [3, (ii) p.122, and Lemma 3.1]). Clearly, H is α-invariant, $K = gp(x^3, y)$ is a retract of H, and $[x^3, y]$, $[(xy)^3, y] \in K'$. Let $G = K/K''$, η be the retraction of H onto K, and $u = [x^3, y][y, x^3]$, $v = [x^3, y][y, (xy)^3] \in K$. Now $v \in \mathfrak{S}(H)$ but $u = v^\eta \notin \mathfrak{S}(K)$. $\square$

Example 4. We illustrate the use of a σ-retract by considering G, a free metabelian group, with at most a countably infinite number of free metabelian generators $\Gamma = \{g_i\}_{i \in I}$, $I \subseteq Z^+$, and let α be an automorphism of order 2, where $\alpha: g_i \mapsto g_{n_i}^{-1}$. Then the α-invariant elements of G are precisely $gp(qq^\alpha; q \in G')$. Indeed, since α has order 2, the sets $\{i, j^i\}$ form a partition of I. We embed Γ into F using $\iota: g_i \mapsto c_{in_i} = [x^i, y^{n_i}]$. Clearly $\iota: g_{n_i} \mapsto c_{n_i i}$. Then $Q = gp(\iota(\Gamma))$ is a σ-invariant retract of $H = F'$, and α is induced by σ. Thus α is strongly switchable via F'. Since $\mathfrak{S}(F/F''') = \mathfrak{S}(F'/(F')'') = 3((F')'/F''')$, we can apply Lemma 2 with $G = F$, $H = F'$, $A = Q$, $V(H) = H'$, $U(H) = H''$ to obtain $\mathfrak{S}(Q)Q'' < smQ'Q''$. Clearly $smQ'Q'' < \mathfrak{S}(Q)Q''$. Hence, $\mathfrak{S}(Q/Q'') = smQ'/Q''$. But G is naturally isomorphic to Q/Q'', with α being induced by σ. Therefore, we have our result. $\square$

The problem of determining which automorphisms of reduced free groups are strongly or weakly switchable is open. It can be shown that, for finitely generated free abelian groups, any automorphism of order 2 with no fixed points is strongly switchable, but for countably generated free abelian groups, the answer is not certain, but would seem to be in the negative.

REFERENCES

1. S.A. Krstič, *On symmetric words in nilpotent groups*, Publications, Institute of Mathematics, Beograd (New Series) **27** (1980), no. 41, 139–142.
2. O. Macedonska, *On symmetric words in nilpotent groups*, Fundamenta Mathematicae **120** (1984), 119–125.
3. W. Magnus, A. Karrass, and D. Solitar, *Combinatorial group theory*, 2nd revised ed., Dover, New York, 1970.
4. E. Plonka, *On symmetric words in free nilpotent groups*, Bull. Acad. Polon. Sci. **18** (1970), 427–429.
5. _____, *Symmetric words in nilpotent groups of class $\leq$ 3*, Fundamenta Mathematicae **97** (1977), 95–103.

SILESIAN TECHNICAL UNIVERSITY, GLIWICE, POLAND AND DEPARTMENT OF MATHEMATICS AND STATISTICS, YORK UNIVERSITY, NORTH YORK, ONTARIO, CANADA M3J 1P3

DEPARTMENT OF MATHEMATICS AND STATISTICS, YORK UNIVERSITY, NORTH YORK, ONTARIO, CANADA M3J 1P3

E-mail address: DSOLITAR@NEXUS.YORKU.CA

Contemporary Mathematics
Volume **169**, 1994

Explicit Matrices for Fuchsian Groups

BERNARD MASKIT

ABSTRACT. We give explicit matrix generators for Fuchsian groups representing all closed Riemann surfaces of genus 2, where the groups are all normalized and all act on the upper half plane. These matrix generators, which in general do not have unit determinant, depend polynomially on six parameters. The parameter space is explicitly defined by very simple inequalities. We also give a generalization to higher genus.

0. Introduction. In this note we write down explicit matrix generators for Fuchsian groups representing closed Riemann surfaces of genus 2, where the entries in these matrices are polynomials in certain parameters. The space of parameters $\mathcal{P}_2$ is explicitly described by simple inequalities. In general, these matrices will not have unit determinant, but will lie in $PGL^+(2, \mathbb{R})$ (we use the superscript "+" to denote a class of matrices of positive determinant). Modulo conjugation in $PGL^+(2, \mathbb{R})$, we explicitly write down all such sets of generators which are "canonical" (i.e., they satisfy the usual one defining relation, and they satisfy an orientation condition). This particular parametrization makes strong use of the fact that if one cuts a surface of genus 2 along three disjoint non-dividing geodesics, then the resulting two pairs of pants are isometric.

We also describe a related and slightly different set of generators for genus 2, as well as related explicit parametrizations for all co-compact purely hyperbolic Fuchsian groups of given genus. Other sets of parameters for certain Teichmüller spaces are well known; see for example Helling [H], Keen [K], and the references given there, and Kra-Maskit [K-M]. For general information about the real-analytic structure of Teichmüller spaces, the reader is referred to Abikoff [A].

We choose some purely hyperbolic Fuchsian group of the first kind F_g representing a closed Riemann surface of genus $g \geq 2$. It is well known that if $\phi : F_g \to$

1991 *Mathematics Subject Classification*. Primary 30F35; Secondary 20H10, 32G15.

The author was supported in part by NSF Grant #DMS 9003361; the author also wishes to thank l'Institut des Hautes Etudes Scientifiques for their hospitality and support.

This paper is in final form and no version of it will be submitted for publication elsewhere.

$PSL(2, \mathbb{R})$ is a faithful representation with discrete image, then there is a home-omorpism of the extended real line inducing ϕ. We call ϕ *orientation preserving* if this homeomorphism of the extended real line is orientation-preserving. The discrete faithful representation space $\mathcal{DF}(F_g, PSL(2, \mathbb{R}))$ is the space of all faithful orientation-preserving representations of F_g into $PSL(2, \mathbb{R}) = PGL^+(2, \mathbb{R})$ with discrete image, modulo conjugation in $PSL(2, \mathbb{R})$. It is well known that $\mathcal{DF}(F_g, PSL(2, \mathbb{R}))$ (or, equivalently, $\mathcal{DF}(F_g, PGL^+(2, \mathbb{R}))$ is real-analytically equivalent to $\mathcal{T}_g$, the Teichmüller space of closed Riemann surfaces of genus g; we also include a proof of this fact for $g = 2$.

Our Fuchsian groups will all be normalized, that is, certain elements will have fixed points at 0, 1, and ∞. With this normalization, the assignment of these generators to a point in our parameter space $\mathcal{P}_2$ defines a real-analytic diffeomorphism between $\mathcal{P}_2$ and $\mathcal{DF}(F_2, PGL^+(2, \mathbb{R}))$. We remark that our parameters are all fixed points of elements of the group; we show that these parameters yield a stratification, in the sense of [K-M], of the corresponding space of quasifuchsian groups; it follows that there is a real-analytic diffeomorphism between $\mathcal{T}_2$ and $\mathcal{P}_2$.

For any $g \geq 2$, we also describe a related procedure for writing down an explicit parameter space $\mathcal{P}_g$, and matrices, $A_1, B_1, \ldots, A_g, B_g$, where these matrices all have positive determinant; for $g \geq 3$, the entries in the matrices are algebraic rather than polynomial functions of the parameters; and the mapping $\Phi_g : \mathcal{P}_g \to \mathcal{DF}(F_g, PGL^+(2, \mathbb{R}))$ is a real-analytic diffeomorphism.

A similar representation for $\mathcal{T}_{1,1}$, the space of tori with one hole and/or puncture, was obtained in [M3]. For any other deformation space of positive dimension, of a group without torsion, one can use these techniques to generate a parameter space, defined by explicit inequalities, and a set of canonical generators for Fuchsian groups of the given signature, where the entries in these generators are real-analytic functions of the parameters, and so obtain a corresponding canonically defined embedding of this (reduced) Teichmüller space as a domain in Euclidean space.

We remark that our result for genus 2 has the following consequence. There are matrices A_1, B_1, A_2, B_2, each with positive determinant and integer entries, so that these are a canonical set of generators for a Fuchsian group of genus 2. This is related to a result of Magnus [Ma], who showed that there exists a genus 2 co-compact Fuchsian subgroup of $PSL(2, \mathbb{Q})$. Later, Takeuchi [T] showed that, for every genus $g \geq 2$, the set of genus g cocompact Fuchsian subgroups of $PSL(2, \mathbb{Q})$ is dense in the corresponding Teichmüller space; see also [W-M], and the references listed there. We also obtain a weaker form of Takeuchi's theorem for genus 2.

THEOREM. *Let $g \geq 2$ be given. Then there is a group $G_g \in GL^+(2, \mathbb{Z})$ so that the projection of G_g into $PGL^+(2, \mathbb{R}) \cong PSL(2, \mathbb{R})$ is a purely hyperbolic Fuchsian group representing a closed Riemann surface of genus g. Further, for $g = 2$, the set of Riemann surfaces, marked with a set of generators for the* *fundamental group, that can be so represented, is dense in the Teichmüller space.*

1. The deformation space. The usual definition of the Teichmüller space starts with a given Fuchsian group F_0, and is defined as the space of normalized quasiconformal deformations of F_0. For groups representing closed surfaces, this is equivalent to $\mathcal{DF}(F_g, PGL^+(2, \mathbb{R}))$. In this case, one can also give a non-variational definition. For genus 2, let $\mathcal{M}_2$ be the set of quadruples of matrices $(A_1, B_1, A_2, B_2) \in PGL^+(2, \mathbb{R})^4$, where $F = \langle A_1, B_1, A_2, B_2 \rangle$ is discrete, and has exactly the one defining relation:

$$A_1 B_1 A_1^{-1} B_1^{-1} A_2 B_2 A_2^{-1} B_2^{-1} = 1.$$

Then $\tilde{\mathcal{M}}_2$, which is $\mathcal{M}_2$ modulo conjugation in $PGL(2, \mathbb{R})$, is essentially the same space as $\mathcal{DF}(F_2, PGL^+(2, \mathbb{R}))$.

It is well known that if A_1, B_1, A_2, B_2 are as above, then the axes of A_i and B_i intersect. Let $\mathcal{M}_2^+$ be the subset of $\mathcal{M}_2$ where these axes intersect positively; that is, if we let (x_i, y_i) be the repelling, respectively, attracting, fixed points of A_i, and we conjugate (in $PGL^+(2, \mathbb{R})$) so that B_i has its repelling fixed point at 0, and its attracting fixed point at ∞, then $x_i < 0$ and $y_i > 0$. Then we can likewise define $\tilde{\mathcal{M}}_2$ to be $\mathcal{M}_2^+$ modulo conjugation in $PGL^+(2, \mathbb{R})$.

2. Basic computations. We start with some basic facts about hyperbolic isometries. Throughout this paper, all references to geometric objects, such as lines, reflections, etc. refer to the hyperbolic plane, $\mathbb{H}^2$.

2.1. Reflections. The canonical isomorphism between the group of orientation preserving isometries of $\mathbb{H}^2$ and $PSL(2, \mathbb{R})$ is well known, as is its extension to an isomorphism between the group of all isometries of $\mathbb{H}^2$ and $PGL(2, \mathbb{R})$.

If one has four distinct real points, $x_{i1}, x_{i2}, x_{j1}, x_{j2}$, then there is a unique involution $R_{ij} \in PGL(2, \mathbb{R})$, where R_{ij} interchanges x_{i1} with x_{i2}, and interchanges x_{j1} with x_{j2}. If the hyperbolic line connecting x_{i1} and x_{i2} intersects the hyperbolic line connecting x_{j1} with x_{j2}, then R_{ij} preserves orientation; otherwise it reverses orientation.

In any case, write $p_k = x_{k1} x_{k2}$, and $s_k = x_{k1} + x_{k2}$. Then we can write, provided none of our parameters is ∞,

$$R_{ij} = \begin{pmatrix} p_i - p_j & s_i p_j - p_i s_j \\ s_i - s_j & p_j - p_i \end{pmatrix}. \tag{2.1}$$

In the special case that $x_{j1} = 0$ and $x_{j2} = \infty$, we obtain

$$R_{ij} = \begin{pmatrix} 0 & p_i \\ 1 & 0 \end{pmatrix}. \tag{2.2}$$

2.2. Roots of transformations. If A is a hyperbolic transformation with fixed points at (x, y), then we can write A in the form:

$$\frac{A(z) - x}{A(z) - y} = \lambda \frac{z - x}{z - y};$$

where either $0 < \lambda < 1$, or $\lambda > 1$. Then, using the real k-th root of λ, the (hyperbolic) k-th root of A can be defined by:

$$\frac{\sqrt[k]{A}(z) - x}{\sqrt[k]{A}(z) - y} = \sqrt[k]{\lambda}\,\frac{z - x}{z - y}.$$

3. **Pants groups.** A *pants group* is a Fuchsian group Υ where $\mathbb{H}^2/\Upsilon$ is a sphere with three (hyperbolic) boundary components. The geodesic lengths of these boundary components are the *sizes* of the generators of Υ.

In what follows, we will write the fixed points of the hyperbolic transformation A as (x, y); this will always mean that A has its repelling fixed point at x, and its attracting fixed point at y.

3.1. A free pants group. Our first goal is to write down a single pants group depending on six parameters, which are the fixed points of the generators. Let $x_{11} < x_{12} < x_{21} < x_{22} < x_{31} < x_{32}$ be given. We permit some x_{i1} to be ∞ only if the corresponding $x_{i2} = 0$, and we do not permit any $x_{i2} = \infty$ (here we consider $\hat{\mathbb{R}} = \mathbb{R} \cup \{\infty\}$ as being cyclically ordered; i.e., $-\infty = \infty$). let R_{ij} denote the reflection that interchanges both x_{i1} with x_{i2} and x_{j1} with x_{j2}. Let $A_1 = R_{12} \circ R_{31}$, $A_2 = R_{23} \circ R_{12}$, $A_3 = R_{31} \circ R_{23}$. Then A_i has its fixed points at (x_{i1}, x_{i2}), and the group $\Upsilon = \langle A_1, A_2, A_3 \rangle$ is a pants group where these three generators satisfy the one relation $A_3 \circ A_2 \circ A_1 = 1$.

An easy proof of the above facts can be obtained as follows. Let L_i be the hyperbolic line connecting x_{i1} to x_{i2}, and let M_{ij} be the common orthogonal to L_i and L_j; then M_{ij} is the fixed point set of R_{ij}. by Poincaré's polygon theorem (see [M1] for a proof), the region bounded by these three lines is a fundamental domain for the discrete group $\tilde{\Upsilon}$ generated by these three reflections, and the only relations in $\tilde{\Upsilon}$ are that the three generators are involutions. The desired results now easily follow.

We will have some number of variations on the above in what follows. In each variation, we will have some of the fixed points as parameters, these will be labelled as x_{ij}, and some of the fixed points as computed variables, these will be labelled as z_{ij}.

3.2. A pants group with one size determined. Our next goal is to write down a pants group, where the size of one hole is determined; specifically, we assume that we are given λ, $0 < \lambda < 1$, and we write down three generators for a pants group, in terms of parameters, where one of the generators is $A_3(z) = \lambda z$.

Assume we are given three points $x_{11} < x_{12} < x_{21} < 0$, subject to the additional inequality

$$\lambda x_{11} x_{12} < x_{21}^2. \tag{3.1}$$

We also use the parameters $(x_{31}, x_{32}) = (\infty, 0)$. Let L_1 be the line joining x_{11} to x_{12}, and let L_3 be the line joining 0 to ∞. Let M_{13} be the common orthogonal to L_1 and L_3. Let $M_{23} = \sqrt{A_3}(M_{13})$. We next construct the line L_2 to have one endpoint at the given point x_{21}, and to have M_{23} as its common orthogonal with L_3. Easy computations show that the other endpoint of L_2 is at $z_{22} = \lambda x_{11} x_{12} / x_{21}$; inequality (3.1) guarantees that $x_{21} < z_{22} < 0$. We then proceed exactly as in 3.1.

3.2.1. Computational note. In the normalized case, as above, where A_3 has its fixed points at $(\infty, 0)$, the endpoint of M_{13} is at the geometric mean of x_{11} and x_{12}; then M_{23} has its endpoint at $\sqrt{\lambda x_{11} x_{12}}$; so R_{23}, the reflection in M_{23} can be written as

$$R_{23} = \begin{pmatrix} 0 & \lambda x_{11} x_{12} \\ 1 & 0 \end{pmatrix};$$

it follows that the attractive fixed point of A_2 is at $z_{22} = \lambda x_{11} x_{12} / x_{21}$.

In the unnormalized case, the analogue of inequality (3.1), and the computation of z_{22} are somewhat more complicated. In particular, the fixed points in inequality (3.1) do not appear as a cross-ratio, so its generalization is not obvious.

We want generators A_1, with fixed points at (x_{11}, x_{12}), A_2, with fixed points at (x_{21}, z_{22}), where z_{22} is still to be determined, and A_3, which is given. We assume also that we know the fixed points of A_3 at (x_{31}, x_{32}). We are also given that the points $x_{31}, x_{11}, x_{12}, x_{21}, x_{32}$ all lie in this order on the circle at infinity. Since we know the endpoints of the axes of A_1 and A_3, we can write down the reflection R_{13} in M_{13}, their common orthogonal. Then $M_{23} = \sqrt{A_3}(M_{13})$ must be the common orthogonal between the axes of A_2 and A_3; hence R_{23}, the reflection in M_{23}, can be written as $R_{23} = \sqrt{A_3} \circ R_{13} \circ (\sqrt{A_3})^{-1}$. We then compute $z_{22} = R_{23}(x_{21})$. The unnormalized form of inequality (3.1) then states that z_{22} lies between x_{21} and x_{32}.

We still need to find a matrix representing $\sqrt{A_3}$; this computation occurs in the next section.

3.2.2. Remarks on multipliers. For various of our computations, we will need to construct a root of a hyperbolic transformation, where we are given the fixed points. This is easily accomplished using the following. Since we know the fixed points, we need to be able to determine the multiplier; the following proposition gives a formula for the multiplier.

PROPOSITION. *Let the hyperbolic transformation A be given as the matrix*

$$A = \begin{pmatrix} \alpha & \beta \\ \gamma & \delta \end{pmatrix},$$

where the fixed points of A are known to be (x, y). Then the multiplier $\lambda = \lambda(A)$ satisfies

$$\lambda = \frac{\alpha - x\gamma}{\alpha - y\gamma}.$$

PROOF. Write the transformation A as

$$\frac{A(z) - x}{A(z) - y} = \lambda \frac{z - x}{z - y},$$

and set $z = \infty$.

3.3. A pants group with equal sized legs and free waist. We start with the parameters $0 < x_{11} < x_{12} < 1$. We adjoin the fixed points $z_{21} = 1/x_{12}$ and $z_{22} = 1/x_{11}$, set $x_{31} = \infty$ and $x_{32} = 0$, and then proceed as in 3.1. Observe that reflection in the unit circle interchanges L_1 with L_2, while keeping L_3 invariant. Hence this orientation reversing transformation interchanges A_1 with A_2 while conjugating A_3 onto its inverse. It follows that A_1 and A_2 have the same trace.

3.3.1. Computational note. We remark that 1 is not a fixed point of any element of $\Upsilon = \langle A_1, A_2, A_3 \rangle$. However, if one uses the commutator $[A, B] = A \circ B - B \circ A$, which is the element of order 2 whose axis in $\mathbb{H}^3$ is orthogonal to the axes of both A and B (see Fenchel [F] or Jørgensen [J]), then 1 is a fixed point of $[[A_1, A_2], A_3]$. We can specify further that it is the fixed point on the same side of the axis of A_3 as the fixed points of A_1 and A_2.

3.4. A pants group with equal sized legs and given waist. Here we assume we are given the transformation $A_3(z) = \lambda(z)$, $0 < \lambda < 1$, and we assume we are given two parameters $x_{11} < x_{12} < 0$, satisfying the additional inequality

$$\lambda < (x_{12}/x_{11})^2. \tag{3.2}$$

Let L_1 be the line with endpoints at x_{11} and x_{12}, and let L_3 be the line with endpoints at 0 and ∞. Let M_{13} be the common orthogonal between L_1 and L_3; let $M = \sqrt[4]{A_3}(M_{13})$; denote reflection in M by R, and let $z_{21} = R(x_{12})$, $z_{22} = R(x_{11})$. We can compute these as follows. Write $p_1 = x_{11}x_{12}$. Then M_{13} has its endpoints at $\pm\sqrt{p_1}$; it follows that

$$R = \begin{pmatrix} 0 & p_1\sqrt{\lambda} \\ 1 & 0 \end{pmatrix};$$

from which it follows that $z_{21} = \sqrt{\lambda}x_{11}$ and $z_{22} = \sqrt{\lambda}x_{12}$. Inequality (3.2), together with our assumption that $x_{11} < x_{12} < 0$, guarantees that these points are in the proper order; i.e., $x_{11} < x_{12} < z_{21} < z_{22} < 0$. Also, as above, we are assured that the reflection in M interchanges the axes L_1 and L_2, while keeping the third axis L_3 invariant. Note also that the (hyperbolic) distance along L_3 between the points of intersection with L_1 and L_2 is exactly $\frac{1}{2}|\log \lambda|$; hence the third generator is the given transformation A_3.

3.4.1. Computational note. Inequality (3.2) as written refers to the normalized situation. However, we can rewrite it as a cross-ratio:

$$\sqrt{\lambda(A_3)} < (x_{12}, x_{11}; x_{32}, x_{31} = \frac{(x_{12} - x_{32})}{(x_{12} - x_{31})} \cdot \frac{x_{11} - x_{31}}{(x_{11} - x_{32})}, \tag{3.2'}$$

where the fixed points of A_3 are at (x_{31}, x_{32}).

Once we are given A_3, together with its fixed points, and we are given the parameters x_{11} and x_{12}, then we can compute $\lambda(A_3)$ and $\sqrt[4]{A_3}$ using 3.2.2.

We still need to compute z_{21} and z_{22} in the unnormalized case. In this case, M_{13} is the common orthogonal between L_1 and L_3, whose endpoints we know; hence we can write down the reflection R_{13} in the common orthogonal M_{13}. Then $R = \sqrt[4]{A_3} \circ R_{13} \circ (\sqrt[4]{A_3})^{-1}$ maps M_{13} onto M_{23}. Hence $z_{21} = R(x_{12})$, and $z_{22} = R(x_{11})$.

4. Boundary half-spaces. Let $\Upsilon = \langle A_1, A_2, A_3 \rangle$ be as constructed in 3.1; let L_i be the axis of A_i; let M_{ij} be the common orthogonal to L_i and L_j; and denote reflection in M_{ij} by R_{ij}. Let $\tilde{\Upsilon} = \langle R_{12}, R_{23}, R_{31} \rangle$ be the extended group generated by these three reflections. Let $\tilde{D}$ be the region bounded by the three lines M_{12}, M_{23}, and M_{31}. Finally let H_i be the half-plane bounded by L_i, where H_i is disjoint from the other two axes; H_i is called a *boundary half-space*.

One can use Poincaré's polygon theorem to conclude that $\tilde{\Upsilon}$ is discrete; that the only relations in $\tilde{\Upsilon}$ are that the three generators are involutions; and that $\tilde{D}$ is a fundamental polygon for $\tilde{\Upsilon}$. Since the sides of $\tilde{D}$ are orthogonal to whichever of the axes L_i they meet, one easily concludes that H_i is precisely invariant under the subgroup of $\tilde{\Upsilon}$ generated by the two R_{ij}, $j \neq i$. It then follows that H_i is precisely invariant under $\langle A_i \rangle$ in Υ.

5. Special case of genus 2. In this section, we use the special fact about surfaces of genus 2 that three disjoint non-dividing geodesics necessarily divide such a surface into two isometric pairs of pants. Using this fact, we write down a parameter space $\mathcal{P}_2 \in \mathbb{R}^6$, and we write down four matrices $A_1, B_1, A_2, B_2 \in PGL^+(2, \mathbb{R})$, depending on the point in $\mathcal{P}_2$, where these matrices generate a purely hyperbolic discrete group representing a closed surface of genus 2. This process is reversible; that is, starting with canonical generators (i.e., they satisfy the usual defining relation and orientation condition), A_1, B_1, A_2, B_2, generating a purely hyperbolic Fuchsian group representing a closed Riemann surface of genus 2, the corresponding point in $\mathcal{P}_2$ is canonically determined.

We start with the parameter space

$$\mathcal{P}_2 = \{x_{21}, x_{32}, x_{41}, x_{42}, y_1, y_2 \in \mathbb{R}^6 : x_{21} < 0 < 1 < y_1 < x_{32} < x_{41} < y_2 < x_{42}\}.$$

We will identify these parameters, along with the normalization points, $0, 1, \infty$, as fixed points of certain elements of our group.

First we use the operation described in 3.1, with $(x_{31} = 1, x_{32})$, (x_{41}, x_{42}) and $(\infty, 0)$ as the three pairs of fixed points. This yields the discrete group $\Upsilon_1 = \langle A, E, C \rangle$, where A has fixed points at $(1, x_{32})$, E has fixed points at (x_{41}, x_{42}), and C has fixed points at $(\infty, 0)$. We will explicitly write down matrices for these transformations as we need them. We note that $C(z) = (x_{32}/p_4)z$, where $p_4 = x_{41}x_{42}$.

Next let Υ_2 be Υ_1 conjugated by the transformation $z \to x_{21}\bar{z}$ (recall that $x_{21} < 0$). This yields $\Upsilon_2 = \langle A', E', C \rangle$, where A' has its fixed points at $(x_{21}, z_{22}) = (x_{21}, x_{21}x_{32})$, E' has its fixed points at $(z_{11}, z_{12}) = (x_{21}x_{41}, x_{21}x_{42})$, and C is the same transformation as above.

The imaginary axis splits the extended complex plane into two closed discs; the one containing the left half-plane is precisely invariant under $\langle C \rangle$ in Υ_1, and the one containing the right half-plane is precisely invariant under $\langle C \rangle$ in Υ_2; also, E maps the imaginary axis onto some circle in the right half-plane. Hence the hypotheses of the first combination theorem (see [M1,VII.C]) are satisfied. We conclude that $\Gamma_1 = \langle A, A', E, E', C \rangle$ is discrete; Γ_1 is the amalgamated free product, $\Gamma_1 = \Upsilon_1 *_{\langle C \rangle} \Upsilon_2$; and that $\mathbb{H}^2/\Gamma_1$ is a sphere with four holes, where the four holes are given by the boundary half-spaces bounded by the axes of A, A', E, and E'.

Rather than say that Γ_1 is the amalgamated free product, one can equivalently give the natural presentation:

$$\Gamma_1 = \langle A, A', E, E', C : CE'A' = CEA = 1 \rangle.$$

Since we obtained Υ_2 from Υ_1 by conjugation by a hyperbolic isometry, A and A' have the same translation length, as do E and E'. Hence, if B is a transformation mapping the fixed points of A onto those of A', then B will necessarily conjugate A onto either A' or $(A')^{-1}$. We let B be the unique transformation mapping the triple of points $(1, y_1, x_{32})$ onto the triple of points $(x_{21}, 0, z_{22})$. Then $A' = B \circ A \circ B^{-1}$.

Let H_1 be the closed disc bounded by the axis of A' and its reflection in the real axis, where H_1 contains the boundary half-space of A'. Similarly, let H_2 be the closed disc bounded by the axis of A and its reflection, and containing its boundary half-space. Then (H_1, H_2) is precisely invariant under $(\langle A' \rangle, \langle A \rangle)$ in Γ_1, and the complement of the union of the translates of these two discs is non-empty — it contains the interior of the convex hull of Γ_1. Hence the second combination theorem (see [M1, VII.E]) is applicable. We conclude that $\Gamma_2 = \langle A, A', B, C, E, E' \rangle$ is discrete; Γ_2 is the HNN-extension of Γ_1 by adjoining B; and $\mathbb{H}^2/\Gamma_2$ is a torus with two equal size holes, where the holes are the projections of the boundary half-spaces of E and E'.

We can translate the above algebraic information about Γ_2 into the presentation:

$$\Gamma_2 = \langle A, A', B, C, E, E' : CEA = CE'A' = A^{-1}B^{-1}A'B = 1 \rangle.$$

We repeat the above process, and adjoin D, mapping the triple $(z_{11}, 0, z_{12})$ onto the triple (x_{41}, y_2, x_{42}), and therefore conjugating E' to E. We thus obtain our final group Γ, where $\mathbb{H}^2/\Gamma$ is a closed Riemann surface of genus 2. We also obtain the following presentation for Γ:

$$\langle A, A', B, C, D, E, E' : CEA = CE'A' = A^{-1}B^{-1}A'B = D^{-1}E^{-1}DE' = 1 \rangle.$$

Now set $A_1 = A$, $B_1 = B$, $A_2 = D^{-1}$, $B_2 = E^{-1}$, and observe that these also generate Γ; they satisfy the usual relation:

$$A_1 \circ B_1 \circ A_1^{-1} \circ B_1^{-1} \circ A_2 \circ B_2 \circ A_2^{-1} \circ B_2^{-1} = 1;$$

one easily sees that the projections of the axes of these generators form a canonical homology basis on $S = \mathbb{H}^2/\Gamma$.

We are now in a position to write down the corresponding matrices; for convenience in writing, we introduce some elementary symmetric functions. Set $p_3 = x_{32}$, $s_3 = 1 + x_{32}$, $p_4 = x_{41}x_{42}$, $s_4 = x_{41} + x_{42}$. With this notation, our matrices appear as:

$$A_1 = \begin{pmatrix} p_4 s_3 - p_3 s_4 & p_3(p_3 - p_4) \\ p_4 - p_3 & p_3(s_3 - s_4) \end{pmatrix},$$

$$B_1 = \begin{pmatrix} x_{21} p_3 & -x_{21} p_3 y_1 \\ y_1 & p_3 - y_1 - p_3 y_1 \end{pmatrix},$$

$$A_2 = \begin{pmatrix} p_4 x_{21} & -x_{21} p_4 y_2 \\ y_2 & p_4 - s_4 y_2 \end{pmatrix},$$

$$B_2 = \begin{pmatrix} p_4 s_3 - p_3 s_4 & p_4(p_3 - p_4) \\ p_4 - p_3 & p_4(s_3 - s_4) \end{pmatrix}.$$

5.1. Stratifications.

Let G be a finitely generated non-elementary torsion-free Kleinian group, normalized so that $0, 1, \infty$ are loxodromic fixed points. Let $T(G)$ denote the deformation space of G; this is the space of equivalence classes of normalized quasiconformal deformations of G, where each quasiconformal deformation preserves the normalization points $0, 1, \infty$, and two such deformations are equivalent if they are equal on the limit set of G. Let d denote the complex dimension of $T(G)$.

A *stratification* of $T(G)$ is a set of d distinct limit points of G, $x_1, \ldots, x_d$, all distinct from $0, 1, \infty$, where the map $\phi \in T(G) \to (\phi(x_1), \ldots, \phi(x_d))$ defines a complex analytic embedding of $T(G)$ into $\mathbb{C}^d$. As remarked in [K-M], in order to check that a set of d fixed points of loxodromic elements of G, together with the choice of fixed points to lie at the normalization points, forms a stratification, it suffices to check that this map is injective.

In our case, $G = F_2$ is some given Fuchsian group representing a closed surface of genus 2 and our $d = 6$ points are the six coordinates of $\mathcal{P}_2$; these, together with the normalization points, are attracting fixed points of hyperbolic elements of G as follows: The points x_{21}, 0, 1, y_1, x_{32}, x_{41}, y_2, x_{42}, ∞ are, in order, the attracting fixed points of $B_1 A_1^{-1} B_1^{-1}$, $A_1^{-1} B_2$, A_1^{-1}, $B_1^{-1} A_1^{-1} B_2 B_1$, A_1, B_2, $A_2^{-1} A_1^{-1} B_2 A_2$, B_2^{-1}, $B_2^{-1} A_1$.

LEMMA 5.1.1. *The mapping* $T(F) \to (x_{21}, y_1, x_{32}, x_{41}, y_2, x_{42})$ *defines a stratification of* $T(F)$.

PROOF. We need to permit our parameters to become complex, and to check that the corresponding group elements, now elements of a quasifuchsian group, are still uniquely determined by the normalization points and the parameters.

An easy computation shows that if one knows that A and C, and the product AC, all have two fixed points; one knows both fixed points of C, normalized to lie at 0 and ∞; one knows both fixed points of A; and one knows the sum and product of the fixed points of $E^{-1} = AC$, then one knows A, C and E as elements of $PSL(2, \mathbb{C})$.

It was remarked in [K-M] that one can continuously and uniquely lift all elements of all the groups of $T(G)$ to $SL(2, \mathbb{C})$. Hence the trace of every matrix is a well defined function on $T(G)$.

Using the above remark, another easy computation shows that if one knows that C, A' and E' all have two fixed points; one knows the matrix for C, one knows both the repelling fixed point of $A' = BAB^{-1}$, and the trace of the matrix for A'; and one knows the trace of the matrix for $E' = BAB^{-1}C = D^{-1}ED$, then one knows the matrices in $SL(2, \mathbb{C})$ for A' and E'.

The elements B and D are defined as mapping three of our normalization or parameter points onto three other such points; hence they are uniquely determined.

We have shown that these nine points determine at most one set of generators, satisfying the corresponding relations, in $T(G)$.

5.2. Real-analytic embeddings. It was shown in [K-M] that if one has a stratification of $T(F)$, where F is a finitely generated non-elementary Fuchsian group, then the real points of this stratification define a real-analytic embedding of the corresponding Teichmüller space into $\mathbb{R}^d$.

THEOREM 5.3. *The mapping $p \in \mathcal{P}_2 \rightarrow (A_1, B_1, A_2, B_2)$ defines a real-analytic diffeomorphism between $\mathcal{P}_2$ and $\tilde{\mathcal{M}}_2$.*

PROOF. We have shown that every point in $\mathcal{P}_2$ determines a unique quadruple A_1, B_1, A_2, B_2 of elements of $PGL^+(2, \mathbb{R})$, where the group generated by these elements is discrete of the first kind and represents a closed Riemann surface of genus 2; i.e., we have a well defined map $\Phi : \mathcal{P}_2 \rightarrow \tilde{\mathcal{M}}_2$. We note that our matrices all have positive determinant, and that the entries are polynomials in our parameters; hence Φ is real-analytic.

We next note that, since there is an orientation-preserving quasiconformal deformation taking any one point of $\mathcal{M}_2$ onto any other such point, these nine fixed points always occur in the same order on the (oriented) extended real line. Hence, if we start with a set of canonical generators for such a Fuchsian group, compute the fixed points of these nine elements, and normalize appropriately, we will obtain a point in $\mathcal{P}_2$. That is, starting with the matrices A_1, B_1, A_2, B_2, not necessarily normalized, but generating a discrete purely hyperbolic Fuchsian group of the first kind, we compute the attracting fixed points of the above nine words in these generators. Since every word in these generators stays hyperbolic throughout our space, the mapping from the generators to the attracting fixed point of any particular word is real-analytic. One can now write each of the coordinates of Φ^{-1} as a cross-ratio of four of these fixed points; this is also real-analytic.

THEOREM 5.4. *There is a real analytic diffeomorphism between T_2 and $\mathcal{DF}(F_2, PGL^+(2, \mathbb{R}))$.*

PROOF. It was remarked in [K-M] that a stratification of a space of quasi-fuchsian groups defines a real-analytic embedding of the corresponding space of Fuchsian groups into $\mathbb{R}^d$. It follows from Lemma 5.1.1 that our map between T_2 and P_2 is a real analytic diffeomorphism.

Theorem 5.3 establishes a real-analytic diffeomorphism between P_2 and $\tilde{M}_2$, which is the same space as $\mathcal{DF}(F_2, PGL^+(2, \mathbb{R}))$, but defined non-variationally.

5.5. Remark. We observe that if we choose our six coordinates to be rational, then, after clearing fractions, we will have represented a dense set of points in T_2 as subgroups of $PGL^+(2, \mathbb{Z})$.

Since every Fuchsian group representing a closed Riemann surface of genus 2 contains a subgroup representing a closed Riemann surface of any genus $g \geq 2$, we know that we can represent a Riemann surface of any genus by a subgroup of $PGL^+(2, \mathbb{Z})$.

6. A different view. It is sometimes convenient to work with a non-canonical set of generators for $\pi_1(S)$, and it is usually pleasanter to work with parameters without subscripts. We rewrite our parameter space as

$$P_2 = \{a, \ldots, f \in \mathbb{R}^6 : a < 0 < 1 < b < c < d < e < f\};$$

i.e., we have set $a = x_{21}$, $b = y_1$, $c = x_{32}$, $d = x_{41}$, $e = y_2$, and $f = x_{42}$. We also note that Γ is generated by A, B, C, D, where the axes of A and B, B and C, and C and D intersect exactly once, positively. These generators have the one defining relation:

$$A \circ B^{-1} \circ C \circ D^{-1} \circ C^{-1} \circ A^{-1} \circ D \circ B = 1.$$

Our matrices for these generators in these parameters are as follows:

$$A = \begin{pmatrix} df(1+c) - c(d+f) & c(c - df) \\ df - c & c(1 + c - d - f) \end{pmatrix},$$

$$B = \begin{pmatrix} ac & -abc \\ b & c - b - bc \end{pmatrix},$$

$$C = \begin{pmatrix} c & 0 \\ 0 & df \end{pmatrix},$$

$$D = \begin{pmatrix} df - e(d+f) & adef \\ -e & adf \end{pmatrix}.$$

6.1. An expanded set of generators. The generators $A, \ldots, D$ appear asymmetrically; that is, the axis of A intersects only the axis of B, while the axis of B intersects the axes of both A and C. We can enlarge this set of generators to include $E = C^{-1} \circ A^{-1}$, and $F = B^{-1} \circ D^{-1}$. Then, writing $A_1 = A$, $A_2 = B$, $A_3 = C$, $A_4 = D$, $A_5 = E$, and $A_6 = F$, and thinking of these as cyclically ordered, the axis of each A_i intersects both the axes of A_{i-1}, which it intersects negatively, and that of A_{i+1}, which it intersects positively, while intersecting the axis of no other A_j. We note that the six points of intersection of these six axes are necessarily the six Weierstrass points on S.

With this expanded set of generators, we obtain the presentation:

$$\Gamma = \langle A, B, C, D, E, F : AB^{-1}CD^{-1}EF^{-1} = ACE = FDB = 1 \rangle.$$

This expanded set of generators has the advantage that it geometrically exhibits the element of order 6 in the mapping class group: $A_i \to A_{i+1}^{-1}$, or, equivalently,

$$(A, \dots, E, F) \to (B^{-1}, \dots, F^{-1}, A^{-1}).$$

This automorphism, together with the twist about A, generate the mapping class group.

7. **The general case.** In this section, we outline a technique for writing down an explicit matrix parametrization for the Teichmüller space of closed Riemann surfaces of any genus; however, both the inequalities describing the parameter space and the entries in the matrices get quite complicated as the genus increases.

There are also significant notational problems as the genus increases; we outline the general procedure here, with the focus on genus 4; this is sufficiently general to meet all the other significant problems. The case of genus 2 is again special; although we do not need another parametrization for genus 2, we work out that case first, as a guide to the general case.

7.1. **Special case of genus** 2. In this case, we again call our parameters by the same names:

$$x_{12}, y_1, x_{11}, x_{41}, y_2, x_{42};$$

but they have different meanings from those given in section 5. These parameters satisfy the following inequalities:

$$x_{12} < y_1 < x_{11} < 0 < 1 < x_{41} < y_2 < x_{42}, \quad x_{41}x_{42} > x_{12}/x_{11}.$$

The construction is quite straightforward. As in 3.3, we use the three pairs of points: $(z_{31}, z_{32}) = (1/x_{42}, 1/x_{41})$, (x_{41}, x_{42}), and $(\infty, 0)$ to construct a first pants group with two equal legs, $\Xi_1 = \langle A, A', C \rangle$, where A has fixed points (z_{31}, z_{32}), A' has fixed points (x_{41}, x_{42}), and C has fixed points $(\infty, 0)$. Note that we have placed a fixed point of a Fenchel-Jørgensen commutator at the point 1.

We now have $C(z) = z/x_{41}^2 x_{42}^2 = z/p_4^2$, and we use the construction given in 3.4 to construct a second pants group Ξ_2 with two equal legs, and waist equal to C. This gives us a second pants group $\Xi_2 = \langle B, B', C \rangle$, where B has fixed points $(z_{21}, z_{22}) = (x_{12}/p_4, x_{11}/p_4)$, and B' has fixed points at (x_{11}, x_{12}). Note that we need the inequality $p_4 > x_{12}/x_{11}$ for this operation.

Exactly as above, the group $\Gamma_1 = \langle \Xi_1, \Xi_2 \rangle$ is formed via the first combination theorem, and has a similar presentation; i.e.,

$$\Gamma_1 = \langle A, A', B, B', C : CA'A = CB'B = 1 \rangle,$$

and this Γ_1 also represents a sphere with four holes; however, now the holes represented by A and A' have equal sizes, and the holes represented by B and B' have equal sizes.

We then use the second combination theorem to adjoin a transformation D mapping the triple $(z_{31}, 0, z_{32})$ onto the triple (x_{42}, y_2, x_{41}), and also adjoin the transformation E mapping the triple $(z_{21}, 0, z_{22})$ onto the triple (x_{12}, y_1, x_{11}). We obtain our final group Γ, representing a closed Riemann surface of genus 2, and having the presentation:

$$\Gamma = \langle A, A', B, B', C, D, E : CA'A = CB'B = AD^{-1}A'D = BE^{-1}B'E = 1 \rangle.$$

Solving the above relations, and checking the orientation, we see that the axes of $A_1 = B^{-1}$, $B_1 = E$, $A_2 = D$, and $B_2 = A^{-1}$ form a canonical homotopy basis for Γ. Using the above computations, one now easily writes down the corresponding matrices; again, we set $p_1 = x_{11}x_{12}$, $s_1 = x_{11} + x_{12}$, $p_4 = x_{41}x_{42}$, and $s_4 = x_{41} + x_{42}$. We obtain:

$$A_1 = \begin{pmatrix} s_1 p_4 & -p_1(1 + p_4) \\ p_4^2(1 + p_4) & -s_1 p_4^2 \end{pmatrix},$$

$$B_1 = \begin{pmatrix} p_1 p_4 - y_1 s_1 p_4 & y_1 p_1 \\ -y_1 p_4 & p_1 \end{pmatrix},$$

$$A_2 = \begin{pmatrix} -p_4 & y_2 \\ y_2 - s_4 & 1 \end{pmatrix},$$

$$B_2 = \begin{pmatrix} s_4 & -(1 + p_4) \\ p_4(1 + p_4) & -p_4 s_4 \end{pmatrix}.$$

7.2. Parameters for the general case. We start with a surface of genus $g > 2$. We use a pants decomposition in which there are g pairs of pants having equal size legs (i.e., two of the holes in the pair of pants correspond to the same non-dividing geodesic on the surface), and $g - 2$ general pairs of pants (i.e., the boundary geodesics for these correspond to dividing geodesics). We start with unnormalized pants groups $\Xi_1, \ldots, \Xi_g$ corresponding to the pants with the equal sized legs, and unnormalized pants groups $\Upsilon_1, \ldots, \Upsilon_{g-2}$ corresponding to the other pants groups.

Each of the groups Υ_i has three distinguished generators; we call these C_{i1}, C_{i2} and C_{i3}. The repelling fixed point of C_{ij} is labelled as s_{ij1} if it a parameter, and it is labelled as t_{ij1} if it is to be computed from the other parameters. Similarly, the attracting fixed point is either s_{ij2}, if it is a parameter, or t_{ij2} if it is to be computed.

Each of the groups Ξ_i likewise has three distinguished generators; these are called A_i, A_i' and D_i. The fixed points of A_i are at (x_{i1}, x_{i2}); these will always be parameters. The fixed points of A_i' are at (z_{i1}, z_{i2}); we will always compute these from the other parameters. We will not need to separately label the fixed points of D_i.

For $i = 1, \ldots, g$, we will also have the transformation B_i mapping the triple of points (x_{i1}, y_i, x_{i2}) onto the triple $(z_{i2}, 0, z_{i1})$, where y_i is another parameter lying between x_{i1} and x_{i2}.

In what follows, we will label the axis of the transformation A as L_A.

We first normalize Υ_1 so that C_{11} has its fixed points at $(\infty, 0)$; C_{12} has its fixed points at $(1, s_{122})$, and C_{13} has its fixed points at (s_{131}, s_{132}). This gives us our first set of parameter inequalities:

$$1 < s_{122} < s_{131} < s_{132}. \tag{7.1}$$

We next normalize Ξ_1 so that its fixed points lie on the other side of $L_{C_{11}}$, and so that $D_1 = C_{11}^{-1}$. Then A_1 has its fixed points at (x_{11}, x_{12}), and A_1' has its fixed points at (z_{11}, z_{12}), where $x_{11} < x_{12} < 0$. We also adjoin B_1 conjugating A_1 onto $(A_1')^{-1}$, and mapping y_1 to 0. We denote reflection in the common orthogonal to L_{A_1} and $L_{C_{11}}$ by $R_{A_1 C_{11}}$. Using 3.4, we can write

$$z_{11} = \sqrt[4]{C_{11}} \circ R_{A_1 C_{11}} \circ (\sqrt[4]{C_{11}})^{-1}(x_{12}),$$

and

$$z_{12} = \sqrt[4]{C_{11}} \circ R_{A_1 C_{11}} \circ (\sqrt[4]{C_{11}})^{-1}(x_{11}).$$

This gives us two new sets of inequalities for our parameters:

$$x_{11} < y_1 < x_{12} < 0; \quad z_{11} = \sqrt[4]{C_{11}} \circ R_{A_1 C_{11}} \circ (\sqrt[4]{C_{11}})^{-1}(x_{12}) > x_{12}. \tag{7.2}$$

We next repeat the process above, and normalize Ξ_2 so that $D_2 = C_{12}^{-1}$, where the fixed points of A_2 lie on the other side of the axis of C_{12} from those of C_{11} and C_{13}. We also choose the parameter y_2, lying between x_{21} and x_{22}, and adjoin the transformation B_2, mapping the triple (x_{21}, y_1, x_{22}) onto the triple $(z_{22}, 0, z_{21})$. For this operation, we need the following inequalities:

$$1 < x_{21} < y_2 < x_{22} < s_{122}; \quad \sqrt[4]{C_{12}} \circ R_{A_2 C_{12}} \circ (\sqrt[4]{C_{12}})^{-1}(x_{22}) > x_{22}. \tag{7.3}$$

If $g = 3$, then we repeat the above process one more time, normalizing Ξ_3 so that $D_3 = C_{13}^{-1}$, and adjoining B_3.

If $g > 3$, which we now assume, then we normalize Υ_2 so that $C_{21}^{-1} = C_{13}$. Then C_{22} has its fixed points at (s_{221}, s_{222}), and C_{23} has its fixed points at (s_{231}, t_{232}). Using 3.2, we solve for t_{232} by letting $M = M_{C_{13} C_{22}}$ be the common orthogonal between $L_{C_{13}}$ and $L_{C_{22}}$, and letting R denote reflection in M; then $t_{232} = \sqrt{C_{13}} \circ R \circ (\sqrt{C_{13}})^{-1}(s_{231})$. Having solved for t_{232}, we need only the one additional set of inequalities:

$$s_{131} < s_{221} < s_{222} < s_{231} < t_{232} = \sqrt{C_{13}} \circ R \circ (\sqrt{C_{13}})^{-1}(s_{231}) < s_{132}. \tag{7.4}$$

We then normalize Ξ_3 so that $D_3 = C_{22}^{-1}$, and we adjoin B_3, as above. If $g = 4$, then we also normalize Ξ_4 so that $D_4 = C_{23}^{-1}$, and adjoin B_4. We note that, for $g = 4$, our parameter space is exactly described by the inequalities (7.1–7.4).

If $g > 4$, then normalize Υ_3 so that $C_{31}^{-1} = C_{23}$, and continue as above.

7.3. Matrices for the general case.

We first observe that the transformations of the form A_i and B_i actually generate the group. Each D_i is a commutator of the corresponding A_i and B_i. Thus, for Υ_{g-2}, two of the generators are known; therefore the third generator is known; then the same statement is true for Υ_{g-3}, etc. Hence, in principle, it would suffice to compute only the A_i and B_i. However, in order to compute A_i, we need to know its fixed points, which are the parameters x_{i1} and x_{i2}; the fixed points of A_i', which are the computed variables z_{i1} and z_{i2}; and the fixed points of D_i, which is the inverse of some C_{jk}. We also need to know this C_{jk} in order to compute z_{i1} and z_{i2}. Hence, we need to compute all the C_{jk} as well.

We need the exact same information for B_i, which is defined as mapping the triple (x_{i1}, y_i, x_{i2}) onto the triple $(z_{i2}, 0, z_{i1})$.

If one puts together the inequalities in (7.1–7.4), then they say exactly that the points $x_{ij}, z_{ij}, s_{ijk}, t_{ijk}$ lie in a given order on the real axis. If one expands the set of parameters to include all the z_{ij} and t_{ijk}, then the parameter space is just defined by the order in which these parameters occur. For $g = 4$, this order is the following:

$$x_{11} < y_1 < x_{12} < z_{11} < z_{12} < 0 = s_{112} < 1 = s_{121} < x_{21} < y_2 < x_{22} <$$

$$< z_{21} < z_{22} < s_{122} < s_{131} < s_{221} < x_{31} < y_3 < x_{32} < z_{31} < z_{32} <$$

$$< s_{222} < s_{231} < x_{41} < y_4 < x_{42} < z_{41} < z_{42} < t_{232} < s_{132} < \infty = s_{111}.$$

Once one views the z_{ij} and t_{ijk} as parameters, then one can also easily write down the matrices A_i and B_i; each B_i is defined as mapping a triple of these parameters onto another such triple; each A_i is a product of two reflections, where each of these reflections is of the form (2.1); that is, it is a reflection in the common orthogonal between two lines whose endpoints are parameters. One can now easily write down the matrices $A_1, B_1, \dots, A_4, B_4$.

REFERENCES

[A]   W. Abikoff, *The Real Analytic Theory of Teichmüller Spaces*, Lecture Notes in Math. number 820, Springer-Verlag, Berlin and New York.

[F]   W. Fenchel, *Elementary Geometry in Hyperbolic Space*, Studies in Mathematics 11, de Gruyter, Berlin and New York.

[H]   H. Helling, *Über den Raum der kompakten Riemannschen Flächen vom Geschlecht 2.*, J. Reine Angew. Math. **268/269** (1974), 286–293.

[J]   T. Jørgensen, *Compact 3-manifolds of constant negative curvature fibering over the circle*, Ann. of Math. **106** (1977), 61–72.

[K]   L. Keen, *On Fricke Moduli*, Advances in the Theory of Riemann Surfaces (L.V. Ahlfors, L. Bers, H.M. Farkas, R. C. Gunning I. Kra, H. E. Rauch, eds.), Annals of Math. Studies Number 66, Princeton University Press, Princeton, 1988, pp. 251–265.

[K-M]   I. Kra and B. Maskit, *The deformation space of a Kleinian group*, Amer. J. Math. **103** (1981), 1065–1102.

[Ma]   W. Magnus, *Rational Representations of Fuchsian Groups and non-parabolic Subgroups of the modular Group*, Nachr. Akad. Wiss. Gottingen Math-Phys. Kl. II (1973), 179–189.

[M1]   B. Maskit, *Kleinian Groups*, Springer-Verlag, Heidelberg and New York, 1988.

[M2]   _____, *Parameters for Fuchsian groups I: Signature (0,4)*, Holomorphic Functions and Moduli II, Math. Sci. Res. Inst. Pub. 11, Springer-Verlag, New York, 1988, pp. 251–265.

[M3]   _____, *Parameters for Fuchsian groups II: topological type (1,1)*, Annal. Acad. Sci. Fenn. Ser. A.I. **14** (1990), 265–275.

[T]   K. Takeuchi, *Fuchsian groups contained in* $SL(2, \mathbb{Q})$, J. Math Soc. Japan **23** (1971), 82–94.

[W-M]   P. L. Waterman and C. Maclachlan, *Fuchsian groups and algebraic number fields*, Trans. Amer. Math. Soc. **287** (1985), 353–364.

MATHEMATICS DEPARTMENT, THE UNIVERSITY AT STONY BROOK, STONY BROOK NY 11794-3651

E-mail address: bernie@math.sunysb.edu

Contemporary Mathematics
Volume **169**, 1994

Levi-Properties Generated by Varieties

ROBERT FITZGERALD MORSE

ABSTRACT. Levi-properties were first introduced by L. C. Kappe and are modeled after groups investigated by F. W. Levi where conjugates commute. Let $\mathfrak{X}$ be a group theoretic class. A group is in the derived class $L(\mathfrak{X})$ if the normal closure of each element in the group is an $\mathfrak{X}$-group. The property of being in the class $L(\mathfrak{X})$ is called the Levi-property generated by $\mathfrak{X}$. In the case where $\mathfrak{X}$ is a variety, we show that $L(\mathfrak{X})$ is also a variety. Given the laws defining any variety $\mathfrak{V}$, the laws defining a variety $\mathfrak{W}$ can be exactly stated such that $L(\mathfrak{V}) \leq \mathfrak{W}$. However, there exists a variety $\mathfrak{V}$ such that $L(\mathfrak{V}) < \mathfrak{W}$. Our investigations show for varieties defined by outer commutator laws, denoted by $\mathfrak{O}$, the varieties $L(\mathfrak{O})$ and $\mathfrak{W}$ coincide.

1. Introduction

Given a group theoretic class $\mathfrak{X}$, we define a derived class of groups $L(\mathfrak{X})$ as the class of those groups in which the normal closure of each element in the group is an $\mathfrak{X}$-group. The property of being in the class $L(\mathfrak{X})$ is called the Levi-property generated by $\mathfrak{X}$. In this paper we investigate Levi-properties in which the generating classes are varieties.

Levi-properties were first introduced by L.C. Kappe in [3]. This characterization of groups is modeled after 2-Engel groups first classified by Levi [7]. These groups are exactly those groups in which the normal closure of each element is abelian. Subsequent investigations considered the problem of determining the Levi-property generated by a specific variety of groups such as nilpotency of a given class [1, 4, 6], n-abelian [5], and n-central [6]. In each case, the Levi-properties generated by these varieties are themselves varieties whose laws are derived from the laws of the generating variety. The goal of this paper is to generalize these results.

1991 *Mathematics Subject Classification.* Primary 20E10 20F12.

Fundamental to this investigation is the result that if $\mathfrak{V}$ is a variety of groups, then $L(\mathfrak{V})$ is also a variety (Theorem 2.1). We call $L(\mathfrak{V})$ the Levi-variety generated by $\mathfrak{V}$. However, not every variety of groups can be presented in the form $L(\mathfrak{V})$ for any $\mathfrak{V}$ (Proposition 2.3).

The variety of all abelian groups, denoted by $\mathfrak{A}$, is defined by the word $[x, y]$. The law $[x^{y_1}, x^{y_2}] = 1$ holds in the Levi-variety $L(\mathfrak{A})$. We generalize this case and show that, for any variety $\mathfrak{V}$ and for each law which holds in $\mathfrak{V}$, the same law with each variable in the law replaced with distinct conjugates of a single variable is a law in the Levi-variety generated by $\mathfrak{V}$ (Corollary 3.3). However, the variety defined by these conjugate laws may strictly contain $L(\mathfrak{V})$.

For the abelian case, $L(\mathfrak{A})$ is exactly characterized by the conjugate law $[x^{y_1}, x^{y_2}] = 1$. A similar characterization is given for Levi-varieties generated by varieties defined by the outer commutator words of P. Hall [2] (Corollary 4.4).

2. Levi-varieties

Our notation is standard (see, for instance, [10]). We fix $x, x_1, \ldots, y, y_1, \ldots$ to be variables and $\mathfrak{w}$ and $\mathfrak{y}$ to be words in these variables. Denote the variety of nilpotent groups of class c by $\mathfrak{N}_c$. The variety of n-Engel groups, defined by the commutator word $[x, \underbrace{y, \ldots, y}_{n}]$, is denoted by $\mathfrak{E}_n$.

The following result is fundamental to our investigation.

THEOREM 2.1. *If $\mathfrak{V}$ is a variety of groups, then $L(\mathfrak{V})$ is also a variety.*

PROOF. A straightforward argument shows that the class $L(\mathfrak{V})$ is closed under the formation of subgroups, homomorphic images, and Cartesian products. It follows by a result of Birkhoff that $L(\mathfrak{V})$ is a variety (see [9, Theorem 15.23]). □

Distinct classes of groups may not generate distinct Levi-properties. The normal closure of each element of a group being an $\mathfrak{X}$-group may put a constraint on the class $\mathfrak{X}$ as it can occur in the normal closures. This constraint may act on two distinct classes of groups such that both will generate the same Levi-property. For example, $\mathfrak{E}_2$ strictly contains $\mathfrak{N}_2$ since there exist 2-Engel groups which are nilpotent of exactly class 3 [7]. However, each of these varieties generate the same Levi-variety, namely $\mathfrak{E}_3$ [4]. That is, 2-Engel groups are constrained as they appear in the normal closures to those 2-Engel groups which are nilpotent of class less than 3. Hence, the varieties $\mathfrak{E}_2$ and $\mathfrak{N}_2$ are equivalent as they occur in the normal closure of each element in a group. We denote by $\overline{\mathfrak{X}}$ those $\mathfrak{X}$-groups which occur in the normal closures of each group in $L(\mathfrak{X})$. Note that $\overline{\mathfrak{X}} \subseteq \mathfrak{X}$. The lemma below describes relations and implications between classes of groups and the Levi-properties they generate.

LEMMA 2.2. *Let $\mathfrak{X}$ and $\mathfrak{Y}$ be classes of groups. Then we have the following implications:*

(i) *If $\mathfrak{X}$ is inherited by normal subgroups, then $\mathfrak{X} \leq L(\mathfrak{X})$.*

(ii) *If $\mathfrak{X} \leq \mathfrak{Y}$, then $L(\mathfrak{X}) \leq L(\mathfrak{Y})$. (But $\mathfrak{X} < \mathfrak{Y}$ does not necessarily imply $L(\mathfrak{X}) < L(\mathfrak{Y})$ as noted above.)*

(iii) *If $L(\mathfrak{X}) \leq L(\mathfrak{Y})$, then $\overline{\mathfrak{X}} \leq \overline{\mathfrak{Y}}$. In particular, $L(\mathfrak{X}) < L(\mathfrak{Y})$ implies $\overline{\mathfrak{X}} < \overline{\mathfrak{Y}}$.*

PROOF. Let $\mathfrak{X}$ and $\mathfrak{Y}$ be classes of groups.

(i) Obvious.

(ii) Let $G \in L(\mathfrak{X})$. Hence, $z^G \in \overline{\mathfrak{X}} \leq \mathfrak{X}$ for all $z \in G$. If $\mathfrak{X} \leq \mathfrak{Y}$, then $z^G \in \mathfrak{Y}$ for all $z \in G$ which implies $z^G \in \overline{\mathfrak{Y}}$ and $G \in L(\mathfrak{Y})$. Thus, $L(\mathfrak{X}) \leq L(\mathfrak{Y})$.

(iii) Assume $L(\mathfrak{X}) \leq L(\mathfrak{Y})$. Let $G \in L(\mathfrak{X})$. Then $z^G \in \overline{\mathfrak{X}}$ for all $z \in G$. Suppose $z_0^G \notin \overline{\mathfrak{Y}}$ for some $z_0 \in G$. This implies $G \notin L(\mathfrak{Y})$, a contradiction. If $L(\mathfrak{X}) < L(\mathfrak{Y})$, then there exists a group $G \in L(\mathfrak{Y})$ such that there exists a $z \in G$ such that z^G is not an $\mathfrak{X}$-group. Hence, we have strict containment, i.e., $\overline{\mathfrak{X}} < \overline{\mathfrak{Y}}$ $\square$.

Every variety generates a Levi-variety. However, not all varieties are Levi-varieties.

PROPOSITION 2.3. *Every proper subvariety of zero exponent of $\mathfrak{E}_2$ is not a Levi-variety.*

PROOF. Let $\mathfrak{V}$ be a proper subvariety of zero exponent of $\mathfrak{E}_2$. Suppose that there exists a variety $\mathfrak{X}$ such that $L(\mathfrak{X}) = \mathfrak{V}$. Then $\mathfrak{X}$ must be of zero exponent and hence $\mathfrak{X}$ contains the variety of abelian groups. Therefore, by Lemma 2.2 (ii), we have $\mathfrak{E}_2 = L(\mathfrak{A}) \leq L(\mathfrak{X}) = \mathfrak{V}$. This contradicts the assumption that $\mathfrak{V}$ is a proper subvariety of $\mathfrak{E}_2$.

3. Laws defining Levi-varieties

In light of Theorem 2.1, can one specifically state all the laws of $L(\mathfrak{V})$ if those of $\mathfrak{V}$ are given? In general, this is an open question. However, laws can be derived from the laws defining $\mathfrak{V}$ which always hold in $L(\mathfrak{V})$ by replacing each variable in these laws with distinct conjugates of a single variable. Moreover, if $\mathfrak{w}$ is an n-variable law of $\mathfrak{V}$, then the law derived from $\mathfrak{w}$ is also an n-variable law.

The following lemma shows a certain amount of freedom is allowed when working with laws concerning conjugates of elements.

LEMMA 3.1. *Let G be any group. Then G satisfies the law*

$$\mathfrak{w}\left(x^{y_1}, \ldots, x^{y_{k-1}}, x, x^{y_{k+1}}, \ldots, x^{y_{n+1}}\right) = 1$$

if and only if G satisfies the law $\mathfrak{w}\left(x^{y_1}, \ldots, x^{y_k}, \ldots, x^{y_{n+1}}\right) = 1$.

THEOREM 3.2. *Let $\mathfrak{w}(x_1, \ldots, x_n)$ be a word and $\mathfrak{V}$ be the variety defined by $\mathfrak{w}$. Then the n-variable word $\mathfrak{w}(x^{y_1}, \ldots, x^{y_{n-1}}, x)$ is a law of $L(\mathfrak{V})$.*

PROOF. Let $\mathfrak{V}$ be the variety defined by the word $\mathfrak{w}(x_1, \ldots, x_n)$. Suppose $G \in L(\mathfrak{V})$. Then for all $z \in G$, $z^G \in \mathfrak{V}$, $\mathfrak{w}$ reduces to the identity element for $z^{g_1}, z^{g_2}, \cdots \in z^G$, where $g_1, g_2, \ldots$ are arbitrary elements of G. Hence, $\mathfrak{w}(x^{y_1}, \ldots, x^{y_n})$ is a law in G. By Lemma 3.1, this law is equivalent to the n-variable law

$$\mathfrak{w}(x^{y_1}, \ldots, x^{y_{n-1}}, x) = 1. \qquad \square$$

The following corollary is now obvious.

COROLLARY 3.3. *Let Λ be an index set and $W = \{\mathfrak{w}_\lambda | \lambda \in \Lambda\}$ be a set of words in variables $\{x_1, x_2, \ldots\}$. If $\mathfrak{V}$ is the variety defined by the set of words W, then $L(\mathfrak{V})$ is a subvariety of the variety $\mathfrak{W}$ defined by the set of words W, where the variables $\{x_1, x_2, \ldots\}$ are replaced by $\{x^{y_1}, x^{y_2}, \ldots\}$.*

We conclude this section with an example for which $L(\mathfrak{V})$ is a strict subvariety of $\mathfrak{W}$.

EXAMPLE 3.4. Let $W = \{[x_1^2, x_2]\}$ and $\mathfrak{V}$ be the variety defined by W. Then $\mathfrak{V}$ is the variety of 2-central groups, and $\mathfrak{W}$ is defined by the law $[x^2, x^y]$. Let G be the infinite dihedral group $G = \langle h, i : i^2 = 1, h^i = h^{-1} \rangle$. We first show that $G \in \mathfrak{W}$. If $x \in G$, then either $x \in \langle h \rangle$, so the normal closure of x is abelian, or $x = h^n i$ for some integer n, in which case $x^2 = 1$. In both cases, $[x^2, x^y] = 1$ for all $y \in G$ as required. However, G is not a 3-Engel group. Therefore, $G \notin L(\mathfrak{V})$ [6, Theorem 8].

4. Outer Commutator Laws

Let $L(\mathfrak{V})$ and $\mathfrak{W}$ be varieties as defined in Corollary 3.3. It is not always the case that $L(\mathfrak{V}) < \mathfrak{W}$. The varieties $L(\mathfrak{V})$ and $\mathfrak{W}$ may coincide when the generating variety $\mathfrak{V}$ is defined by certain words. For example, the two varieties coincide when the generating variety is defined by single variable words, i.e., $\mathfrak{w}(x) = x^n$. This is an immediate consequence of Lemma 3.1. In this section, we show that this is also true for generating varieties defined by outer commutator words. Our treatment of outer commutator words follows Robinson [10].

We define the weight of a word as the number of distinct variables it contains. The unique outer commutator word of weight one is Θ_0 where $\Theta_0(x_1) = x_1$. The word Θ is an outer commutator word of weight $n > 1$ if $\Theta = [\Theta_1, \Theta_2]$, where Θ_1 and Θ_2 are outer commutator words of weight n_1 and n_2 respectively and $n = n_1 + n_2$. Since each variable of an outer commutator word appears at most once, we see the lower central words are outer commutator words whereas the n-Engel words $(n > 1)$ are not.

Let G be a group and $z \in G$. Every element $u \in z^G$ can be expressed as a finite product of $z^{\pm g_i}$, $g_i \in G$. Using the length of this product, we define the function $\eta : z^G \to \{\mathbb{N} \cup 0\}$ as follows:

$$\eta(u) = \begin{cases} 0, & \text{for } u = 1 \\ k, & \text{for } u = z^{\pm g_1} \ldots z^{\pm g_k}, \ g_i \in G, \text{ and } k \text{ minimal} \end{cases}$$

We observe that the function η is not affected by conjugation:

$$(1) \qquad \eta\left(u^g\right) = \eta\left(u\right) \qquad u \in z^G, g \in G.$$

Let $z \in G$ and $u_i \in z^G$. We define the total length $\mathcal{L}$ of a word $\mathfrak{w}$ with entries $u_1, \ldots, u_j$ as

$$\mathcal{L}\left(\mathfrak{w}(u_1, \ldots, u_j)\right) = \sum_{i=1}^{j} \eta\left(u_i\right).$$

Two words with equal weight and identical entries from z^G have equal total length.

For any word $\mathfrak{w}\left(x_1, \ldots, x_j\right)$, we have

$$(2) \qquad \mathfrak{w}\left(z_1, \ldots, z_j\right)^h = \mathfrak{w}\left(z_1^h, \ldots, z_j^h\right) \qquad z_i, h \in G.$$

With this observation, it follows from (1) that the total length of a word $\mathfrak{w}$ is not affected by conjugation, that is, for $z, g \in G$ and $u_i \in z^G$,

$$(3) \qquad \mathcal{L}\left(\mathfrak{w}\left(u_1, \ldots, u_j\right)^g\right) = \mathcal{L}\left(\mathfrak{w}\left(u_1^g, \ldots, u_j^g\right)\right) = \mathcal{L}\left(\mathfrak{w}\left(u_1, \ldots, u_j\right)\right).$$

The following two lemmas will facilitate the proof of our main result.

LEMMA 4.1. *Let Θ be an outer commutator word of weight n, and let G be a group.*

(i) *Let $z, g_i \in G$, $i = 1, \ldots, n$. Then there exist $h_i \in G$, $i = 1, \ldots, n$, such that*

$$\Theta\left(z^{g_1}, \ldots, z^{-g_k}, \ldots, z^{g_n}\right) = \Theta\left(z^{h_1}, \ldots, z^{h_k}, \ldots, z^{h_n}\right)^{-1}.$$

(ii) *If G satisfies the law $\Theta\left(x^{y_1}, \ldots, x^{y_n}\right) = 1$, then G also satisfies the law*

$$\Theta\left(x^{\delta_1 y_1}, \ldots, x^{\delta_n y_n}\right) = 1$$

where $\delta_i = \pm 1$, $i = 1, \ldots, n$.

PROOF.

(i) We prove this statement by induction on n, the weight of Θ. Let G be a group, and let $z, g_i \in G$, $i = 1, \ldots, n$. For $n = 1$, we have $\Theta_0\left(z^{-g_1}\right) = z^{-g_1}$. Relabeling g_1 as h_1, we see (i) is true for $n = 1$.

Assume (i) is true for all outer commutator words of weight less than n. Suppose Θ is an outer commutator of weight n. Then by definition, we have outer commutators Φ and Γ of weight $i > 0$ and $n-i$ respectively such that $\Theta = [\Phi, \Gamma]$.

Without loss of generality, assume $1 \leq k \leq i$. We apply the induction hypothesis to the outer commutator Φ, and hence, there exist $\bar{h}_j \in G$, $j = 1, \ldots, i$, such that

$$\Phi\left(z^{g_1}, \ldots, z^{-g_k}, \ldots, z^{g_i}\right) = \Phi\left(z^{\bar{h}_1}, \ldots, z^{\bar{h}_i}\right)^{-1}.$$

Hence,

$$\Theta\left(z^{g_1}, \ldots, z^{-g_k}, \ldots, z^{g_n}\right) = \left[\Phi\left(z^{\bar{h}_1}, \ldots, z^{\bar{h}_k}, \ldots, z^{\bar{h}_i}\right)^{-1}, \Gamma\left(z^{g_{i+1}}, \ldots, z^{g_n}\right)\right].$$

By standard commutator expansion and relabeling, we obtain the desired result. A similar argument holds if $i + 1 \le k \le n$.

(ii) Suppose the law $\Theta\left(x^{y_1}, \ldots, x^{y_n}\right) = 1$ holds in G. For $z, g_i \in G$ and $\delta_i = \pm 1$, $i = 1, \ldots, n$, we obtain

$$\Theta\left(z^{\delta_1 g_1}, \ldots, z^{\delta_n g_n}\right) = \Theta\left(z^{h_1}, \ldots, z^{h_n}\right)^{\pm 1}$$

by applying (i) to each $\delta_i = -1$, $i = 1, \ldots, n$. By hypothesis, we have

$$\Theta\left(z^{h_1}, \ldots, z^{h_n}\right)^{\pm 1} = 1.$$

Therefore, G satisfies the desired law. $\square$

LEMMA 4.2. *Let G be a group, $z \in G$ and Θ be an outer commutator word of weight n. Let $1 \ne u_i \in z^G$, $i = 1, \ldots, n$, and $\eta\left(u_k\right) \ge 2$ for some $k \in \{1, \ldots, n\}$. Then there exist $v_i, w_i \in z^G$, $i = 1, \ldots, n$, with*

$$\Theta\left(u_1, \ldots, u_n\right) = \Theta\left(v_1, \ldots, v_n\right) \Theta\left(w_1, \ldots, w_n\right)$$

and

$$0 < \mathcal{L}\left(\Theta\left(v_1, \ldots, v_n\right)\right) < \mathcal{L}\left(\Theta\left(u_1, \ldots, u_n\right)\right),$$
$$0 < \mathcal{L}\left(\Theta\left(w_1, \ldots, w_n\right)\right) < \mathcal{L}\left(\Theta\left(u_1, \ldots, u_n\right)\right).$$

PROOF. We prove the lemma by induction on n, the weight of Θ. Let G be a group and $z \in G$. A straightforward argument shows that the lemma is true for $n = 1$.

Assume the result is true for all outer commutator words of weight less than n. Suppose Θ is an outer commutator of weight n. Then by definition, we have outer commutators Φ and Γ of weight less than n such that for variables $x_1, \ldots, x_n$

$$\Theta\left(x_1, \ldots, x_n\right) = \left[\Phi\left(x_1, \ldots, x_i\right), \Gamma\left(x_{i+1}, \ldots, x_n\right)\right].$$

Let $1 \ne u_j \in z^G$ for $j = 1, 2, \ldots, n$, and assume $\eta\left(u_k\right) \ge 2$ for some $1 \le k \le i$. We apply the induction hypothesis to the outer commutator Φ, and hence, there exist $\bar{v}_j, w_j \in z^G$, $j = 1, \ldots, i$, such that

$$\Phi\left(u_1, \ldots, u_i\right) = \Phi\left(\bar{v}_1, \ldots, \bar{v}_i\right) \Phi\left(w_1, \ldots, w_i\right)$$

and

(4) $$0 < \mathcal{L}\left(\Phi\left(\bar{v}_1, \ldots, \bar{v}_i\right)\right) < \mathcal{L}\left(\Phi\left(u_1, \ldots, u_i\right)\right),$$

(5) $$0 < \mathcal{L}\left(\Phi\left(w_1, \ldots, w_i\right)\right) < \mathcal{L}\left(\Phi\left(u_1, \ldots, u_i\right)\right).$$

Set $\Phi\left(w_1, \ldots, w_i\right) = c$ whenever an element is conjugated by $\Phi\left(w_1, \ldots, w_i\right)$. By standard commutator expansion, we obtain

$$\Theta(u_1, \ldots, u_n) = [\Phi(\bar{v}_1, \ldots, \bar{v}_i) \cdot \Phi(w_1, \ldots, w_i), \Gamma(u_{i+1}, \ldots, u_n)]$$
$$= [\Phi(\bar{v}_1, \ldots, \bar{v}_i), \Gamma(u_{i+1}, \ldots, u_n)]^c [\Phi(w_1, \ldots, w_i), \Gamma(u_{i+1}, \ldots, u_n)].$$

By (2) we see

(6) $\quad \Theta(u_1, \ldots, u_n) = [\Phi(\bar{v}_1^c, \ldots, \bar{v}_i^c), \Gamma(u_{i+1}^c, \ldots, u_n^c)] [\Phi(w_1, \ldots, w_i), \Gamma(u_{i+1}, \ldots, u_n)].$

By (3) we have

$$0 < \mathcal{L}\left([\Phi(\bar{v}_1^c, \ldots, \bar{v}_i^c), \Gamma(u_{i+1}^c, \ldots, u_n^c)]\right) = \mathcal{L}([\Phi(\bar{v}_1, \ldots, \bar{v}_i), \Gamma(u_{i+1}, \ldots, u_n)]),$$

and hence, by (4)

$$0 < \mathcal{L}\left([\Phi(\bar{v}_1^c, \ldots, \bar{v}_i^c), \Gamma(u_{i+1}^c, \ldots, u_n^c)]\right) < \mathcal{L}([\Phi(u_1, \ldots, u_i), \Gamma(u_{i+1}, \ldots, u_n)]).$$

It follows by (5) that

$$0 < \mathcal{L}([\Phi(w_1, \ldots, w_i), \Gamma(u_{i+1}, \ldots, u_n)]) < \mathcal{L}([\Phi(u_1, \ldots, u_i), \Gamma(u_{i+1}, \ldots, u_n)]).$$

The desired decomposition is obtained by relabeling the group elements of $\Theta(u_1, \ldots, u_n)$. Setting $\bar{v}_j^c = v_j$ and $u_j = w_j$ for $j = 1, \ldots, i$, and $u_j^c = w_j$ for $j = i+1, \ldots, n$, in (6) proves the lemma for $1 \le k \le i$. A similar argument holds if $i+1 \le k \le n$. $\square$

From the following theorem, we see that if $\mathfrak{D}$ is a variety defined by an n-variable outer commutator word, then $L(\mathfrak{D})$ is a variety defined by an n-variable word.

THEOREM 4.3. Let $\Theta(x_1, \ldots, x_n)$ be an outer commutator word and $\mathfrak{D}$ be the variety defined by Θ. Then $L(\mathfrak{D})$ is the variety defined by the n-variable word $\Theta(x^{y_1}, \ldots, x^{y_{n-1}}, x)$.

PROOF. Let G be a group. Suppose $G \in L(\mathfrak{D})$. Then, by Theorem 3.2, the law $\Theta(x^{y_1}, \ldots, x^{y_{n-1}}, x) = 1$ holds in G.

Now suppose the group G is in the variety defined by the word

$$\Theta(x^{y_1}, \ldots, x^{y_{n-1}}, x).$$

It suffices to show that, for every $z \in G$, the law $\Theta(x_1, \ldots, x_n) = 1$ holds in z^G. It follows that $z^G \in \mathfrak{D}$ and consequently $G \in L(\mathfrak{D})$.

By induction on the total length of Θ, we will show $\Theta(u_1, \ldots, u_n) = 1$ for $z \in G$ and every $u_1, \ldots, u_n \in z^G$. If $\eta(u_i) = 0$ for some $i \in \{1, \ldots, n\}$, then u_i is the identity element. It follows that $\Theta(u_1, \ldots, u_n) = 1$ since Θ is a commutator word. Suppose $\mathcal{L}(\Theta(u_1, \ldots, u_n)) < n$. Then $\eta(u_i) = 0$ for at least one $i \in \{1, \ldots, n\}$ and hence $\Theta(u_1, \ldots, u_n) = 1$. From now on, we assume, without loss of generality, that $\eta(u_i) > 0$ for $i = 1, \ldots, n$. Suppose $\mathcal{L}(\Theta(u_1, \ldots, u_n)) = n$.

Then $\eta(u_i) = 1$ for each $i = 1, \ldots, n$. Hence, it follows that $u_i = z^{\pm g_i}$. Since G satisfies the law $\Theta(x^{y_1}, \ldots, x^{y_n}) = 1$ by hypothesis, we can apply Lemma 4.1(ii) and thus $\Theta(u_1, \ldots, u_n) = 1$.

Suppose the claim is true when the total length of Θ is less than m, $m > n$. We now show the result for $\mathcal{L}(\Theta(u_1, \ldots, u_n)) = m$. Since m is larger than n, there is at least one u_i, $i \in \{1, \ldots, n\}$, such that $\eta(u_i) \geq 2$. By Lemma 4.2 there exist $v_i, w_i \in z^G$, $i = 1, \ldots, n$, with $\Theta(u_1, \ldots, u_n) = \Theta(v_1, \ldots, v_n)\Theta(w_1, \ldots, w_n)$ such that

$$0 < \mathcal{L}(\Theta(v_1, \ldots, v_n)) < \mathcal{L}(\Theta(u_1, \ldots, u_n)) = m,$$
$$0 < \mathcal{L}(\Theta(w_1, \ldots, w_n)) < \mathcal{L}(\Theta(u_1, \ldots, u_n)) = m.$$

By the induction hypothesis, $\Theta(v_1, \ldots, v_n) = 1$ and $\Theta(w_1, \ldots, w_n) = 1$. Hence $\Theta(u_1, \ldots, u_n) = 1$. Therefore, for any $u_1, u_2, \ldots, u_n \in z^G$, we have $\Theta(u_1, \ldots, u_n) = 1$ as desired. $\square$

COROLLARY 4.4. *Let Λ be an index set and $W = \{\Theta_\lambda | \lambda \in \Lambda\}$ be a set of outer commutator words in variables $\{x_1, x_2, \ldots\}$. If $\mathfrak{D}$ is the variety defined by the set of words W, then $L(\mathfrak{D})$ is the variety defined by the set of words W in variables $\{x^{y_1}, x^{y_2}, \ldots\}$.*

5. Acknowledgements

The author would like to thank his dissertation advisor Professor Luise-Charlotte Kappe for her guidance and encouragement in preparing this paper. The author would also like to thank Professor Samuel M. Vovsi for reviewing this paper. His suggestions directly led to the formulation of Proposition 2.3.

REFERENCES

1. N. Gupta and F. Levin, *On soluble Engel groups and Lie algebras*, Arch. Math. **34** (1980), 289-295..
2. P. Hall, *The Edmonton notes on nilpotent groups*, Queen Mary College Mathematics Notes, 1969.
3. L. C. Kappe, *On Levi-formations*, Arch. Math. **23** (1972), 561-572.
4. L. C. Kappe and W. P. Kappe, *On 3-Engel groups*, Bull. Austral. Math. Soc. **7** (1972), 391-405.
5. L. C. Kappe and R. F. Morse, *Groups with 3-abelian normal closures*, Arch. Math. **51** (1988), 104-110.
6. L. C. Kappe and R. F. Morse, *Levi-properties in metabelian groups*, Contemporary Mathematics **109** (1990), 59-72.
7. F. W. Levi, *Groups in which the commutator operation satisfies certain algebraic conditions*, J. Indian Math. Soc. **6** (1942), 87-97.
8. F. W. Levi, *Notes on Group Theory I, II*, J. Indian Math. Soc. **8** (1944), 1-9.
9. H. Neumann, *Varieties of groups*, Springer-Verlag, Berlin, 1967.
10. D. J. S. Robinson, *Finiteness conditions and generalized soluble groups (2 vols.)*, Springer-Verlag, Berlin, 1972.

IBM FEDERAL SYSTEMS COMPANY, ROUTE 17C, OWEGO, NEW YORK 13827

E-mail address: rfmorse@owgvm0.vnet.ibm.com

Contemporary Mathematics
Volume **169**, 1994

Chains of Primitive Ideals

D. S. PASSMAN

Dedicated to the Memory of Professor Wilhelm Magnus

ABSTRACT. In this brief note, we offer two types of examples to show that the intersection of a descending chain of primitive ideals in a ring need not be primitive. The "torsion-free" examples are based on free rings; the "torsion" examples use group rings of locally finite groups.

§1. Free Rings

If R is a ring and X is a set, then we let $R\langle X \rangle$ and $R[X]$ denote the free ring and the polynomial ring, respectively, generated over R by the variables in X. We start with a special case of a result of [**L**].

LEMMA 1. *Let S be a countable domain and let X be a countably infinite set of variables. Then $R = S\langle X \rangle$ is primitive.*

PROOF. Say $X = \{x_1, x_2, \dots\}$. Since R is countable, we can enumerate its nonzero elements as $r_1, r_2, \dots$ and, for each $n \geq 1$, we set $\alpha_n = x_n r_n + 1$. If $\sum_n \alpha_n R = R$, then we can write $1 \in R$ as a finite sum

$$1 = \sum_{n=1}^{k} \alpha_n \beta_n = \sum_{n=1}^{k} (x_n r_n + 1)\beta_n$$

for suitable $\beta_n \in R$. But if $m = \max_n \{ \deg r_n + \deg \beta_n \}$ occurs at $n = t$, then we see that the monomials in $x_t r_t \beta_t$ of degree $m + 1$ all start with x_t and hence cannot be cancelled in the above equation, a contradiction.

Thus $\sum_n \alpha_n R \neq R$ and we can choose a maximal right ideal M of R containing this sum. Then $V = R/M$ is an irreducible R-module which we claim is faithful. Indeed, if I is the kernel of the action of R on V, then $I \subseteq M$ and $I \triangleleft R$. Finally, if $I \neq 0$, then $r_j \in I$ for some j, so $\alpha_j - 1 = x_j r_j \in I \subseteq M$. But $\alpha_j \in M$, so $1 \in M$, a contradiction. Thus $I = 0$ and R is a primitive ring. □

As a consequence, we have

1991 *Mathematics Subject Classification.* Primary 16D60, 16S10, 16S34.

COROLLARY 2. *Let K be a countable field, and let X and Y be countable sets with X infinite. Then $R = K[Y]\langle X \rangle$ is a primitive ring.*

This is clear since $S = K[Y]$ is a countable domain. On the other hand, if Y is allowed to be uncountable, then just the opposite occurs.

LEMMA 3. *Let K be a field and let X be a countable set. If Y is uncountable, then $R = K[Y]\langle X \rangle$ is not primitive.*

PROOF. Suppose by way of contradiction that R has a faithful irreducible right module V. Since $S = K[Y]$ is central in R, it follows that the rational function field $F = K(Y)$ is contained in the endomorphism ring of V. Since F is a field, there is an F-epimorphism $\theta : V \to F$ and, in particular, θ is an S-epimorphism. But R is a countably generated S-module, so V_S is countably generated and hence so is $\theta(V_S) = F_S$. On the other hand, Y is uncountable, so it is clear that F_S cannot be countably generated and the lemma is proved. □

By combining the preceding two results, we obtain

THEOREM 4. *Let K be a countable field, and let X be a countably infinite set. If Y has the cardinality of the continuum, then $R = K[Y]\langle X \rangle$ has a descending chain of primitive ideals with intersection equal to the nonprimitive ideal 0.*

PROOF. Since Y has the cardinality of the continuum, it is easy to see that $K[Y]$ is residually a polynomial ring in finitely many variables. Indeed, label the variables in Y by the real numbers in the interval $[0, 1]$, and for each n let $\pi_n : K[Y] \to K[Y]$ be the endomorphism which truncates the decimal expansion for the label of each variable to n decimal places. Then certainly the image of π_n is a polynomial ring in finitely many variables $Y_n \subseteq Y$ and the kernels of the various π_n decrease and have intersection 0. It follows that R has a descending chain of ideals P_n with $R/P_n \cong K[Y_n]\langle X \rangle$, so each P_n is primitive by Corollary 2. Finally, $\bigcap_n P_n = 0$ and 0 is not a primitive ideal by the preceding lemma. □

Thus we have obtained appropriate "torsion-free" examples. We close this section by taking a closer look at the argument of Lemma 3. Here S plays the role of $K[Y]$ and R corresponds to $K(Y)$.

LEMMA 5. *Let S be a commutative ring and let V be a right S-module. Assume that R is an S-algebra acting faithfully on V and that R is a right Artinian ring.*

   i. *If V_S is finitely generated, then so is R_S.*
   ii. *Let κ be an infinite cardinal. If V_S has κ generators, then so does R_S.*

PROOF. Let $N = \operatorname{rad} R$ and let W be any irreducible right R-module. Choose e to be an idempotent in R such that $\bar{e} = e + N \in R/N$ is the central idempotent corresponding to W. Then eRe has radical $eNe = N \cap eRe$ and its unique irreducible module is the restriction W_{eRe}.

Since R acts faithfully on V, we have $Ve \neq 0$ and hence Ve properly contains $V(eNe)$. Thus $Ve/V(eNe)$ is a nonzero (unital) module for eRe/eNe. Indeed, since eRe/eNe is a simple Artinian ring, $Ve/V(eNe)$ is a nontrivial direct sum of copies of W_{eRe}. It follows that we have abelian group epimorphisms

$$V \to Ve \to W_{eRe}$$

and, in fact, these are S-module epimorphisms. In particular, since V_S has κ generators (or is finitely generated), we conclude that W_S also has this property.

Finally, R has a finite composition series with factors W_i and, by the above, each such factor has κ generators (or is finitely generated) as an S-module. Thus the same is clearly true for R_S. $\square$

§2. Group Rings

If R is a ring and G is a multiplicative group, then we let $R[G]$ denote the group ring of G over R. We begin with the lovely result of [**D**] which is not as well-known as it should be. It appears in [**P**], but not in the original edition of that book.

THEOREM 6. *Let K be a field and let A be a normal abelian subgroup of the group G. If $K[A]$ contains more than $2^{|G/A|}$ idempotents, then $K[G]$ is not primitive.*

PROOF. Suppose $K[G]$ is primitive, and let its faithful irreducible module be $K[G]/M$ with M a maximal right ideal. Then $M \cap K[A] \neq K[A]$, so we can choose a maximal right ideal L of $K[A]$ with $L \supseteq M \cap K[A]$. Since $K[A]$ is commutative, $K[A]/L$ is a field and we let $\xi\colon K[A] \to K[A]/L$ be the natural map. Note that ξ takes idempotents to 0 or 1.

Let e be a nonzero idempotent of $K[A]$ and set $I = \sum_{x \in G} K[G]e^x$. Then I is a nonzero G-stable left ideal of $K[G]$, so $I \lhd K[G]$. Hence, since $K[G]$ acts faithfully on $K[G]/M$, we have $I \not\subseteq M$ and thus $M + I = K[G]$. Write

$$1 = \mu + \sum_{x \in S} \alpha_x e^x$$

with $\mu \in M$, $\alpha_x \in K[G]$ and S a finite subset of G. We multiply this equation on the right by $\prod_{x \in S}(1 - e^x)$ and, since $K[A]$ is commutative, we obtain

$$\prod_{x \in S}(1 - e^x) = \mu \prod_{x \in S}(1 - e^x) \in K[A] \cap M \subseteq L.$$

But L is a maximal ideal of $K[A]$, so this implies that $1 - e^x \in L$ for some $x \in S$ and therefore $e^x \notin L$.

Write $\bar{G} = G/A$, and to each idempotent $e \in K[A]$ associate the *sequence* $\prod_{\bar{x} \in \bar{G}} \xi(e^{\bar{x}})$ of 0's and 1's. There are $2^{|\bar{G}|}$ such sequences and, by assumption, $K[A]$ has more than this many idempotents. Thus there exist distinct idempotents e, f of $K[A]$ with $\prod_{\bar{x} \in \bar{G}} \xi(e^{\bar{x}}) = \prod_{\bar{x} \in \bar{G}} \xi(f^{\bar{x}})$. It follows that the idempotent $e - ef$ satisfies $\xi((e - ef)^x) = 0$ for all $x \in G$, and hence, by the observation of the preceding paragraph, this idempotent must be zero. Thus $e = ef$ and similarly $f = ef$, a contradiction since $e \neq f$. $\square$

In the past, there was some question as to whether the above inequality of cardinals is sharp. We show now that this is indeed the case. While the following argument applies to more general situations, we will restrict our attention to one small example.

LEMMA 7. *Let p be a prime, let K be a field containing a primitive pth root of 1 and let A be an elementary abelian p-group with $|A| = \mathbf{c}$, the continuum. Then there exists a countable set Λ of linear characters $\lambda\colon K[A] \to K$ with the following properties.*

   i. *$\bigcap_{\lambda \in \Lambda} \ker \lambda = 0$.*
   ii. *If Λ' is any cofinite subset of Λ, then $\bigcap_{\lambda \in \Lambda'} \ker \lambda = 0$.*
   iii. *Let n be any integer, let A^n denote the direct product of n copies of A, and let Λ_n be the set of all linear characters*

$$\mu = \lambda_1 \otimes \lambda_2 \otimes \cdots \otimes \lambda_n \colon K[A^n] \to K$$

   with the λ_i distinct elements of Λ. Then $\bigcap_{\mu \in \Lambda_n} \ker \mu = 0$.

PROOF. (i) For this, we can assume that $A = \prod_{i=1}^{\infty} C_i$ is the complete direct product of countably many cyclic groups C_i of order p. For each integer n, let $\pi_n\colon A \to \prod_{i=1}^{n} C_i = C^n$ be the projection of A onto the product of the first n factors. Then certainly $\bigcap_n \ker \pi_n = 1$. Furthermore, if we extend π_n to $\pi_n^*\colon K[A] \to K[C^n]$, then clearly $\bigcap_n \ker \pi_n^* = 0$. Now note that each $K[C^n]$ is semisimple and hence has a set Ω_n of p^n linear characters $\lambda\colon K[C^n] \to K$ with $\bigcap_{\lambda \in \Omega_n} \ker \lambda = 0$. When we view each λ as a linear character of $K[A]$ via composition with π_n^*, then it is clear that $\Lambda = \bigcup_{n=1}^{\infty} \Omega_n$ has the required property.

(ii) Say $\Lambda = \Lambda' \cup \{\lambda_1, \lambda_2, \dots, \lambda_m\}$, and let $I = \bigcap_{\lambda \in \Lambda'} \ker \lambda$ and $J = \bigcap_{j=1}^{m} \ker \lambda_j$. By (i), $I \cap J = 0$ and hence $IJ = 0$. Furthermore, J has finite codimension in $K[A]$. If $\alpha \in I$, then there exists a finite subgroup $B \subseteq A$ with $\alpha \in K[B]$. Let $\{x_i \mid i \in \mathcal{I}\}$ be a complete set of coset representatives for B in A. Since

$$\mathrm{r.ann}_{K[A]}\, \alpha = \oplus \sum_{i \in \mathcal{I}} (\mathrm{r.ann}_{K[B]}\, \alpha) x_i,$$

it follows easily that either $\alpha = 0$ or $\mathrm{r.ann}_{K[A]}\, \alpha$ has uncountable codimension in $K[A]$. But $\alpha \in I$, so $\alpha J = 0$. Thus, since J has finite codimension, we conclude that $\alpha = 0$ and hence that $I = 0$.

(iii) We proceed by induction on n. Write $A^n = A^{n-1} \times A^1$ and let $0 \neq \alpha \in K[A^n]$. Then $\alpha = \sum_i \alpha_i a_i$, where $\alpha_i \in K[A^{n-1}]$ and the a_i are distinct elements of A^1. If $\alpha_0 \neq 0$, then by induction there exists $\mu = \lambda_1 \otimes \cdots \otimes \lambda_{n-1}$, with the λ_j distinct elements of Λ and with $\mu(\alpha_0) \neq 0$. Now $\mu \otimes 1\colon K[A^n] \to K[A^1]$ and $(\mu \otimes 1)(\alpha) = \sum_i \mu(\alpha_i) a_i \neq 0$. Furthermore, by (ii), there exists $\lambda_n \in \Lambda$ such that $\lambda_n \neq \lambda_1, \dots, \lambda_{n-1}$ and $\lambda_n(\sum_i \mu(\alpha_i) a_i) \neq 0$. Thus $\mu' = \mu \otimes \lambda_n = \lambda_1 \otimes \cdots \otimes \lambda_n \in \Lambda_n$ satisfies $\mu'(\alpha) \neq 0$ and the lemma is proved. $\square$

PROPOSITION 8. *Let p be a prime, let K be a field with a primitive pth root of unity and let A be an elementary abelian p-group with $|A| = \mathbf{c}$, the continuum. If Sym_∞ is the countably infinite, locally finite symmetric group and if $G = A \wr \mathrm{Sym}_\infty$ is the permutation group wreath product, then $K[G]$ is primitive.*

PROOF. We will construct a concrete faithful irreducible module for $K[G]$. To start with, the base group B of G is the direct sum of countably many copies of A and $G = B \rtimes H$ where $H = \mathrm{Sym}_\infty$ permutes the summands via its natural permutation action. Let $\Lambda = \{\lambda_1, \lambda_2, \ldots\}$ be the countable set of linear characters of A given by the preceding lemma. Then the tensor product $\mu = \otimes_{i=1}^\infty \lambda_i$ defines a linear character $\mu \colon K[B] \to K$ and hence μ corresponds to a 1-dimensional irreducible $K[B]$-module W. Let $V = W \otimes_{K[B]} K[G]$ be the induced $K[G]$-module. We claim that V is faithful and irreducible.

First, note that the restriction $V_B = \oplus \sum_{x \in H} W \otimes x$ is a direct sum of 1-dimensional irreducible $K[B]$-modules. Indeed, $W \otimes x$ corresponds to the linear character $\mu^x \colon K[B] \to K$ given by $\mu^x(b) = \mu(xbx^{-1})$ for all $b \in B$, and we note that $\mu^x = \otimes_{i=1}^\infty \lambda_{ix}$ where ix is the image of i under the permutation $x \in \mathrm{Sym}_\infty$. Since all λ_i are distinct, it follows that all μ^x are distinct and hence the $W \otimes x$ are nonisomorphic $K[B]$-modules. We conclude therefore that V_B is a completely reducible $K[B]$-module whose simple submodules are precisely the modules $W \otimes x$ with $x \in H$. In particular, if $0 \neq U$ is a $K[G]$-submodule of V, then $U_B \supseteq W \otimes y$ for some y. But then $U \supseteq (W \otimes y)K[G] = V$ and we conclude that V is irreducible.

Next, we observe that $K[B]$ acts faithfully on V. To this end, let $0 \neq \beta \in K[B]$. Since B is the direct sum of copies of A, it follows that $\beta \in K[A^n]$, where A^n denotes the direct product of the first n copies of A for some n. Hence, by part (iii) of the preceding lemma, there exist distinct linear characters $\lambda_1', \ldots, \lambda_n' \in \Lambda$ such that $(\lambda_1' \otimes \cdots \otimes \lambda_n')(\beta) \neq 0$. Now Sym_∞ is n-transitive in its permutation action, so there exists $z \in H = \mathrm{Sym}_\infty$ with $\lambda_i' = \lambda_{iz}$ for $i = 1, 2, \ldots, n$. Hence $\mu^z(\beta) \neq 0$, so β acts nontrivially on $W \otimes z$ and therefore on V.

Finally, suppose $\alpha \in K[G]$ acts trivially on V and write $\alpha = \sum_{t \in H} \alpha_t t$ with $\alpha_t \in K[B]$. Then, for any $x \in H$ and $0 \neq w \in W$, we have

$$0 = (w \otimes x)\alpha = \sum_{t \in H} w\mu^x(\alpha_t) \otimes xt,$$

so $\mu^x(\alpha_t) = 0$ for all x, t. It follows that each α_t acts trivially on V, so $\alpha_t = 0$ and hence $\alpha = 0$. Thus V is indeed a faithful irreducible $K[G]$-module and $K[G]$ is therefore a primitive ring. $\square$

Notice that, in the preceding example, B is a normal abelian subgroup of G with $|G/B| = \mathbf{a}$, the countably infinite cardinal, and with $|B| = \mathbf{c}$. Furthermore, since each subgroup of B of order p gives rise to an idempotent of $K[B]$, it follows easily that $K[B]$ has precisely $\mathbf{c}$ idempotents. But $\mathbf{c} = 2^{\mathbf{a}}$ and $K[G]$ is primitive, so we conclude that Theorem 6 does indeed require that the number of idempotents in $K[A]$ be strictly larger than $2^{|G/A|}$. Finally, we combine Theorem 6 and

Proposition 8 to obtain our "torsion" examples.

THEOREM 9. *Let p be a prime, let K be a field containing a primitive pth root of unity and let A be an elementary abelian p-group with $|A| = 2^{\mathbf{c}}$. If G is the locally finite group $G = A \wr \mathrm{Sym}_\infty$, then $K[G]$ contains a directed family of primitive ideals with intersection equal to the nonprimitive ideal 0.*

PROOF. Since $|A| = 2^{\mathbf{c}}$, we can assume that $A = \prod_{i \in \mathcal{I}} C_i$ is the complete direct product of $|\mathcal{I}| = \mathbf{c}$ many cyclic groups C_i of order p. For each countably infinite subset $\mathcal{S} \subseteq \mathcal{I}$, let $\pi_{\mathcal{S}} \colon A \to \prod_{i \in \mathcal{S}} C_i = C^{\mathcal{S}}$ be the natural projection and extend this to an epimorphism

$$\pi_{\mathcal{S}} \colon G = A \wr \mathrm{Sym}_\infty \to C^{\mathcal{S}} \wr \mathrm{Sym}_\infty = H_{\mathcal{S}}.$$

It is clear that $\ker \pi_{\mathcal{S}} \cap \ker \pi_{\mathcal{T}} = \ker \pi_{\mathcal{S} \cup \mathcal{T}}$ and that $\bigcap_{\mathcal{S}} \ker \pi_{\mathcal{S}} = 1$. Next, extend $\pi_{\mathcal{S}}$ to the algebra epimorphism $\pi_{\mathcal{S}}^* \colon K[G] \to K[H_{\mathcal{S}}]$ and set $P_{\mathcal{S}} = \ker \pi_{\mathcal{S}}^*$. Then it follows from Proposition 8 and the above that each $P_{\mathcal{S}}$ is a primitive ideal of $K[G]$ and that $P_{\mathcal{S}} \cap P_{\mathcal{T}} \supseteq P_{\mathcal{S} \cup \mathcal{T}}$. In other words, $\{ P_{\mathcal{S}} \mid \mathcal{S} \subseteq \mathcal{I}, |\mathcal{S}| = \mathbf{a} \}$ is a directed set of primitive ideals of $K[G]$, and of course $\bigcap_{\mathcal{S}} P_{\mathcal{S}} = 0$. Since $K[G]$ is not primitive, by Theorem 6, the result follows. $\square$

REFERENCES

[D] O. I. Domanov, *A prime but not primitive regular ring*, in Russian, Uspekhi Mat. Nauk **32** (1977), 219–220.

[L] A. I. Lichtman, *The primitivity of free products of associative algebras*, J. Algebra **54** (1978), 153–158.

[P] D. S. Passman, *The Algebraic Structure of Group Rings*, Reprint Edition, Krieger, Malabar, Florida, 1985.

DEPARTMENT OF MATHEMATICS, UNIVERSITY OF WISCONSIN-MADISON, MADISON, WISCONSIN 53706

E-mail address: passman@math.wisc.edu

Contemporary Mathematics
Volume **169**, 1994

Families of Closed Geodesics on Hyperbolic Surfaces with Common Self-intersections

THEA PIGNATARO AND HANNA SANDLER

ABSTRACT. It is shown that every intersection point of two distinct closed geodesics on an orientable hyperbolic surface is simultaneously a double point of an arbitrarily large number of closed geodesics of equal length. Moreover, this configuration persists under arbitrary deformations of the surface. An analogous result is given for certain pairs of nonintersecting closed geodesics.

1. Introduction and Statement of Results

On a hyperbolic surface, a *double point* is a point through which a single closed geodesic passes in at least two different directions. The notion of a *quadruple point* is defined similarly. We refer to a property of a given hyperbolic surface as being *permanent* if it persists under arbitrary deformations of the surface. It was shown in [**7**] that on any orientable hyperbolic surface a point which is an intersection point of two distinct closed geodesics is either a permanent simultaneous double point of two closed geodesics of permanently equal length or a permanent quadruple point. In this paper, we offer the following further result:

THEOREM 1. *On any orientable hyperbolic surface, a point which is an intersection point of two distinct closed geodesics is a permanent double point of an arbitrarily large number of closed geodesics of permanently equal length.*

For pairs of nonintersecting closed geodesics, we have the related result:

1991 *Mathematics Subject Classification.* Primary 51M10; Secondary 30F35.

Research of the first author was supported by the Alfred P. Sloan Foundation and by NSF grant #DMS-9204533.

Research of the second author was supported in part by NSF grant #DMS-9009521.

This paper is in final form and no version of it will be submitted for publication elsewhere.

THEOREM 2. *Let S be an open orientable hyperbolic surface arising from a representation of the free group on n generators, $F_n = \langle X_1, X_2, \ldots \rangle$, $2 \leq n \leq \aleph_0$, in $SL(2, \mathbb{C})$. Assume that the closed geodesics on S associated to X_1 and X_2 do not intersect. Then their common perpendicular ℓ is a permanent common perpendicular of an arbitrarily large number of closed geodesics of permanently equal length each of which permanently passes through ℓ twice at right angles.*

The proofs of Theorems 1 and 2 appear at the end of the paper, after several necessary preliminary results. These theorems encompass a result of Horowitz [4], which states that in any free group of rank at least two there exists an arbitrarily large number of inconjugate elements which have the same trace under all representations of the group in $SL(2, \mathbb{C})$. The geometric interpretation of this is that on any open hyperbolic surface there exists an arbitrarily large number of closed geodesics of equal length which remain equal in length under arbitrary deformations of the surface. Both this result and our Theorem 2 can be easily adapted to closed hyperbolic surfaces S of genus g. More specifically, if $G = \langle X_1, Y_1, \ldots, X_g, Y_g; \prod [X_i, Y_i] \rangle$ is such a surface group, then the elements of the free subgroup $F_g = \langle X_1, \ldots, X_g \rangle$ are conjugate in F_g precisely when they are conjugate in G [9]. It follows that the statement of Theorem 2 holds when applied to the geodesics on S associated to two of the generators of F_g.

It seems appropriate to mention that Horowitz pursued his result in response to a question of Magnus regarding the largest number of inconjugate elements in a free group which may be chosen with the property that the elements have equal trace under all representations in $SL(2, \mathbb{C})$.

We also remark that, for certain special surfaces, constructions of closed geodesics with multiple self-intersections were first given by Jørgensen [6] using Schottky groups and by Birman-Series [2] for surfaces with certain symmetries. These constructions give self-intersections of arbitrarily large multiplicity.

The authors are grateful to Troels Jørgensen for suggesting the work presented here. We would like to express our gratitude to Martin Lustig for his thoughtful suggestions, which greatly improved the statements of our results. We would also like to thank Ara Basmajian for his helpful advice.

2. Results on Palindromes

Recall that a *palindrome* is a word or sentence which reads the same forwards as backwards. Mathematically, a palindrome is a finite word, in the letters of an alphabet, with this same property. Elements of a free group on a certain set of generators may be viewed as freely reduced words whose letters are either generators or inverses of generators. This gives rise to the notion of a palindrome in any free group, once a set of generators has been specified.

Let us describe the relevance of palindromes in free groups to Theorem 1. Given a point P on a hyperbolic surface which is the intersection of two distinct closed geodesics, palindromes may be used to produce an infinite supply of closed geodesics which pass through P (see [5] and §4 below). Among all of these geodesics, we will select an arbitrarily large number which have permanently equal length and which have permanent self-intersections at P. This corresponds to constructing a family of palindromes which satisfy certain algebraic compatibility relations. A similar procedure with palindromes will be used in Theorem 2. We need the following:

PROPOSITION 1. *Let* $F_n = \langle X_1, X_2, \ldots \rangle$ *be a free group on* n *generators, where* $2 \leq n \leq \aleph_0$. *Every conjugacy class in* F_n *contains at most two (freely reduced) palindromes in the generators* $X_1, X_2, \ldots$.

PROOF. Our proof uses a simple geometric construction. Let m be any positive integer. Suppose w is a freely reduced palindrome of length m in the generators $X_1, X_2, \ldots$. In other words, $w = L_1 L_2 \cdots L_m$, with $L_1 = L_m = X_{i_1}^{\epsilon_1}$, $L_2 = L_{m-1} = X_{i_2}^{\epsilon_2}, \ldots$, where the indices $i_1, i_2, \ldots$ are positive integers no greater than n and $\epsilon_j = \pm 1$.

The cases of $m = 1$ and 2 can be checked easily. Let us therefore assume that $m \geq 3$. Consider a regular m-gon with the edges labelled consecutively by the letters $L_1, \ldots, L_m$ of w. The cyclically reduced conjugates of w are the cyclic permutations $L_2 L_3 \cdots L_m L_1, L_3 \cdots L_m L_1 L_2, \ldots$ (see e.g. [8], Theorem 1.3). Suppose the r-th cyclic permutation $L_r \cdots$ is a palindrome. Consider the line which passes through the center of the m-gon and the vertex between L_{r-1} (or L_m if $r = 1$) and L_r. Reflection through this line does not affect the labels on the edges. The collection of all such axes slices the m-gon into several pieces. Since there is reflective symmetry about each of these axes, the pieces must be equal in size. We may as well assume that there are more than two pieces, since we are done otherwise. Let $w_0 = L_1 \cdots L_{r-1}$, where r is the smallest integer greater than 1 such that the cyclic permutation $L_r \cdots L_{r-1}$ of w is a palindrome. Let $\tilde{w}_0 = L_{r-1} L_{r-2} \cdots L_1$. Symmetry about the axis which passes between L_{r-1} and L_r implies $\tilde{w}_0 = L_r \cdots L_{2r-2}$. Similarly, we see that $w = w_0 \tilde{w}_0 w_0 \tilde{w}_0 \cdots$.

Let t denote the ratio of the lengths of w and w_0. Suppose v is a freely reduced palindrome which is conjugate to w. There are only a few possibilities. If $w_0 = \tilde{w}_0$, then $v = w_0^t = w$. Otherwise, $w_0 \neq \tilde{w}_0$, t is even, and either $v = (w_0 \tilde{w}_0)^{t/2} = w$ or $v = (\tilde{w}_0 w_0)^{t/2}$. Our claim follows. $\square$

Let X and Y be generators of a free group of rank two. We will construct certain families of palindromes in X and Y. However, instead of using X and Y directly, we construct palindromes in the conjugate elements $A = XY^2X$ and $B = YX^2Y$. Observe that A and B are palindromes in X and Y and, consequently, every palindrome in A and B may be regarded as a palindrome in X and Y. The elements A and B have the added feature that they have equal trace under all representations in $SL(2, \mathbb{C})$, and this will be useful when proving our assertions about equality of lengths of closed geodesics.

Palindromes in two letters may be described simply in terms of the operator σ which is defined by $\sigma W(U, V) = W^\sigma(U, V) = W(U^{-1}, V^{-1})$, where W is a word in two letters. Letting $W^{-\sigma}$ denote the inverse of W^σ, we see that a palindrome is any word W with the property that $W = W^{-\sigma}$. We define a word $W_0(U,V) = UV^2U^{-1}$ and note that W_0^{-1} is obtained by inverting the second argument. Associated to W_0 is the palindrome $W = W_0 W_0^\sigma W_0^{-1} W_0^{-\sigma}$.

We define recursively a sequence $\mathcal{D}_0, \mathcal{D}_1, \ldots$ of sets of unordered pairs of palindromes in A and B. The geometric meaning of these sets will become clear when we prove the main theorems. For example, in the case of Theorem 1, it will turn out that each pair in $\mathcal{D}_n$ corresponds to a geodesic with a double point at the intersection point given in the hypothesis of the theorem. Let $\mathcal{D}_0$ denote the set which contains the single pair $\{A, B\}$. For positive values of n, take $\mathcal{D}_n$ to be the set of all pairs of the form $\{W(U, V), W(U, V^{-1})\}$, where $\{U, V\}$ is a pair in $\mathcal{D}_{n-1}$. By our choice of W, the elements of each pair in $\mathcal{D}_n$, for each n, are conjugate.

Note that each pair in $\mathcal{D}_{n-1}$ has two orderings, each of which gives rise to a pair in $\mathcal{D}_n$. For example, $\{U, V\}$ produces the pairs $\{W(U, V), W(U, V^{-1})\}$ and $\{W(V, U), W(V, U^{-1})\}$. Suppose that U and V generate a free group of rank two or, equivalently, that U and V do not commute. Then, by inspection, no two of the elements $W(U, V)$, $W(U, V^{-1})$, $W(V, U)$ and $W(V, U^{-1})$ commute (hence no two are the same or mutually inverse) and all have the same length viewed as words in U and V. An inductive argument shows that our assumption that U and V generate a free group of rank two actually holds for all pairs $\{U, V\} \in \mathcal{D}_{n-1}$.

It will be convenient to say that the elements $W(U, V)$, $W(U, V^{-1})$, $W(V, U)$ and $W(V, U^{-1})$ "come from" the pair $\{U, V\}$. Using Proposition 1, we see that all four of these elements cannot lie in a single conjugacy class.

PROPOSITION 2. *For every positive integer n, each element of each pair in $\mathcal{D}_n$ comes from only one pair in $\mathcal{D}_{n-1}$. Furthermore, no two elements of distinct pairs in $\mathcal{D}_n$ are the same or mutually inverse.*

PROOF. Define a sequence $\{l_n\}$ of positive integers by the recursion: $l_n = 16l_{n-1} - 8l_{n-2}$, where $l_0 = 1$, $l_1 = 16$. It is easily verified that $l_n > 15l_{n-1}$ for $n \geq 1$. In addition to proving the assertions above, we will show that each element of each pair in $\mathcal{D}_n$ has length l_n when viewed as a freely reduced word in A and B. Throughout the proof, by "letters" we mean either A or B or their inverses. When we refer to lengths of words in these letters we mean lengths measured after free reduction in A and B. Our assertions are proven simultaneously by induction on n. By inspection $W(A, B)$, $W(A, B^{-1})$, $W(B, A)$, $W(B, A^{-1})$ all have length 16. By the above remarks, no two of these words are the same or mutually inverse. Hence the case of $n = 1$ is clear.

Assume $n > 1$. Suppose $W_1(A, B)$ is a freely reduced word in A and B which belongs to a pair in $\mathcal{D}_n$. According to the construction of $\mathcal{D}_n$, there exists at least one pair $\{U, V\} \in \mathcal{D}_{n-1}$ such that $W_1(A, B)$ equals either $W(U, V)$ or $W(U, V^{-1})$. Fix such a pair $\{U, V\}$. We will show that, in fact, there is a unique such pair, and we will give an algorithm for determining it.

The algorithm for recovering $\{U, V\}$ from $W_1(A, B)$ is rather technical. Before attacking it rigorously, we give an overview. It will be shown that the length of W_1 is necessarily a multiple of 4. Thus we may let W_2 denote the segment which consists of the leftmost one fourth of W_1. Let W_1', W_2', W_3' and W_4' be the segments of W_2 defined as follows: W_1' is the first l_{n-1} letters, W_2' is the second l_{n-1} letters, W_3' is the second-to-last l_{n-1} letters, W_4' is the last l_{n-1} letters. (It will turn out that the words W_2' and W_3' overlap.) We will show that either W_1' or W_4' is a palindrome, and the other is not. Then $\{U, V\}$ may be recovered from $W_1(A, B)$ as follows: if W_1' is a palindrome, then $\{U, V\} = \{W_1'(A, B), W_2'(A, B)\}$, and if W_4' is a palindrome, then $\{U, V\} = \{W_4'^{-1}(A, B), W_3'(A, B)\}$.

Let us now verify that the algorithm just described is well-defined and that it recovers $\{U, V\}$ from $W_1(A, B)$. We consider the case where $W_1(A, B) = W(U, V)$. An analogous argument handles the case $W_1(A, B) = W(U, V^{-1})$. Let $\widetilde{W}_2$ be the freely reduced word such that $\widetilde{W}_2(A, B) = UV^2U^{-1}$. Then

$$W_1(A, B) = W(U, V) = \widetilde{W}_2(A, B)\, \widetilde{W}_2^\sigma(A, B)\, \widetilde{W}_2^{-1}(A, B)\, \widetilde{W}_2^{-\sigma}(A, B).$$

The four factors on the right have equal length and are freely reduced as words in A and B. We claim their product is also freely reduced. Consider, for example, the first two factors: $\widetilde{W}_2(A, B) = UV^2U^{-1}$ and $\widetilde{W}_2^\sigma(A, B) = U^{-1}V^{-2}U$. The last letter of $\widetilde{W}_2(A, B)$, be it A, A^{-1}, B or B^{-1}, must coincide with the first letter of $\widetilde{W}_2^\sigma(A, B)$. Therefore, there is no cancellation between the first two factors. The other factors are handled similarly. It follows that the length of W_1 is a multiple of 4, and $\widetilde{W}_2$ coincides with W_2 of the algorithm.

There exist words U_0 and V_0 in A and B such that $U = W(U_0, V_0)$ and $V = W(U_0, V_0^{-1})$ and either $\{U_0, V_0\}$ or $\{U_0, V_0^{-1}\}$ belongs to $\mathcal{D}_{n-2}$. By the inductive assumption, or inspection if $n = 2$, the lengths of U_0 and V_0 are l_{n-2}. The product UV has length $2l_{n-1}$. This becomes obvious once one replaces U and V by $W(U_0, V_0)$ and $W(U_0, V_0^{-1})$, respectively, and expands W as $W_0 W_0^\sigma W_0^{-1} W_0^{-\sigma}$. The same type of expansion shows that VU^{-1} has length $2l_{n-1} - 2l_{n-2}$. Indeed, the only cancellation of A's and B's occurs when taking the product of two consecutive factors $U_0^{-1} V_0^{-2} U_0$ and this results in the loss of exactly $2l_{n-2}$ letters. Therefore, $W_2(A, B) = (UV)(VU^{-1})$ has length $4l_{n-1} - 2l_{n-2}$.

It is now clear that $W_1(A, B)$ has length $16l_{n-1} - 8l_{n-2} = l_n$. In addition, $W_1'(A, B) = U$, $W_2'(A, B) = V$ and we will now show that W_4' is not a palindrome. By definition, W_4' consists of the rightmost l_{n-1} letters of W_2. Equivalently, this is just the rightmost l_{n-1} letters of VU^{-1}. Expanding U and V as in the previous paragraph, we see that there exist unique freely reduced words α, β and γ satisfying $U = \alpha\gamma$, $V = \beta\gamma$ and such that α and β have length $l_{n-1} - l_{n-2}$ and γ has length l_{n-2}. If δ denotes the rightmost l_{n-2} letters of β, then $W_4' = \delta\alpha^{-1}$. Examining lengths on both sides of the equation $V = \beta\gamma$, we see that there is no cancellation in the product $\beta\gamma$. It follows that $\delta \neq \gamma^{-1}$ or, equivalently, $W_4' \neq U^{-1}$. On the other hand, if W_4' is a palindrome, then one can produce a contradiction by showing that $W_4' = U^{-1}$. Indeed, $W_4' = \delta\alpha^{-1}$ and $U^{-1} = \gamma^{-1}\alpha^{-1}$ would be palindromes of length l_{n-1} whose rightmost $l_{n-1} - l_{n-2}$ letters agree. Hence, W_4' cannot be a palindrome.

The case where $W_1(A, B) = W(U, V^{-1})$ is entirely analogous. In this case, $W_4'^{-1}(A, B) = U$, $W_3'(A, B) = V$ and W_1' is not a palindrome. Therefore, each element of each pair in $\mathcal{D}_n$ comes from a unique pair in $\mathcal{D}_{n-1}$.

Suppose there is an element which lies in two different pairs in $\mathcal{D}_n$. According to the algorithm, the elements of these two pairs come from a unique pair in $\mathcal{D}_{n-1}$. However, the remarks preceding Proposition 2 imply that when the elements of two different pairs in $\mathcal{D}_n$ come from a common pair in $\mathcal{D}_{n-1}$, these pairs in $\mathcal{D}_n$ must be disjoint. This shows that no two elements of distinct pairs in $\mathcal{D}_n$ are the same.

Finally, we verify our assertion concerning inverses. Suppose that both $W(U, V)$ and its inverse $W(U^{-1}, V^{-1})$ belong to pairs in $\mathcal{D}_n$ and $\{U, V\} \in \mathcal{D}_{n-1}$. (A similar argument will work when $\{U, V^{-1}\} \in \mathcal{D}_{n-1}$.) Then, applying the above algorithm, we see that $W(U^{-1}, V^{-1})$ comes from $\{U^{-1}, V^{-1}\}$ and this pair must belong to $\mathcal{D}_{n-1}$. The inductive assumption implies $\{U, V\} = \{U^{-1}, V^{-1}\}$. Hence $U = V^{-1}$, which is impossible since, according to the remarks preceding Proposition 2, U and V cannot commute. Therefore no two elements of distinct pairs in $\mathcal{D}_n$ are mutually inverse. $\square$

We observe that it follows from Proposition 2 and the remarks preceding it that the set $\mathcal{D}_n$ has cardinality 2^n for each n.

3. Trace Equivalence

In this section, we present our main tool for establishing the equality of lengths of the closed geodesics in Theorems 1 and 2.

Fix a characteristic zero commutative ring K with identity. If ρ is an arbitrary representation of the free group $\langle X, Y \rangle$ in $SL(2, K)$ and U belongs to $\langle X, Y \rangle$, we let $\tau_U(\rho)$ denote the trace of $\rho(U)$. In this way, we associate to each U a K-valued function τ_U on the space of representations of $\langle X, Y \rangle$ in $SL(2, K)$. We make use of the fact that there exists a polynomial P_U in three variables with integer coefficients such that $\tau_U = P_U(\tau_X, \tau_Y, \tau_{XY})$. (This fact, for $K = R$, was known to Fricke and Klein [3]. A proof for general K may be found in [4].)

PROPOSITION 3. *Suppose n is a nonnegative integer and $\{U_1, V_1\}$ and $\{U_2, V_2\}$ are pairs in $\mathcal{D}_n$. Then $\tau_{U_1} = \tau_{V_1} = \tau_{U_2} = \tau_{V_2}$ and $\tau_{U_1 V_1} = \tau_{U_2 V_2}$. In addition, no two of the elements U_1, V_1, U_2 and V_2 commute unless $\{U_1, V_1\} = \{U_2, V_2\}$.*

PROOF. We use induction on n. The case of $n = 0$ is obvious. Assume our claim is true for $n - 1$. Let $U_1 = W(\tilde{U}_1, \tilde{V}_1)$, $V_1 = W(\tilde{U}_1, \tilde{V}_1^{-1})$, $U_2 = W(\tilde{U}_2, \tilde{V}_2)$ and $V_2 = W(\tilde{U}_2, \tilde{V}_2^{-1})$, where $\{\tilde{U}_1, \tilde{V}_1\}$ and $\{\tilde{U}_2, \tilde{V}_2\}$ belong to $\mathcal{D}_{n-1}$. The inductive assumption implies that $\tau_{\tilde{U}_1} = \tau_{\tilde{V}_1} = \tau_{\tilde{U}_2} = \tau_{\tilde{V}_2}$ and $\tau_{\tilde{U}_1 \tilde{V}_1} = \tau_{\tilde{U}_2 \tilde{V}_2}$.

There exist polynomials P_1 and P_2 with integer coefficients such that $\tau_{W(U,V)} = P_1(\tau_U, \tau_V, \tau_{UV})$ and $\tau_{W(U,V)W(U,V^{-1})} = P_2(\tau_U, \tau_V, \tau_{UV})$, for arbitrary U and V. We have $\tau_{U_1} = P_1(\tau_{\tilde{U}_1}, \tau_{\tilde{V}_1}, \tau_{\tilde{U}_1 \tilde{V}_1}) = P_1(\tau_{\tilde{U}_2}, \tau_{\tilde{V}_2}, \tau_{\tilde{U}_2 \tilde{V}_2}) = \tau_{U_2}$. Similarly, using P_2, we may show that $\tau_{U_1 V_1} = \tau_{U_2 V_2}$. In addition, $\tau_{U_1} = \tau_{V_1}$ and $\tau_{U_2} = \tau_{V_2}$, since U_1 and V_1 are conjugate, as are U_2 and V_2.

We have observed in the previous section that the elements of a given pair in $\mathcal{D}_n$ do not commute. In particular, this applies to the pairs $\{U_1, V_1\}$ and $\{U_2, V_2\}$. Now suppose U_1 and U_2 commute. Since $\tau_{U_1} = \tau_{U_2}$, this implies that either $U_1 = U_2$ or $U_1 = U_2^{-1}$. Both of these cases contradict Proposition 2. Hence U_1 and U_2 do not commute. In this manner, one can show that no two of the elements U_1, V_1, U_2 and V_2 commute unless $\{U_1, V_1\} = \{U_2, V_2\}$. $\square$

4. Geometric Interpretation

We will consider hyperbolic surfaces which are orientable. Such a surface can be thought of as the quotient space of a hyperbolic plane by a certain group of orientation-preserving isometries of hyperbolic 3-space which leaves the plane invariant and preserves its orientation. The group acts freely and discontinuously on the invariant plane and may be identified with a discrete subgroup of $SL(2, \mathbb{C})$. On a hyperbolic surface, the closed geodesics correspond to (conjugacy classes of) axes of the associated covering transformations. Recall that the trace of $X \in SL(2, \mathbb{C})$, denoted τ_X, is twice the hyperbolic cosine of half the length of the associated geodesic. (More details may be found in [5], for example.)

Let X and Y be elements of $SL(2, \mathbb{C})$ with no common fixed points on the Riemann sphere. The Lie product $XY - YX$ determines a Möbius transformation whose axis ℓ is the common perpendicular of the axes of X and Y. The axis of the Lie product is perpendicular to the axes of all nontrivial palindromes in X and Y. In the case where the axes of X and Y intersect, this means that their intersection point is a point which is common to the axes of all nontrivial palindromes in X and Y (see [5]).

Our algebraic results have an immediate geometric interpretation. Let X and Y be elements of $SL(2, \mathbb{C})$ with no common fixed points on the Riemann sphere whose axes intersect at some point $\widetilde{P}$ on a hyperbolic plane. Assume that X and Y generate a free group of rank two which acts discontinuously. Then $\langle X, Y \rangle$ may be viewed as the covering group of an open surface of genus one with one hole. Let P denote the point on this surface corresponding to $\widetilde{P}$.

Now fix n. The 2^{n+1} elements which belong to pairs in $\mathcal{D}_n$ have distinct axes, according to Proposition 3. Since these elements are palindromes in X and Y, their axes all pass through $\widetilde{P}$. The elements are conjugate in pairs, and Proposition 1 implies there are no other conjugacy relations between these elements. Thus the 2^n pairs in $\mathcal{D}_n$ determine 2^n distinct closed geodesics. The total intersection multiplicity at $\widetilde{P}$ is 2^{n+1}. Since a neighborhood of $\widetilde{P}$ is mapped homeomorphically onto a neighborhood of P, we see that the collection of geodesics passes through P at least 2^{n+1} times in at least 2^{n+1} different directions. Each geodesic passes through P at least twice in at least two different directions. Therefore P is realized as a simultaneous double point of 2^n distinct closed geodesics. This phenomenon is a consequence of abstract properties of the group $\langle X, Y \rangle$, namely conjugacy and properties of palindromes. Therefore, varying the representation ρ of $\langle X, Y \rangle$ in $SL(2, \mathbb{C})$, and thereby deforming the surface, does not destroy the configuration of simultaneous double points. Now if $\{U, V\}$ and $\{U', V'\}$ are distinct pairs in $\mathcal{D}_n$, then $\tau_U(\rho) = \tau_{U'}(\rho)$, according to Proposition 3. Since the trace determines the length of the associated closed geodesic, we see that the closed geodesics arising from $\mathcal{D}_n$ have permanently equal length. Therefore, we conclude that P may be realized as a permanent simultaneous double point of an arbitrary number of closed geodesics of permanently equal length.

We now make use of the fact that the one-holed open surface of genus one covers every orientable hyperbolic surface to obtain Theorem 1.

PROOF OF THEOREM 1. Let G be a surface group represented in $SL(2, \mathbb{C})$ and S the associated surface. Suppose P is an intersection point of two distinct closed geodesics on S, and let X and Y be elements of G whose axes correspond to these geodesics and intersect in a point which covers P. We may as well assume that X and Y generate a free group of rank two. Indeed, this becomes true after replacing X and Y by sufficiently high powers, and such a change of variables does not alter the associated axes.

Let $\widetilde{G} = \langle X, Y \rangle$, let $\widetilde{S}$ denote the associated genus one open surface, and let $\widetilde{P} \in \widetilde{S}$ be the point corresponding to the intersection of the axes of X and Y. Our basic construction realizes $\widetilde{P}$ as a permanent simultaneous double point of an arbitrarily large number of closed geodesics of permanently equal length. Since the covering map $\widetilde{S} \to S$ is a local homeomorphism, the family of closed geodesics on $\widetilde{S}$ projects to a family of closed geodesics on S which (collectively) pass through P at least 2^{n+1} times in different directions.

Presumably, the family of 2^n closed geodesics on $\widetilde{S}$ associated to $\mathcal{D}_n$ projects to a family of fewer geodesics on S. Let $c(n)$ denote the number of such geodesics. Algebraically, this collapsing would correspond to inconjugate elements in $\widetilde{G}$ becoming conjugate in G, and $c(n)$ measures the number of conjugacy classes in G represented by elements of pairs in $\mathcal{D}_n$.

Let us show that $c(n)$ is unbounded. (We thank Martin Lustig for pointing out this fact.) The quotient of the closure of the Nielsen region of $\widetilde{G}$ by the action of $\widetilde{G}$ is a compact surface $\widetilde{S}_0 \subset \widetilde{S}$ which is a torus with a geodesic boundary curve. Since all of the axes of the hyperbolic elements in $\widetilde{G}$ lie in the closure of the Nielsen region, $\widetilde{S}_0$ contains all of the closed geodesics of $\widetilde{S}$ (see [1]). The compactness of $\widetilde{S}_0$ implies that there is a constant k such that every point in S has less than k preimage points in $\widetilde{S}_0$. But then for any closed geodesic in S, the number of preimage geodesics in $\widetilde{S}_0$ is also bounded by k. Therefore $c(n) \geq 2^n/k$, and hence $c(n)$ is unbounded.

Now suppose N is a positive integer. We may choose n for which $c(n) \geq 2^N$. The pairs in $\mathcal{D}_n$ must give at least 2^N distinct closed geodesics each of which has a double point at P. Theorem 1 follows once we observe that the desired permanence properties follow from the corresponding properties on $\widetilde{S}$ and the fact that every representation of G in $SL(2, \mathbb{C})$ restricts to a representation of $\widetilde{G}$. $\square$

We note that the geodesics corresponding to the pairs in $\mathcal{D}_1$ also appear in the basic construction in [**7**].

PROOF OF THEOREM 2. The proof of Theorem 2 is similar to the proof of Theorem 1. Let S be an open orientable hyperbolic surface associated to a representation of the free group $F_n = \langle X_1, X_2, \ldots \rangle$, $2 \leq n \leq \aleph_0$, in $SL(2, \mathbb{C})$. Let $X = X_1$ and $Y = X_2$. We assume the axes of X and Y do not intersect. The surface $\widetilde{S}$ corresponding to the group $\widetilde{G} = \langle X, Y \rangle$ is a three-holed sphere. Then the Lie product gives a common perpendicular to these axes and this projects to a geodesic $\tilde{\ell}$ on $\widetilde{S}$. Once again we perform our basic construction on X and Y. This yields a family of 2^n closed geodesics on $\widetilde{S}$. Using the techniques of [**5**], we see that each of the geodesics in this family intersects $\tilde{\ell}$ at least twice at right angles. Projection from $\widetilde{S}$ onto S does not reduce the number of distinct geodesics in this family, since inconjugate elements in $\widetilde{G}$ remain inconjugate in F_n. Moreover, the number of intersection points of each of these geodesics with the common perpendicular does not decrease after projecting to S, since primitive elements in $\widetilde{G}$ remain primitive in F_n. The permanence properties are established exactly as in the proof of Theorem 1. $\square$

REFERENCES

1. A. Beardon, *The Geometry of Discrete Groups*, Graduate Texts in Mathematics #91, Springer-Verlag, New York, 1983.
2. J. Birman and C. Series, *Geodesics with multiple self-intersections and symmetries*, Low-dimensional Topology and Kleinian Groups (D.B.A. Epstein, ed.), Cambridge University Press, 1984, pp. 3–11.
3. R. Fricke and F. Klein, *Vorlesungen über die Theorie der Automorphen Funktionen*, vol. 1, B.G. Teubner, Leipzig, 1897.
4. R. Horowitz, *Characters of free groups represented in the two dimensional special linear group*, Commun. Pure and Appl. Math. **25** (1972), 635–649.
5. T. Jørgensen, *Closed geodesics on Riemann surfaces*, Proc. Am. Math. Soc. **72, # 1** (1978), 140–142.
6. T. Jørgensen, unpublished result described in *On Lengths and Self-Intersections of Closed Geodesics on Riemann Surfaces*, Ph.D. thesis of H. Sandler (1989), Columbia University, New York.
7. T. Jørgensen and H. Sandler, *Double points on hyperbolic surfaces*, Proc. Amer. Math. Soc. (to appear).
8. W. Magnus, A. Karrass and D. Solitar, *Combinatorial Group Theory*, Second Revised Edition, Dover Publications Inc, New York, 1976.
9. B. Randol, *The length spectrum of a Riemann surface is always of unbounded multiplicity*, Proc. Am. Math. Soc. **78, # 3** (1980), 455–456.

DEPARTMENT OF MATHEMATICS, THE CITY COLLEGE OF NEW YORK, NEW YORK, NY 10031

DEPARTMENT OF MATHEMATICS & STATISTICS, THE AMERICAN UNIVERSITY, WASHINGTON, DC 20016

E-mail address: sandler@american.edu

Contemporary Mathematics
Volume **169**, 1994

On the Isometry Groups
of Hyperbolic Manifolds

JOHN G. RATCLIFFE

ABSTRACT. In this paper, it is proved that the group of isometries of a generic, complete, geometrically finite, hyperbolic n-manifold is finite.

1. Introduction

In 1885, Poincaré [5] proved that the group of isometries of a closed hyperbolic surface is finite. This was generalized by Löbell [3] in 1930 when he proved that the group of isometries of a complete, geometrically finite, hyperbolic surface is finite. In 1972, Lawson and Yau [2] proved that the group of isometries of a compact hyperbolic n-manifold is finite. This was generalized by Avérous and Kobayashi [1] in 1976 when they proved that the group of isometries of a complete hyperbolic n-manifold of finite volume is finite. In this paper, the following theorem is proved that generalizes all these theorems.

THEOREM 1. *Let $M = H^n/\Gamma$ be a nonelementary, geometrically finite, hyperbolic space-form such that Γ leaves no m-plane of H^n invariant for $m < n-1$. Then the group $\mathrm{I}(M)$ of isometries of M is finite.*

Before proving the above theorem, examples will be given that show that none of the hypotheses: (1) nonelementary, (2) geometrically finite, nor (3) leaves no m-plane invariant for $m < n-1$ can be dropped from Theorem 1. As a reference for geometrically finite hyperbolic manifolds, see Chapter 12 of Ratcliffe [6].

1991 *Mathematics Subject Classification.* Primary 30F40, 53C25; Secondary 20H10, 22E40, 51M10, 57S17.

This paper is in final form and no version of it will be submitted for publication elsewhere

2. Examples

Before we consider our examples, we need the following lemma:

LEMMA 1. *Let $M = H^n/\Gamma$ be a hyperbolic space-form, and let* N *be the normalizer of* Γ *in* $I(H^n)$. *Then* $I(M)$ *is isomorphic to* N/Γ.

PROOF. An isometry ϕ of M lifts to an isometry $\tilde{\phi}$ of H^n such that $\tilde{\phi}\Gamma\tilde{\phi}^{-1} = \Gamma$. Moreover $\tilde{\phi}$ is unique up to composition with an element of Γ. Conversely, if ψ is an isometry of H^n such that $\psi\Gamma\psi^{-1} = \Gamma$, then ψ induces an isometry of M. We conclude that $I(M)$ is isomorphic to N/Γ.

EXAMPLE 1. Let U^n be the upper half-space model of hyperbolic n-space, and let Γ be the group generated by $n-1$ independent horizontal translations of U^n. Then Γ is an elementary, geometrically finite, discrete subgroup of $I(U^n)$. The group Γ leaves no m-plane of U^n invariant for $m < n-1$, since the cohomological dimension of Γ is $n-1$. The normalizer N of Γ in $I(U^n)$ contains the group of all horizontal translations of U^n. Therefore N/Γ is infinite, and so $I(U^n/\Gamma)$ is infinite by Lemma 1.

EXAMPLE 2. Let B^n be the conformal ball model of hyperbolic n-space. In 1976, Millson [4] proved that for each $n \geq 2$ there is a compact hyperbolic space-form B^n/Γ such that the abelianization of Γ is infinite. The commutator subgroup Γ' is nonelementary, since it has the same limit set as Γ. Moreover Γ' leaves no m-plane of B^n invariant for $m < n$, since Γ is of the first kind. The group Γ' is a normal subgroup of Γ of infinite index. Therefore, Γ' is not geometrically finite, since Γ' is of the first kind and B^n/Γ' has infinite volume. The normalizer N of Γ' in $I(B^n)$ contains Γ. Therefore N/Γ' is infinite, and so $I(B^n/\Gamma')$ is infinite by Lemma 1.

EXAMPLE 3. Suppose that $1 < m < n-1$. We regard B^m to be an m-plane of B^n. Let Γ be a nonelementary, geometrically finite, discrete subgroup of $I(B^m)$. Then Γ extends to a nonelementary, geometrically finite, discrete subgroup of $I(B^n)$ by Poincaré extension. The normalizer N of Γ in $I(B^n)$ contains the group of all rotations of B^n that fix each point of B^m. Therefore N/Γ is infinite, and so $I(B^n/\Gamma)$ is infinite by Lemma 1.

3. Main Results

For the remainder of the paper, we shall work in the conformal ball model B^n of hyperbolic n-space, and we shall identify the group $I(B^n)$ of isometries of B^n with the group $M(B^n)$ of Möbius transformations of $E^n \cup \{\infty\}$ that leave B^n invariant. Compare the next lemma with Lemma 3.2 of Wang [8].

LEMMA 2. *Let* Γ *be a finitely generated, nonelementary, discrete subgroup of* $M(B^n)$ *which leaves no m-plane of B^n invariant for $m < n - 1$. Then the normalizer* N *of* Γ *in* $M(B^n)$ *is discrete.*

PROOF. Let $\{g_1, \ldots, g_m\}$ be a set of generators for Γ with $g_1 = 1$. Let x be a point of B^n that is fixed only by the identity element of Γ. Set

$$s = \text{dist}(x, \Gamma x - \{x\}).$$

Let

$$U = \{\phi \in \text{M}(B^n) : d(\phi(g_i x), g_i x) < s/2 \text{ for } i = 1, \ldots, m\}.$$

Then U is an open neighborhood of the identity in $\text{M}(B^n)$.

Suppose that h is an element of $\text{N} \cap U$. Then we have

$$\begin{aligned}
d(g_i^{-1} h^{-1} g_i h x, x) &= d(g_i h x, h g_i x) \\
&\leq d(g_i h x, g_i x) + d(g_i x, h g_i x) \\
&= d(h x, x) + d(g_i x, h g_i x) \\
&< s.
\end{aligned}$$

Hence $g_i^{-1} h^{-1} g_i h x = x$, and so $g_i^{-1} h^{-1} g_i h = 1$. Therefore h and g_i commute for each $i = 1, \ldots, m$. As $g_1, \ldots, g_m$ generate Γ, we have that h commutes with every element of Γ.

Now let y be an arbitrary point of the limit set $L(\Gamma)$ of Γ. Then there is a sequence $\{f_i\}$ of elements of Γ such that $f_i x \to y$. Observe that for each i, we have that

$$d(f_i x, h f_i x) = d(f_i x, f_i h x) = d(x, h x).$$

Consequently, we have that

$$\lim_{i \to \infty} |f_i x - h f_i x| = 0.$$

Therefore $hy = y$. Thus h is the identity on $L(\Gamma)$.

Let m be the least integer such that $L(\Gamma)$ is contained in an $(m-1)$-sphere of S^{n-1}. By conjugating Γ, we may assume that $L(\Gamma) \subset S^{m-1}$. As Γ leaves the convex hull $C(\Gamma)$ of $L(\Gamma)$ invariant, Γ also leaves $\overline{B}^m$ invariant, since $\overline{B}^m$ is the affine hull of $C(\Gamma)$. By hypothesis, $m = n - 1$ or n.

Assume first that $m = n$. Then we can choose points $y_0, \ldots, y_n$ of $L(\Gamma)$ that are the vertices of an ideal n-simplex. As $hy_i = y_i$ for each i, we deduce that $h = 1$. Hence $\text{N} \cap U = \{1\}$. Now assume that $m = n - 1$. Then we can conclude as above that h is the identity on B^{n-1}. Therefore h is either the identity or the reflection ρ of B^n in the hyperplane B^{n-1}. Therefore, we have that $\text{N} \cap U \subset \{1, \rho\}$. Hence, the identity is open in N, and therefore N is discrete.

THEOREM 1. *Let $M = B^n/\Gamma$ be a nonelementary, geometrically finite, hyperbolic space-form such that Γ leaves no m-plane of B^n invariant for $m < n - 1$. Then the group $\text{I}(M)$ of isometries of M is finite.*

PROOF. The group Γ is finitely generated, since Γ is geometrically finite. Therefore N is discrete by Lemma 2. Now since Γ is an infinite normal subgroup of N, we have that $L(\Gamma) = L(\mathrm{N})$. Therefore N leaves $L(\Gamma)$ invariant. Hence N also leaves invariant the set

$$B(\Gamma) = C(\Gamma) \cap B^n.$$

Therefore N leaves invariant the neighborhood $N(B(\Gamma), 1)$ of $B(\Gamma)$ in B^n of radius one.

Since the set $N(B(\Gamma), 1)$ is open, there is a point x of $N(B(\Gamma), 1)$ that is not fixed by any $g \neq 1$ in N. Let D be the Dirichlet domain for N centered at x. Set

$$E = D \cap N(B(\Gamma), 1).$$

Then E is a fundamental domain for the action of N on $N(B(\Gamma), 1)$. Let $\{h_i\}$ be a set of Γ-coset representatives in N. Then $F = \cup h_i E$ is a fundamental region for the action of Γ on $N(B(\Gamma), 1)$. Let $\partial_N(F)$ be the boundary of F in $N(B(\Gamma), 1)$. As D is a locally finite fundamental domain for N, we have

$$\partial_N F \subset \cup h_i \partial D.$$

Therefore $\mathrm{Vol}(\partial_N F) = 0$. Hence we have

$$\mathrm{Vol}(F) = \mathrm{Vol}(N(B(\Gamma), 1)/\Gamma).$$

Now let

$$C(M) = B(\Gamma)/\Gamma$$

be the convex core of M. Then we have

$$N(B(\Gamma), 1)/\Gamma = N(C(M), 1).$$

As Γ is geometrically finite, we have

$$\mathrm{Vol}(F) = \mathrm{Vol}(N(C(M), 1)) < \infty.$$

Now since

$$[\mathrm{N} : \Gamma] = \mathrm{Vol}(F)/\mathrm{Vol}(E),$$

we deduce that N/Γ is finite, and therefore $\mathrm{I}(M)$ is finite.

COROLLARY. *The group of isometries of any nonelementary, geometrically finite, 3-dimensional, hyperbolic space-form is finite.*

The next theorem extends the above corollary.

THEOREM 2. *If $M = B^3/\Gamma$ is a nonelementary, 3-dimensional, hyperbolic space-form such that Γ is finitely generated and of the second kind, then the group of isometries of M is finite.*

PROOF. An isometry of M extends to a conformal automorphism of the ideal boundary of M. By Ahlfors' finiteness theorem, the ideal boundary of M is a finite union of Riemann surfaces of finite hyperbolic type. Therefore, the group of conformal automorphisms of the ideal boundary of M is finite. Now since a Möbius transformation of B^3 is determined by its action on a nonempty open subset of S^2, we conclude that the group of isometries of M is finite.

We end with an example that shows that the hypothesis "of the second kind" cannot be dropped from Theorem 2.

EXAMPLE 4. In 1975, Riley [7] proved that there is a 3-dimensional hyperbolic space-form B^3/Γ of finite volume which is homeomorphic to the figure-eight knot space. The commutator subgroup Γ' is nonelementary, since it has the same limit set as Γ. Therefore Γ' is of the first kind. The group Γ' is finitely generated, since it is a free group of rank two. The normalizer N of Γ' in $I(B^3)$ contains Γ. Therefore N/Γ' is infinite, and so $I(B^3/\Gamma')$ is infinite by Lemma 1.

REFERENCES

1. G. Avérous and S. Kobayashi, *On automorphisms of spaces of nonpositive curvature with finite volume*, Differential Geometry and Relativity, D. Reidel, Dordrecht, 1976, pp. 20–26.
2. H. B. Lawson and S. T. Yau, *Compact manifolds of nonpositive curvature*, J. Diff. Geom. **7** (1972), 211–228.
3. F. Löbell, *Ein Satz über die eindeutigen Bewegungen Clifford-Kleinscher Flächen in sich*, J. Reine Angew. Math. **162** (1930), 114–124.
4. J. J. Millson, *On the first Betti number of a constant negatively curved manifold*, Ann. Math. **104** (1976), 235–247.
5. H. Poincaré, *Sur un théorème de M. Fuchs*, Acta Math. **7** (1885), 1–32.
6. J. G. Ratcliffe, *Foundations of Hyperbolic Manifolds,*, Graduate Texts in Math., vol. 149, Springer-Verlag, Berlin, Heidelberg, and New York, 1994.
7. R. Riley, *A quadratic parabolic group*, Math. Proc. Camb. Phil. Soc. **77** (1975), 281–288.
8. H-C. Wang, *On a maximality property of discrete subgroups with fundamental domain of finite measure*, Amer. J. Math. **89** (1967), 124–132.

DEPARTMENT OF MATHEMATICS, VANDERBILT UNIVERSITY, NASHVILLE, TN 37240

E-mail address: RATCLIFJ@VUCTRVAX.VANDERBILT.EDU

Contemporary Mathematics
Volume 169, 1994

A Generalization of Lazard's Theorem on Modular Dimension Subgroups

VLADIMIR TASIĆ

ABSTRACT. We prove that the description of dimension subgroups mod p^e given by Lazard for free groups remains valid in the case of groups with torsion-free lower central factors.

1. Introduction

Given a group G, the subgroup defined by $D_n(G, R) = G \cap (1 + \Delta^n)$ where Δ is the augmentation ideal of the group ring RG is called the nth dimension subgroup over R. In the case $R = Z$, these subgroups are called integral dimension subgroups, and in the case $R = Z_{p^e}$, they are called dimension subgroups modulo p^e. Magnus' result which states that integral dimension subgroups of a free group coincide with its lower central subgroups has its modular analogue in Lazard's theorem which gives a formula for the nth dimension subgroup modulo p^e of a free group in terms of the lower central subgroups. Specifically, Lazard identified the nth dimension subgroup mod p^e of the free group F as

$$F_{n,p^e} = \prod_{ip^{(j-e+1)^+} \geq n} \gamma_i(F)^{p^j}$$

where $k^+ = \max\{k, 0\}$. In general, the subgroup $G_{n,p^e} = \prod_{ip^{(j-e+1)^+} \geq n} \gamma_i(G)^{p^j}$ may be properly contained in $D_n(G, Z_{p^e})$ (see [4]). In certain cases, however, the equality holds: for instance if $n \leq p$ or $e = 1$ (see [4] and [3] repectively). We shall prove the modular analogue of that part of Jennings' theorem which asserts that Magnus' statement remains true for groups with torsion-free lower central factors. The result, which generalizes Lazard's theorem, is as follows:

THEOREM 1. *Let G be a finitely generated group with torsion-free lower central factors. Then the nth dimension subgroup modulo p^e of G is equal to G_{n,p^e}.*

2. Notation and preliminary results

Recall from [3] that with the lower central series of G we can associate a Lie algebra, denoted by $L(G)$, with the set $\oplus_n(\gamma_n(G)/\gamma_{n+1}(G))$ as its carrier; similarly with the series of rational dimension subgroups we can associate a Lie algebra $L^*(G)$ on the set $\oplus_n(D_n(G, Q)/D_{n+1}(G, Q))$. Finally, let $\mathrm{gr}(A)$ denote the associated graded algebra of the algebra A.

1991 Mathematics Subject Classification. Primary 20F14.
This paper is in final form and no version of it will be submitted for publication elsewhere.

THEOREM A. *(Quillen [5])* $\mathrm{gr}(QG)$ *is the universal graded enveloping algebra of the Lie algebra* $L^*(G) \otimes Q$.

THEOREM B. *(cf. Jennings [2]) If G is a group with torsion-free lower central factors, then* $D_n(G, Q) = \gamma_n(G)$.

As an immediate consequence of these results, we have

THEOREM C. *Let G be a group with torsion-free lower central factors. Then* $\mathrm{gr}(QG)$ *is the universal enveloping algebra of the Lie algebra* $L(G) \otimes Q$.

We have found it convenient to use the language of free group rings in the proof; this method is described in [1], and we refer the reader to that monograph for the details. Briefly, if G is given by the presentation $G = F/R$, $\mathbf{r}$ is the kernel of the natural map $ZF \to Z(F/R)$ and $\mathbf{f}$ is the augmentation ideal of ZF, then if we succeed in demonstrating that in ZF

$$F \cap (1 + \mathbf{r} + \mathbf{f}^n + p^e \mathbf{f}) \subseteq \prod_{ip^{(j-e+1)^+} \geq n} \gamma_i(F)^{p^j} R,$$

Theorem 1 will follow. We shall prove this inclusion in the next paragraph.

3. Proof of Theorem 1

Let $G = F/R$, and let $\{z_{k,j}\}_{k,j \geq 1}$ be the sequence of commutators in F such that $\{z_{k,j} R \gamma_{k+1}(F)\}_{j \geq 1}$ is a basis for $\gamma_k(F) R / \gamma_{k+1}(F) R$. We must show that if $w - 1 \in \mathbf{r} + \mathbf{f}^n + p^e \mathbf{f}$ and $w \in F$, then $w \in F_{n,p^e} R$. Certainly w can be written as $w = u \prod z_{k,j}^{b_{k,j}}$ for some $u \in R$. Thus $w - 1 \in \mathbf{r} + \mathbf{f}^n + p^e \mathbf{f}$ is equivalent to

$$\prod z_{k,j}^{b_{k,j}} - 1 \in \mathbf{r} + \mathbf{f}^n + p^e \mathbf{f}.$$

Repeated application of the standard identities

$$uv - 1 = (u - 1)(v - 1) + (u - 1) + (v - 1),$$

$$u^m - 1 = \sum_{j \geq 1} \binom{m}{j} (u - 1)^j$$

yields

$$w - 1 = \sum_{k=1}^{n-1} \sum \binom{b_{i_1,j_1}}{l_1} \cdots \binom{b_{i_s,j_s}}{l_s} (z_{i_1,j_1} - 1)^{l_1} \cdots (z_{i_s,j_s} - 1)^{l_s} \equiv 0$$

mod $\mathbf{r} + \mathbf{f}^n + p^e \mathbf{f}$, where the inner sum is being taken over the sequences (i_r, j_r), l_r such that $(1,1) \leq (i_1, j_1) < \cdots < (i_s, j_s)$ lexicographically and $i_1 l_1 + \cdots + i_s l_s = k$.

Modulo $\mathbf{r}$, every element of $\mathbf{f}$ is a linear combination of ordered products, say π_t, in the $(z_{i,j} - 1)$ so that the relation above can be rewritten as

$$\sum_{k=1}^{n-1}\sum \binom{b_{i_1,j_1}}{l_1}\cdots\binom{b_{i_s,j_s}}{l_s}(z_{i_1,j_1} - 1)^{l_1}\cdots(z_{i_s,j_s} - 1)^{l_s} + \sum_t p^e\alpha_t\pi_t \equiv 0$$

modulo $\mathbf{r}+\mathbf{f}^n$, for some $\alpha_t \in Z$. Looking at this relation successively modulo $\mathbf{r}+\mathbf{f}^{k+1}$ for $k = 1,\ldots,n-1$, we obtain a set of relations among the ordered products $(z_{i_1,j_1} - 1)^{l_1}\cdots(z_{i_s,j_s} - 1)^{l_s}$ and π_t in the graded ring $\mathrm{gr}(QG) \cong \mathrm{gr}(QF/\mathbf{r})$. Observe that these ordered products are basic elements of the graded algebra $\mathrm{gr}(QG)$, it being the universal envelope of the Lie algebra $L(G)\otimes Q$ (by Theorem C and the Poincaré-Birkhoff-Witt theorem); consequently, each (basic) ordered product $(z_{i_1,j_1} - 1)^{l_1}$, occurring when $s = 1$ and $i_1 l_1 = k < n$ in the sum above, must 'cancel' with some π_t. Hence its coefficient must be divisible by p^e (or be 0) and we obtain the relations

$$\binom{b_{i,j}}{l} \equiv 0 \bmod p^e$$

for each $il = k < n$. Let $b_{i,j} = p^\alpha m$ where m is prime to p. Without loss of generality, we can assume that $i < n$ and $\alpha \geq e$. Thus $(\alpha - e + 1)^+ = \alpha - e + 1$ so that it suffices to show that $ip^\alpha \geq np^{e-1}$. Now let β be the largest integer with the property $ip^\beta < n$, and put $l = p^\beta$; then as p^e divides $\binom{b_{i,j}}{p^\beta}$, it follows that $b_{i,j}$ is divisible by $p^{\beta+e}$. On the other hand, by the definition of β, $ip^{\beta+e} \geq p^{e-1}n$ and hence $ip^\alpha \geq p^{e-1}n$. Consequently, $z_{i,j}^{b_{i,j}} \in F_{n,p^e}$. This proves the theorem.

ACKNOWLEDGEMENT. I would like to thank the referee for suggestions which helped me improve the presentation.

REFERENCES

1. Narain Gupta, *Free Group Rings*, Contemporary Math. 66, AMS, Providence, RI (1987).

2. S.A.Jennings, The group ring of a class of infinite nilpotent groups, *Canad. J. Math.* **7** (1950) 169-187.

3. M.Lazard, Sur les groupes nilpotentes et les algebres de Lie, *Ann. École Norm. Sup.* **71** (1954) 101-190.

4. Siegfried Moran, Dimension subgroups mod n, *Proc. Cambridge Phil. Soc.* **68** (1970) 579-582.

5. D.G.Quillen, On the associated graded ring of a group ring, *J. Algebra* **10** (1968) 411-418.

MATHEMATICAL INSTITUTE
UNIVERSITY OF OXFORD
24-29 St.GILES
OXFORD OX1 3LB, ENGLAND
E-mail: vtasic@maths.ox.ac.uk

Recent Titles in This Series

(*Continued from the front of this publication*)

139 **Vinay Deodhar, Editor,** Kazhdan-Lusztig theory and related topics, 1992

138 **Donald St. P. Richards, Editor,** Hypergeometric functions on domains of positivity, Jack polynomials, and applications, 1992

137 **Alexander Nagel and Edgar Lee Stout, Editors,** The Madison symposium on complex analysis, 1992

136 **Ron Donagi, Editor,** Curves, Jacobians, and Abelian varieties, 1992

135 **Peter Walters, Editor,** Symbolic dynamics and its applications, 1992

134 **Murray Gerstenhaber and Jim Stasheff, Editors,** Deformation theory and quantum groups with applications to mathematical physics, 1992

133 **Alan Adolphson, Steven Sperber, and Marvin Tretkoff, Editors,** p-adic methods in number theory and algebraic geometry, 1992

132 **Mark Gotay, Jerrold Marsden, and Vincent Moncrief, Editors,** Mathematical aspects of classical field theory, 1992

131 **L. A. Bokut', Yu. L. Ershov, and A. I. Kostrikin, Editors,** Proceedings of the International Conference on Algebra Dedicated to the Memory of A. I. Mal'cev, Parts 1, 2, and 3, 1992

130 **L. Fuchs, K. R. Goodearl, J. T. Stafford, and C. Vinsonhaler, Editors,** Abelian groups and noncommutative rings, 1992

129 **John R. Graef and Jack K. Hale, Editors,** Oscillation and dynamics in delay equations, 1992

128 **Ridgley Lange and Shengwang Wang,** New approaches in spectral decomposition, 1992

127 **Vladimir Oliker and Andrejs Treibergs, Editors,** Geometry and nonlinear partial differential equations, 1992

126 **R. Keith Dennis, Claudio Pedrini, and Michael R. Stein, Editors,** Algebraic K-theory, commutative algebra, and algebraic geometry, 1992

125 **F. Thomas Bruss, Thomas S. Ferguson, and Stephen M. Samuels, Editors,** Strategies for sequential search and selection in real time, 1992

124 **Darrell Haile and James Osterburg, Editors,** Azumaya algebras, actions, and modules, 1992

123 **Steven L. Kleiman and Anders Thorup, Editors,** Enumerative algebraic geometry, 1991

122 **D. H. Sattinger, C. A. Tracy, and S. Venakides, Editors,** Inverse scattering and applications, 1991

121 **Alex J. Feingold, Igor B. Frenkel, and John F. X. Ries,** Spinor construction of vertex operator algebras, triality, and $E_8^{(1)}$, 1991

120 **Robert S. Doran, Editor,** Selfadjoint and nonselfadjoint operator algebras and operator theory, 1991

119 **Robert A. Melter, Azriel Rosenfeld, and Prabir Bhattacharya, Editors,** Vision geometry, 1991

118 **Yan Shi-Jian, Wang Jiagang, and Yang Chung-chun, Editors,** Probability theory and its applications in China, 1991

117 **Morton Brown, Editor,** Continuum theory and dynamical systems, 1991

116 **Brian Harbourne and Robert Speiser, Editors,** Algebraic geometry: Sundance 1988, 1991

115 **Nancy Flournoy and Robert K. Tsutakawa, Editors,** Statistical multiple integration, 1991

114 **Jeffrey C. Lagarias and Michael J. Todd, Editors,** Mathematical developments arising from linear programming, 1990

113 **Eric Grinberg and Eric Todd Quinto, Editors,** Integral geometry and tomography, 1990

112 **Philip J. Brown and Wayne A. Fuller, Editors,** Statistical analysis of measurement error models and applications, 1990

(See the AMS catalog for earlier titles)